APPLIED REGRESSION ANALYSIS
AND OTHER MULTIVARIABLE METHODS

Applied Regression Analysis
and
Other Multivariable Methods

DAVID G. KLEINBAUM
LAWRENCE L. KUPPER

The University of North Carolina
at Chapel Hill

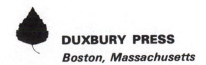

DUXBURY PRESS
Boston, Massachusetts

PWS PUBLISHERS

Prindle, Weber & Schmidt •♣• Duxbury Press • ♦ • PWS Engineering •⟁• Breton Publishers •⚙
20 Park Plaza • Boston, Massachusetts 02116

ISBN 0-87150-355-7

(previously ISBN 0-87872-139-8)

Applied Regression Analysis and Other Multivariable Methods was edited and prepared for composition by Service to Publishers, Inc. Interior design was provided by Service to Publishers, Inc., and the cover was designed by Oliver Kline.

L.C. Catalog Card Number 77-22327

Printed in the United States of America

86 87 88 89 — 11

LIBRARY OF CONGRESS CATALOGING IN PUBLICATION DATA

Kleinbaum, David G
  Applied regression analysis and other multivariable methods.

  Bibliography: p. 486
  Includes index.
  1. Multivariable methods. 2. Regression analysis.
I. Kupper, Lawrence L., joint author. II. Title.
QA278.K58   519.5'3   77–22327
ISBN 0-87150-355-7
(previously ISBN 0-87872-139-8)

To the memory of John C. Cassel

CONTENTS

PREFACE

This is a second-level statistics text intended primarily for advanced undergraduates, graduate students, and working professionals in the health, social, biological, and behavioral sciences who engage or plan to engage in applied research in their fields. We also hope that the book will provide the professional statistician with some new insights concerning the application of advanced statistical techniques to real-life data-analysis problems.

Our primary purpose in writing this text is to fill what we perceive to be the communications gap faced by the application-oriented reader who has a limited mathematical background, but would like to continue from where most beginning statistics texts end. In our view, workers in quantitative fields other than statistics are becoming more and more aware of the need to use advanced statistical techniques for the analysis of their data. This book is designed to help such researchers extend the scope of their knowledge of advanced statistical methods so that they will be aware of the advantages and disadvantages of the various procedures and thus will be able to interact much more efficiently with professional statisticians. We hasten to emphasize that a working knowledge of the material in this book does not make one a statistician; indeed, we are of the strong opinion that no research study should be undertaken without the involvement of a professional statistician.

The contents are based on our class notes for a second-level biostatistics service course designed for nonstatistics majors in the health and related fields. Both of us have been teaching and consulting with students and professionals in the public health area as one of our major responsibilities for the past six years. Most of the examples described in this book, particularly those used to introduce and illustrate the concepts in the body of the text, are based on this experience and reflect an orientation toward research involving a wide range of the social and behavioral aspects of health problems. However, some of the exercises dealing with biological applications were adapted from data presented in two textbooks on statistical methods for biologists, those by Bliss (1967, 1970) and by Sokal and Rohlf (1969).

The development of topics emphasizes the intuitive logic and assumptions that underlie the techniques covered, the purposes for which these techniques are designed, the advantages and disadvantages of the techniques, the interpretations that can validly be made based on the use of the techniques, and the form of the statistical calculations required. The use of the mathematical formulas presented involves no more than simple algebraic manipulations. Proofs are of secondary importance and are generally omitted. Neither calculus nor matrix algebra is used anywhere in the main text, although we have included an appendix on matrices for the interested reader.

The text is not intended to be a general reference work dealing with *all* statistical techniques available for analyzing data on several variables. Instead, what we feel to be the most important advanced topics are covered in considerable detail. After becoming proficient with the material in this text, the reader should be able to benefit from more specialized discussions dealing with applied topics of interest not specifically covered in this text.

Among the notable features of this book are the following:

1. It covers the topics of regression analysis and analysis of variance in considerably more detail than do other applied texts, and there are numerous worked examples and exercises, reflecting a wide variety of applications.

2. It relates several advanced topics involving the use of linear models, and it highlights the relationship between analysis of variance and regression analysis.

3. It provides a detailed discussion of the connection between multiple regression analysis and multiple and partial correlation analysis.

4. It presents several advanced topics in a unique, nonmathematical manner; these topics include factor analysis, discriminant analysis, and the analysis of categorical data using linear models.

5. It discusses applications based on recent studies in a variety of disciplines.

6. It uses representative computer results from packaged programs to illustrate concepts in the body of the text, as well as to provide a basis for exercises for the reader. (We have avoided providing thorough descriptions of available packaged programs, however, because of their frequently changing nature.)

7. The *complete* set of data for most exercises is provided, along with related computer results. This allows an instructor using this book to assign computer work based on available packaged programs, if desired. However, if such programs are not conveniently available or if the instructional objectives involve a minimum of computer work, the computer results given can still provide the student with practical experience in interpreting computer output from statistical programs.

The first four chapters of this text are introductory in nature. Chapter 1 describes several examples of recent studies involving the use of the statistical techniques discussed in this book. Chapter 2 reviews different methods of variable classification and their importance in guiding the researcher to the choice of an appropriate method of analysis. Chapter 3 reviews the basic statistical concepts needed to facilitate the reading of this book and establishes much of the notation used throughout the text. Chapter 4 provides a general overview of regression analysis.

Chapters 5 through 8 are concerned with straight-line regression analysis, and Chapter 9 deals with polynomial regression.

Chapters 10 and 11 provide general descriptions of multiple regression analysis and correlation analysis; Chapter 12 reviews by example the concepts presented in these two chapters and discusses the notion of statistical interaction. Chapters 13 through 15 deal with specialized regression topics concerning the use of dummy variables, analysis of covariance, and the selection of the best regression model. Chapter 16 provides a general overview of the techniques of residual analysis, data transformations, and weighted-least-squares analysis.

Chapters 17 through 20 discuss one-way and two-way analysis-of-variance procedures. These chapters are somewhat unique with regard to their descriptions of the relationship between regression and analysis of variance, random and mixed models, interaction, and ANOVA procedures when there are unequal cell numbers.

Chapters 21 through 23 cover the topics of factor analysis, two-group discriminant analysis, and categorical data analysis in a novel, nonmathematical, applied-oriented manner.

For formal classroom instruction, the chapters fall naturally into three clusters: Chapters 4 through 16, on regression analysis; Chapters 17 through 20, on analysis of variance; and, Chapters 21 through 23, on miscellaneous advanced topics.

There have been a number of contributors, both direct and indirect, to the preparation of this text. Dr. Kupper's work was supported by Research Career Development Award #1-K04-ES00003 from the National Institute of Environmental Health Sciences. We especially want to thank Anna Kleinbaum for help in editing, typing, and proofing the earlier versions of the manuscript and for providing overall advice on the instructional aspects of this text. We also wish to thank Cindy Kupper for the many hours that she helped in proofreading. We greatly appreciate the efforts of Steve Cohen and Agam Sinha in developing exercises and their solutions for many of the chapters. For typing and checking the final manuscript, we thank Ann Thomas, Gwen Entsminger, and Jackie O'Neal. We also thank those many students and colleagues at the University of North Carolina (most notably, Sherman James, Tony Lachenbruch, and Ken Read), who have read and suggested improvements in the various drafts of the manuscript. We are indebted to John Cassel, Bernard Greenberg, and James Grizzle, who have provided the professional and administrative guidance that has enabled us to gain the broad experience necessary to write this book.

DAVID G. KLEINBAUM
LAWRENCE L. KUPPER

APPLIED REGRESSION ANALYSIS
AND OTHER MULTIVARIABLE METHODS

CHAPTER 1 SOME MULTIVARIABLE EXAMPLES

Multivariable[1] techniques deal with problems that involve describing the relationship between two or more variables. The choice of an appropriate technique depends upon the general characteristics of the variables under investigation.

In Chapter 2 we shall discuss the classification of variable characteristics and the use of such a classification in choosing a method of analysis. For now, it seems important and appropriate to acquaint you with a few examples. The examples that follow concern *real* problems involving variables cutting across a variety of disciplines to which the methods described in this book are applicable. We shall return to these examples later when illustrating various methods of multivariable analysis.

EXAMPLE 1.1   *Study of the association among the physician–patient relationship,*
perceptions of pregnancy, and the outcomes of pregnancy,
illustrating the use of regression analysis, discriminant analysis,
and factor analysis

Thompson (1972) and Hulka et al. (1971) looked at both the process and the outcomes of medical care in a cohort of 107 pregnant married women in North Carolina. The data were obtained through questionnaires completed by physicians, through patient interviews, and by a review of medical record data. Several variables were recorded for each patient.

One research goal of primary interest was to determine what association, if any, existed between SATISfaction[2] with medical care and a number of variables meant to describe patient perception of pregnancy and the physician–patient relationship. Three perception-of-pregnancy variables measured the patient's WORRY during pregnancy, her desire (WANT) for the baby, and her concern about childBIRTH. Two other variables measured the physician–patient relationship in terms of informational communication (INFCOM) concerning prescriptions and affective communication (AFFCOM)

---

[1] The term "multivariable" is preferred to "multivariate." The distinction will be explained in Chapter 2.

[2] Capital letters denote the abbreviated variable "name."

1

concerning perceptions. Several other variables considered were age (AGE), social class (SOCLS), education (EDUC), and parity (PAR).

The perception-of-pregnancy variables leading to scores for each patient were developed by use of a method called *factor analysis*, which summarized the information obtained from several questions on this subject asked of all patients. *Regression analysis* was used to describe the relationship between scores measuring patient satisfaction with medical care and the variables above. From this analysis, variables found not related to satisfaction could be eliminated, while those found to be associated with satisfaction could be ranked in order of importance. Also, the effects of confounding variables such as AGE and SOCLS could be considered, so that any associations found could not be attributed solely to such variables; measures of the strength of the relationship between satisfaction and other variables could be obtained; and a functional equation predicting level of patient satisfaction in terms of the other variables found to be important in describing satisfaction could be developed.

Another question of interest in this study was whether or not patient perception of pregnancy and/or the physician–patient relationship was associated with COMPlications of pregnancy. A variable describing complications was defined and determined for each patient to take the value 1 if the patient experienced one or more complications of pregnancy and 0 if no complications were experienced. The method of *discriminant analysis* was used to evaluate the relationship between complications of pregnancy and other variables. This method, like regression analysis, allows the researcher to determine and rank important variables that can distinguish patients who have complications from patients who do not. With this method the researcher can also obtain an equation (called a *logistic function*) which provides an estimate of the probability that a patient with a given set of symptoms will have complications of pregnancy. We will see later that the computational aspects of discriminant analysis and regression analysis are intimately related.

EXAMPLE 1.2   *Study of the relationship between water hardness and sudden death, illustrating the use of regression analysis*

Hamilton's (1971) study of the effects of environmental factors on mortality used 88 North Carolina *counties* as the observational units, in contrast to *individual patients*, who were the observational units for the preceding study. Hamilton's primary goal was to determine the relationship, if any, between the sudden death rate of residents of a county and the measure of water hardness for that county. However, it was also of interest to compare the mortality/water hardness relationship among four regions of the state and to determine whether this relationship was affected by other variables, such as the habits of the county coroner in recording deaths, the distance from the county seat to the main hospital, the per capita income, and the population per physician.

Regression analysis was used in this study. The four regions could have been compared by doing a separate analysis for each region and then comparing results. But it was also possible to perform the *comparative analysis* in a single step, by defining a number of artificial or *dummy* variables to represent the regions.

EXAMPLE 1.3   *Comparative study of the effects of two instructional designs for teaching statistics, illustrating the use of analysis of covariance*

Kleinbaum and Kleinbaum (1976a) compared through a classroom experiment two alternative approaches for teaching probability to graduate students taking an introductory course in biostatistics. A class of 52 students was randomly split into two groups stratified on the basis of the students' fields of study. Both groups were taught in the lecture format by the same instructor. However, one group—the control—was tested in terms of the standard lecture method involving an instructor (using chalk and a blackboard), following distribution of a previously prepared handout. The classroom procedure for the experimental group, on the other hand, used a systematically designed (to fit a carefully defined set of objectives) unit of 16 transparencies and a set of practice problems to be done and reviewed in class. Both groups were given the same pre- and post-tests as well as questionnaires to measure attitudes. They both received a copy of the objectives, the handout, and a homework assignment. The primary question of interest for this study was whether or not the instructional design for the experimental group was more effective than that for the control group in terms of both cognitive learning and attitudes. With regard to cognitive learning, it was important to take into account pretest scores so that any possible difference found in post-test scores could not be attributable to differing levels of ability or knowledge between the two groups at the beginning of the experimental period. The appropriate method of analysis for this situation was *analysis of covariance*, which showed that post-test scores, when adjusted for pretest scores, were significantly higher for the experimental group. It was further shown, through use of ordinary *t* tests, that students gave significantly higher ratings to the experimental teaching approach than to the standard approach.

EXAMPLE 1.4   *Study of race and social influence in cooperative problem-solving dyads, illustrating the use of analysis of variance and analysis of covariance*

James (1973) conducted an experiment on 140 seventh- and eighth-grade males to investigate the effects of two factors, race of the experimenter (E) and race of comparison norm (N), on social influence behaviors in three types of dyads: white–white, black–black, and white–black. Subjects played a game of strategy called Kill the Bull, which required that 14 separate decisions for proceeding toward a defined goal be made on a game board. In the game, each two players (dyad) must reach a consensus on a direction at each decision step, after which they signal the E, who then rolls a die to determine how far they can advance along their chosen path of six squares. Photographs of the current champion players (N) [either two black youths (black norm) or two white youths (white norm)] were placed above the game board.

Four measures of social influence activity were used as outcome variables of interest. One of these, for example, was called *performance output*, which was a measure of the number of times a given subject attempted to influence his dyad to move in a particular direction.

The major research question focused on the outcomes for biracial dyads. Previous research of this type had used only white investigators and implicit white comparison norms, with results indicating that the white partner tended to dominate the influencing of decisions. In James's study, it was of interest to see if such an "interaction disability," previously attributable to blacks, would be maintained, removed, or reversed if either the comparison norm and/or the experimenter were black. One approach to analysis of this problem was to perform a *two-way analysis of variance* on social-influence-activity difference scores between black and white partners, to determine whether such differences were affected by either the race of E or the race of N. No such significant effects were found, however, implying that neither E nor N influenced biracial dyad interaction. Nevertheless, through use of *analysis of covariance*, it could be shown that, controlling for factors such as age, height, grade, and verbal and mathematical test scores, there was no statistical evidence of white dominance in any of the experimental conditions.

Furthermore, when looking at combined output scores for both subjects in "same race" dyads (white–white or black–black), using a *three-way analysis of variance* (the three factors being race of dyad, race of *E*, and race of *N*), it was seen that subjects in all black dyads were more verbally active (i.e., exhibited more of a tendency to influence decisions) under black *E's* than under white *E*'s; the same distinction was found for white dyads. This property would generally be referred to in the statistical jargon as a "race of dyad/race of *E* interaction." This property was also found to hold up after *analysis of covariance* was used to control for the effects of age, height, and verbal and mathematical test scores.

EXAMPLE 1.5   *Study of the relationship of culture change to health, illustrating the use of factor analysis and analysis of variance*

Patrick et al. (1974) have studied the effects of cultural change on health in the U.S. Trust Territory island of Ponape. Medical and sociological data were obtained on a sample of about 2,000 people by means of physical exams and a sociological questionnaire. This Micronesian island has experienced rapid westernization and modernization during its American occupation since 1945. The question of primary interest was whether or not rapid social and cultural change caused a rise in the level of blood pressure and in the prevalence of coronary heart disease. A specific hypothesis guiding the research was that persons with high levels of cultural ambiguity and incongruity and low levels of supportive affiliations with others will have high levels of blood pressure and high risk for coronary heart disease.

A preliminary step in the evaluation of this hypothesis involved determining three variables, measuring an individual's attitude toward modern life, preparation for modern life, and involvement in modern life. Each of these variables was created by isolating specific questions from a sociological questionnaire. Then a *factor analysis*, based on the method of *principal components*, determined how best to combine the scores on specific questions into a single overall score that defined the variable under consideration. Two cultural incongruity variables were then defined. One involved the discrepancy between attitude toward modern life and involvement in modern life. The other cultural incongruity variable was defined as the discrepancy between preparation for modern life and involvement in modern life.

These variables were then analyzed to determine their relationship, if any, to blood pressure and coronary heart disease. Individuals with large positive or negative scores on either of the two incongruity variables were hypothesized to have high blood pressure and high risk for coronary heart disease.

One approach toward analysis was to categorize both discrepancy scores into high and low groups. Then a *two-way analysis of variance* could be done using blood pressure as the outcome variable. We will see later that this problem can also be described as a regression problem.

EXAMPLE 1.6   *Study of the effect of neighborhood racial characteristics on the health of low-income housing residents, illustrating the use of analysis of variance and analysis of covariance*

Daly (1973) explored the possible health effects that racial-minority status might have on residents, almost all of whom were black, of four Turnkey III housing developments in Winston-Salem, N.C. Data were collected from the female adult of the household by means of a questionnaire. The information gathered included personal and demographic characteristics, residential history, indices of perceived power over the environment, indices of social integration, and a measure of self-perceived general medical and psychiatric health called the Cornell Medical Index (CMI).

The main study hypothesis was that the general health of Turnkey residents, as measured by the CMI, would improve as the percentage of black households in the surrounding neighborhood increased. The study sites were chosen in part so that the racial composition of the surrounding neighborhood showed a gradient of 100%, 65%, 36%, and 17% black.

One direct approach toward evaluating the study hypothesis was to compare the mean CMI scores for the four neighborhoods by *analysis of variance*. Furthermore, additional variables, such as a measure of perceived power over one's environment and length of residence in the Turnkey development, could be controlled for by *analysis of covariance* or *two-way analysis of variance*, depending on how these variables were defined. A significant difference in the mean CMI scores among the four developments would indicate that the study hypothesis was true, provided that the differences in mean scores between developments were in the right direction and that such differences held up when accounting for the contribution of other variables strongly associated with CMI scores.

The six examples described should indicate the variety of research questions to which multivariable statistical techniques are applicable. In Chapter 2 we shall provide a broader view of such techniques, and the remaining chapters will be devoted to discussing each technique in detail.

CHAPTER 2 VARIABLE CLASSIFICATION AND THE CHOICE OF ANALYSIS

2.1 VARIABLE CLASSIFICATION

Variables can be classified in a number of ways. Such classifications are useful for determining the method of data analysis. In this section we describe three methods of classification: by gapiness, by descriptive orientation, and by level of measurement.

2.1.1 GAPPINESS

The classification scheme we shall call *gappiness* seeks to determine whether or not there are gaps between successively observable values of a variable (Figure 2.1). If there are gaps between observations, the variable is said to be *discrete*; if there are no gaps between observations, the variable is said to be *continuous*. To be more precise, a variable is discrete if, between any two potentially observable values, there is a value that is not possibly observable. A variable is continuous if, between any two potentially observable values, there is always another potentially observable value.

Examples of continuous variables are age, blood pressure, cholesterol level, height, and weight. Examples of discrete variables are sex (e.g., 0 if male and 1 if female), number of deaths, group identification (e.g., 1 if group *A* and 2 if group *B*), and state of disease (e.g., 1 if a coronary case and 0 if not a coronary case).

FIGURE 2.1   *Discrete versus continuous variables*

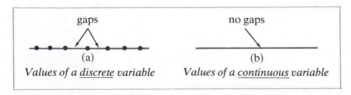

FIGURE 2.2   *Sample frequency distributions of continuous and discrete variables*

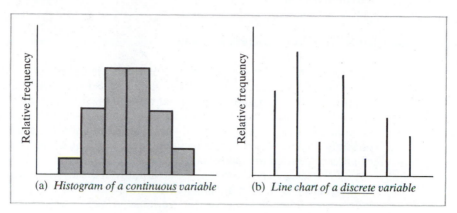

(a) *Histogram of a <u>continuous</u> variable*   (b) *Line chart of a <u>discrete</u> variable*

When dealing with actual data, the sampling frequency distributions for continuous variables are represented differently than for discrete variables. Data on a continuous variable are usually *grouped* into class intervals and a frequency distribution is determined by counting observations in each interval. Such a distribution is usually represented by a histogram, as shown in Figure 2.2a. Data on a discrete variable, on the other hand, are not usually grouped but are represented instead by a line chart, as shown in Figure 2.2b.

It is important to note that discrete variables can sometimes be treated as continuous variables for the purposes of analysis. This is particularly useful when the possible values of such a variable, even though discrete, are not very far apart and cover a wide range of numbers. In such a case, the possible values, although technically gappy, show such small gaps between values that a visual representation would approximate an interval (Figure 2.3).

Furthermore, a line chart such as Figure 2.2b, representing the frequency distribution of data on such a variable, would probably show few frequencies greater than 1, and consequently would be uninformative. As an example, the variable "social class" is usually measured as discrete; one popular social class measure[1] takes on the integer values between 11 and 77. When data on this variable are grouped into classes (e.g., 11–15, 16–20, etc.), the resulting frequency histogram will give a clearer picture of the characteristics of the variable than a line chart would. Thus, in this case, treating social class as a continuous variable would be more useful than treating it as discrete.

Just as it may often be useful to treat a discrete variable as continuous, some variables that are fundamentally continuous may be grouped into categories and treated as discrete variables in a given analysis. For example, the variable "age" can be

FIGURE 2.3   *Discrete variable that may be treated as continuous*

[1] Hollingshead's "Two-Factor Index of Social Position," a description of which can be found in Green (1970).

made discrete by grouping its values into two categories, "young" and "old." Similarly, blood pressure becomes a discrete variable if it is categorized into low, medium, and high groups, or into deciles.

2.1.2 DESCRIPTIVE ORIENTATION

A second scheme for classification of variables is based on whether a variable is *to describe or to be described by* other variables. Such a classification depends on the study objectives and orientation rather than on the inherent mathematical structure of the variable itself. If the variable under investigation is to be described in terms of other variables, we call it a *dependent variable*. If the variable is one that we are using possibly in conjunction with other variables to describe a given dependent variable, we call it an *independent variable*.

For example, in Thompson's (1972) study of the relationship of patient perception of pregnancy to patient satisfaction with medical care, the perception variables are independent and the satisfaction variable is dependent (Figure 2.4). Similarly, in studying the relationship of water hardness to sudden death rate in North Carolina counties, the water hardness index measured in each county is an independent variable, and the sudden death rate for that county, the dependent variable (Figure 2.5).

FIGURE 2.4 *Descriptive orientation for Thompson's 1972 study of satisfaction with medical care*

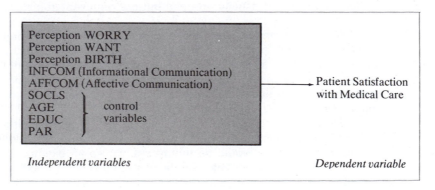

FIGURE 2.5 *Descriptive orientation for Hamilton's 1971 study of water hardness and sudden death*

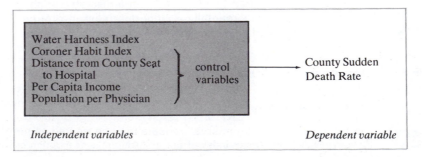

Usually, the distinction between independent and dependent variables is clear, as it is in the examples we have given. Nevertheless, a variable considered as dependent for evaluating one study objective may be considered as independent for evaluating a different objective. For example, in Thompson's study, in addition to determining the relationship of perceptions as independent variables to patient satisfaction, it was also of interest to determine the relationship of social class, age, and education to perceptions treated as *dependent* variables.

2.1.3 LEVELS OF MEASUREMENT

A third classification scheme deals with the level of mathematical preciseness of measurement of the variable. There are three such levels: nominal, ordinal, and interval.

The weakest level of measurement is the *nominal*. At this level the numbers the variable takes on are used simply to classify or label different categories. The variable "sex," for example, is nominal, since by assigning the numbers 1 and 0 to denote male and female, respectively, we can distinguish the two sex categories. A variable that describes treatment group is also nominal, provided that the treatments involved cannot be ranked according to some criterion (e.g., dosage level).

A somewhat higher level of measurement is one that allows not only *grouping* into separate categories, but also *ordering* of categories. This level is called *ordinal*. The treatment group may be considered ordinal if, for example, different treatments differ by dosage. In this case we could tell not only which treatment group an individual falls into, but also who experienced a heavier dose of the treatment. Also, social class is an ordinal variable, since an ordering can be made among its different categories. For example, all members of the upper middle class are "higher" in some sense than all members of the lower middle class.

A limitation, perhaps debatable, in the preciseness of a measurement such as social class is the amount of information that it supplies about the magnitude of the differences among different categories. Thus, although upper middle class is higher than lower middle class, it is debatable how *much* higher.

A variable that can give not only an ordering but also a meaningful measure of the distance between categories is called an *interval* variable. To be interval, a variable must have some sort of standard or well-accepted physical unit of measurement. Height, weight, blood pressure, and number of deaths all satisfy this requirement, whereas subjective measures such as perception of pregnancy, personality type, prestige, and social stress must be utilized with caution in this regard.

As with the other classification schemes, the same variable may be considered to be at one level of measurement in one analysis and at a different level in another analysis. Thus, age may be considered as interval in a regression analysis or may be grouped into categories and considered nominal in an analysis of variance.

It should also be pointed out that the various levels of mathematical preciseness are cumulative. An ordinal scale possesses all the properties of a nominal scale plus ordinality. An interval scale is also nominal and ordinal. The cumulativeness of these levels allows the researcher to drop back one or more levels of measurement in analyzing the data. So, an interval variable may be treated as nominal or ordinal for purposes of a particular analysis, and an ordinal variable may be analyzed as nominal.

2.2 OVERLAPPING NATURE OF CLASSIFICATION SCHEMES

It is important to realize that the three classification schemes described overlap in the sense that any variable can be labeled according to each scheme. Social class, for example, may be considered as ordinal, discrete, and independent in a given study; blood pressure may be considered interval, continuous, and dependent in the same or another study.

The overlap between the level of measurement classification and the gappiness classification is shown in Figure 2.6. The diagram does not include classification into dependent or independent variables, because that is entirely a function of the study objectives and not of the variable structure itself. In reading the diagram one should consider any variable as being represented by some point within the triangle. If the point falls below the dashed line within the triangle, it is classified as discrete; if it falls above this line, it is continuous. Also, a point that falls in the area marked "interval" is classified as an interval variable; and similarly for the other two levels of measurement.

It can be observed from Figure 2.6 that any nominal variable must be discrete, although a discrete variable may be nominal, ordinal, or interval. Also, a continuous variable must be either ordinal or interval, although there may be ordinal or interval variables which are not continuous. For example, "sex" is nominal and discrete; "age" may be considered interval and continuous, or, if grouped into categories, nomial and discrete. "Social class," depending on the way it is measured and on the viewpoint of the researcher, may be considered ordinal and continuous, ordinal and discrete, or nominal and discrete.

2.3 CHOICE OF ANALYSIS

Any researcher faced with the need to analyze data requires a rationale for choosing between alternative methods of analysis. Several considerations should enter into such a choice. These include (1) the purpose of the investigation, (2) the general mathematical characteristics of the variables involved, (3) the statistical assumptions

FIGURE 2.6   *Variable classification overlap*

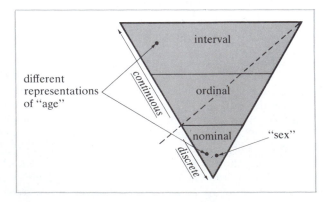

TABLE 2.1   *Rough guide to multivariable methods*

NAME OF METHOD	DEPENDENT-VARIABLE CLASSIFICATION	INDEPENDENT-VARIABLE CLASSIFICATION	GENERAL PURPOSE
Multiple regression analysis	Continuous	Classically all continuous, but, practically, any type(s) can be used	To describe the extent, direction, and strength of the relationship between several independent variables and a continuous dependent variable
Analysis of variance	Continuous	All nominal	To describe the relationship between a continuous dependent variable and one or more nominal independent variables
Analysis of covariance	Continuous	A mixture of nominal variables and continuous variables (the latter to be used as control variables)*	To describe the relationship between a continuous dependent variable and one or more nominal independent variables, controlling for the effect of one or more continuous independent variables
Discriminant analysis	Nominal	Classically all continuous, but, practically, can have a mixture of various types as long as some are continuous	To determine how one or more independent variables can be used to discriminate among different categories of a nominal dependent variable
Factor analysis	(The variables used in a factor analysis are classically continuous, but, practically, may be of any type. These variables are not clearly identifiable as being either dependent or independent, although the resulting factors may be used as dependent or independent variables in a later analysis.)		Variable construction or reduction using several variables to define one or more new composite variables called *factors*
Categorical data analysis using linear models	Nominal	Mostly nominal, but sometimes ordinal	To describe the relationship between a nominal dependent variable and several nominal or ordinal independent variables, although applications to situations involving only dependent variables are possible

* Generally speaking, a "control" variable is a variable that has to be considered before any relationships of interest can be quantified; this is because a control variable is one that is possibly related to the variables of primary interest and must be taken into account in studying the relationships among the primary variables. For example, in describing the relationship between blood pressure and physical activity, we would probably consider age and sex as control variables because they are related to blood pressure and physical activity and so, unless taken into account, could confound any conclusions regarding the primary relationship of interest.

made about these variables, and (4) the manner in which the data are collected (e.g., the sampling procedure). Knowledge of the first two is generally sufficient to guide the researcher toward the appropriate analyses. The researcher must consider the latter two items before finalizing initial recommendations.

In this section we focus on the use of variable classification, as it relates to the first two items above, in choosing an appropriate method of analysis. Table 2.1 provides a rough guide to help the researcher in this choice when several variables are involved. This guide distinguishes among various multivariable methods. It considers the types of variable sets usually associated with each method and gives a general description of the purposes of each method. The table is meant to point the researcher in the right direction in the choice of method of analysis. However, in addition to using the table, a careful check must be made of the statistical assumptions being used. These assumptions will be described fully later in the text. Table 2.2 shows how these guidelines can be applied to the examples given in Chapter 1.

TABLE 2.2   *Application of the guide to multivariable methods using examples in Chapter 1*

STUDY	NAME OF MULTIVARIABLE METHOD	DEPENDENT VARIABLE	INDEPENDENT VARIABLES	PURPOSE
Example 1.1	Multiple regression analysis	Patient satisfaction with medical care (SATIS), a continuous variable	WANT; WORRY; BIRTH; INFCOM; AFFCOM; AGE; EDUC; SOCLS; PAR	To describe the relationship between SATIS and the variables WANT, WORRY, etc.
Example 1.1	Discriminant analysis	Complications of pregnancy (0 = no, 1 = yes), a nominal variable	WANT; WORRY; BIRTH; INFCOM; AFFCOM; AGE; EDUC; SOCLS; PAR	To determine whether or not and to what extent the independent variables can be used to discriminate between mothers having and not having pregnancy complications
Example 1.1	Factor analysis	None	Several scores from questions on patient perceptions of pregnancy	To reduce the information in the scores to a small set of composite variables describing the essential characteristics of patient perceptions of pregnancy
Example 1.2	Multiple regression analysis	Sudden death rate for a county in North Carolina during a 5-year period, a continuous variable	Water hardness index for county; distance between hospital and county seat; per capita income; population per physician	To describe the relationship between sudden death rate and the independent variables
Example 1.3	Analysis of covariance	Post-test score, a continuous variable	Pretest score; group designation (e.g., 1 = experimental, 0 = control)	To compare post-test scores for experimental and control groups, adjusting for the possible effect of pretest scores

TABLE 2.2   (*continued*)

STUDY	NAME OF MULTIVARIABLE METHOD	DEPENDENT VARIABLE	INDEPENDENT VARIABLES	PURPOSE
Example 1.4	Analysis of covariance	Social-influence-activity score, a continuous variable	Race of subject (e.g., 1 = white, 2 = black), age, height, etc.	To determine whether or not one racial group dominates the other in biracial dyads, after controlling for age, height, etc.
Example 1.4	Two-way analysis of variance	Social influence activity difference score between black and white partners in biracial dyads, a continuous variable	Race of experimenter (e.g., 1 = white, 2 = black); race of comparison norm (e.g., 1 = white, 2 = black)	To determine whether there is any effect of experimenter's race and of the comparison norm's race on the difference score
Example 1.5	Two-way analysis of variance	Systolic blood pressure (SBP), a continuous variable	Discrepancy between attitude toward and involvement in modern life, categorized as high and low; discrepancy between preparation for and involvement in modern life, categorized as high and low	To describe the relationship between nominal discrepancy scores and SBP
Example 1.5	Factor analysis	None	Several scores from questions on attitudes of Ponapeans toward modern life	To reduce the information in these scores to a single composite variable summarizing an individual's attitudes toward modern life
Example 1.6	Analysis of covariance	CMI score, a continuous variable	Three dummy variables describing neighborhood; length of residence in Turnkey development; measure of individual power over environment	To compare the mean CMI scores for different neighborhoods, controlling for length of residence and perceived power over environment

Some comments are appropriate concerning Table 2.1. To begin with, there are several methods for dealing with multivariable problems which are *not* included in this list or in this text. Among those not included are nonparametric methods of analysis of variance, multivariate multiple regression and multivariate analysis of variance (which are extensions of the corresponding methods given here, allowing for *several* dependent variables), and methods of cluster analysis. In this book, we will not cover all multivariable techniques, but simply those used most often by health and social researchers.

[Statisticians generally use the term *multivariate analysis* to describe a method whose theoretical framework allows for the simultaneous consideration of several *dependent* variables. On the other hand, researchers in the biomedical and health sciences who are not statisticians view this term as describing any statistical technique involving several variables, even if only one dependent variable is considered at a time. In this text we prefer to avoid these contrasting viewpoints by using the term "multivariable analysis" to denote the latter more general description.]

It should also be pointed out that multiple regression analysis is a general technique which can be utilized with all kinds of variables. In fact, analysis of variance and analysis of covariance may be considered as special cases of regression analysis. Furthermore, even the method of discriminant analysis can be implemented using regression analysis, although the theoretical bases of the two methods are different.

Finally, factor analysis has been included in Table 2.1, since it is often used in conjunction with one or more of the other methods. Often, as when dealing with a sociological or psychological questionnaire, the researcher is confronted with a large set of basic variables which he wishes to reduce to a much smaller set of meaningful variables. In this case factor analysis can be used to combine the basic variables into one or more composite variables which summarize the essential information contained in all the variables.

CHAPTER 3

BASIC STATISTICS: A REVIEW

3.1 PREVIEW

The purpose of this chapter is to provide a review of fundamental statistical concepts and methods which are needed for an understanding of the more sophisticated multivariable techniques to be discussed in this text. Also, through this review, we shall introduce the statistical notation (using conventional symbols whenever possible) that we employ throughout the text.

The broad subject-matter area associated with the word *statistics* concerns the variety of methods and procedures for collecting, classifying, summarizing, and analyzing data, the latter two of which we focus on here. The primary goal of most statistical-analysis efforts is to make *statistical inferences*, that is, to draw valid conclusions about a *population* of measurements based upon information contained in a *sample* from that population.

Once sample data have been collected, it is useful, prior to analysis, to examine the data using tables, graphs, and *descriptive statistics*, such as the sample mean and the sample variance. Such descriptive efforts are important for representing the essential features of the data in easily interpretable terms.

Following such examination, statistical inference-making procedures are considered through two related aspects: *estimation* and *hypothesis testing*. The techniques involved in each aspect are based on certain assumptions about the probability pattern (or *distribution*) of the (*random*) variables being studied.

The key terms above—descriptive statistics, random variables, probability distribution, estimation, and hypothesis testing—will each be reviewed in the sections to follow.

3.2 DESCRIPTIVE STATISTICS

A descriptive statistic may be defined to be any single numerical measure computed from a set of data, which is designed to describe a particular aspect or characteristic of the data set. The most common types of descriptive statistics are measures of *central tendency* and *variability* (or *dispersion*).

The central tendency in a sample of data refers to what is often called the "average value" of the variable being observed. Of the several measures of central tendency, the most commonly used is the sample mean, which we denote by $\bar{X}$ whenever our underlying variable is called X. The formula for the sample mean is given by

$$\bar{X} = \frac{\sum\limits_{i=1}^{n} X_i}{n}$$

where n denotes the sample size; $X_1, X_2, \ldots, X_n$ denote the n independent measurements on X; and $\sum$ denotes summation. The sample mean $\bar{X}$, in contrast to such other measures of central tendency as the median or mode, uses in its computation all the observations in the sample. This property means that $\bar{X}$ is necessarily affected by the presence of extreme X values, in which case it may be preferable to use the median instead of the mean. A remarkable property of the sample mean, which makes it particularly desirable for use in making statistical inferences, follows from the *Central Limit Theorem*, which states that *whenever n is moderately large, $\bar{X}$ has approximately a normal distribution, regardless of the distribution of the underlying variable X*.

Measures of central tendency (such as $\bar{X}$) do not, however, completely summarize all the features of the data. It is obvious, for example, that two sets of data with the same mean can differ widely in appearance (e.g., an $\bar{X}$ of 4 can result both from the values 4, 4, 4 and from the values 0, 4, 8). Thus, we customarily consider, in addition to $\bar{X}$, measures of variability, which tell us the extent to which the values of the measurements in the sample differ from one another.

The two measures of variability most often considered are the *sample variance* and the *sample standard deviation*. These are given by the following formulas when considering observations $X_1, X_2, \ldots, X_n$ on a single variable X:

$$\text{sample variance} = S^2 = \frac{1}{n-1} \sum_{i=1}^{n} (X_i - \bar{X})^2$$

$$\text{sample standard deviation} = S = \sqrt{\frac{1}{n-1} \sum_{i=1}^{n} (X_i - \bar{X})^2}$$

The formula for S^2 describes variability in terms of an "average" of squared deviations from the sample mean, although $n-1$ is used as the divisor instead of n. This use of $n-1$ is due essentially to considerations that make S^2 a good estimator of the variability in the entire population.

A drawback to the use of S^2 is that it is in squared units of the underlying variable X. To have a measure of dispersion which is in the same units as X, then, we simply take the square root of S^2 and call it the sample standard deviation S. A convenient computational formula for S is given by

$$S = \sqrt{\frac{\sum\limits_{i=1}^{n} X_i^2 - \left(\sum\limits_{i=1}^{n} X_i\right)^2 \big/ n}{n-1}}$$

Using S in combination with $\bar{X}$ thus gives a fairly succinct picture of both the amount of spread and the center of the data, respectively.

When more than one variable is being considered in the same analysis, as will be the case throughout this text, we shall use different letters and/or different subscripts to differentiate among the variables, and we will modify the notations for mean and variance accordingly. For example, if we are using X to stand for age and Y to stand for systolic blood pressure, we would denote the sample mean and sample standard deviation for each variable as $(\bar{X}, S_X)$ and $(\bar{Y}, S_Y)$, respectively.

3.3 RANDOM VARIABLES AND DISTRIBUTIONS

The term *random variable* is used to denote a variable whose observed values may be considered as outcomes of a stochastic or random experiment (e.g., the drawing of a random sample). The values of such a variable in a particular sample, then, cannot be anticipated with certainty before the sample is gathered. Thus, if we select a random sample of persons from some community and determine the systolic-blood-pressure level (W), cholesterol level (X), race (Y), and sex (Z) of each person, then W, X, Y, and Z are four random variables whose particular realizations (or observed values) for a given person in the sample could not have been known for sure beforehand. In this text we shall denote random variables by capital Roman letters.

The probability pattern that gives the relative frequencies associated with all the possible values of a random variable in the population is generally called the *probability distribution* of the random variable. We represent such a distribution by a table, graph, or mathematical expression which provides the probabilities corresponding to the different values or ranges of values taken on by a random variable.

Discrete random variables (such as the number of deaths in a sample of patients or the number of arrivals at a clinic), whose possible values are countable, have (gappy) distributions which are graphed as a series of lines, the heights of these lines representing the probabilities associated with the various possible discrete outcomes (Figure 3.1). *Continuous* random variables (such as blood pressure and weight), whose possible values are uncountable, have (nongappy) distributions which are graphed as smooth curves, and an *area* under such a curve represents the probability associated with a *range of values* of the continuous variable (Figure 3.1). We note in passing that the probability of a continuous random variable taking on one particular value is zero, because there is no area above a single point.

FIGURE 3.1   *Discrete versus continuous distributions* [$P(X = a)$ *is read*: "*probability that X takes the value a*"]

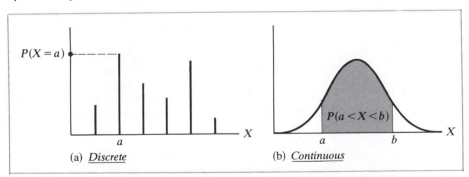

(a) *Discrete*   (b) *Continuous*

The next two sections will discuss two particular distributions of enormous practical importance: the binomial (which is discrete) and the normal (which is continuous).

3.3.1 THE BINOMIAL DISTRIBUTION

A *binomial* random variable describes the number of occurrences of a particular event in a series of n trials under the following four conditions:

1. The n trials are identical.
2. The outcome of any one trial is independent of (i.e., is not affected by) the outcome of any other trial.
3. There are two possible outcomes on each trial: "success" (i.e., the event of interest occurs) or "failure" (i.e., the event of interest does *not* occur), with probabilities p and $q = 1 - p$, respectively.
4. The probability of success, p, remains the same for all trials.

For example, the distribution of the number of lung cancer deaths in a random sample of $n = 400$ persons would be considered to be binomial if the four conditions were satisfied, as would the number of persons in a sample of $n = 70$ who favor a certain form of legislation.

The two *parameters* of the binomial distribution that one must specify to determine the precise shape of the probability distribution and to compute binomial probabilities are n and p. The usual notation used for this distribution is, therefore, $B(n, p)$, and, if X has a binomial distribution, it is customary to write

$$X \frown B(n, p)$$

where "$\frown$" stands for "is distributed as." The probability formula for this discrete random variable X is given by the expression

$$P(X = j) = {}_nC_j p^j (1 - p)^{n-j}, \qquad j = 0, 1, \ldots, n$$

where ${}_nC_j = n!/j!(n - j)!$ denotes the number of combinations of n things taken j at a time. We shall return to a consideration of the binomial distribution and its extension to multinomial data in our discussion of analysis of categorical data in Chapter 23.

3.3.2 THE NORMAL DISTRIBUTION

The *normal distribution*, denoted as $N(\mu, \sigma)$, where μ and σ are the two parameters, is described by the well-known bell-shaped curve (Figure 3.2). The parameters μ (the mean) and σ (the standard deviation) characterize the center and the spread of the distribution, respectively. We generally attach a subscript to the parameters μ and σ to distinguish among variables; that is, we often write

$$X \frown N(\mu_X, \sigma_X)$$

to denote a normally distributed X.

FIGURE 3.2   *The normal distribution*

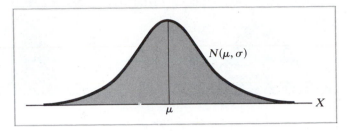

An important property of any normal curve is its *symmetry*, which distinguishes it from some other continuous distributions which we shall discuss. This symmetry property is quite helpful when using tables to determine probabilities or percentiles of the normal distribution.

Probability statements concerning a normally distributed random variable X which are of the form $P(a \leq X \leq b)$ require for computation the use of a single table (Table A-1 in Appendix A). This table gives the probabilities (or areas) associated with the *standard normal distribution*, which is a normal distribution with $\mu = 0$ and $\sigma = 1$. It is customary to denote a standard normal random variable by the letter Z, so that we write

$$Z \frown N(0, 1)$$

To compute the probability $P(a \leq X \leq b)$ for an X that is $N(\mu, \sigma)$, we must transform (i.e., *standardize*) X to Z by applying the conversion formula

$$Z = \frac{X - \mu}{\sigma} \tag{3.1}$$

to each of the elements in the probability statement about X as follows:

$$P(a \leq X \leq b) = P\left(\frac{a - \mu}{\sigma} \leq Z \leq \frac{b - \mu}{\sigma}\right)$$

We then look up the equivalent probability statement concerning Z in the $N(0, 1)$ tables.

This rule also applies to the sample mean $\bar{X}$ whenever the underlying variable X is normally distributed or whenever the sample size is moderately large (by the Central Limit Theorem). However, since the standard deviation of $\bar{X}$ is $\sigma/\sqrt{n}$, the conversion formula has the form

$$Z = \frac{\bar{X} - \mu}{\sigma/\sqrt{n}}$$

An inverse procedure to computing a probability for a range of values of X, as illustrated above, is to find a percentile of the distribution of X. A *percentile* is a value of the variable X *below which* the area under the probability distribution has a certain specified value. We denote the ($100p$)th percentile of X by X_p and picture it as in Figure 3.3, where p is the amount of area under the curve to the left of X_p. In determining X_p for

FIGURE 3.3   *The (100p)th percentiles of X and Z*

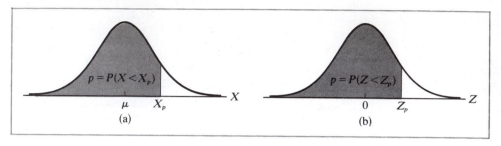

a given p we must again use the conversion formula (3.1). However, since the procedure requires that we first determine Z_p and then convert back to X_p, we generally rewrite the conversion formula as

$$X_p = \mu + \sigma Z_p \qquad\qquad (3.2)$$

For example, if $\mu = 140$, $\sigma = 40$, and we wish to find $X_{0.95}$, the $N(0, 1)$ tables first give us $Z_{0.95} = 1.645$, with which we convert back to $X_{0.95}$ as follows:

$$X_{0.95} = 140 + (40)Z_{0.95} = 140 + 40(1.645) = 205.8$$

Formulas (3.1) and/or (3.2) can also be used to approximate probabilities and percentiles for the binomial distribution $B(n, p)$ whenever n is moderately large (e.g., $n > 20$). The usual conditions for this approximation to be accurate require that both $np > 5$ and $n(1 - p) > 5$. Under such conditions the mean and standard deviation of the approximating normal distribution are

$$\mu = np \quad \text{and} \quad \sigma = \sqrt{np(1-p)}$$

3.4   THE SAMPLING DISTRIBUTIONS OF t, χ^2, AND F

The Student's t, chi-square (χ^2), and Fisher's F distributions are particularly important for their use in statistical inference-making procedures.

The *(Student's) t distribution*, which is symmetric about zero, as is the normal distribution (Figure 3.4a), was originally developed to describe the behavior of the random variable

$$T = \frac{\bar{X} - \mu}{S/\sqrt{n}} \qquad\qquad (3.3)$$

which represents an alternative to

$$Z = \frac{\bar{X} - \mu}{\sigma/\sqrt{n}}$$

FIGURE 3.4   *The t, χ^2, and F distributions*

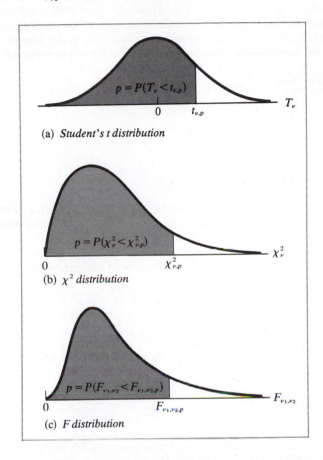

(a) *Student's t distribution*

(b) χ^2 *distribution*

(c) *F distribution*

whenever the population variance σ^2 is unknown and is estimated by S^2. The denominator of (3.3), $S/\sqrt{n}$, is the *estimated standard error of $\bar{X}$*. When the underlying distribution of X is normal, and when $\bar{X}$ and S^2 are calculated from a random sample from that distribution, then (3.3) has the *t distribution with $n-1$ degrees of freedom*, where $n-1$ is the parameter that must be specified in order to look up tabulated percentiles of this distribution. We denote all this by writing

$$T = \frac{\bar{X} - \mu}{S/\sqrt{n}} \frown t_{n-1}$$

It has generally been shown by mathematical statisticians that the t distribution is appropriate for describing the behavior of any random variable of the general form

$$T = \frac{\hat{\theta} - \mu_{\hat{\theta}}}{S_{\hat{\theta}}}$$

(3.4)

where $\hat{\theta}$ is any random variable that is approximately normally distributed with mean $\mu_{\hat{\theta}}$ and standard deviation $\sigma_{\hat{\theta}}$, where $S_{\hat{\theta}}$ is the estimated standard error of $\hat{\theta}$, and where $\hat{\theta}$ and $S_{\hat{\theta}}$ are statistically independent. For example, when random samples are taken from two normally distributed populations with the same standard deviation [e.g., from $N(\mu_1, \sigma)$ and $N(\mu_2, \sigma)$] and we consider $\hat{\theta} = \bar{X}_1 - \bar{X}_2$ in (3.4), we can write

$$T = \frac{(\bar{X}_1 - \bar{X}_2) - (\mu_1 - \mu_2)}{S_p \sqrt{1/n_1 + 1/n_2}} \frown t_{n_1 + n_2 - 2}$$

where

$$S_p^2 = \frac{(n_1 - 1)S_1^2 + (n_2 - 1)S_2^2}{n_1 + n_2 - 2} \tag{3.5}$$

estimates the common variance σ^2 in the two populations. The quantity S_p^2 is called a *pooled sample variance*, since it is calculated by pooling the data from both samples in order to estimate the common variance σ^2.

The *chi-square* (or χ^2) *distribution* is a nonsymmetrical distribution (Figure 3.4b) and describes, for example, the behavior of the nonnegative random variable

$$\frac{(n-1)S^2}{\sigma^2} \tag{3.6}$$

where S^2 is the sample variance based on a random sample of size n from a normal distribution. The variable given by (3.6) has the chi-square distribution with $n-1$ degrees of freedom:

$$\frac{(n-1)S^2}{\sigma^2} \frown \chi_{n-1}^2$$

Because of the nonsymmetry of the χ^2 distribution, both upper and lower percentage points of the distribution need to be tabulated, and such tabulations are solely a function of the degrees of freedom associated with the particular χ^2 distribution of interest. The chi-square distribution has widespread application in the analyses of categorical data (see Chapter 23).

The *F distribution*, which is also skewed to the right, like the χ^2 distribution (Figure 3.4c), is appropriate for modeling the probability distribution of the ratio of independent estimators of two population variances. For example, given random samples of sizes n_1 and n_2 from $N(\mu_1, \sigma_1)$ and $N(\mu_2, \sigma_2)$, respectively, so that estimates S_1^2 and S_2^2 of σ_1^2 and σ_2^2 can be calculated, it can be shown that

$$\frac{S_1^2 / \sigma_1^2}{S_2^2 / \sigma_2^2} \tag{3.7}$$

has the F distribution with the two parameters $n_1 - 1$ and $n_2 - 1$, which are called the *numerator* and *denominator* degrees of freedom, respectively. We write this as

$$\frac{S_1^2 \sigma_2^2}{S_2^2 \sigma_1^2} \frown F_{n_1 - 1, n_2 - 1}$$

The F distribution may also be related to the t distribution when the numerator degrees of freedom equals *one*; that is, the square of a variable distributed as Student's t with ν degrees of freedom has the F distribution with 1 and ν degrees of freedom. In other words,

$$T^2 \frown F_{1,\nu} \quad \text{if and only if} \quad T \frown t_\nu$$

Percentiles of the t, χ^2, and F distributions may be obtained from tables A-2, A-3, and A-4 in Appendix A. The shapes of the curves that describe these probability distributions, together with the notation that we will use to denote their percentile points, are given in Figure 3.4.

3.5 STATISTICAL INFERENCE: ESTIMATION

Two general categories of statistical inference, estimation and hypothesis testing, may be distinguished from one another by their differing purposes:

estimation: concerned with estimating the specific value of an unknown population parameter

hypothesis testing: concerned with making a decision concerning a hypothesized value of an unknown population parameter

In estimation, which we focus on in this section, we wish to estimate an unknown parameter θ using a random variable $\hat{\theta}$ ("theta hat," called a *point estimator* of θ). This point estimator is in the form of a formula or rule, for example,

$$\bar{X} = \frac{1}{n}\sum_{i=1}^{n} X_i \quad \text{or} \quad S^2 = \frac{1}{n-1}\sum_{i=1}^{n}(X_i - \bar{X})^2$$

which tells us how to calculate a specific point estimate given a particular set of data.

To estimate a parameter of interest (e.g., a population mean μ; a binomial proportion p; the difference between two population means, $\mu_1 - \mu_2$; or the ratio of two population standard deviations, σ_1/σ_2), the usual procedure is to select a random sample from the population or populations of interest, calculate the point estimate of the parameter, and then associate with this estimate a measure of its variability, which usually takes the form of a confidence interval for the parameter of interest.

As its name implies, a *confidence interval* (CI) consists of two boundary points between which we have a certain specified *level of confidence* that the population parameter lies. For example, a 95% CI for a parameter θ would consist of lower and upper limits determined so that, in repeated sets of samples each of the same size, 95% of all such intervals so calculated would be expected to contain the parameter θ. Care must be taken when interpreting such a CI not to consider θ as a random variable which either falls or does not fall in the calculated interval; rather, θ is a fixed (unknown) constant and the random quantities are the lower and upper limits of the CI, which vary from sample to sample.

We shall illustrate the procedure for computing a confidence interval with two examples, one concerning estimation of a single population mean μ and one concerning estimation of the difference between two population means $\mu_1 - \mu_2$. In each case we shall find that the appropriate CI required has the general form

$$\left(\begin{array}{c}\text{point estimate of}\\ \text{the parameter}\end{array}\right) \pm \left(\begin{array}{c}\text{a percentile of}\\ \text{the } t \text{ distribution}\end{array}\right) \cdot \left(\begin{array}{c}\text{estimated standard}\\ \text{error of the estimate}\end{array}\right) \qquad (3.8)$$

This general form also carries over to confidence intervals for other parameters which are to be considered in the remainder of the text (e.g., those considered in multiple regression analysis).

EXAMPLE 1   Suppose that we have determined the quantitative graduate record exam scores (i.e., QGRE scores) for a random sample of nine student applicants to a certain graduate department in a university, and we have found that $\bar{X} = 520$ and $S = 50$. If we wish to estimate with 95% confidence the population mean QGRE score (μ) for all such applicants to the department, and we are willing to assume that the population of such scores from which our random sample was selected is approximately normally distributed, then the confidence interval for μ is given by the general formula

$$\bar{X} \pm t_{n-1,1-\alpha/2} S/\sqrt{n} \qquad (3.9)$$

which gives the $100(1-\alpha)\%$ (small-sample) confidence interval for μ when σ is unknown. In our problem $\alpha = 1 - 0.95 = 0.05$ and $n = 9$, so by substitution of the given information into (3.9), we obtain

$$520 \pm t_{8,0.975}(50/\sqrt{9})$$

Since $t_{8,0.975} = 2.3060$, this formula becomes

$$520 \pm 2.3060(50/\sqrt{9})$$

or

$$520 \pm 38.43$$

Our 95% confidence interval for μ is thus given by

$$481.57 \leq \mu \leq 558.43$$

One final point is worth mentioning. If we wanted to use this CI to help us decide whether 600 is a likely value for μ or not (i.e., we are interested in making a decision concerning a specific value for μ), we would conclude that 600 is not a likely value, since it is not contained in the 95% CI for μ just developed. This helps to illustrate the connection between estimation and hypothesis testing.

EXAMPLE 2   Suppose that we wish to compare the change in health status of two groups of mental patients undergoing different forms of treatment for the same disorder. Suppose that we have a measure of change in health status based on a questionnaire given to each patient at two different times, and we are willing to assume that this measure of change in health status is approximately normally distributed with the same variance for the populations of patients from which we select our independent random samples. The data obtained are summarized as follows:

Group 1: $n_1 = 15$, $\bar{X}_1 = 15.1$, $S_1 = 2.5$

Group 2: $n_2 = 15$, $\bar{X}_2 = 12.3$, $S_2 = 3.0$

where the underlying variable X denotes the change in health status between time 1 and time 2.

A 99% CI for the true mean difference $(\mu_1 - \mu_2)$ in health-status change between these two groups would be given by the following formula (which assumes equal population variances; i.e., $\sigma_1^2 = \sigma_2^2$):

$$(\bar{X}_1 - \bar{X}_2) \pm t_{n_1+n_2-2, 1-\alpha/2} S_p \sqrt{\frac{1}{n_1} + \frac{1}{n_2}} \tag{3.10}$$

where S_p^2 is the pooled sample variance given by (3.5). Here we have

$$S_p^2 = \frac{(15-1)(2.5)^2 + (15-1)(3.0)^2}{15 + 15 - 2} = 7.625$$

so

$$S_p = \sqrt{7.625} = 2.76$$

Since $\alpha = 0.01$, our percentile in (3.10) is given by $t_{28, 0.995} = 2.7633$. So, the 99% CI for $\mu_1 - \mu_2$ is then given by

$$(15.1 - 12.3) \pm 2.7633(2.76) \sqrt{\frac{1}{15} + \frac{1}{15}}$$

which reduces to

$$2.80 \pm 2.78$$

yielding the following 99% confidence interval for $(\mu_1 - \mu_2)$:

$$0.02 \leq \mu_1 - \mu_2 \leq 5.58$$

Since the value zero is not contained in this interval, we would conclude (with 99% confidence) that there is a significant difference in health-status change between the two groups.

3.6 TESTING HYPOTHESES

Although closely related to confidence-interval estimation, hypothesis testing has a slightly different orientation. When developing a confidence interval, we use our sample data to estimate what we think is a *likely* set of values for the parameter of interest; when performing a statistical test of a null hypothesis concerning a certain parameter, on the other hand, we use our sample data to *test* whether our estimated value for the parameter is *different enough* from the hypothesized value to conclude that the null hypothesis is *unlikely* to be true.

The general procedure used in testing a statistical null hypothesis is basically the same regardless of the parameter being considered. This procedure (which we will illustrate by example) consists of the following seven steps:

1. Check the assumptions concerning the properties of the underlying variable(s) being measured, which are needed to justify the use of the testing procedure being considered.
2. State the null hypothesis H_0 and the alternative hypothesis H_A.
3. Specify the significance level α.
4. Specify the test statistic to be used and its distribution under H_0.
5. Form the decision rule for rejecting and not rejecting H_0 (i.e., specify the rejection and acceptance regions for the test).
6. Compute the value of the test statistic from the observed data.
7. Draw your conclusions concerning rejection or nonrejection of H_0.

EXAMPLE   Suppose that we again consider the random sample of nine student applicants with mean QGRE score $\bar{X} = 520$ and standard deviation $S = 50$. Suppose, further, that the department chairperson, because of the declining reputation of the department, suspects that this year's applicants are not quite as good quantitatively as in the previous 5 years, for which the average QGRE score of applicants was 600. Under the assumption that the population of QGRE scores from which our random sample has been selected is normally distributed, it is possible to test the null hypothesis that the population mean score associated with this year's applicants is 600 versus the alternative hypothesis that it is less than 600. The *null hypothesis*, in mathematical terms, is then $H_0: \mu = 600$, which asserts that the population mean μ for this year's applicants is not different from what has been generally found in the past. The *alternative hypothesis* is stated as $H_A: \mu < 600$, since there is interest in detecting whether the QGRE scores, on the average, have gotten worse. We have thus far considered the first two steps of our testing procedure:

1. Assumptions: The variable QGRE score has a normal distribution, from which a random sample has been selected.
2. Hypotheses: $H_0: \mu = 600$, $H_A: \mu < 600$.

The next step is to decide what error (or probability) we are willing to tolerate with regard to the possibility of incorrectly rejecting H_0 (i.e., making a type I error). We call this probability of making a type I error the *significance level* α.

[There are actually two types of errors that can be made when performing a statistical test: a type II error occurs if we fail to reject H_0 when H_0 is actually false. We denote the probability of a type II error by β and call $(1 - \beta)$ the *power* of the test. For a fixed sample size, α and β for a given test are inversely related in the sense that lowering one has the effect of increasing the other. However, both α and β can be simultaneously lowered by increasing the sample size.]

We usually specify α to have a value such as 0.10, 0.05, 0.025, or 0.01. Suppose, for now, that we choose $\alpha = 0.025$, so that step 3 is:

3.   Use $\alpha = 0.025$.

Step 4 requires specification of the test statistic that will be used to test H_0. In this case, with H_0: $\mu = 600$, we can state that:

4. $T = (\bar{X} - 600)/(S/\sqrt{9}) \frown t_8$ under H_0.

Step 5 requires us to specify the decision rule that we will use to accept or reject H_0. In determining this rule, we divide the possible values of T into two sets: one of these sets is called the *rejection region* (or *critical region*), which consists of those values of T for which we reject H_0; the other set, called the *acceptance region*, consists of those values for which we do not reject H_0. The idea here is that if our computed value of T falls in the rejection region, we conclude that our observed results have deviated enough from H_0 to cast considerable doubt on the validity of the null hypothesis.

In our example we determine the critical region by choosing a point from t tables, called the *critical point*, which defines the boundary between the acceptance and rejection regions. The value we choose is

$$-t_{8, 0.975} = -2.306$$

in which case the probability that the test statistic takes values less than -2.306 under H_0 is exactly $\alpha = 0.025$ (the significance level; Figure 3.5). We thus have the following decision rule:

5.   *Reject H_0 if* $T = \dfrac{\bar{X} - 600}{S/\sqrt{9}} < -2.306$; *do not reject H_0 otherwise.*

We now simply apply the decision rule to our data by computing the observed value of T. In our example, since $\bar{X} = 520$ and $S = 50$, our computed T is

6. $T = \dfrac{\bar{X} - 600}{S/\sqrt{9}} = \dfrac{520 - 600}{50/3} = -4.8.$

FIGURE 3.5   *The critical point for our example*

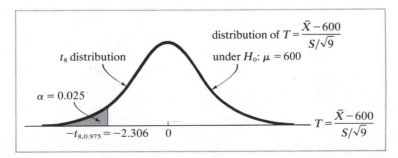

distribution of $T = \dfrac{\bar{X} - 600}{S/\sqrt{9}}$ under H_0: $\mu = 600$

t_8 distribution

$\alpha = 0.025$

$-t_{8, 0.975} = -2.306$ 0

$T = \dfrac{\bar{X} - 600}{S/\sqrt{9}}$

The last step is to make the decision about H_0 based on the rule given in step 5:

7. Since $T = -4.8$, which lies below -2.306, we reject H_0 at significance level 0.025, and conclude that there is evidence that students currently applying to the department have *significantly lower* QGRE scores than 600.

In addition to the procedure above, it is often of interest to compute a *P value*, which quantifies *exactly how unusual the observed results are if H_0 were true*. An equivalent way of describing the P value is as follows: *The P value gives the probability of obtaining a value of the test statistic "at least" as unfavorable to H_0 as the observed value* (Figure 3.6).

To get an idea of the approximate size of the P value in this example, the approach is to determine from the table of the distribution of T under H_0 the two percentiles that bracket the observed value of T. In this case the two percentiles are

$$-t_{8,0.995} = -3.355 \quad \text{and} \quad -t_{8,0.9995} = -5.041$$

Since the observed value of T lies between these two values, we can conclude that the area we seek lies between the two areas corresponding to these two percentiles,

$$0.0005 < P < 0.005$$

In interpreting this inequality, we observe that the P value is *quite small*. This small P indicates that we have observed a highly unusual result if, indeed, H_0 is true; in fact, this P value is so small as to lead us to reject H_0. Furthermore, the size of this P value means that even for an α as small as 0.005 we would reject H_0.

For the general computation of a P value, it is important to point out that the appropriate P value for a two-tailed test will be twice that for the corresponding one-tailed test. Furthermore, if an investigator wishes to draw conclusions about a test on the basis of the P value (e.g., in lieu of specifying α a priori), the following guidelines are recommended:

1. If P is small (less than 0.01), reject H_0.·
2. If P is large (greater than 0.10), do not reject H_0.
3. If $0.01 < P < 0.10$, the significance is borderline; that is, we reject H_0 for $\alpha = 0.10$ but do not reject H_0 for $\alpha = 0.01$.

Note that if we actually do specify α a priori, we reject H_0 when $P < \alpha$.

FIGURE 3.6   *The P value*

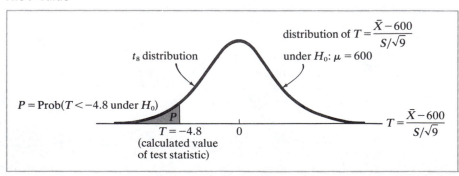

To complete this review we will look at one more worked example about hypothesis testing, this time involving a comparison of two means, μ_1 and μ_2. Consider the following health-status-change data, which were discussed earlier:

Group 1: $n_1 = 15$, $\bar{X}_1 = 15.1$, $S_1 = 2.5$

$(S_p = 2.76)$

Group 2: $n_2 = 15$, $\bar{X}_2 = 12.3$, $S_2 = 3.0$

Suppose that we wish to test at significance level 0.01 whether or not the true average change in health status differs between the two groups. The steps required to perform this test would then be carried out as follows:

1. Assumptions: We have independent random samples from two normally distributed populations. The population variances are assumed to be equal.
2. $H_0: \mu_1 = \mu_2$, $H_A: \mu_1 \neq \mu_2$.
3. $\alpha = 0.01$.
4. $T = \dfrac{(\bar{X}_1 - \bar{X}_2) - 0}{S_p\sqrt{1/n_1 + 1/n_2}} \frown t_{28}$ under H_0.
5. Reject H_0 if $|T| \geq t_{28,0.995} = 2.763$; do not reject H_0 otherwise (Figure 3.7).
6. $T = \dfrac{(\bar{X}_1 - \bar{X}_2) - 0}{S_p\sqrt{1/n_1 + 1/n_2}} = \dfrac{15.1 - 12.3}{2.76\sqrt{\frac{1}{15} + \frac{1}{15}}} = 2.78$.
7. Since $T = 2.78$ exceeds $t_{28,0.995} = 2.763$, we would reject H_0 at $\alpha = 0.01$ and conclude that the true average change in health status is different in the two groups.

The P value for this test is given by the shaded area in Figure 3.8. For the t distribution with 28 degrees of freedom, we find that $t_{28,0.995} = 2.763$ and $t_{28,0.9995} = 3.674$. Thus, $P/2$ is given by the inequality

$$(1 - 0.9995) < P/2 < (1 - 0.995)$$

so

$$0.001 < P < 0.01$$

FIGURE 3.7   *Critical region for the health-status-change example*

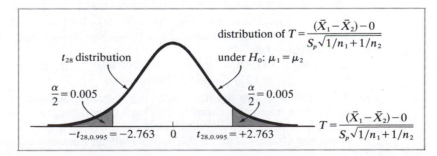

FIGURE 3.8   *P value for the health-status-change example*

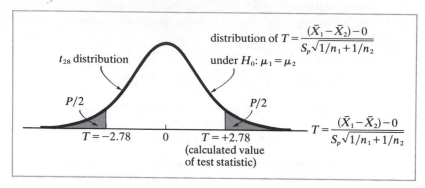

1. Suppose that 175 out of a sample of 625 pregnant married women developed complications from pregnancy. Let X be a random variable which takes the value 0 for a woman without any complications and 1 for a woman with at least one complication. The sample mean $(\bar{X})$ for these data is
(a) 28.0   (b) 0.0   (c) 1.0   (d) 0.62   (e) 0.28

2. The sample median of X values for the data in Problem 1 is
(a) 313   (b) 1   (c) 0   (d) 0.28   (e) 0.62

3. The sample standard deviation S_X for the data on X in Problem 1 is
(a) 0.4494   (b) 0.2019   (c) 126   (d) 0.7200   (e) None of these

4. Which of the following statements about descriptive statistics is correct?
(a) The median uses *all* the data for its computation.
(b) The mean should be preferred to the median as a measure of central tendency if the data are noticeably skewed.
(c) The variance has the same units of measurement as the original observations.
(d) The variance can never be zero.
(e) The variance is like an average of squared deviations from the mean.

5. Suppose that the weight (W) of male patients registered at a diet clinic has the normal distribution with mean 190 and variance 100. For a random sample of patients of size $n = 25$, which of the following statements is equivalent to the statement $P(\bar{W} > 195)$, where $\bar{W}$ denotes the mean weight of the sample?
(a) $P(Z < -0.5)$ (b) $P(Z < 2.5)$ (c) $P(Z > -2.5)$
(d) $P(Z > -0.5)$ (e) $P(Z < -2.5)$
(*Note*: Z is a standard normal random variable.)

6. Find the interval (a, b) such that $P(a < \bar{W} < b) = 0.90$, where $\bar{W}$ is, as in Problem 5, the mean of a random sample of size 25 from a normal distribution with mean 190 and variance 100.
(a) (186.08, 193.92)   (b) (173.55, 206.45)   (c) (170.40, 209.60)
(d) (186.71, 193.29)   (e) None of these

7. For workers in the tire-manufacturing industry, suppose it is known that the population mean vital capacity (FEV) is 3.91 and the population variance is 4. If a random sample of 49 workers from a certain plant is chosen, the probability that the mean vital capacity for this sample will be between 2.5 and 4.5 is:

(a) 0.9806   (b) 0.3760   (c) 0.4213   (d) 0.8417
(e) None of these

8. The limits of a 99% confidence interval for the mean μ of a normal population with unknown variance are found by adding to and subtracting from the sample mean a certain multiple of the estimated standard error of the sample mean. If the sample size on which this confidence interval is based is 28, the *multiple* referred to in the previous sentence is:
(a) 2.467   (b) 2.473   (c) 2.7633   (d) 2.5758   (e) None of these

9. A random sample of 25 persons attending a certain diet clinic was found to have lost (over a 3-week period) an average of 30 pounds, with a sample standard deviation of 10. For these data, a 99% confidence interval for the true mean weight loss for all patients attending the clinic would have the following limits:
(a) (15.05, 44.95)   (b) (13.22, 46.78)   (c) (14.54, 45.46)
(d) (26.64, 33.36)   (e) None of these

10. A random sample of cities under 25,000 persons is to be taken, the purpose being to estimate the true mean air-pollution index for such cities. If the standard deviation is assumed to be 0.0150 and if we require 99% confidence that the sample mean index is to be within 0.0027 of the population mean for such cities, what is an approximate value for the minimum sample size (i.e., the minimum number of cities) required?
(a) 37   (b) 167   (c) 205   (d) 72   (e) None of these
(*Hint:* The width of the confidence interval is to be at most 0.0054.)

11. From two normal populations assumed to have the same variance, there were drawn independent random samples of sizes 21 and 10. The first sample (with $n_1 = 21$) yielded mean and standard deviation 116.4 and 10.0, respectively, while the second sample ($n_2 = 10$) gave mean and standard deviation 105.9 and 11.0, respectively. The estimated standard error of the difference in sample means is given by:
(a) 15.7240   (b) 3.9653   (c) 1.2888   (d) 10.3207
(e) None of these

12. For the data of Problem 11, suppose that a test of $H_0: \mu_1 = \mu_2$ versus $H_A: \mu_1 > \mu_2$ yielded a computed value of the appropriate test statistic equal to 2.65. Which of the following conclusions may then be drawn?
(a) Reject H_0 for $\alpha = 0.05$ but do not reject H_0 for $\alpha = 0.01$.
(b) Reject H_0 since $0.005 < P < 0.01$.
(c) Reject H_0 since $0.01 < P < 0.05$.
(d) Conclude borderline significance since $0.01 < P < 0.05$.
(e) None of the above conclusions is appropriate.

13. Independent random samples are drawn from two normal populations, which are assumed to have the same variance. One sample (of size 5) yields mean 86.4 and standard deviation 8.0, and the other sample (of size 7) has mean 78.6 and standard deviation 10. The limits of a 95% confidence interval for the difference in population means are found by adding to and subtracting from the difference in sample means a certain multiple of the estimated standard error of this difference. This *multiple* is:
(a) 2.201   (b) 2.228   (c) 1.812   (d) 1.960   (e) None of these

14. If a 99% confidence interval for $(\mu_1 - \mu_2)$ is given by $4.3 < (\mu_1 - \mu_2) < 8.1$, which of the following conclusions can be drawn *based on this interval*?

(a) Accept H_0: $\mu_1 = \mu_2$ at $\alpha = 0.05$ if your alternative is H_A: $\mu_1 \neq \mu_2$.
(b) Reject H_0: $\mu_1 = \mu_2$ at $\alpha = 0.01$ if your alternative is H_A: $\mu_1 \neq \mu_2$.
(c) Reject H_0: $\mu_1 = \mu_2$ at $\alpha = 0.01$ if your alternative is H_A: $\mu_1 < \mu_2$.
(d) Accept H_0: $\mu_1 = \mu_2$ at $\alpha = 0.01$ if your alternative is H_A: $\mu_1 \neq \mu_2$.
(e) Accept H_0: $\mu_1 = \mu_2 + 3$ at $\alpha = 0.01$ if your alternative is H_A: $\mu_1 \neq \mu_2 + 3$.

15. Suppose that the critical region for a certain test of hypothesis is of the form $|T| \geq 3.5$ and the computed value of T from the data is -2.75. Which of the following statements is correct?
(a) H_0 should be rejected.
(b) The significance level α is the probability that, under H_0, either T is greater than 2.75 or less than -2.75.
(c) The acceptance region is given by $-3.5 < T < 3.5$.
(d) The acceptance region consists of those values of T above 3.5 or below -3.5.
(e) The P value of this test is given by the area to the right of $T = 3.5$ for the distribution of T under H_0.

16. Suppose that $\bar{X}_1 = 125.2$ and $\bar{X}_2 = 125.4$ are the mean blood pressures for two samples of workers from different plants in the same industry. Suppose, further, that a test of H_0: $\mu_1 = \mu_2$ using these sample means is rejected for $\alpha = 0.001$. Which of the following conclusions is most reasonable?
(a) There is a meaningful difference (clinically speaking) in population means but not a statistically significant difference.
(b) The difference in population means is both statistically and meaningfully significant.
(c) There is a statistically significant difference but not a meaningfully significant difference in population means.
(d) There is neither a statistically significant nor a meaningfully significant difference in population means.
(e) The sample sizes used must have been quite small.

17. The choice of an alternative hypothesis (H_A) should depend primarily on:
(a) The data obtained from the study.
(b) What the investigator is interested in determining.
(c) The critical region.
(d) The significance level.
(e) The power of the test.

18. Suppose that H_0: $\mu_1 = \mu_2$ is the null hypothesis, that $0.005 < P < 0.01$, and that $\alpha = 0.005$. Our conclusion would be to:
(a) Reject H_0 since P is small.
(b) Accept H_0 since P is small.
(c) Reject H_0 since α is less than P.
(d) Accept H_0 since α is less than P.
(e) Reject H_0 since α is large.

19. If the P value for a certain test satisfies the inequality $P > 0.25$, which of the following conclusions would you make?
(a) You would not reject H_0.
(b) You would reject H_0 for $\alpha = 0.05$.
(c) You would reject H_0 for $\alpha = 0.10$.
(d) Your acceptance region has a lower limit of 0.25.
(e) Your significance level α is greater than 0.25.

20. Suppose that H_0: $\mu_1 = \mu_2$ is the null hypothesis, and that $0.005 < P < 0.01$. Which of the following conclusions is most appropriate?
   (a) Accept H_0 because P is small.
   (b) Reject H_0 because P is small.
   (c) Accept H_0 because P is large.
   (d) Reject H_0 because P is large.
   (e) Accept H_0 at $\alpha = 0.01$.

CHAPTER 4 INTRODUCTION TO REGRESSION ANALYSIS

4.1 PREVIEW

Regression analysis is a statistical tool for evaluating the relationship of one or more independent variables $X_1, X_2, \ldots, X_k$ to a single continuous dependent variable Y. It is most often used when the independent variables are not controllable, as when collected in a sample survey or other observational study. Nevertheless, it is equally applicable to more controlled experimental situations.

In practical applications there are several *possibly overlapping* situations for which a regression analysis would be appropriate. Among these are the following:

APPLICATION 1   You wish to *characterize the relationship* between the dependent and independent variables in the sense of determining the extent, direction, and strength of the association among these variables. For example ($k = 2$), in Thompson's (1972) study described in Chapter 1, one of the primary questions was to describe the extent, direction, and strength of the association between "patient satisfaction with medical care (Y)" and the variables "affective communication between patient and physician (X_1)" and "informational communication between patient and physician (X_2)."

APPLICATION 2   You seek a *quantitative formula* or equation to describe (e.g., predict) the dependent variable Y as a function of the independent variables $X_1, X_2, \ldots, X_k$. For example ($k = 1$), a quantitative formula may be desired for a study dealing with the effect of dosage of a blood-pressure-reducing treatment (X_1) on blood pressure change (Y).

APPLICATION 3   You want to describe quantitatively or qualitatively the relationship between $X_1, X_2, \ldots, X_k$ and Y. You wish to *control* for the effects of still other variables, $C_1, C_2, \ldots, C_p$, which you believe have an important relationship with the dependent variable. For example ($k = 2$, $p = 2$), a study in the area of chronic disease epidemiology might involve describing the relationship of blood pressure (Y) to smoking habits (X_1) and social class (X_2), controlling for age (C_1) and weight (C_2).

APPLICATION 4 You want to *determine which of several independent variables are important and which are not* for describing or predicting a dependent variable. You may want to control for other variables. You may also want to *rank* independent variables in their order of importance. In Thompson's (1972) study, for example ($k = 4$, $p = 2$), it was of interest to determine for the dependent variable "satisfaction with medical care (Y)" which of the following independent variables are important descriptors: WORRY (X_1), WANT (X_2), INFCOM (X_3), AFFCOM (X_4). It was also considered necessary to control for AGE (C_1) and EDUC (C_2).

APPLICATION 5 You want to determine the *best interpretive mathematical model* for describing the relationship between a dependent and one or more independent variables. Any of the previous examples can be used to illustrate this.

APPLICATION 6 You wish to *compare* several derived regression relationships. An example would be a study to determine if smoking (X_1) is related to blood pressure (Y) in the same way for males as for females, controlling for age (C_1).

4.2 REGRESSION AND CAUSALITY

It is important to be cautious about the results obtained from a regression analysis. A strong relationship (i.e., an association) found between variables does not necessarily prove or even imply that the independent variables are *causes* of the dependent variable. In order to make such causal inferences, both *additional methodology* [see Blalock (1964)] and *experimentation* are required, as in a carefully controlled clinical trial.

Although causality cannot necessarily be inferred from a regression analysis, a meaningful interpretation of the relationship between variables can be described in a *statistical* sense. It is through statistical techniques, such as the use of confidence intervals and tests of hypotheses, that the investigator can infer the extent to which changes in the independent variables are related to changes in the dependent variable. However, such statistical procedures stop short of permitting the conclusion that the relationship is causal.

For example, if blood pressure and age were determined for a sample of individuals, regression analysis could help demonstrate the generally accepted statistical relationship that, on the average, blood pressure increases with age, and could also help quantify the strength of the relationship. However, it could not be further inferred without consideration of additional variables and further experimentation that the aging process actually *causes* increased blood pressure.

Also, the *statistical statements* that we will be able to make based on regression or other multivariable analyses need to be distinguished from *deterministic statements*. The law of falling bodies, for example, is a deterministic statement which assumes an ideal setting, the values of the dependent variable varying in a completely prescribed way so as to trace a perfect (error-free) mathematical function of the values of the independent variables.

Statistical statements, on the other hand, allow for the possibility of error in the description of a relationship. For example, in relating blood pressure to age, persons of the same age are unlikely to have exactly the same blood pressure. Nevertheless, we could conclude that, *on the average*, blood pressure increases with age. We could even predict what blood pressure to *expect* for a given age and associate a measure of accuracy with that prediction. Such statements, it should be added, although not deterministic, are precise. Through the use of probability and statistical theory, they take into account the irregularities of the real world that are associated with measurement error and individual variability. But, because they deal with the real world, they require careful interpretation. This is the part that is often quite difficult!

CHAPTER 5 STRAIGHT-LINE REGRESSION ANALYSIS

5.1 PREVIEW

The simplest (but by no means trivial) form of the general regression problem deals with one dependent variable Y and one independent variable X. We have previously described the general problem in terms of k independent variables $X_1, X_2, \ldots, X_k$. Let us now restrict our attention to the special case $k = 1$, but denote X_1 as X in order to keep our notation as simple as possible. We find it useful to begin with a single independent variable in order to clarify the basic concepts and assumptions of regression analysis. Furthermore, in this regard, researchers often find it sensible to begin by looking at one independent variable even when other independent variables will also be considered eventually.

5.2 REGRESSION WITH A SINGLE INDEPENDENT VARIABLE

We shall begin this section by describing the statistical problems of finding the *curve* (e.g., straight line, parabola, etc.) that *best fits* the data in such a way as to closely approximate the true (but unknown) relationship between X and Y.

5.2.1 THE PROBLEM

Given a sample of n individuals (or other study units such as geographical locations, time points, or pieces of physical material), we observe for each a value of X and a value of Y. We thus have n pairs of observations which can be denoted by $(X_1, Y_1), (X_2, Y_2), \ldots, (X_n, Y_n)$, where the subscripts now refer to the different individuals rather than to different variables. These pairs may be considered as points in two-dimensional space, so that we may plot them on a graph. Such a graph is called a

FIGURE 5.1   *Scatter diagram of age and systolic blood pressure*

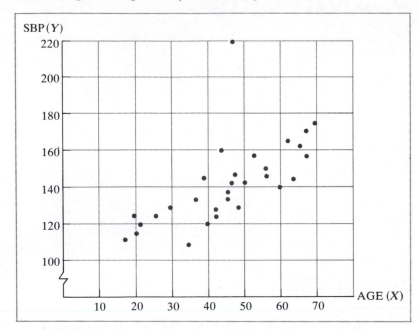

scatter diagram. For example, measurements of age and systolic blood pressure (SBP) for 30 individuals might yield the scatter diagram given in Figure 5.1.

5.2.2 BASIC QUESTIONS TO BE ANSWERED

There are two basic questions to be dealt with in any regression analysis:

1. What is the most appropriate mathematical model to use? In other words, should we use a straight line, a parabola, a log function, or what?
2. Given a specific model, what do we mean by and how do we determine the best-fitting model for the data? In other words, if our model is a straight line, how do we find the best-fitting line?

5.2.3 GENERAL STRATEGY

There are several general strategies for studying the relationship between two variables by means of regression analysis. The most common of these is called the *forward method*. This strategy begins with a simply structured model, usually a straight line, and then adds more complexity to the model in successive steps, if necessary. Another strategy is called the *backward method*. This strategy begins with a complicated model, such as a high-degree polynomial, and then successively simplifies the model, if possible, by eliminating unnecessary terms. A third approach uses a *model suggested from experience or theory* and allows for revision of this model either toward or away from complexity, as dictated by the data.

The choice of strategy depends essentially on the type of problem and on the data that we have; there are no hard-and-fast rules in this regard. Of course, we might try more than one of these approaches and then use those results which describe the relationship between the dependent and independent variables in the most reasonably interpretable way.

Because it is the most reasonable strategy to use in the absence of experience or theory to indicate otherwise, the *forward* method will now be described:

1. Begin by assuming that a straight line is the appropriate model. Later, the validity of this assumption can be investigated.

2. Find the best-fitting straight line as that line, among all possible straight lines, which best agrees (in some well-defined way to be outlined below) with the data.

3. Determine whether or not the straight line found in step 2 significantly helps to describe the dependent variable *Y*. Inherent in this step is the necessity to check that certain basic statistical assumptions (e.g., normality) are met. These assumptions will be discussed in detail below.

4. Examine whether or not the assumption of a straight-line model is correct. One approach for doing this is called *testing for lack of fit*, although other approaches can be used instead.

5. If step (4) finds the assumption of a straight line to be invalid, fit a new model (e.g., a parabola) to the data, determine how well it describes *Y* (i.e., repeat step 3), and then decide whether or not the new model is appropriate (i.e., repeat step 4).

6. Continue to try new models until an appropriate one is found.

A flow diagram for this strategy is given in Figure 5.2.

FIGURE 5.2   *Flow diagram of the forward method*

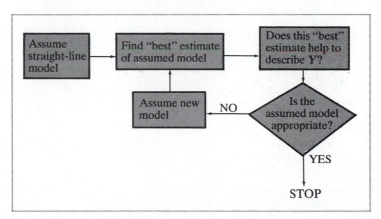

Since the usual (forward) approach to regression analysis with a single independent variable begins with the assumption of a straight-line model, we will first focus attention on this model. Before describing the *statistical* methodology for this special case, it is appropriate to review some basic straight-line mathematics. You may wish to omit the next section if already familiar with its contents.

5.3 MATHEMATICAL PROPERTIES OF A STRAIGHT LINE

Mathematically, a straight line can be described by an equation of the form

$$y = \beta_0 + \beta_1 x \tag{5.1}$$

We have used lowercase letters y and x in this equation instead of capital letters to emphasize that we are treating these variables in a purely mathematical, rather than in a statistical, context. The symbols β_0 and β_1 have constant values for a given line and are, therefore, not considered variables; β_0 is called the *y intercept* of the line and β_1 is called the *slope*. Thus, $y = 5 - 2x$ describes a straight line with intercept 5 and slope -2, whereas $y = -4 + 1x$ describes a different line with intercept -4 and slope 1. These two lines are shown in Figure 5.3.

The intercept β_0 is the value of y when $x = 0$. For the line $y = 5 - 2x$, $y = 5$ when $x = 0$. For the line $y = -4 + 1x$, $y = -4$ when $x = 0$. The slope β_1 is the amount of change in y for each $1 -$ unit change in x. For any given straight line, this rate of change is always constant. Thus, for the line $y = 5 - 2x$, when x changes 1 unit from 3 to 4, y changes -2 units (the value of the slope) from $5 - 2(3) = -1$ to $5 - 2(4) = -3$. When x changes from 1 to 2, also 1 unit, y changes from $5 - 2(1) = 3$ to $5 - 2(2) = 1$, also -2 units.

The properties of any straight line can be viewed graphically as in Figure 5.3. To plot a given line, all that is needed is to plot any two points on the line and then connect them with a ruler. One of the two points often used is the point corresponding to the intercept. This point is given by $x = 0$, $y = 5$ for the line $y = 5 - 2x$ and by $x = 0$, $y = -4$ for $y = -4 + 1x$. The other point for each line may be determined by arbitrarily selecting an x and finding the corresponding y. An x of 3 was used in our two examples. Thus, for $y = 5 - 2x$, an x of 3 yields a y of $5 - 2(3) = -1$. For $y = -4 + 1x$, an x of 3 yields a y of

FIGURE 5.3   *Straight-line plots*

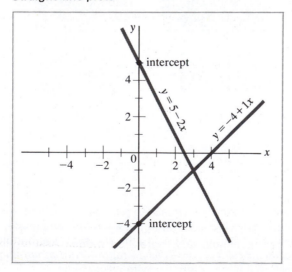

$-4 + 1(3) = -1$. The line $y = 5 - 2x$ can then be drawn by connecting the points ($x = 0$, $y = 5$) and ($x = 3$, $y = -1$). The line $y = -4 + 1x$ can be drawn from the points ($x = 0$, $y = -4$) and ($x = 3$, $y = -1$).

It can also be seen from Figure 5.3 that for the equation $y = 5 - 2x$, y decreases as x increases. Such a line is said to have *negative* slope. Indeed, this definition agrees with the sign of the slope -2 in the equation. Similarly, the line $y = -4 + 1x$ is said to have *positive* slope, since y increases as x increases.

5.4 STATISTICAL ASSUMPTIONS FOR A STRAIGHT-LINE MODEL

Suppose that we have tentatively assumed a straight-line model as the first step in the forward method for determining the best model describing the relationship between X and Y. We now wish to determine the best-fitting line. Certainly, we will have no trouble in deciding what is meant by "best fitting" if the data are such as to allow a single straight line to pass through each and every point in the scatter diagram. Unfortunately, this will never happen with real-life data, as we mentioned in Chapter 4. For example, persons of the same age are unlikely to have the same blood pressure, height, or weight.

Thus, we must realize that the straight line we seek is an approximation to the true state of affairs, and cannot be expected to predict precisely each individual's Y from that individual's X. In fact, this would be so even if we measured X and Y on the whole population of interest instead of on just a sample from that population. In addition, the fact that the line is to be determined from the sample data and not from the population requires us to consider statistical problems concerning the estimation of unknown population parameters.

What are these parameters? The ones of primary concern at this point are the intercept β_0 and slope β_1 of the straight line of the general mathematical form of (5.1) that best fits the X-Y data for the entire population. In order to make inferences from the sample about this population line, we need to make certain statistical assumptions. These assumptions are as follows.

ASSUMPTION 1 For any fixed value of the variable X, Y is a random variable with a certain probability distribution. The (population) mean of this distribution will be denoted as $\mu_{Y|X}$ and the (population) variance as $\sigma^2_{Y|X}$. The notation "$Y|X$" indicates that the mean and variance of the random variable Y depend on the value of X.

This assumption will apply to any regression model, whether a straight line or not. Figure 5.4 pictorially illustrates this assumption. The different distributions are drawn vertically to correspond to different values of X. The dots denoting the mean values $\mu_{Y|X}$ at different X's have been connected to form the *regression equation*, which is the population model to be estimated from the data.

ASSUMPTION 2 *Independence*

The Y values are statistically independent of one another. As with Assumption 1, this assumption applies to almost any regression model. Assumption 2 is sometimes violated when different observations are made on the same individual at different

FIGURE 5.4   *General regression equation*

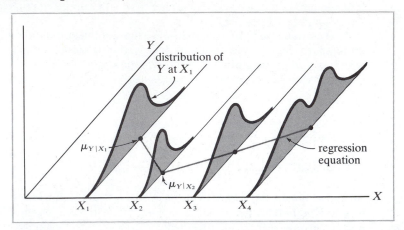

points in time. For example, if weight was measured on an individual at different times, it is to be expected that weight at one time would be related to weight at a later time. As another example, if blood pressure is measured on a given individual over time, one would expect the blood pressure value at one time to be approximately in the same range as the blood pressure value at an adjacent time point.

When observations are not independent, special methods can be used to find the best-fitting curve. One such method which incorporates information reflecting the extent of statistical dependence among observations on the same individual is called *growth-curve analysis* [e.g., see Allen and Grizzle (1969)].

ASSUMPTION 3   *The straight-line assumption*

The mean value of Y, $\mu_{Y|X}$, is a straight-line function of X; that is, if you connect the dots denoting the different mean values $\mu_{Y|X}$, you will obtain a straight line. This assumption is represented pictorially in Figure 5.5.

FIGURE 5.5   *Straight-line assumption*

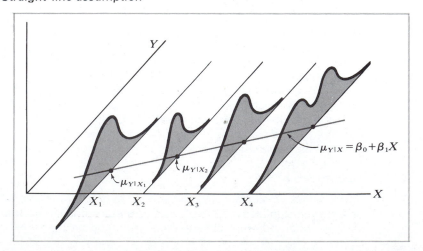

Using mathematical symbols, we can describe Assumption 3 by the equation

$$\mu_{Y|X} = \beta_0 + \beta_1 X \qquad (5.2)$$

where β_0 and β_1 are the intercept and slope of this (population) straight line, respectively. Equivalently, we can express (5.2) in the form

$$Y = \beta_0 + \beta_1 X + E \qquad (5.3)$$

where E denotes a random variable which has mean 0 at fixed X (i.e., $\mu_{E|X} = 0$ for any X). To be more specific, since X is fixed and not random, (5.3) represents the dependent variable Y as the sum of a constant term $(\beta_0 + \beta_1 X)$ and a random variable (E). Thus, the probability distributions of Y and E differ only in the value of this constant term; that is, since E has mean 0, so Y must have mean $(\beta_0 + \beta_1 X)$.

[The statement "X is fixed and not random" is often associated with the statement "X is measured without error." For our purposes, the practical implication of either statement with regard to making statistical inferences from sample to population is that the *only* random component on the right-hand side of (5.3) is E.]

Equations (5.2) and (5.3) describe a *statistical* model. These equations should be distinguished from the *mathematical* model for a straight line described by (5.1), which does not consider Y as a random variable.

The variable E describes how far away an individual's response is from the population regression line (Figure 5.6). In other words, what we observe at a given X (namely, Y) is in *error* from that expected on the average (namely, $\mu_{Y|X}$) by an amount E, which is random and varies from individual to individual. For this reason, E is commonly referred to as the *error component* in the model (5.3); mathematically, E is given by the formula

$$E = Y - (\beta_0 + \beta_1 X)$$

or by

$$E = Y - \mu_{Y|X}$$

FIGURE 5.6   *Error component E*

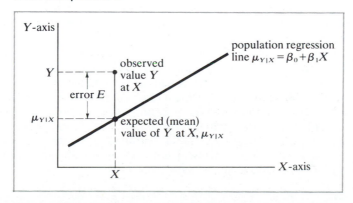

This concept of an error component is particularly important for defining a good-fitting line since, as we will see in the next section, a line that fits the data well ought to have small deviations (or errors) between what is observed and what is predicted by the fitted model.

ASSUMPTION 4   *The variance of Y is the same for any X*

This assumption is called the *assumption of homoscedasticity*, where "homo" means "the same" and "scedastic" means "scatter." An example that illustrates the violation of this assumption (called *heteroscedasticity*) is given in Figure 5.5. In this figure the distribution of Y at X_1 can be seen to have considerably more spread than the distribution of Y at X_2. This means that $\sigma^2_{Y|X_1}$, the variance of Y at X_1, is greater than $\sigma^2_{Y|X_2}$, the variance of Y at X_2.

In mathematical terms, the homoscedastic assumption can be written

$$\sigma^2_{Y|X} \equiv \sigma^2 \qquad \text{for all } X$$

This formula is a shortcut way of saying that since $\sigma^2_{Y|X_i} = \sigma^2_{Y|X_j}$ for *any* two values (say, X_i and X_j) of X, we might as well simplify our notation by giving this common variance a single name, say σ^2, which does not involve X at all.

A number of techniques of varying statistical sophistication can be used to determine whether or not the homoscedastic assumption is satisfied. Some of these procedures will be discussed in Chapter 16.

ASSUMPTION 5   *For any fixed value of X, Y has a normal distribution*

This assumption makes it convenient to evaluate the statistical significance (e.g., by means of confidence intervals and tests of hypotheses) of the relationship between X and Y as reflected by the fitted line.

Figure 5.5 provides an example when this assumption is violated. In addition to the variances not being all equal in this figure, it can also be seen that the distributions of Y at X_3 and of Y at X_4 are not normal. The distribution at X_3 is skewed, whereas the normal distribution is symmetric. The distribution at X_4 is bimodal (two humps), whereas the normal distribution is unimodal (one hump). Methods for determining whether or not the normality assumption is tenable are described in Chapter 16.

It is important to emphasize that if the normality assumption is not *badly* violated, the conclusions reached by a regression analysis assuming normality will generally be reliable and accurate. This stability property with respect to deviations from normality is a type of *robustness*. Consequently, we recommend that considerable leeway be given before deciding that the normality assumption is so badly violated as to require alternative inference-making procedures.

If the normality assumption is deemed unsatisfactory, an attempt may be made to transform the observations using a log, square root, or other function to make the new set of observations approximately normal. Care must be taken when using such transformations to see that other assumptions, such as variance homogeneity, are not violated for the transformed variable. Fortunately, it often turns out in practice that such transformations usually help both the normality and variance homogeneity assumptions to be more closely satisfied.

5.5 DETERMINING THE BEST-FITTING STRAIGHT LINE

By far the simplest and quickest method for determining a straight line is to choose that line which can best be drawn by eye. Although such a method paints a reasonably good picture, this procedure is much too subjective and imprecise and is worthless for purposes of statistical inference. Two analytical approaches for finding the best-fitting straight line will now be described.

5.5.1 THE LEAST-SQUARES METHOD

The *least-squares method* determines the best-fitting straight line as that line which *minimizes* the sum of squares of the lengths of the vertical-line segments (Figure 5.7) drawn from the observed data points on the scatter diagram to the fitted line. The idea here is that the smaller the deviations of observed values from this line (and conse-quently the smaller the sum of squares of these deviations), the "closer" or "snugger" the best-fitting line will be to the data.

Using mathematical notation, the least-squares method is described as follows. Let $\hat{Y}_i$ denote the estimated response at X_i based on the fitted regression line; in other words, $\hat{Y}_i = \hat{\beta}_0 + \hat{\beta}_1 X_i$, where $\hat{\beta}_0$ and $\hat{\beta}_1$ are the intercept and slope of the fitted line. The vertical distance between the observed point (X_i, Y_i) and the corresponding point $(X_i, \hat{Y}_i)$ on the fitted line is given by the absolute value $|Y_i - \hat{Y}_i|$ or $|Y_i - (\hat{\beta}_0 + \hat{\beta}_1 X_i)|$. The sum of squares of all such distances is then given by

$$\sum_{i=1}^{n} (Y_i - \hat{Y}_i)^2 = \sum_{i=1}^{n} (Y_i - \hat{\beta}_0 - \hat{\beta}_1 X_i)^2$$

The least-squares solution is defined to be that choice of $\hat{\beta}_0$ and $\hat{\beta}_1$ for which the sum of squares above is a minimum. It is standard jargon to refer to $\hat{\beta}_0$ and $\hat{\beta}_1$ as the least-squares estimates of the parameters β_0 and β_1, respectively, in the statistical model given by (5.3).

The *minimum sum of squares* corresponding to the least-squares estimates $\hat{\beta}_0$ and $\hat{\beta}_1$ is usually called the *sum of squares about the regression line*, the *residual sum*

FIGURE 5.7   *Deviations of observed points from the fitted regression line*

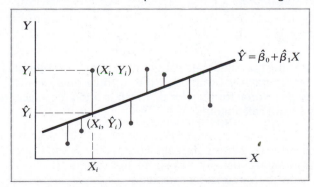

of squares, or the *sum of squares due to error* (SSE). The measure SSE is of great importance in assessing the quality of the straight-line fit, and its interpretation will be discussed in Section 5.6.

Mathematically, the essential property of the measure SSE can be stated in the following way. If β_0^* and β_1^* denote any other possible estimators of β_0 and β_1, we must have

$$SSE = \sum_{i=1}^{n} (Y_i - \hat{\beta}_0 - \hat{\beta}_1 X_i)^2 \le \sum_{i=1}^{n} (Y_i - \beta_0^* - \beta_1^* X_i)^2$$

5.5.2 THE MINIMUM-VARIANCE METHOD

The *minimum-variance method* is more classically statistical in nature than the method of least squares, which can be looked upon as a purely mathematical algorithm. It deals with the problem of defining "best fit" as a statistical estimation problem. The goal is to find point estimators of β_0 and β_1 with good statistical properties. In this regard, under the previous assumptions, the best line is determined by those estimators $\hat{\beta}_0$ and $\hat{\beta}_1$ which are *unbiased* for their unknown population counterparts β_0 and β_1, respectively, and which, in addition, have minimum variance among all unbiased (linear) estimators of β_0 and β_1.

5.5.3 THE SOLUTION TO THE BEST-FIT PROBLEM

Fortunately, both the least-squares and minimum-variance methods yield exactly the same solution, which we will state without proof.

Let $\bar{Y}$ denote the sample mean of the observations on Y and $\bar{X}$ denote the sample mean of the values of X. Then the best-fitting straight line is determined by the formulas

$$\hat{\beta}_1 = \frac{\sum_{i=1}^{n} (X_i - \bar{X})(Y_i - \bar{Y})}{\sum_{i=1}^{n} (X_i - \bar{X})^2} \tag{5.4}$$

$$\hat{\beta}_0 = \bar{Y} - \hat{\beta}_1 \bar{X} \tag{5.5}$$

In performing the actual calculations of $\hat{\beta}_0$ and $\hat{\beta}_1$, one should first compute $\bar{Y}$ and $\bar{X}$, then $\hat{\beta}_1$, and, finally, $\hat{\beta}_0$. Although the formula given in (5.4) is mathematically correct, a computational formula for $\hat{\beta}_1$ which is often more accurate and convenient when using a hand calculator is

$$\hat{\beta}_1 = \frac{\sum_{i=1}^{n} X_i Y_i - \left(\sum_{i=1}^{n} X_i\right)\left(\sum_{i=1}^{n} Y_i\right) \big/ n}{\sum_{i=1}^{n} X_i^2 - \left(\sum_{i=1}^{n} X_i\right)^2 \big/ n} \tag{5.6}$$

The least-squares line may be generally represented either by

$$\hat{Y} = \hat{\beta}_0 + \hat{\beta}_1 X \tag{5.7}$$

or, equivalently, by

$$\hat{Y} = \bar{Y} + \hat{\beta}_1 (X - \bar{X}) \tag{5.8}$$

Either (5.7) or (5.8) may be used to determine predicted Y's corresponding to X's actually observed or to other X values in the region of experimentation.

[Equation (5.8) can be seen, through some simple algebra, to be equivalent to (5.7). The right-hand side of (5.8), $\bar{Y} + \hat{\beta}_1 (X - \bar{X})$, can be alternatively written as $\bar{Y} + \hat{\beta}_1 X - \hat{\beta}_1 \bar{X}$, which in turn equals $\bar{Y} - \hat{\beta}_1 \bar{X} + \hat{\beta}_1 X$, which is equivalent to (5.7) since $\hat{\beta}_0 = \bar{Y} - \hat{\beta}_1 \bar{X}$.]

EXAMPLE   Table 5.1 lists observations on systolic blood pressure (SBP) and age for a sample of 30 individuals. The scatter diagram for this sample was presented in Figure 5.1. For this data set, the best-fitting line may be calculated as follows:

$$\bar{Y} = 142.53$$

$$\bar{X} = 45.13$$

$$\sum X_i Y_i - (\sum X_i)(\sum Y_i)/n = 199,576 - (1,354)(4,276)/30 = 6,585.87$$

$$\sum X_i^2 - (\sum X_i)^2/n = 67,894 - (1,354)^2/30 = 6,783.47$$

$$\hat{\beta}_1 = 6,585.87/6,783.47 = 0.97$$

$$\hat{\beta}_0 = \bar{Y} - \hat{\beta}_1 \bar{X} = 142.53 - (0.97)(45.13) = 98.71$$

TABLE 5.1   *Observations on systolic blood pressure and age for a sample of 30 individuals*

INDIVIDUAL (i)	SBP (Y)	AGE (X)	INDIVIDUAL (i)	SBP (Y)	AGE (X)
1	144	39	16	130	48
2	220	47	17	135	45
3	138	45	18	114	17
4	145	47	19	116	20
5	162	65	20	124	19
6	142	46	21	136	36
7	170	67	22	142	50
8	124	42	23	120	39
9	158	67	24	120	21
10	154	56	25	160	44
11	162	64	26	158	53
12	150	56	27	144	63
13	140	59	28	130	29
14	110	34	29	125	25
15	128	42	30	175	69

FIGURE 5.8   *Best-fitting straight line to age–systolic blood pressure data of Table 5.1*

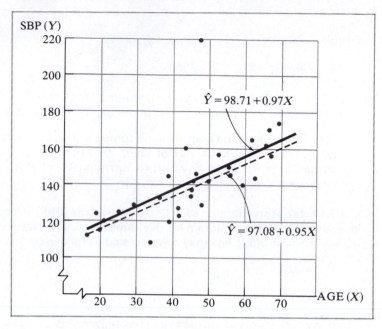

The equation for this line is thus given by

$$\hat{Y} = 98.71 + 0.97X \tag{5.9}$$

or, equivalently, by

$$\hat{Y} = 142.53 + (0.97)(X - 45.13)$$

The line (5.9) is graphed as the solid line in Figure 5.8 and reflects the clear trend that systolic blood pressure (SBP) increases as age increases. Notice that one point, (47, 220), seems quite out of place with the other data, and such an observation is often called an *outlier*. Because the presence of an outlier can affect the least-squares estimates, it is important to make an evaluation as to whether or not an outlier should be removed from the data. Usually, this decision can be made only after a thorough evaluation of the experimental conditions, the data-collection process, and the data itself. (See Chapter 16 for further discussion concerning the treatment of outliers.) If the decision is difficult, one can always determine the effect of removing the outlier by refitting the model to the remaining data. If we do this, the resulting least-squares line is

$$\hat{Y} = 97.08 + 0.95X$$

and is shown on the graph as the dashed line. As might be expected, this line is slightly below the one obtained using all the data.

5.6 MEASURE OF THE QUALITY OF THE STRAIGHT-LINE FIT AND ESTIMATE OF σ^2

Once the least-squares line is determined, we would like to evaluate whether or not the fitted line actually aids in predicting Y, and, if so, to what extent. A measure that helps to answer these questions is provided by

$$SSE = \sum_{i=1}^{n} (Y_i - \hat{Y}_i)^2$$

where $\hat{Y}_i = \hat{\beta}_0 + \hat{\beta}_1 X_i$. Clearly, if $SSE = 0$, the straight line fits perfectly; that is, $Y_i = \hat{Y}_i$ for each i, and every observed point lies on the fitted line. Furthermore, as the fit gets worse, SSE gets larger, since the deviations of points from the regression line will be larger.

Two possible factors contribute to the inflation of SSE. The first factor is that there may be a lot of variation in the data; that is, σ^2 may be large. The second factor is that the assumption of a straight-line model may not be appropriate. It is important therefore to determine the separate effects of each of these components since they address decidedly different issues with regard to the fit of the model. For the time being, we will assume that the second factor is not at issue. Thus, assuming that the straight-line model is appropriate, we can obtain an estimate of σ^2 using SSE. Such an estimate is needed for making statistical inferences concerning the true (i.e., population) straight-line relationship between X and Y. This estimate of σ^2 is given by the formula

$$S_{Y|X}^2 = \frac{1}{n-2} \sum_{i=1}^{n} (Y_i - \hat{Y}_i)^2 = \frac{1}{n-2} SSE \qquad (5.10)$$

A convenient computational formula for $S_{Y|X}^2$ is

$$S_{Y|X}^2 = \frac{n-1}{n-2}(S_Y^2 - \hat{\beta}_1^2 S_X^2) \qquad (5.11)$$

where

$$S_Y^2 = \frac{\sum_{i=1}^{n} Y_i^2 - \left(\sum_{i=1}^{n} Y_i\right)^2 \Big/ n}{n-1} \qquad \text{(i.e., the sample variance of the observed } Y\text{'s)}$$

and

$$S_X^2 = \frac{\sum_{i=1}^{n} X_i^2 - \left(\sum_{i=1}^{n} X_i\right)^2 \Big/ n}{n-1} \qquad \text{(i.e., the sample variance of the } X\text{'s)}$$

EXAMPLE   For the data of Table 5.1, $S^2_{Y|X}$ is calculated as follows:

$$S^2_Y = 509.91$$

$$S^2_X = 233.91$$

$$\hat{\beta}_1 = 0.97$$

$$S^2_{Y|X} = \tfrac{29}{28}[509.91 - (0.97)^2(233.91)] = 299.77$$

Some explanation is appropriate as to why $S^2_{Y|X}$ estimates σ^2, especially since, on first glance, (5.10) looks different from the formula usually used for the sample variance. The usual formula for the sample variance is given by $\sum_{i=1}^{n}(Y_i - \bar{Y})^2/(n-1)$, which is appropriate when the Y's are independent with the same mean μ and variance σ^2. Since, in this case, μ is unknown (its estimate being, of course, $\bar{Y}$), we must divide by $n-1$ instead of n in order for the sample variance to be an unbiased estimator of σ^2. To put it another way, we subtract 1 from n because to estimate σ^2 we had to estimate *one* other parameter first, μ.

If a straight-line model is appropriate, the population mean response $\mu_{Y|X}$ changes with X. For example, using the least-squares line (5.9) as an approximation to the population line for the age–SBP data, the estimated mean of the Y's at $X = 40$ is approximately 138, whereas the estimated mean of the Y's at $X = 70$ is close to 167. Therefore, instead of subtracting $\bar{Y}$ from each Y_i when estimating σ^2, we should actually subtract $\hat{Y}_i$ from Y_i because $\hat{Y}_i = \hat{\beta}_0 + \hat{\beta}_1 X_i$ is the estimate of $\mu_{Y|X_i}$. Furthermore, we subtract 2 from N in the denominator of our estimate since the determination of $\hat{Y}_i$ requires the estimation of two parameters, β_0 and β_1.

When we discuss testing for lack of fit of the assumed model, we will show how to get an estimate of σ^2 which does not assume that the straight-line model is the correct one.

5.7 INFERENCES CONCERNING THE SLOPE AND INTERCEPT

To assess whether or not, using the fitted line, X helps to predict Y, and to take into account the uncertainties of using a sample, it is standard practice to compute confidence intervals and/or test statistical hypotheses about the unknown parameters in the assumed straight-line model. Such confidence intervals and tests require, as previously described in Section 5.4, the assumption that the random variable Y has a normal distribution at each fixed value of X. Under this assumption it can be deduced that the estimators $\hat{\beta}_0$ and $\hat{\beta}_1$ are each normally distributed, with respective means β_0 and β_1 when (5.2) holds, and with easily derivable variances. These estimators, together with estimates of their variances, can then be used to form confidence intervals and test statistics based on the t *distribution*.

[An important property that allows the normality assumption on Y to carry over to $\hat{\beta}_0$ and $\hat{\beta}_1$ is that these estimators are *linear functions* of the Y's.

Such a function is defined by a formula of the structure

$$L = \sum_{i=1}^{n} c_i Y_i$$

or, equivalently, $L = c_1 Y_1 + c_2 Y_2 + \cdots + c_n Y_n$, where the c_i's are constants not involving the Y's. A simple example of a linear function is $\bar{Y}$, which can be written as

$$\sum_{i=1}^{n} \frac{1}{n} Y_i$$

Here the c_i's equal $1/n$ for each i. The normality of $\hat{\beta}_0$ and $\hat{\beta}_1$ derives from a statistical theorem which states that linear functions of independent normally distributed observations are themselves normally distributed.]

More specifically, to test the hypothesis $H_0: \beta_1 = \beta_1^{(0)}$, where $\beta_1^{(0)}$ is some hypothesized value for β_1, the test statistic used is

$$T = \frac{\hat{\beta}_1 - \beta_1^{(0)}}{S_{Y|X}/S_X\sqrt{n-1}} \tag{5.12}$$

This test statistic has the t distribution with $n-2$ degrees of freedom (df) when H_0 is true. In this formula $S_{Y|X}^2$ denotes the sample estimate of σ^2 defined by (5.10), and S_X^2 is the sample variance of the X's used in (5.11) for calculating $S_{Y|X}^2$. The denominator is an estimate of the unknown standard error of the estimator $\hat{\beta}_1$, given by

$$\sigma_{\hat{\beta}_1} = \frac{\sigma}{S_X\sqrt{n-1}}$$

Thus, the test statistic (5.12) is the ratio of a normally distributed random variable divided by an estimate of its standard deviation. Such a statistic will have a t distribution for the kinds of situations encountered in this text.

Similarly, to test the hypothesis $H_0: \beta_0 = \beta_0^{(0)}$, the test statistic used is

$$T = \frac{\hat{\beta}_0 - \beta_0^{(0)}}{S_{Y|X}\sqrt{1/n + \bar{X}^2/(n-1)S_X^2}} \tag{5.13}$$

which also has the t distribution with $n-2$ degrees of freedom when $H_0: \beta_0 = \beta_0^{(0)}$ is true. The denominator here estimates the standard deviation of $\hat{\beta}_0$, given by

$$\sigma_{\hat{\beta}_0} = \sigma\sqrt{1/n + \bar{X}^2/(n-1)S_X^2}$$

The reason that both test statistics above have $n-2$ degrees of freedom is that both involve $S_{Y|X}^2$, which itself has $n-2$ degrees of freedom and is the only random component in the denominator of each of the test statistics.

TABLE 5.2 *Confidence intervals, tests of hypotheses, and prediction intervals for straight-line regression analysis**

PARAMETER	$100(1-\alpha)\%$ CONFIDENCE INTERVAL	H_0	TEST STATISTIC (T)	DISTRIBUTION UNDER H_0
β_1	$\hat{\beta}_1 \pm t_{n-2,1-\alpha/2}\dfrac{S_{Y\mid x}}{S_X\sqrt{n-1}}$	$\beta_1 = \beta_1^{(0)}$	$T = \dfrac{(\hat{\beta}_1 - \beta_1^{(0)})S_X\sqrt{n-1}}{S_{Y\mid x}}$	t_{n-2}
β_0	$\hat{\beta}_0 \pm t_{n-2,1-\alpha/2}S_{Y\mid x}\sqrt{\dfrac{1}{n}+\dfrac{\bar{X}^2}{(n-1)S_X^2}}$	$\beta_0 = \beta_0^{(0)}$	$T = \dfrac{\hat{\beta}_0 - \beta_0^{(0)}}{S_{Y\mid x}\sqrt{\dfrac{1}{n}+\dfrac{\bar{X}^2}{(n-1)S_X^2}}}$	t_{n-2}
$\mu_{Y\mid X_0}$	$\bar{Y}+\hat{\beta}_1(X_0-\bar{X})$ $\pm t_{n-2,1-\alpha/2}S_{Y\mid x}\sqrt{\dfrac{1}{n}+\dfrac{(X_0-\bar{X})^2}{(n-1)S_X^2}}$	$\mu_{Y\mid X_0} = \mu_{Y\mid X_0}^{(0)}$	$T = \dfrac{\bar{Y}+\hat{\beta}_1(X_0-\bar{X})-\mu_{Y\mid X_0}^{(0)}}{S_{Y\mid x}\sqrt{\dfrac{1}{n}+\dfrac{(X_0-\bar{X})^2}{(n-1)S_X^2}}}$	t_{n-2}
Y_{X_0} (single observation, not a parameter)	$\bar{Y}+\hat{\beta}_1(X_0-\bar{X})$ $\pm t_{n-2,1-\alpha/2}S_{Y\mid x}\sqrt{1+\dfrac{1}{n}+\dfrac{(X_0-\bar{X})^2}{(n-1)S_X^2}}$ (prediction interval for a new individual's Y)		(not relevant)	

* $\mu_{Y\mid X} = \beta_0 + \beta_1 X$ is the assumed true regression model.

$\hat{\beta}_1 = \sum_1^n (X_i - \bar{X})(Y_i - \bar{Y}) \Big/ \sum_1^n (X_i - \bar{X})^2 = \left[\sum_1^n X_i Y_i - \left(\sum_1^n X_i\right)\left(\sum_1^n Y_i\right)\Big/ n\right] \Big/ \left[\sum_1^n X_i^2 - \left(\sum_1^n X_i\right)^2 \Big/ n\right].$

$\hat{\beta}_0 = \bar{Y} - \hat{\beta}_1 \bar{X}.$

$\hat{Y} = \hat{\beta}_0 + \hat{\beta}_1 X = \bar{Y} + \hat{\beta}_1 (X - \bar{X}).$

$t_{n-2,1-\alpha/2}$ is the $100(1-\alpha/2)\%$ point of the t distribution with $n-2$ df.

$S_{Y\mid x}^2 = \dfrac{1}{n-2} \sum_1^n (Y_i - \hat{Y}_i)^2 = \dfrac{n-1}{n-2}(S_Y^2 - \hat{\beta}_1^2 S_X^2).$

$S_Y^2 = \dfrac{1}{n-1} \sum_1^n (Y_i - \bar{Y})^2 = \left[\sum_1^n Y_i^2 - \left(\sum_1^n Y_i\right)^2 \Big/ n\right] \Big/ (n-1).$

$S_X^2 = \dfrac{1}{n-1} \sum_1^n (X_i - \bar{X})^2 = \left[\sum_1^n X_i^2 - \left(\sum_1^n X_i\right)^2 \Big/ n\right] \Big/ (n-1).$

TABLE 5.3   *Sample calculations of confidence intervals, tests of hypotheses, and prediction intervals for the age–SBP data of Table 5.1**

PARAMETER	$100(1-\alpha)\%$ CONFIDENCE INTERVAL	H_0	TEST STATISTIC (T)
β_1	($\alpha = 0.05$) $0.97 \pm 2.0484\left(\dfrac{17.31}{15.29\sqrt{30-1}}\right)$ or $0.54 < \beta_1 < 1.40$	$\beta_1 = 0$	$T = \dfrac{(0.97-0)(15.29)\sqrt{30-1}}{17.31}$ $= 4.62$ Reject H_0 at $\alpha = 0.05$ (two-tailed test), since $t_{28,0.975} = 2.0484 \ (P < 0.001)$
β_0	($\alpha = 0.05$) $98.71 \pm (2.0484)(17.31)\sqrt{\dfrac{1}{30} + \dfrac{(45.13)^2}{(30-1)(15.29)^2}}$ or $78.23 < \beta_0 < 119.20$	$\beta_0 = 75$	$T = \dfrac{98.71 - 75}{17.31\sqrt{\dfrac{1}{30} + \dfrac{(45.13)^2}{(30-1)(15.29)^2}}}$ $= 2.37$ Reject H_0 at $\alpha = 0.05$ (two-tailed test), since $t_{28,0.975} = 2.0484 \ (0.02 < P < 0.05)$
$\mu_{Y\mid 65}$ (i.e., $X_0 = 65$)	($\alpha = 0.10$) $142.53 + (0.97)(65-45.13)$ $\pm (1.7011)(17.31)$ $\times \sqrt{\dfrac{1}{30} + \dfrac{(65-45.13)^2}{(30-1)(15.29)^2}}$ or $153.09 < \mu_{Y\mid 65} < 171.07$	$\mu_{Y\mid 65} = 147$	$T = \dfrac{142.53 + (0.97)(65-45.13) - 147}{17.31\sqrt{\dfrac{1}{30} + \dfrac{(65-45.13)^2}{(30-1)(15.29)^2}}}$ $= 2.86$ Reject H_0 at $\alpha = 0.10$ (two-tailed test), since $t_{28,0.95} = 1.7011 \ (0.0001 < P < 0.001)$
Y_{65} (prediction interval at $X_0 = 65$)	($\alpha = 0.10$) $142.53 + (0.97)(65-45.13)$ $\pm (1.7011)(17.31)$ $\times \sqrt{1 + \dfrac{1}{30} + \dfrac{(65-45.13)^2}{(30-1)(15.29)^2}}$ or $131.29 < Y_{65} < 192.87$		(not relevant)

* $n = 30$, $\hat{\beta}_0 = 98.71$, $\beta_1 = 0.97$, $\bar{Y} = 142.53$, $\bar{X} = 45.13$, $S_{Y\mid X} = 17.31$, $S_X = 15.29$.

In testing either of the preceding hypotheses at significance level α, H_0 should be rejected whenever any of the following occur:

$$\begin{cases} T \geq t_{n-2,1-\alpha} \text{ for an upper one-tailed test} & (\text{i.e., } H_a: \beta_1 > \beta_1^{(0)} \text{ or } H_a: \beta_0 > \beta_0^{(0)}) \\ T \leq -t_{n-2,1-\alpha} \text{ for a lower one-tailed test} & (\text{i.e., } H_a: \beta_1 < \beta_1^{(0)} \text{ or } H_a: \beta_0 < \beta_0^{(0)}) \\ |T| \geq t_{n-2,1-\alpha/2} \text{ for a two-tailed test} & (\text{i.e., } H_a: \beta_1 \neq \beta_1^{(0)} \text{ or } H_a: \beta_0 \neq \beta_0^{(0)}) \end{cases}$$

where $t_{n-2,1-\alpha}$ denotes the $100(1-\alpha)\%$ point of the t distribution with $n-2$ degrees of freedom. As an alternative to using a specified significance level, P values may be computed based on the calculated value of the test statistic T.

Table 5.2 contains an easy-to-use summary of the formulas needed for performing statistical tests and computing confidence intervals for β_0 and β_1. Also given in this table are formulas for inference-making procedures concerned with prediction using the fitted line, and these formulas are described in Sections 5.9 and 5.10. Table 5.3 gives worked examples illustrating each of the formulas in Table 5.2 using the age-SBP data previously considered.

5.8 INTERPRETATIONS OF TESTS FOR SLOPE AND INTERCEPT

Special care must be taken to correctly interpret the results of tests concerning the slope and intercept, especially since, in practice, errors are often made in this regard. In this section we discuss the conclusions that can be made based on acceptance[1] or rejection of the null hypotheses most commonly tested concerning the slope and intercept. In the discussion we shall assume that the usual assumptions concerning normality, independence, and variance homogeneity are not violated. If these assumptions do not hold, any conclusions based on testing procedures developed under such assumptions are suspect.

5.8.1 TEST FOR ZERO SLOPE

The most important test of the hypothesis dealing with the parameters of the straight-line model concerns whether or not the slope of the regression line is significantly different from zero, or equivalently, whether or not X helps to predict Y using a straight-line model. The appropriate null hypothesis for this test is $H_0: \beta_1 = 0$. Care must be taken in interpreting the result of the test of this hypothesis.

If we ignore for now the ever-present possibility of making a type I error (i.e., rejecting a true H_0) or a type II error (i.e., accepting a false H_0), we can make the following interpretations concerning accepting or rejecting $H_0: \beta_1 = 0$:

1. If $H_0: \beta_1 = 0$ is accepted (i.e., not rejected), this means that either:
    (a) For a true underlying straight-line model, X provides little or no help in predicting Y; that is, $\bar{Y}$ is essentially as good as $\bar{Y} + \hat{\beta}_1(X - \bar{X})$ for predicting Y (see Figure 5.9, case 1a, for an example).

---

[1] The reader is reminded that, statistically speaking, "acceptance of H_0" really means that "there is insufficient evidence to reject H_0."

FIGURE 5.9   *Interpreting the test for zero slope*

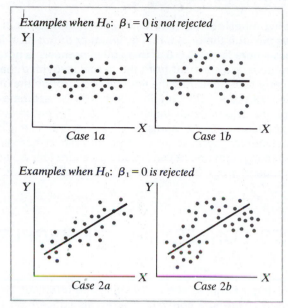

or

(b) The true underlying relationship between *X* and *Y* is *not* linear; that is, the true model may involve quadratic, cubic, or other more complex functions of *X* (see Figure 5.9, case 1b, for an example).

Combining (a) and (b), we can say that accepting H_0: $\beta_1 = 0$ *implies that a straight-line model in X is not the best model to use and does not provide much help for predicting Y.*

2. If H_0: $\beta_1 = 0$ is rejected, this means that
    (a)   *X* provides significant information for the prediction of *Y*; that is, the model $\bar{Y} + \hat{\beta}_1(X - \bar{X})$ is far better than the naive model $\bar{Y}$ for predicting *Y* (see Figure 5.9, case 2a, for an example).
and perhaps
    (b) A better model, for example, might have a curvilinear term (e.g., see case 2b of Figure 5.9), although there is a definite linear component.

Combining (a) and (b), we can say that *rejecting* H_0: $\beta_1 = 0$ *implies that a straight-line model in X is a better model than one which does not include X at all*, but at the same time it may very well represent only a "linear approximation" to a truly "nonlinear relationship."

An important point implied by the above interpretations is that *whether or not the hypothesis H_0: $\beta_1 = 0$ is rejected, it may be that a straight-line model is not appropriate and that, instead, some other curve better describes the relationship between X and Y.*

5.8.2 TEST FOR ZERO INTERCEPT

Another specific hypothesis sometimes tested concerns whether or not the population straight line goes through the origin, that is, whether or not its Y intercept β_0 is zero. The null hypothesis here is $H_0: \beta_0 = 0$. If this null hypothesis is not rejected, it may be appropriate to remove the constant from the model, provided that there is previous experience or theory to suggest that the line may go through the origin and that there are observations taken around the origin to improve the estimate of β_0. To force the fitted line through the origin merely because $H_0: \beta_0 = 0$ is not rejected can sometimes give a spurious appearance to the regression line. In any case this hypothesis is rarely of interest in most studies, because data are not usually gathered near the origin. For example, when dealing with age (X) and blood pressure (Y), we are not interested in knowing what happens at $X = 0$ and we rarely choose values of X near 0.

5.9 INFERENCES CONCERNING THE REGRESSION LINE $\mu_{Y|X} = \beta_0 + \beta_1 X$

In addition to making inferences about the slope and intercept, we may also want to perform tests and/or compute confidence intervals concerning the regression line itself. More specifically, we may want to find for a given $X = X_0$ a confidence interval for $\mu_{Y|X_0}$, the mean value of Y at X_0.[2] We may also be interested in testing the hypothesis $H_0: \mu_{Y|X_0} = \mu_{Y|X_0}^{(0)}$, where $\mu_{Y|X_0}^{(0)}$ is some hypothesized value of interest.

The test statistic to use for the hypothesis $H_0: \mu_{Y|X_0} = \mu_{Y|X_0}^{(0)}$ is given by the formula

$$T = \frac{\hat{Y}_{X_0} - \mu_{Y|X_0}^{(0)}}{S_{\hat{Y}_{X_0}}} \tag{5.14}$$

where $\hat{Y}_{X_0} = \hat{\beta}_0 + \hat{\beta}_1 X_0 = \bar{Y} + \hat{\beta}_1 (X_0 - \bar{X})$ is the predicted value of Y at X_0 and

$$S_{\hat{Y}_{X_0}} = S_{Y|X} \sqrt{1/n + (X_0 - \bar{X})^2/(n-1)S_X^2} \tag{5.15}$$

This test statistic, like those for slope and intercept, has the t distribution with $n - 2$ degrees of freedom when H_0 is true. The denominator $S_{\hat{Y}_{X_0}}$ is an estimate of the standard deviation of $\hat{Y}_{X_0}$, which is given by

$$\sigma_{\hat{Y}_{X_0}} = \sigma \sqrt{1/n + (X_0 - \bar{X})^2/(n-1)S_X^2}$$

The corresponding confidence interval for $\mu_{Y|X_0}$ is given by the formula

$$\hat{Y}_{X_0} \pm t_{n-2,1-\alpha/2} S_{\hat{Y}_{X_0}} \tag{5.16}$$

In addition to inferences about specific points on the regression line, it is often useful to construct a confidence interval for the regression line over the entire range of X values. The most convenient way to do this is to plot the upper and lower confidence

[2] The point $(X_0, \mu_{Y|X_0})$, of course, lies on the population regression line.

FIGURE 5.10   *90% Confidence and prediction bands for age–systolic blood pressure data of Table 5.1*

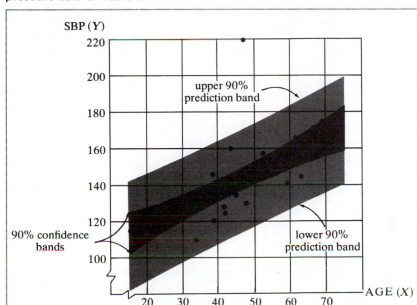

limits obtained for several specified values of X and then to sketch the two curves that connect these points. Such curves are called *confidence bands* for the regression line. The confidence bands for the data of Table 5.1 are indicated in Figure 5.10.

    Sketching confidence bands by hand can be a painful job. However, it can be made reasonably easy if a number of helpful hints are followed:

1.   Simplify the confidence-interval formula (5.16) by substitution of all known numbers except X_0.

EXAMPLE   For 90% confidence bands for our age–SBP data, this simplified expression is given by

$$142.53 + (0.97)(X_0 - 45.13) \pm 29.45\sqrt{0.033 + (X_0 - 45.13)^2/6,783.48}$$

2.   Calculate the confidence limits for $X_0 = \bar{X}$. These are the easiest limits to calculate because two potentially messy terms in the preceding formula disappear. It is important to note that the minimum-width confidence interval is always obtained when $X_0 = \bar{X}$, since the second term in the square-root expression of the formula above is then zero. Furthermore, as can be seen in Figure 5.10, the farther X_0 is away from $\bar{X}$, the wider the confidence interval is. Thus, *we do a better job of estimating $\mu_{Y|X_0}$ near the center of the region of experimentation than in the periphery.*

EXAMPLE   At $X_0 = \bar{X} = 45.13$, the formula simplifies to $142.53 \pm 29.45\sqrt{0.033}$, which yields a lower limit of 137.16 and an upper limit of 147.91.

3. Calculate the confidence limits for values of X_0 of the form $\bar{X} \pm k$, $\bar{X} \pm 2k$, $\bar{X} \pm 3k$, . . . , for a suitable k chosen so that the region of experimentation is uniformly covered. The advantage of adopting this approach is that the confidence-interval calculations become more systematic and more simple to perform; for example, the term under the square-root sign in the confidence-interval formula is the same for $\bar{X} + k$ as it is for $\bar{X} - k$ (namely, $\sqrt{0.033 + k^2/6{,}783.48}$), and similarly for the other values.

EXAMPLE For the value $X_0 = \bar{X} + 10 = 55.13$, the formula becomes $142.53 + (0.97)(10) \pm 29.45\sqrt{0.033 + 100/6{,}783.48}$, which yields limits of 145.78 and 158.70. For $X_0 = \bar{X} - 10 = 35.13$, the formula becomes $142.53 - (0.97)(10) \pm$ [same right-hand side as for $(\bar{X} + 10)$]. The limits obtained here are 126.37 and 139.28. The computations for other values are performed similarly.

4. Plot all limits on a graph, as in Figure 5.10, and then sketch the two confidence bands (which should be parabolic in shape).

5.10 PREDICTION OF A NEW VALUE OF Y AT X_0

We have just dealt with estimating the mean $\mu_{Y|X_0}$ at $X = X_0$. In practice, we may wish instead to estimate the response Y of a single individual based on the fitted regression line; that is, we may want to predict an individual's Y given his $X = X_0$. Note that the obvious point estimate to use in this case is $\hat{Y}_{X_0} = \hat{\beta}_0 + \hat{\beta}_1 X_0$. Thus, $\hat{Y}_{X_0}$ is used to estimate both the mean $\mu_{Y|X_0}$ and an individual's response Y at X_0.

It is, of course, necessary to put some bounds (i.e., limits) on this estimate to take into account the variability of the estimate. Here, however, we could not say that we are constructing a confidence interval for Y, since Y is not a parameter, nor could we perform a test of hypothesis for the same reason. The term used to describe the "hybrid limits" we require is the *prediction interval* (PI), which is given by

$$\bar{Y} + \hat{\beta}_1(X_0 - \bar{X}) \pm t_{n-2,1-\alpha/2} S_{Y|X}\sqrt{1 + 1/n + (X_0 - \bar{X})^2/(n-1)S_X^2} \tag{5.17}$$

We first note that an estimate of an individual's response should naturally have more variability than an estimate of a group's mean response. This is reflected by the extra term 1 under the square-root sign in (5.17), which is not found in the square-root part of the confidence-interval formula for $\mu_{Y|X}$ [see (5.15)]. To be more specific, in predicting an actual observed Y for a given individual, there are two sources of error operating: individual error as measured by σ^2 and the error in estimating $\mu_{Y|X_0}$ using $\hat{Y}_{X_0}$. More precisely, this can be expressed by the equation

$$\underbrace{Y - \hat{Y}_{X_0}}_{\substack{\text{error in} \\ \text{predicting} \\ \text{an individual's} \\ Y \text{ at } X_0}} = \underbrace{(Y - \mu_{Y|X_0})}_{\substack{\text{deviation of} \\ \text{individual's} \\ Y \text{ from true} \\ \text{mean at } X_0}} + \underbrace{(\mu_{Y|X_0} - \hat{Y}_{X_0})}_{\substack{\text{deviation of} \\ \hat{Y}_{X_0} \text{ from true} \\ \text{mean at } X_0}}$$

This representation allows us to write the variance of an individual's predicted response at X_0 as

$$\text{Var } Y + \text{Var } \hat{Y}_{X_0} = \sigma^2 \left[1 + \frac{1}{n} + \frac{(X_0 - \bar{X})^2}{(n-1)S_X^2} \right]$$

This variance expression is estimated by replacing σ^2 by its estimate $S_{Y|X}^2$. This accounts for the expression on the right-hand side of the prediction interval in (5.17).

In order to describe individual predictions over the *entire* range of X values, *prediction bands* may be determined in a manner analogous to that used for the computation of confidence bands. Figure 5.10 gives 90% prediction bands for the age–SBP data. As expected, the 90% prediction bands in this figure are wider than the corresponding 90% confidence bands.

PROBLEMS   1.   The table gives the dry weights (Y) of 11 chick embryos ranging in age from 6 to 16 days (X). Also given in the table are the values of the common logarithms of the weights (Z).

AGE (X) (days)	6	7	8	9	10	11
DRY WEIGHT (Y) (grams)	0.029	0.052	0.079	0.125	0.181	0.261
LOG$_{10}$ DRY WEIGHT (Z)	−1.538	−1.284	−1.102	−0.903	−0.742	−0.583

AGE (X) (days)	12	13	14	15	16
DRY WEIGHT (Y) (grams)	0.425	0.738	1.130	1.882	2.812
LOG$_{10}$ DRY WEIGHT (Z)	−0.372	−0.132	0.053	0.275	0.449

(a)   Draw two scatter diagrams on graph paper, one showing the ages and dry weights of the chick embryos and the other showing the ages and log dry weights.

(b)   Sketch straight lines by eye on both scatter diagrams. Which line seems to give the better fit?

(c)   Calculate the least-squares estimates of the parameters of the regression line for each set of data, and draw the estimated lines on the respective scatter diagrams. The following summations will be useful (all summations go from $i = 1$ to $i = 11$):

$$\sum X_i = 121, \qquad \sum Y_i = 7.714, \qquad \sum Z_i = -5.879$$

$$\sum X_i^2 = 1{,}441, \qquad \sum Y_i^2 = 13.578, \qquad \sum Z_i^2 = 7.375$$

$$\sum X_i Y_i = 110.712, \qquad \sum X_i Z_i = -43.12$$

(d)   For each of the estimated lines, find a 95% confidence interval for the true slope and then interpret the interval in each case with regard to the null hypothesis that the true slope is zero.

(e) For each of the estimated lines, calculate and sketch the 90% confidence and prediction bands.

(f) In your opinion, which of the two regression lines has the better fit? Explain.

2. The table gives the systolic blood pressure (SBP), body size (QUET),[3] age (AGE), and smoking history (SMK = 0 if nonsmoker, SMK = 1 if a current smoker or a previous smoker) for a hypothetical sample of 32 white males over 40 years old from the town of Angina.

PERSON No.	SBP	QUET	AGE	SMK
1	135	2.876	45	0
2	122	3.251	41	0
3	130	3.100	49	0
4	148	3.768	52	0
5	146	2.979	54	1
6	129	2.790	47	1
7	162	3.668	60	1
8	160	3.612	48	1
9	144	2.368	44	1
10	180	4.637	64	1
11	166	3.877	59	1
12	138	4.032	51	1
13	152	4.116	64	0
14	138	3.673	56	0
15	140	3.562	54	1
16	134	2.998	50	1
17	145	3.360	49	1
18	142	3.024	46	1
19	135	3.171	57	0
20	142	3.401	56	0
21	150	3.628	56	1
22	144	3.751	58	0
23	137	3.296	53	0
24	132	3.210	50	0
25	149	3.301	54	1
26	132	3.017	48	1
27	120	2.789	43	0
28	126	2.956	43	1
29	161	3.800	63	0
30	170	4.132	63	1
31	152	3.962	62	0
32	164	4.010	65	0

2.1. Draw scatter diagrams on graph paper for each of the following variable pairs:
(a) SBP (Y) vs. QUET (X)
(b) SBP (Y) vs. SMK (X)
(c) QUET (Y) vs. AGE (X)
(d) SBP (Y) vs. AGE (X)

2.2. For scatter diagrams (a), (c), and (d), sketch by eye a line that fits the data reasonably well. Comment on the relationships described.

2.3. (a) Determine the least-squares estimates of slope (β_1) and intercept (β_0) for the straight-line regression of SBP (Y) on SMK (X). A computer printout which provides the essential information for these data is provided.

[3] QUET stands for "quetelet index," a measure of size defined by QUET = 100 (weight/height^2).

Computer printout for Problem 2[4]

SBP (Y) on SMK (X)

MEANS AND STANDARD DEVIATIONS

X: 0.53125 +- .50693   Y: 144.53125 +- 14.39755

CORRELATIONS[5]

PEARSON: 0.247333   SPEARMAN: 0.213763   KENDALL: 0.176227

LINEAR REGRESSION OF Y ON X

SLOPE: 7.02353 +- 5.02360
INTERCEPT: 140.80000 +- 3.66147
S(Y|X):   14.18032

ANALYSIS OF VARIANCE: F(1,30) = 1.95478

SBP (Y) on QUET (X)

MEANS AND STANDARD DEVIATIONS

X: 3.44109 +- 0.49708   Y: 144.53125 +- 14.39755

CORRELATIONS

PEARSON: 0.742004   SPEARMAN: 0.746928   KENDALL: 0.572019

LINEAR REGRESSION OF Y ON X

SLOPE: 21.49167 +- 3.54515
INTERCEPT: 70.57640 +- 12.32187
S(Y|X):   9.81160

ANALYSIS OF VARIANCE: F(1,30) = 36.75122

QUET (Y) on AGE (X)

MEANS AND STANDARD DEVIATIONS

X: 53.25000 +- 6.95608   Y: 3.44109 +- 0.49708

CORRELATIONS

PEARSON: 0.802751

LINEAR REGRESSION OF Y ON X

SLOPE: 5.73642E - 02 +- 7.77991E - 03
INTERCEPT: 0.38645 +- 0.41769
S(Y|X):   0.30131

ANALYSIS OF VARIANCE: F(1,30) = 54.36664

SBP (Y) on AGE (X)

MEANS AND STANDARD DEVIATIONS

X: 53.25000 +- 6.95608   Y: 144.53125 +- 14.39755

CORRELATIONS

PEARSON: 0.775204   SPEARMAN: 0.762119   KENDALL: 0.603708

LINEAR REGRESSION OF Y ON X

SLOPE: 1.60450 +- 0.23872.
INTERCEPT: 59.09163 +- 12.81626
S(Y|X):   9.24543

ANALYSIS OF VARIANCE: F(1,30) = 45.17692

[4]This computer printout has been chosen to typify the output obtained via the use of a standard straight line regression program, and, in this regard, one major goal of these exercises is to help the reader learn to be able to interpret and utilize efficiently such output.
[5]The Pearson correlation coefficient is discussed in detail in Chapter 6. The Spearman and Kendall correlations are *nonparametric* measures of association [see Siegel (1956)].

(b) (1) Compare the value of $\hat{\beta}_0$ with the mean SBP for nonsmokers.
    (2) Compare the value of $(\hat{\beta}_0 + \hat{\beta}_1)$ with the mean SBP for smokers.
    (3) Explain the results of these comparisons.
(c) Test the hypothesis that the true slope (β_1) is zero. (Make sure that you compute the P value for your test.)
(d) Is the test in part (c) equivalent to the usual two-sample t test for the equality of two population means assuming equal but unknown variances? Demonstrate your answer numerically.

2.4. (a) Determine the least-squares estimates of slope and intercept for the straight-line regression of SBP (Y) on QUET (X). Refer to the computer printout if you wish.
(b) Sketch the estimated regression line on the scatter diagram involving SBP and QUET.
(c) Test the hypothesis of zero slope (again, make sure to compute the P value).
(d) Find a 95% confidence interval for $\mu_{Y|\bar{X}}$.
(e) Calculate 95% prediction bands.
(f) Based on the above, would you conclude that blood pressure increases as body size increases?
(g) Are any of the assumptions for straight-line regression clearly not satisfied in this example?

2.5. Answer questions (a) to (c) in Problem 2.4 for the regression of QUET (Y) on AGE (X).

2.6. Answer questions (a) to (c) in Problem 2.4 for the regression of SBP (Y) on AGE (X).

3. For married couples with one or more offspring, a demographic study was conducted to determine the effect of husband's annual income (at marriage) on the time (in months) between marriage and birth of the first child. The table gives the husband's annual income (INC), and time between marriage and birth of the first child (TIME), for a hypothetical sample of 20 couples.
(a) Draw a scatter diagram on graph paper for the variables TIME (Y) and INC (X).
(b) Attempt to sketch by eye a line that fits the data reasonably well.
(c) What does this tell you about the relationship described?
(d) Use the computer printout to determine the least-squares estimates of slope (β_1) and intercept (β_0) for the straight-line regression of TIME (Y) on INC (X).

INC	TIME	INC	TIME
$ 5,775	16.2	$ 4,608	9.7
9,800	35.0	24,210	20.0
13,795	37.2	19,625	38.2
4,120	9.0	18,000	41.25
25,015	24.4	13,000	44.0
12,200	36.75	5,400	9.2
7,400	31.75	6,440	20.0
9,340	30.0	9,000	40.2
20,170	36.0	18,180	32.0
22,400	30.8	15,385	39.2

Computer printout for Problem 3

```
MEANS AND STANDARD DEVIATIONS

X: 13193.15000 +- 6824.87806    Y: 29.04250 +- 11.31785

CORRELATIONS

PEARSON: 0.430411   SPEARMAN: 0.428733   KENDALL: 0.311347

LINEAR REGRESSION OF Y ON X

SLOPE:        7.13761E - 04 +- 3.52813E - 04
INTERCEPT:   19.62575 +- 5.21291
S(Y|X):      10.49580

ANALYSIS OF VARIANCE: F(1,18) = 4.09277
```

(e) Draw the estimated regression line on the scatter diagram for TIME on INC.

(f) Are any of the assumptions for straight-line regression clearly not satisfied in this example?

(g) Test the null hypothesis that the true slope β_1 is zero at the $\alpha = 0.05$ level. Interpret the results of this test.

(h) Can you suggest a model other than a straight line which would better describe the TIME–INC relationship?

4. A sociologist assigned to a correctional institution was interested in studying the relationship between intelligence and delinquency. A delinquency index (ranging from 0 to 50) was formulated to account for both the severity and frequency of crimes committed, while intelligence was measured by IQ. The table displays the delinquency index (DI) and IQ of a sample of 18 convicted minors.

DI (Y)	IQ (X)	DI (Y)	IQ (X)
26.2	110	22.1	92
33.0	89	18.6	116
17.5	102	35.5	85
25.25	98	38.0	73
20.3	110	30.0	90
31.9	98	19.7	104
21.1	122	41.1	82
22.7	119	39.6	134
10.7	120	25.15	114

(a) Draw a scatter diagram on graph paper for the variable pair (DI, IQ).

(b) Given that $\hat{\beta}_1 = -0.249$ and $\hat{\beta}_0 = 52.273$, draw the estimated regression line on the scatter diagram.

(c) How do you account for the fact that when IQ $= 0$, $\hat{Y} = 52.273$, even though the delinquency index goes no higher than 50?

(d) Find a 95% confidence interval for the true slope β_1 using the fact that $S_{Y|X} = 7.704$ and $S_X = 16.192$.

(e) Interpret this confidence interval with regard to testing the null hypothesis of zero slope at the $\alpha = 0.05$ level.

(f) Notice that the convicted minor with an IQ $= 134$ and DI $= 39.6$ appears to be quite out of place in the data. Such an observation is called an outlier.

Decide whether or not the outlier has any effect with regard to estimating the IQ–DI relationship by looking at the graph for the fitted line based on discarding the outlier. (Note that $\hat{\beta}_0 = 70.846$ and $\hat{\beta}_1 = -0.444$ when the outlier is removed.)

(g) Test the null hypothesis of zero slope when the outlier is removed given that $S_{Y|X} = 4.933$, $S_X = 14.693$, $n = 17$. (Use $\alpha = 0.05$.)

(h) For these data would you conclude that the delinquency index decreases as IQ increases?

5. Following the last congressional election, a political scientist attempted to investigate the relationship between campaign expenditures on televised advertisements and subsequent voter turnout. The table presents the percent of total campaign expenditures relegated to televised advertisements (TVEXP) and the percent of registered voter turnout (VOTE) for a hypothetical sample of 20 congressional districts.

VOTE (Y)	TVEXP (X)	VOTE (Y)	TVEXP (X)
35.4	28.5	40.8	31.3
58.2	48.3	61.9	50.1
46.1	40.2	36.5	31.3
45.5	34.8	32.7	24.8
64.8	50.1	53.8	42.2
52.0	44.0	24.6	23.0
37.9	27.2	31.2	30.1
48.2	37.8	42.6	36.5
41.8	27.2	49.6	40.2
54.0	46.1	56.6	46.1

Computer printout for Problem 5

MEANS AND STANDARD DEVIATIONS

X: 36.99000 +- 8.76758   Y: 45.71000 +- 10.81670

CORRELATIONS

PEARSON: 0.954000   SPEARMAN: 0.955181   KENDALL: 0.858743

LINEAR REGRESSION OF Y ON X

SLOPE: 1.17696 +- 8.71805E - 02
INTERCEPT: 2.17407 +- 3.30974
S(Y|X):   3.33177

ANALYSIS OF VARIANCE: F(1,18) = 182.25892

(a) Draw a scatter diagram on graph paper for the observation pairs on VOTE (Y) and TVEXP (X).

(b) Use the printout to determine the least-squares line of VOTE on TVEXP.

(c) Draw the estimated line on the scatter diagram constructed in part (a).

(d) Are any of the assumptions for straight-line regression clearly *not* satisfied in this example?

(e) Test the hypothesis that the slope β_1 is zero using $\alpha = 0.05$, and interpret your result.

(f)   Test the hypothesis H_0: $\mu_{Y|X_0} = 45$ for $X_0 = \bar{X} = 36.99$ at $\alpha = 0.05$.

(g)   Calculate the corresponding 95% confidence interval for $\mu_{Y|36.99}$.

6.   A group of 13 children and adolescents (considered healthy) participated in a psychological study designed to analyze the relationship between age and average total sleep time (ATST). To obtain a measure for ATST (in minutes), recordings were taken on each subject on three consecutive nights and then averaged. The results obtained are displayed in the table.

ATST (min/24 hr)	AGE (years)	ATST (min/24 hr)	AGE (years)
586.0	4.4	515.2	8.9
461.75	14.0	493.0	11.1
491.1	10.1	528.3	7.75
565.0	6.7	575.9	5.5
462.0	11.5	532.5	8.6
532.1	9.6	530.5	7.2
477.6	12.4		

Computer printout for Problem 6

```
MEANS AND STANDARD DEVIATIONS

X: 9.05769 +- 2.77518    Y: 519.30385 +- 40.95056


CORRELATIONS

PEARSON: -0.951547   SPEARMAN: -0.928571   KENDALL: -0.820513


LINEAR REGRESSION OF Y ON X

SLOPE:        -14.04105 +- 1.36812
INTERCEPT:    646.48334 +- 12.91773
S(Y|X):        13.15238


ANALYSIS OF VARIANCE: F(1,11) = 105.33021
```

(a)   Draw a scatter diagram on graph paper for the variable pair ATST (Y), AGE (X).

(b)   Calculate the least-squares estimates of slope and intercept for the straight-line regression of ATST (Y) on AGE (X). Check your results with the computer printout.

(c)   Draw the estimated regression line on the scatter diagram.

(d)   Are any of the assumptions for straight-line regression obviously violated?

(e)   Obtain a 95% confidence interval for β_1.

(f)   Would you reject the null hypothesis H_0: $\beta_1 = 0$ based on the confidence interval obtained in part (e)?

(g)   Calculate and graph 95% confidence bands based on the estimated line.

7.   Several research workers associated with the office of highway safety were evaluating the relationship between driving speed (MPH) and the distance a vehicle travels once brakes are applied (DIST). The results of 19 experimental tests are displayed in the table.

(a)   Draw two scatter diagrams: one for the variable pair DIST (Y_1) and MPH (X) and the other for the pair $\sqrt{\text{DIST}}$ (Y_2) and MPH (X).

MPH (X)	DIST (Y_1)	$\sqrt{\text{DIST}}$ (Y_2)	MPH (X)	DIST (Y_1)	$\sqrt{\text{DIST}}$ (Y_2)
25	37.4	6.12	50	170.0	13.04
35	57.7	7.60	20	20.0	4.47
60	337.6	18.37	15	13.5	3.67
45	142.5	11.94	27.5	40.8	6.39
50	182.4	13.51	55	207.8	14.42
37.5	67.5	8.22	40	105.0	10.25
30	37.5	6.12	45	132.6	11.52
55	225.0	15.00	17.5	19.1	4.37
60	258.1	16.07	22.5	25.0	5.00
65	297.4	17.25			

Computer printout for Problem 7

MEANS AND STANDARD DEVIATIONS

X: 39.73684 +- 15.76509 Y_1: 125.10000 +- 102.87605

CORRELATIONS

PEARSON: 0.954260 SPEARMAN: 0.993850

LINEAR REGRESSION OF Y_1 ON X

```
SLOPE:         6.22708 +- 0.47319
INTERCEPT:   -122.34459 +- 20.15624
S(Y₁|X):       31.64950
```

ANALYSIS OF VARIANCE: F(1,17) = 173.18128

MEANS AND STANDARD DEVIATIONS

X: 39.73684 +- 15.76509 Y_2: 10.17526 +- 4.77577

CORRELATIONS

PEARSON: 0.986289 SPEARMAN: 0.992528

LINEAR REGRESSION OF Y_2 ON X

```
SLOPE:         0.29878 +- 1.21247E-02
INTERCEPT:    -1.59712 +- 0.51647
S(Y₂|X):       0.81097
```

ANALYSIS OF VARIANCE: F(1,17) = 607.2

(b)   Use the printout to determine and graph the least-squares lines for each variable pair given in part (a).

(c)   Which of the two variable pairs seems to be better suited for straight-line regression?

For the variable pair $\sqrt{\text{DIST}}$ (Y_2), MPH (X):

(d)   Test the hypothesis that the true slope β_1 is equal to 1. (Use $\alpha = 0.01$.)

(e)   Construct a 99% confidence interval for β_1.

(f)   Calculate and graph 95% confidence and prediction bands.

8. The table presents the starting annual salaries (SAL) of a group of 30 college graduates who have recently entered the job market, along with their cumulative-grade-point averages (CGPA).

SAL (Y)	CGPA (X)	SAL (Y)	CGPA (X)	SAL (Y)	CGPA (X)
$		$		$	
10,455	2.58	12,500	3.55	13,255	3.55
9,680	2.31	13,310	3.64	13,004	3.55
7,300	2.47	12,105	3.72	8,000	2.47
9,388	2.52	6,200	2.24	8,224	2.47
12,496	3.22	11,522	2.70	10,750	2.78
11,812	3.37	8,000	2.30	11,669	2.78
9,224	2.43	12,548	2.83	12,322	2.98
11,725	3.08	7,700	2.37	11,002	2.58
11,320	2.78	10,028	2.52	10,666	2.58
12,000	2.98	13,176	3.22	10,839	2.58

Computer printout for Problem 8

MEANS AND STANDARD DEVIATIONS

X: 2.83833 +- 0.44841   Y: 10740.66667 +- 1967.65248

CORRELATIONS

PEARSON: 0.827368   SPEARMAN: 0.933515   KENDALL: 0.799219

LINEAR REGRESSION OF Y ON X

SLOPE: 3630.56128 +- 465.76874
INTERCEPT: 435.92357 +- 1337.85967
S(Y|X):   1124.71499

ANALYSIS OF VARIANCE: F(1,28) = 60.75847

(a)–(c) Answer the same questions as in Problem 6(a)–(c) for the variable pair: SAL (Y), CGPA (X). Refer to the printout when necessary to answer the following:

(d) Obtain a 95% confidence interval for β_1.

(e) Would you reject the null hypothesis $H_0: \beta_1 = 4,000$ at the $\alpha = 0.05$ level?

(f) Calculate 95% confidence and prediction bands.

(g) Graph the respective 95% confidence and prediction bands.

(h) Would you reject the hypothesis $H_0: \mu_{Y|X_0} = 11,500$ at $X_0 = 2.75$? (Use $\alpha = 0.05$.)

9. In an experiment designed to describe the dose–response curve for vitamin K, individual rats were depleted of their vitamin K reserves and then fed dried liver for 4 days at different dosage levels.[6] The response of each rat was measured as the concentration of a clotting agent needed to clot a sample of its blood in 3 minutes. The results of the experiment involving 12 rats are given in the table in terms of common logarithms of both dose and response.

[6] Adapted from a study by Schønheyder (1936).

RAT No.	LOG$_{10}$ CONCENTRATION (Y)	LOG$_{10}$ DOSE (X)
1	2.65	0.18
2	2.25	0.33
3	2.26	0.42
4	1.95	0.54
5	1.72	0.65
6	1.60	0.75
7	1.55	0.83
8	1.32	0.92
9	1.13	1.01
10	1.07	1.04
11	0.95	1.09
12	0.88	1.15

Computer printout for Problem 9

MEANS AND STANDARD DEVIATIONS

X: 0.7425 +- 0.3200   Y: 1.6108 +- 0.5736

CORRELATION

PEARSON: -.99568

LINEAR REGRESSION OF Y ON X

SLOPE: -1.78501 +- 0.05267
INTERCEPT: 2.93620 +- 0.04230
S(Y|X):   0.05589

ANALYSIS OF VARIANCE: F(1,10) = 1148.762

(a) Draw a scatter diagram of these data.
(b) Determine the least-squares estimates of slope and intercept for the straight-line regression of Y on X.
(c) Plot the estimated regression line on the scatter diagram.
(d) Determine and sketch 99% confidence bands based on the estimated regression line.
(e) Convert the fitted straight line to an equation in the original units $Y' = 10^Y$ and $X' = 10^X$.
(f) For the converted equation obtained in part (e), determine 99% confidence intervals for the true mean responses at the maximum and minimum doses used in the experiment. [*Note*: $\hat{Y}(\text{max}) = 0.8834$, $\hat{Y}(\text{min}) = 2.6149$.]
(g) If each of the values for X and Y on each rat are converted to their original units X' and Y', the following fitted straight line is obtained: $\hat{Y}' = 237.16095 - 21.32117X'$. How would you evaluate whether or not using the variables X', Y' was better or worse than using X, Y in doing the regression analysis?

10. The susceptibility of catfish to a certain chemical pollutant was determined by immersing individual fish in 2 liters of an emulsion containing the pollutant and measuring the survival time in minutes.[7] The data in the table give the common

[7] Adapted from a study by Nagasawa et al. (1964).

log of survival time (Y) and the common log concentration (X) of the pollutant in parts per million for 18 fish.

FISH No.	LOG$_{10}$ SURVIVAL TIME (Y)	LOG$_{10}$ CONCENTRATION (X)
1	2.516	5.0
2	2.572	5.0
3	2.438	5.0
4	2.621	4.8
5	2.742	4.8
6	2.689	4.8
7	2.830	4.6
8	2.910	4.6
9	2.983	4.6
10	3.175	4.4
11	3.056	4.4
12	3.095	4.4
13	3.332	4.2
14	3.221	4.2
15	3.293	4.2
16	3.447	4.0
17	3.523	4.0
18	3.551	4.0

Computer printout for Problem 10

MEANS AND STANDARD DEVIATIONS

X: 4.500 +- 0.3515   Y: 2.997 +- 0.3550

CORRELATION

PEARSON: -.98823

LINEAR REGRESSION OF Y ON X

SLOPE: -0.99810 +- 0.03863
INTERCEPT: 7.49110 +- 0.17429
S(Y|X):   .05597

ANALYSIS OF VARIANCE: F(1,16) = 667.728

(a) Plot the scatter diagram for these data and draw the estimated straight line of Y on X on your scatter diagram.
(b) Test for the significance of the straight-line regression.
(c) Determine 95% confidence intervals for the true mean survival time $\mu_{Y'|X}$ (where $Y = 10^{Y}$) at values of $X = 5, 4.5,$ and 4.

11. An experiment was conducted to determine the extent to which the growth rate of a certain fungus could be affected by filling test tubes containing the same medium at the same temperature with different inert gases.[8] Three such experiments were performed for each of six gases and the average growth rate over these three tests was used as the response. The table gives the molecular weight (X) of each gas used and the average growth rate (Y) in milliliters per hour for the three tests made for each gas.

[8] Adapted from a study by Schreiner et al. (1962).

GAS	AVERAGE GROWTH RATE, Y	MOLECULAR WEIGHT, X
A	3.85	4.0
B	3.48	20.2
C	3.27	28.2
D	3.08	39.9
E	2.56	83.8
F	2.21	131.3

(a) Find the least-squares estimates of slope and intercept for the straight-line regression of Y on X and draw the estimated straight line on the scatter diagram for this data set.

(b) Test for significant slope of the fitted straight line.

(c) What information has not been used which might improve the sensitivity of the analysis?

(d) What is the 90% confidence interval for the true average growth rate when the gas used has a molecular weight of 100?

(e) Why would it be inappropriate to use the fitted line to estimate the growth rate for a gas with a molecular weight of 200?

(f) Based on the choice of X values used in this study, how would you criticize the experiment in terms of the accuracy of prediction obtained using the fitted straight line?

CHAPTER 6
THE CORRELATION COEFFICIENT AND ITS RELATIONSHIP TO STRAIGHT-LINE REGRESSION ANALYSIS

6.1 DEFINITION OF r

The correlation coefficient is an often-used statistic that not only provides a measure of how two random variables are associated in a sample but has properties that relate it closely to straight-line regression. We define the *sample correlation coefficient r* for two variables X and Y by the formula

$$r = \frac{\sum_{i=1}^{n} (X_i - \bar{X})(Y_i - \bar{Y})}{\left[\sum_{i=1}^{n} (X_i - \bar{X})^2 \sum_{i=1}^{n} (Y_i - \bar{Y})^2 \right]^{1/2}} \tag{6.1}$$

For hand-calculating purposes, another version of this formula is given by

$$r = \frac{\sum_{i=1}^{n} X_i Y_i - \left(\sum_{i=1}^{n} X_i \right)\left(\sum_{i=1}^{n} Y_i \right) / n}{\left\{ \left[\sum_{i=1}^{n} X_i^2 - \left(\sum_{i=1}^{n} X_i \right)^2 / n \right]\left[\sum_{i=1}^{n} Y_i^2 - \left(\sum_{i=1}^{n} Y_i \right)^2 / n \right] \right\}^{1/2}}$$

An equivalent formula for r which illustrates its mathematical relationship to the least-squares estimate of the slope of a fitted regression line is

$$r = \frac{S_X}{S_Y} \hat{\beta}_1 \tag{6.2}$$

EXAMPLE   For the age–SBP data based on the data of Table 5.1, r is calculated as follows:

$$r = \frac{199,576 - (1,354)(4,276)/30}{\{[67,894 - (1,354)^2/30][624,260 - (4,276)^2/30]\}^{1/2}} = 0.66$$

Or, alternatively, from (6.2) we have

$$r = \frac{15.29}{22.58}(0.97) = 0.66$$

6.1.1 SOME MATHEMATICAL PROPERTIES OF r

Three important mathematical properties of r are:

1. The range of possible values of r is from -1 to 1.
2. r is a dimensionless quantity; that is, r is independent of the units of measurement of X and Y.
3. r is positive, negative, or zero according as $\hat{\beta}_1$ is positive, negative, or zero, and vice versa. This property follows directly, of course, from (6.2).

6.2 r AS A MEASURE OF ASSOCIATION

In the statistical assumptions for straight-line regression analysis discussed earlier, we did not consider the variable X to be random. Nevertheless, it turns out that it often makes sense to view the regression problem as one where *both* X and Y are random variables. The measure r, in this context, is then interpretable as an *index of association* between X and Y in the following sense:

1. The more positive r is, the more positive the association is. This means that when r is close to 1, an individual with a high value for one variable will likely have a high value for the other and an individual with a low value for one variable will likely have a low value for the other (Figure 6.1a).
2. The more negative r is, the more negative the association is; that is, an individual with high (low) value for one variable will likely have a low (high) value for the other when r is close to -1 (Figure 6.1b).
3. If r is close to zero, there is little, if any, *linear* association[1] between X and Y (Figure 6.1c).

FIGURE 6.1 *Correlation coefficient as a measure of association*

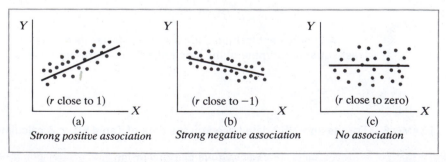

$(r$ close to 1$)$	$(r$ close to $-1)$	$(r$ close to zero$)$
(a)	(b)	(c)
Strong positive association	*Strong negative association*	*No association*

[1] Later we will see that a value of r close to zero does not rule out a possible *nonlinear* association.

[By association, we mean the lack of statistical independence between X and Y. More loosely, the lack of an association means that the value of one variable cannot be reasonably anticipated by knowing the value of the other variable.]

Since r is an index obtained from a *sample* of n observations, it follows that it can be considered as an estimate of an unknown population parameter. This unknown parameter is called the *population correlation coefficient* and is generally denoted by the symbol ρ_{XY}, or simply by ρ, if it is clearly understood which two variables are being considered. We shall agree to use ρ unless the possibility of confusion exists.

[The parameter ρ_{XY} is defined as $\rho_{XY} = \sigma_{XY}/\sigma_X\sigma_Y$, where σ_X and σ_Y denote the population standard deviations of the random variables X and Y and where σ_{XY} is called the *covariance* between X and Y. The covariance σ_{XY} is a population parameter describing the average amount that two variables "covary." In actuality, it is the population mean of the random variable $(X - \bar{X}) \cdot (Y - \bar{Y})$.]

6.3 THE BIVARIATE NORMAL DISTRIBUTION[2]

Another way of looking at straight-line regression arises if we consider X and Y as random variables having the bivariate normal distribution, which is a generalization of the *univariate normal distribution*. Just as the univariate normal distribution is described by a density function which is a bell-shaped curve when plotted in two dimensions, so the bivariate normal distribution is described by a *joint density function* whose plot looks like a bell-shaped surface in three dimensions (Figure 6.2).

FIGURE 6.2 *The bivariate normal distribution*

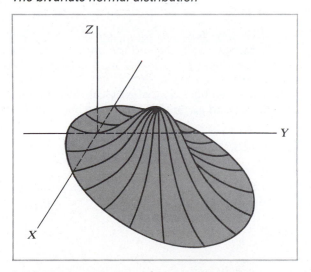

[2] This section is not essential for an understanding of the correlation coefficient as it relates to regression analysis.

One property of the bivariate normal distribution that has implications with regard to straight-line regression analysis is the following. If the bell-shaped surface is cut by a plane *parallel* to the Y-Z plane and passing through a specific X value, the curve or trace that results is a normal distribution. In other words, the distribution of Y for fixed X is univariate-normal. We call such a distribution the *conditional distribution of Y at X* and we denote the corresponding random variable as Y_X. Let us also denote the mean of this distribution as $\mu_{Y|X}$ and the variance as $\sigma^2_{Y|X}$. Then it follows from statistical theory that the mean and variance of Y_X can be written in terms of μ_X, μ_Y, σ^2_X, σ^2_Y, and ρ_{XY} as follows:

$$\mu_{Y|X} = \mu_Y + \rho_{XY}\frac{\sigma_Y}{\sigma_X}(X - \mu_X) \tag{6.3}$$

and

$$\sigma^2_{Y|X} = \sigma^2_Y (1 - \rho^2_{XY}) \tag{6.4}$$

Now suppose that we let $\beta_1 = \rho_{XY}(\sigma_Y/\sigma_X)$ and $\beta_0 = \mu_Y - \beta_1\mu_X$. We can then see that (6.3) has been transformed into the familiar expression for a straight-line model; that is, $\mu_{Y|X} = \beta_0 + \beta_1 X$. Furthermore, if we substitute the estimators $\bar{X}$, $\bar{Y}$, S_X, S_Y, and r for their respective parameters μ_X, μ_Y, σ_X, σ_Y, and ρ_{XY} in (6.3), we obtain the formula

$$\hat{\mu}_{Y|X} = \bar{Y} + r\frac{S_Y}{S_X}(X - \bar{X})$$

The right-hand side is exactly equivalent to the expression for the least-squares straight line given by (5.8), since

$$\hat{\beta}_1 = r\frac{S_Y}{S_X}$$

Thus, *the least-squares formulas for $\hat{\beta}_0$ and $\hat{\beta}_1$ can be developed by assuming that X and Y are random variables having the bivariate normal distribution and by substituting the usual estimates of μ_X, μ_Y, σ_X, σ_Y, and ρ_{XY} into the expression for $\mu_{Y|X}$, the conditional mean of Y given X.*

Note also that our estimate $S^2_{Y|X}$ of $\sigma^2_{Y|X}$ can be obtained by substituting the estimates S^2_Y and r for σ^2_Y and ρ_{XY} in (6.4). This is so because, from (5.11) and (6.2),

$$S^2_{Y|X} = \frac{n-1}{n-2}(S^2_Y - \hat{\beta}^2_1 S^2_X) = \frac{n-1}{n-2}S^2_Y(1 - r^2)$$

which approximates $S^2_Y(1 - r^2)$ even when n is just moderately large. Finally, (6.4) can be algebraically manipulated into the form

$$\rho^2_{XY} = \frac{\sigma^2_Y - \sigma^2_{Y|X}}{\sigma^2_Y} \tag{6.5}$$

This equation describes the square of the population correlation coefficient as the proportionate reduction in the variance of Y due to conditioning on X. The importance of (6.5) in describing the strength of the straight-line relationship will be discussed in the next section.

6.4 *r* AS A MEASURE OF THE STRENGTH OF THE STRAIGHT-LINE RELATIONSHIP

In order to quantify what we mean by the *strength* of the linear relationship between X and Y, we should first consider what would be our predictor of Y if we were not to use X at all. The best predictor in this case would simply be $\bar{Y}$, the sample mean of the Y's. The sum of squares of deviations associated with the naive predictor $\bar{Y}$ would then be given by the formula

$$\text{SSY} = \sum_{i=1}^{n} (Y_i - \bar{Y})^2$$

Now, if the variable X is of any value in predicting the variable Y, the residual sum of squares given by

$$\text{SSE} = \sum_{i=1}^{n} (Y_i - \hat{Y}_i)^2$$

should be considerably less than SSY. If so, the least-squares model $\hat{Y} = \hat{\beta}_0 + \hat{\beta}_1 X$ fits the data better than the horizontal line $\hat{Y} = \bar{Y}$ (Figure 6.3). A quantitative measure of the improvement in the fit obtained by using X is given by the *square of the sample correlation coefficient r*, which can be written in the suggestive form

$$r^2 = \frac{\text{SSY} - \text{SSE}}{\text{SSY}} \tag{6.6}$$

This quantity naturally varies between 0 and 1 since r itself varies between -1 and 1.

Now, what interpretation can be given to the quantity r^2? To answer this question, we first note that the difference or reduction in SSY due to using X to predict

FIGURE 6.3   *Prediction of Y using and not using X*

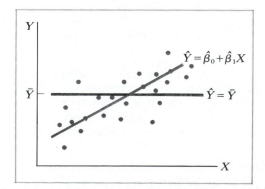

Y may be measured by (SSY − SSE), which is always nonnegative. Furthermore, the *proportionate reduction* in SSY due to using X to predict Y is this difference divided by SSY. Thus, r^2 measures the strength of the linear relationship between X and Y in the sense that it gives the proportionate reduction in the sum of squares of vertical deviations obtained using the least-squares line $\hat{Y} = \hat{\beta}_0 + \hat{\beta}_1 X$ relative to the naive model $\hat{Y} = \bar{Y}$, the predictor of Y if X is ignored. The larger the value of r^2, the greater is the reduction in SSE relative to $\sum_{i=1}^{n} (Y_i - \bar{Y})^2$, and the stronger is the linear relationship between X and Y.

The largest value that r^2 can attain is 1, which occurs when $\hat{\beta}_1$ is nonzero and when SSE = 0 (i.e., when there is a "perfect" positive or negative straight-line relationship between X and Y). By "perfect" we mean that *all* the data points lie on the fitted straight line. In other words, when $Y_i = \hat{Y}_i$ for all i, we must have

$$SSE = \sum_{i=1}^{n} (Y_i - \hat{Y}_i)^2 = 0$$

so that

$$r^2 = \frac{SSY - SSE}{SSY} = \frac{SSY}{SSY} = 1$$

Figure 6.4 illustrates examples of perfect positive and perfect negative linear association.

The smallest value that r^2 may take, of course, is 0. When this occurs, this means that there is no improvement in predictive power using X; that is, SSE = SSY. Furthermore, appealing to (6.2), we see that a correlation coefficient of zero implies a zero estimated slope and, consequently, the absence of any linear relationship (although the possibility of a nonlinear relationship still exists).

Finally, one should *not* be led to a false sense of security by considering the magnitude of r, rather than of r^2, when assessing the strength of the linear association between X and Y. For example, when r is 0.5, r^2 is only 0.25, and it takes an $r > 0.7$ to make $r^2 > 0.5$. Also, when r is 0.3, r^2 is 0.09, which indicates that only 9% of the variation in Y has been explained with the help of X. For the age–SBP data, r^2 is 0.44, as compared with an r of 0.66.

6.5 WHAT r DOES NOT MEASURE

There are two common misconceptions about r (or, equivalently, about r^2) that occasionally lead a researcher to make spurious interpretations about the relationship between X and Y:

1. r^2 *is not a measure of the magnitude of the slope of the regression line*; that is, if the value of r^2 is high (i.e., close to 1), this does not necessarily mean that the magnitude of the slope $\hat{\beta}_1$ is large. This phenomenon is illustrated in Figure 6.4. Notice that r^2 equals 1 in both parts, despite the fact that the slopes are different. Another way to look at this follows from the fact that, using (6.2),

$$\hat{\beta}_1^2 = \frac{S_Y^2}{S_X^2} \qquad \text{when } r^2 = 1$$

FIGURE 6.4   *Examples of perfect linear association*

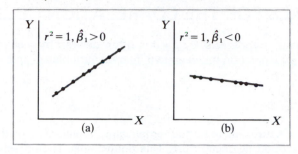

Thus, if two different sets of data have the same amount of X variation but the first set has less Y variation than the second set, the magnitude of the slope for the first set will be smaller than that for the second data set.

2. r^2 *is not a measure of the appropriateness of the straight-line model*. Notice that $r^2 = 0$ in parts (a) and (b) of Figure 6.5, even though there is no evidence of any association between X and Y in (a) and there is strong evidence of a nonlinear association in (b). Also, notice that r^2 is high in parts (c) and (d), even though a straight-line model is quite appropriate in (c) but is not completely appropriate in (d).

FIGURE 6.5   *Examples showing that r^2 is not a measure of the appropriateness of the straight-line model*

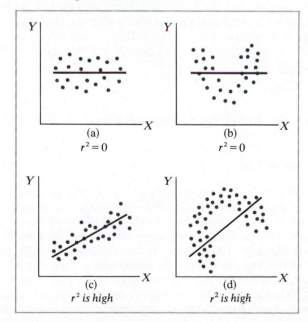

6.6 TESTS OF HYPOTHESES AND CONFIDENCE INTERVALS FOR THE CORRELATION COEFFICIENT

A test of the null hypothesis $H_0: \rho = 0$ is often desired by researchers interested in evaluating the association between two interval variables X and Y.

6.6.1 TEST OF $H_0: \rho = 0$

A test of $H_0: \rho = 0$ turns out to be mathematically equivalent to the test of the hypothesis $H_0: \beta_1 = 0$ described in Section 5.7. This equivalence is suggested by the formulas $\beta_1 = \rho \sigma_Y / \sigma_X$ and $\hat{\beta}_1 = r S_Y / S_X$, which tell us, for example, that β_1 is positive, negative, or zero according as ρ is positive, negative, or zero, and that an analogous relationship exists between $\hat{\beta}_1$ and r. It is possible to write the test statistic for the hypothesis $H_0: \rho = 0$ entirely in terms of r and n, so that one can perform the test without having to fit the straight line. This test statistic is given by the formula

$$T = \frac{r\sqrt{n-2}}{\sqrt{1-r^2}} \tag{6.7}$$

which has the t distribution with $n-2$ degrees of freedom when the null hypothesis $H_0: \rho = 0$ (or, equivalently, $H_0: \beta_1 = 0$) is true. Formula (6.7) yields exactly the same numerical answer as does (5.12), given by

$$T = \frac{\hat{\beta}_1 S_X \sqrt{n-1}}{S_{Y|X}}$$

EXAMPLE   For the age–SBP data of Table 5.1 for which $r = 0.66$, the statistic in (6.7) is calculated as follows:

$$T = \frac{0.66\sqrt{30-2}}{\sqrt{1-(0.66)^2}} = 4.62$$

which is the same value as that obtained for the test for slope in Table 5.3.

6.6.2 TEST OF $H_0: \rho = \rho_0, \rho_0 \neq 0$

A test concerning the null hypothesis $H_0: \rho = \rho_0 \, (\neq 0)$ cannot be directly related to a test concerning β_1; also, the hypothesis $H_0: \rho = \rho_0 \, (\neq 0)$ is not equivalent to the hypothesis $H_0: \beta_1 = \beta_1^{(0)}$ for some value $\beta_1^{(0)}$. Nevertheless, a test of $H_0: \rho = \rho_0 \, (\neq 0)$ is meaningful when previous experience or theory suggests a particular value to use for ρ_0.

The test statistic in this case can be obtained through consideration of the distribution of the sample correlation coefficient r. This distribution happens to be symmetric, like the normal distribution, *only* when ρ is 0. When ρ is nonzero, the distribution of r is skewed. This lack of normality prevents us from using the usual form of test statistic, which has a normally distributed estimator in the numerator and an estimate of its standard deviation in the denominator. However, it turns out that through an appropriate transformation, r can be changed to a statistic which is approximately normal. This transformation is called *Fisher's Z transformation*.[3] The formula for this transformation is

$$\frac{1}{2} \log_e \frac{1+r}{1-r} \tag{6.8}$$

This quantity $\frac{1}{2} \log_e (1+r)/(1-r)$ has approximately the normal distribution with mean $\frac{1}{2} \log_e (1+\rho)/(1-\rho)$ and variance $1/(n-3)$ when n is not too small (e.g., $n \geq 20$). In testing the hypothesis $H_0: \rho = \rho_0 (\neq 0)$, we can then use the test statistic

$$Z = \frac{\frac{1}{2} \log_e (1+r)/(1-r) - \frac{1}{2} \log_e (1+\rho_0)/(1-\rho_0)}{1/\sqrt{n-3}} \tag{6.9}$$

This test statistic has approximately the standard normal distribution [i.e., $Z \frown N(0, 1)$] under H_0. To test $H_0: \rho = \rho_0 (\neq 0)$, we therefore use one of the following critical regions for significance level α:

$Z \geq z_{1-\alpha}$ \qquad (upper one-tailed alternative $H_A: \rho > \rho_0$)

$Z \leq -z_{1-\alpha}$ \qquad (lower one-tailed alternative $H_A: \rho < \rho_0$)

$|Z| \geq z_{1-\alpha/2}$ \qquad (two-tailed alternative $H_A: \rho \neq \rho_0$)

where $z_{1-\alpha}$ denotes the $100(1-\alpha)\%$ point of the standard normal distribution. Computation of Z can be aided by using Table A-5, which gives values of $\frac{1}{2} \log_e (1+r)/(1-r)$ for given values of r.

EXAMPLE   Suppose that from previous experience it can be hypothesized that the true correlation between age and SBP is $\rho_0 = 0.85$. To test the hypothesis $H_0: \rho = 0.85$ against the two-sided alternative $H_A: \rho \neq 0.85$, we perform the following calculations using $r = 0.66$, $\rho_0 = 0.85$, and $n = 30$:

$$\frac{1}{2} \log_e \frac{1+\rho_0}{1-\rho_0} = \frac{1}{2} \log_e \frac{1+0.85}{1-0.85} = 1.2561 \qquad \text{(from Table A-5)}$$

$$\frac{1}{2} \log_e \frac{1+r}{1-r} = \frac{1}{2} \log_e \frac{1+0.66}{1-0.66} = 0.7928 \qquad \text{(from Table A-5)}$$

$$Z = \frac{0.7928 - 1.2561}{1/\sqrt{30-3}} = -2.41$$

For $\alpha = 0.05$, the critical region is

$$|Z| \geq z_{0.975} = 1.96$$

[3] Named after R. A. Fisher, who introduced it in 1925.

Since $|Z| = 2.41$ exceeds 1.96, the hypothesis $H_0: \rho_0 = 0.85$ is rejected at the 0.05 significance level. Further calculations show that the P value for this test is $P = 0.0151$, which tells us that the result is not significant at $\alpha = 0.01$.

6.6.3 CONFIDENCE INTERVAL FOR ρ

A $100(1-\alpha)\%$ confidence interval for ρ can be obtained using Fisher's Z transformation (6.8) as follows. First, compute a $100(1-\alpha)\%$ confidence interval for the parameter $\frac{1}{2}\log_e (1+\rho)/(1-\rho)$ using the formula

$$\frac{1}{2}\log_e \frac{1+r}{1-r} \pm z_{1-\alpha/2}/\sqrt{n-3} \tag{6.10}$$

where $z_{1-\alpha/2}$ is as defined previously.

Denote the lower limit of the confidence interval (6.10) by L_z and the upper limit by U_z; then use Table A-5 (in reverse) to determine the lower and upper confidence limits L_ρ and U_ρ for the confidence interval for ρ. That is, determine L_ρ and U_ρ from the following formulas:

$$L_z = \frac{1}{2}\log_e \frac{1+L_\rho}{1-L_\rho} \quad \text{and} \quad U_z = \frac{1}{2}\log_e \frac{1+U_\rho}{1-U_\rho}$$

The $100(1-\alpha)\%$ confidence interval is then of the form

$$L_\rho < \rho < U_\rho$$

EXAMPLE Suppose that we seek a 95% confidence interval for ρ using the AGE–SBP data for which $r = 0.66$ and $n = 30$. A 95% confidence interval for $\frac{1}{2}\log_e (1+\rho)/(1-\rho)$ is then given by

$$\frac{1}{2}\log_e \frac{1+0.66}{1-0.66} \pm 1.96/\sqrt{30-3}$$

which is equal to

$$0.793 \pm 0.377$$

providing a lower limit of $L_z = 0.416$ and an upper limit of $U_z = 1.170$; that is,

$$0.416 < \frac{1}{2}\log_e \frac{1+\rho}{1-\rho} < 1.170$$

To transform this confidence interval to one for ρ, we determine those values of L_ρ and U_ρ which satisfy

$$0.416 = \frac{1}{2}\log_e \frac{1+L_\rho}{1-L_\rho} \quad \text{and} \quad 1.170 = \frac{1}{2}\log_e \frac{1+U_\rho}{1-U_\rho}$$

Using Table A-5 we see that a value of 0.416 corresponds to an r of about 0.394, so that $L_\rho = 0.394$. Similarly, a value of 1.170 corresponds to an r of about 0.824, so that $U_\rho = 0.824$. The 95% confidence interval for ρ is then given by

$$0.394 < \rho < 0.824$$

Notice that this confidence interval does not contain the value 0.85, which agrees with the conclusion of the previous section that $H_0: \rho = 0.85$ is to be rejected at the 5% level (two-tailed test).

PROBLEMS

1. (a) For the data set of Problem 1 of Chapter 5, compute the correlation coefficients of (1) dry weight with age and (2) log dry weight with age.
   (b) Using Fisher's Z transformation, obtain a 95% confidence interval for ρ based on each of the correlations obtained in part (a).
   (c) For each straight-line regression, calculate r^2 directly by squaring the r obtained in part (a), and also calculate r^2 using the formula $r^2 = (SSY - SSE)/SSY$.
   (d) Based on the results above, which of the two regression lines do you believe provides the better fit? Explain. Does this agree with your earlier conclusion?

2. Examine the five pairs of data points given in the table.

i	1	2	3	4	5
X_i	-2	-1	0	1	2
Y_i	4	1	0	1	4

   (a) What is the mathematical relationship between X and Y?
   (b) Show by computation that, for the straight-line regression of Y on X, $\hat{\beta}_1 = 0$.
   (c) Show by computation that $r = 0$.
   (d) Why is there apparently "no relationship" between X and Y as indicated by the estimates of β_1 and ρ?

3. Consider the data in the table.

i	1	2	3	4	5	6	7	8	9	10
X_i	1	1	1	2	2	2	3	3	3	20
Y_i	1	2	3	1	2	3	1	2	3	20

   (a) Find the sample correlation coefficient r.
   (b) Show that the test statistic $T = \hat{\beta}_1/(S_{Y|X}/S_X\sqrt{n-1})$ for testing $H_0: \beta_1 = 0$ (based on a straight-line regression relationship between Y and X) is exactly equivalent to the test statistic $T' = r\sqrt{n-2}/\sqrt{1-r^2}$ for testing $H_0: \rho = 0$. [Hint: Use $\hat{\beta}_1 = rS_Y/S_X$ and $S_{Y|X}^2 = [(n-1)/(n-2)](S_Y^2 - \hat{\beta}_1^2 S_X^2)$.]

(c) Using T', test $H_0: \rho = 0$ versus $H_A: \rho \neq 0$.

(d) Despite the conclusion obtained in part (c), why should it bother you to conclude that the two variables are linearly related? ("A graph is worth a thousand words.")

4–6. Answer the following questions with regard to the straight-line regressions of Y on X referred to in Problems 2.4, 2.5, and 2.6 of Chapter 5:

(a) Compute r and r^2, and interpret your results.

(b) Find a 99% CI for ρ, and interpret your result with regard to the test of $H_0: \rho = 0$ versus $H_A: \rho \neq 0$ at $\alpha = 0.01$.

7–12. Answer the following questions for each of the data sets of Problems 3–8 in Chapter 5.

(a) Compute r and r^2 for each variable pair, and interpret your results.

(b) Test $H_0: \rho = 0$ versus $H_A: \rho \neq 0$, and interpret your findings.

(c) Find a 95% confidence interval for ρ.

13. Suppose that in a study regarding geographical variation in a certain species of beetle,[4] the mean tibia length (U) and the mean tarsus length (V) were obtained for samples of size 50 from each of 10 different regions spanning five southern states. Suppose further that the results were as given in the table.

Region:	1	2	3	4	5	6	7	8	9	10
U	7.500	7.164	7.512	8.544	7.380	7.860	7.836	8.100	7.584	7.344
V	1.680	1.596	1.680	1.908	1.632	1.752	1.776	1.860	1.692	1.680

(a) Compute the correlation coefficient between tarsus length and tibia length.

(b) Find a 99% confidence interval for ρ.

[4] Adapted from a study by Sokal and Thomas (1965).

CHAPTER 7 THE ANALYSIS-OF-VARIANCE TABLE AND ITS RELATIONSHIP TO REGRESSION ANALYSIS

7.1 PREVIEW

An overall summary of the results of any regression analysis, whether straight line or not, can be provided by a table called an *analysis-of-variance (ANOVA) table*. The name given to this type of table derives primarily from the fact that the basic information in an ANOVA table consists of several estimates of variance. These estimates, in turn, can be used to answer the principal inferential questions of regression analysis. In the straight-line case, these questions are: (1) Is the true slope β_1 zero? (2) What is the strength of the straight-line relationship? (3) Is the straight-line model appropriate?

It should also be pointed out that, historically, the name "analysis-of-variance table" was coined to describe the overall summary table for the statistical procedure known as *analysis of variance*. As we mentioned in Chapter 2 and will see later when discussing the ANOVA method, regression analysis and analysis of variance are very closely related. More precisely, analysis-of-variance problems can be expressed in a regression framework. Thus, it should not be surprising to find that such a table can be used to summarize the results obtained from either method.

7.2 ANOVA TABLE FOR STRAIGHT-LINE REGRESSION

Several slightly different ways of presenting the ANOVA table associated with straight-line regression analysis are given by various textbooks, researchers, and computer program printouts. This section will describe those forms which are most commonly used.

The simplest version of the ANOVA table for straight-line regression is given in Table 7.1, as applied to the age—SBP data of Table 5.1. The mean-square term is obtained by dividing the sum of squares by its degrees of freedom. The variance ratio, or *F* statistic, is obtained by dividing the regression mean square by the residual mean

TABLE 7.1 *ANOVA table for age–SBP data of Table 5.1*

SOURCE	DEGREES OF FREEDOM (*df*)	SUM OF SQUARES (SS)	MEAN SQUARE (MS)	VARIANCE RATIO (*F*)
Regression (*X*)	1	SSY – SSE = 6,394.02	6,394.02	21.33 (*P*<0.001)
Residual	28	SSE = 8,393.44	299.77	
Total (*corrected* for the mean) ($r^2 = 0.43$)	29	SSY = 14,787.46		

square. And r^2 is obtained by dividing the regression sum of squares by the total (corrected) sum of squares.

To explain how this table summarizes the regression results, let us recall that in describing the correlation coefficient, we observed in (6.6) that

$$r^2 = \frac{SSY - SSE}{SSY}$$

where $SSY = \sum_{i=1}^{n} (Y_i - \bar{Y})^2$ is the sum of the squares of deviations of the observed Y's from the mean $\bar{Y}$, and $SSE = \sum_{i=1}^{n} (Y_i - \hat{Y}_i)^2$ is the sum of squares of deviations of observed Y's from the fitted regression line. Since SSY represents the total variation of Y *before* accounting for the linear effect of the variable X, we usually call SSY the *total unexplained variation* or the *total sum of squares about (or corrected for) the mean*. Because SSE measures the amount of variation in the observed Y's that is left *after* accounting for the linear effect of the variable X, we usually call (SSY – SSE) the *sum of squares due to, or explained by, regression*. It turns out, in addition, that (SSY – SSE) is mathematically equivalent to the expression

$$\sum_{i=1}^{n} (\hat{Y}_i - \bar{Y})^2$$

which represents the sum of squares of deviations of the predicted values from the mean $\bar{Y}$. We thus have the following mathematical result:

> total unexplained variation = variation due to regression
>
> + unexplained residual variation

or

$$\sum_{i=1}^{n} (Y_i - \bar{Y})^2 = \sum_{i=1}^{n} (\hat{Y}_i - \bar{Y})^2 + \sum_{i=1}^{n} (Y_i - \hat{Y}_i)^2$$

(7.1)

FIGURE 7.1 *Variation explained and unexplained by (straight-line) regression*

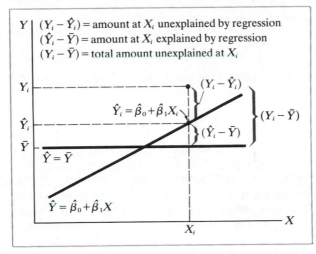

Equation (7.1) is often called the *fundamental equation of regression analysis*, and it holds for any general regression situation. Figure 7.1 illustrates this equation.

It should be pointed out that the mean-square residual term is simply the estimate $S^2_{Y|X}$ presented earlier. If the true regression model is a straight line, then, as mentioned previously in Section 5.5, $S^2_{Y|X}$ is an estimate of σ^2. On the other hand, the mean-square regression term (SSY − SSE) provides an estimate of σ^2 *only* if the variable X does not help to predict the dependent variable Y, that is, if the hypothesis $H_0: \beta_1 = 0$ is true. If, actually, $\beta_1 \neq 0$, the mean-square regression term will be inflated in proportion to the magnitude of β_1 and will correspondingly overestimate σ^2.

Now, it can be shown that the mean-square residual and mean-square regression terms are *statistically* independent of one another. Thus, if $H_0: \beta_1 = 0$ is true, the ratio of these terms represents the ratio of two independent estimates of the same variance σ^2. Under the normality assumption on the Y's, such a ratio has the F distribution, and this F statistic (with the value 21.33 in Table 7.1) can be used to test the hypothesis H_0: *no significant straight-line relationship of Y on X* (i.e., $H_0: \beta_1 = 0$ or $H_0: \rho = 0$).

Fortunately, this way of testing H_0 is *equivalent* to the use of the two-sided t test previously discussed. This is so because of the mathematical result that, for ν degrees of freedom,

$$F_{1,\nu} = T^2_\nu \tag{7.2}$$

so that

$$F_{1,\nu,1-\alpha} = t^2_{\nu,1-\alpha/2} \tag{7.3}$$

The expression in (7.3) states that the $100(1-\alpha)\%$ point of the F distribution with 1 and ν degrees of freedom is exactly the same as the square of the $100(1-\alpha/2)\%$ point of the t distribution with ν degrees of freedom.

To illustrate the equivalence of the F and t tests, we can see from our age–SBP example that $F = 21.33$ and $T^2 = (4.62)^2 = 21.33$, where 4.62 was the figure obtained for

T at the end of Section 6.6.1. Also, it can be seen that $F_{1,28,0.95} = 4.20$ and that $t^2_{28,0.975} = (2.05)^2 = 4.20$.

As a result of these equalities, the critical region

$$|T| > t_{28,0.975} = 2.05$$

for testing $H_0: \beta_1 = 0$ against the two-sided alternative $H_A: \beta_1 \neq 0$ is exactly the same as the critical region

$$F > F_{1,28,0.95} = 4.20$$

That is, if $|T|$ exceeds 2.05, so then will F exceed 4.20. Similarly, if F exceeds 4.20, so will $|T|$ exceed 2.05. Thus, the null hypothesis $H_0: \beta_1 = 0$ or equivalently "H_0: no significant straight-line relationship of Y on X" is rejected at the $\alpha = 0.05$ level of significance.

An alternative but less common representation of the ANOVA table is given in Table 7.2. This table differs from Table 7.1 only in that it splits up the total sum of squares corrected for the mean, SSY, into its two components, the *total uncorrected sum of squares*, $\sum_{i=1}^{n} Y_i^2$, and the *correction factor*, $(\sum_{i=1}^{n} Y_i)^2/n$. The relationship between these components is given by the equation

$$\sum_{i=1}^{n} (Y_i - \bar{Y})^2 = \sum_{i=1}^{n} Y_i^2 - \frac{\left(\sum_{i=1}^{n} Y_i\right)^2}{n}$$

The total (uncorrected) sum of squares $\sum_{i=1}^{n} Y_i^2$ considers the n observations on Y before any estimation of the population mean of Y is considered. The term "regression $(\bar{Y})$" in Table 7.2 refers to the variability explained by using a model involving only β_0 (which is estimated by $\bar{Y}$). This is necessarily the same amount of variability as is explained by using only $\bar{Y}$ to predict Y, without attempting to account for the linear contribution of X to the prediction of Y. The term "regression $(X|\bar{Y})$" describes the contribution of the variable X to predicting Y over and above that contributed by $\bar{Y}$

TABLE 7.2 *Alternative ANOVA table for AGE–SBP data of Table 5.1*

SOURCE	DEGREES OF FREEDOM (df)	SUM OF SQUARES (SS)	MEAN SQUARE (MS)	VARIANCE RATIO (F)
Regression $\{\ \bar{Y}$	1	$\dfrac{\left(\sum_{i=1}^{n} Y_i\right)^2}{n} = 609{,}472.53$		
$\qquad\quad\ X|\bar{Y}$	1	6,394.02	6,394.02	21.33 ($P < 0.001$)
Residual	28	8,393.45	299.77	
Total (*uncorrected* for the mean) ($r^2 = 0.432$)	30	$\sum_{i=1}^{n} Y_i^2 = 624{,}260.00$		

alone. Usually, "regression $(X|\bar{Y})$" is written simply as "regression (X)," the "given $\bar{Y}$" part being suppressed for notational simplicity. We will see more of this notation when we discuss multiple regression in subsequent chapters.

7.3 ASSESSING THE APPROPRIATENESS OF THE STRAIGHT-LINE MODEL

In Section 5.1 it was pointed out that the usual strategy adopted regarding regression with a single independent variable *begins* with the assumption that the straight-line model is the appropriate one to use. This strategy provides for the rejection of the straight-line-model assumption if the data indicate that the use of a more complex model is warranted. In this section we shall describe a method useful in determining whether or not the straight-line-model assumption is reasonable.

The method we shall discuss is based on the use of a *test for lack of fit* of the assumed straight-line model. The key statistic involved is SSE, the residual sum of squares. As we mentioned in Section 5.5, there are two possible reasons for a large value of SSE: the first is that there is a lot of variability in the data itself (that is, σ^2 is large); the second is that the assumed straight-line model is not completely appropriate. In other words, the residual sum of squares is made up of a component describing *pure error* (i.e., a measure of σ^2 unrelated to regression using X) and a component describing the extent of *lack of fit* of the assumed straight-line model.

How do we estimate these two components? First, we find the pure error estimate of σ^2 by means of the use of "replicate observations" (i.e., *replicates* are repeated observations taken at the same value of X). We use repeated observations because our assumptions of Section 5.3 tell us that Y values observed at the same X value are independent observations from a distribution with variance σ^2, so that, for any given X, an estimate of σ^2 can be obtained by applying the usual sample variance formula to the Y observations taken at that X. Such an estimate is considered *pure* because it does not depend on the model under consideration (i.e., it does not depend on X). Note that only X's associated with at least two Y observations can be used to obtain a pure estimate of σ^2, since one Y observation by itself contributes no information concerning variability.

To illustrate what is meant by *replicate observations*, Table 7.3 presents such data for the age–SBP example of Table 5.1. The first two columns of this table give the X values at which replicates are available and the corresponding replicate values of Y associated with each X.

To obtain a pure estimate of σ^2 based on the two observations at age 42, say, one first calculates the usual sum of squares,

$$(124)^2 + (128)^2 - \frac{(124+128)^2}{2} = 8$$

and then divides by the degrees of freedom associated with this sum of squares (namely, $2-1=1$) to obtain the estimate $8/1 = 8$. The other sums of squares at the other X values are found similarly.

In Table 7.3 notice that the sum of squares determined from the data on the two persons of age 47 is extremely large. Upon observing the scatter diagram of these data given in Figure 5.1, one can see that the person of age 47 with blood pressure 220

TABLE 7.3   *Pure error estimates from repeated observations for Age–SBP data of Table 5.1*

X	Y	SS	df
39	144;120	288.0	1
42	124;128	8.0	1
45	138;135	4.5	1
47	220;145	2,812.5	1
56	154;150	8.0	1
67	170;158	72.0	1
		3,193.0	6

seems to stand quite apart from the rest of the data. We previously referred to this particular observation as an outlier and suggested that one might consider the possibility of eliminating it from the data. If, for the moment, we do not eliminate the observation on this person, we would calculate the *pure estimate* of σ^2 by obtaining the total of all the individual sums of squares and then dividing the sum of all the individual degrees of freedom. Thus, the *pure error estimate of* σ^2 is a "pooled" estimate, which in this case is given by

$$S_{pe}^2 = \frac{3,193.0}{6} = 532.17$$

[Such "pooling" of sums of squares is legitimate, of course, only when the assumption of variance homoscedasticity discussed in Section 5.3 holds. The reader is reminded of the two-sample t test as another instance where sums of squares are pooled.]

Notice that if we do not use the data on the person age 47 with SBP 220, we would obtain a value of S_{pe}^2 given by

$$S_{pe}^2 = \frac{3,193.0 - 2,812.5}{5} = 76.10$$

which is quite different from the value obtained before. For now we will carry on our analysis concerning "lack of fit" without dropping this observation, but we will check later to see what happens if we do drop it. In this regard variance homogeneity considerations alone should alert the reader about this particular observation.

We now provide the general formula for obtaining the pure error sum of squares and the corresponding pure error estimate of σ^2. Suppose that

$Y_{11}, Y_{12}, \ldots, Y_{1n_1}$ are n_1 replicate observations at X_1

$Y_{21}, Y_{22}, \ldots, Y_{2n_2}$ are n_2 replicate observations at X_2

$\quad\vdots\qquad\vdots\qquad\qquad\vdots\qquad\vdots\qquad\qquad\vdots$

$Y_{k1}, Y_{k2}, \ldots, Y_{kn_k}$ are n_k replicate observations at X_k

so that we have k sets of replicate observations. Then the contribution to the pure error sum of squares at X_1 and the corresponding estimate of σ^2 (based on $n_1 - 1$ df) are, respectively, given by

$$\sum_{u=1}^{n_1} (Y_{1u} - \bar{Y}_1)^2 \quad \text{and} \quad \frac{1}{n_1 - 1} \sum_{u=1}^{n_1} (Y_{1u} - \bar{Y}_1)^2$$

From this we see that the *pure error sum of squares* is given by

$$\text{pure error sum of squares } (SS_{pe}) = \sum_{i=1}^{k} \sum_{u=1}^{n_i} (Y_{iu} - \bar{Y}_i)^2 \qquad (7.4)$$

and the *pure error estimate of σ^2* is given by

$$S_{pe}^2 = \frac{1}{n_1 + n_2 + \cdots + n_k - k} SS_{pe} = \frac{1}{n_1 + n_2 + \cdots + n_k - k} \sum_{i=1}^{k} \sum_{u=1}^{n_i} (Y_{iu} - \bar{Y}_i)^2 \qquad (7.5)$$

The lack-of-fit sum of squares is obtained very simply by subtracting SS_{pe} from SSE. For our age–SBP data (using all 30 observations) in Table 5.1, we thus obtain

$$\text{SS(lack of fit)} = \text{residual sum of squares} - \text{pure error sum of squares} \qquad (7.6)$$

$$= 8,393.44 - 3,193.00$$
$$= 5,200.44$$

The degrees of freedom for SS(lack of fit) is obtained by subtracting the pure error degrees of freedom from the residual degrees of freedom. We thus have the formula

$$\text{df(lack of fit)} = \text{df(residual)} - \text{df(pure error)} \qquad (7.7)$$

For the age–SBP example, we then obtain

$$\text{df(lack of fit)} = 28 - 6 = 22$$

The next step is to determine the pure error mean square (MS_{pe}) and the mean square for lack of fit (MS_{lof}) by dividing each sum of squares by the corresponding degrees of freedom. The MS_{pe} is simply S_{pe}^2, the pure error estimate of σ^2. For our example we get

$$MS_{pe} = S_{pe}^2 = 532.17 \quad \text{and} \quad MS_{lof} = \frac{5,200.44}{22} = 236.38$$

The final step involves comparing the F statistic

$$F = \frac{MS_{lof}}{MS_{pe}}$$

(7.8)

with the $100(1-\alpha)\%$ point of the F distribution with the *lack-of-fit degrees of freedom as the numerator df* and the *pure error degrees of freedom as the denominator df*. For our example we get

$$F = \frac{236.38}{532.17} = 0.44$$

which does not exceed the 95% point of the F distribution given by $F_{22,6,0.95} = 3.86$. Furthermore, the P value for this test satisfies the relationship $P > 0.25$. Thus, basing this lack-of-fit test on the complete data for all 30 individuals, we would *not* reject the *null hypothesis "H_0: the straight-line model is appropriate"* in favor of the *alternative hypothesis "H_A: the straight-line model is not appropriate."* In other words, we may conclude that the straight-line assumption is reasonable.

Figure 7.2 presents a general flow diagram illustrating the procedure to use for testing the appropriateness of the straight-line model. It should be noted here that this flow diagram applies not only to a test of fit of the straight-line model, but to a test of fit of any regression model.

FIGURE 7.2 *Flow diagram for test for lack of fit*

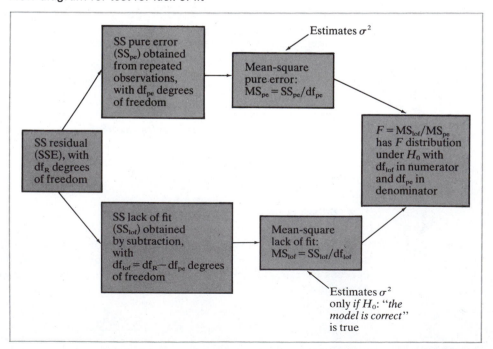

7.4 EXTENDED VERSION OF THE ANOVA TABLE FOR AGE–SBP DATA

The test for lack of fit can be incorporated directly into the ANOVA table for straight-line regression, as shown in Table 7.4. In this table the residual has been partitioned into lack of fit and pure error, giving the essential information concerning the test for lack of fit.

TABLE 7.4 *ANOVA table including test for lack of fit*

SOURCE		df	SS	MS	F
Regression (X)		1	6,394.02	6,394.02	21.33 ($P<0.001$)
Residual	lack of fit	22	5,200.44	236.38	0.44 ($P>0.25$)
	pure error	6	3,193.00	532.17	
Total (corrected) ($r^2 = 0.432$)		29	14,787.46		

7.5 ANOVA TABLE WITH OUTLIER REMOVED

Let us now return to the problem posed earlier concerning our outlier observation, (47; 220). There are two reasonable procedures to use to adjust for the presence of this outlier. First, it might not be a good idea to drop this observation completely, since our outlier possibly contains some information concerning lack of fit. However, its effect on the pure error sum of squares is considerable. A compromise, then, would be to retain the outlier but not to make use of its contribution to the pure error sum of squares. This certainly makes sense in view of the variance homogeneity assumption. Obviously, this causes an increase in the lack-of-fit mean square and a decrease in the pure error mean square (see Table 7.5).

The F test for lack of fit yields an F value of 4.58, which is considerably larger than the 0.44 given in Table 7.4. Since $F_{23,5,0.95} = 4.54$, the null hypothesis of no lack of fit of the straight-line model is barely rejected at the $\alpha = 0.05$ level, in contrast to our earlier

TABLE 7.5 *ANOVA table adjusted for outlier*

SOURCE		df	SS	MS	F
Regression (X)		1	6,394.02	6,394.02	21.33 ($P<0.001$)
Residual	lack of fit	23	8,012.94	348.39	4.58 ($0.025<P<0.05$)
	pure error	5	380.50	76.10	
Total		29	14,787.46		

decision. Note that H_0 would not be rejected at $\alpha = 0.025$ since $F_{23,5,0.975} = 6.29$. Thus, we can say that this approach to dealing with the outlier contradicts, although not very strongly, the previous conclusion of no lack of fit. Nevertheless, we should wonder whether this first approach is too one-sided in the sense that we did everything we could with the outlier to demonstrate lack of fit (i.e., we increased MS_{lof} and decreased MS_{pe} as much as possible). This leads us to consider a second procedure: discard the outlier and reanalyze the data. Table 7.6 is the ANOVA table with the outlier removed.

It is clear from this analysis that we *do not have significant lack of fit*. We therefore conclude that adding additional terms to the model is not really necessary, and that a straight-line model is appropriate to describe the relationship between age and SBP based on the available data.

One other observation can be made from a comparison of Table 7.6 and Table 7.4: the straight-line relationship appears considerably stronger when the outlier is removed. This can be seen in two ways: the F test for slope is more strongly significant when the outlier is removed (e.g., $F = 66.81$ and $P < 0.0001$ in this case), and r^2 is 0.712 when the outlier is removed as compared with only 0.432 when it is retained. Thus, it is clear that the outlier has a strong influence on the results of the analysis, and so whether to drop it or not becomes quite a difficult decision for the researcher. In our example, considerations of variance homogeneity and of lack of fit strongly suggest discarding the outlier.

One other point is worthy of mention here. It is often difficult to decide whether or not a given value of r^2 should be considered "high" in a practical sense, even though a statistical test of $H_0: \rho = 0$ leads to rejection of H_0. The answer depends primarily on the purpose of the investigation and on previous experience or theory. If the investigation is exploratory in nature, a small r^2 value like 0.4 or even 0.2 might be considered high in terms of helping to pinpoint a potentially important independent variable. Also, previous studies might have resulted in much smaller r^2 values than those achieved in a current investigation.

An r^2 value of 0.712 as achieved in the age–SBP example would be considered high in most situations. Nevertheless, it may still be possible to envision further improvement through the addition of other independent variables. For example, if weight and height were considered in addition to age, the r^2 might increase to over 0.90. We shall learn in later chapters, however, that adding more variables will always increase the r^2 value somewhat. The main concern here would be whether or not the increase in r^2 is sufficient to compensate for the increase in the number of variables and for the associated increase in the complexity of the model. The addition of new independent variables and the questions that relate to such additions will be dealt with in the later chapters on multiple regression.

TABLE 7.6   *ANOVA table with outlier removed*

SOURCE		df	SS	MS	F
Regression (X)		1	6,110.10	6,110.10	66.81 ($P < 0.0001$)
Residual {	lack of fit	22	2,088.85	94.95	1.25 ($P > 0.25$)
	pure error	5	380.50	76.10	
Total ($r^2 = 0.712$)		28	8,579.45		

PROBLEMS

1. (a) Determine the ANOVA table for the regression of dry weight (Y) on age (X) for the data of Problem 1 in Chapter 5.

   (b) Perform the F test for the significance of the straight-line regression of Y on X. Verify that this test is equivalent to the t test for zero slope (and for zero correlation).

   (c) Determine the ANOVA table for the regression of log dry weight (Z) on age (X) for the data of Problem 1 in Chapter 5.

   (d) Perform the F test for the significance of the straight-line regression of Z on X.

2–4. Answer the same questions as in Problem 1(a) and (b) for each of the regressions of Y on X using the data of Problems 2.4, 2.5, and 2.6 in Chapter 5. Also, for each of these data sets, do the following:

   (c) Perform a test for lack of fit of the straight-line model (provided that such a test can be performed) and interpret your findings.

5. (a) Determine the ANOVA table for the regression of TIME (Y) on INC (X) in Problem 3 of Chapter 5.

   (b) Test the hypothesis "H_0: no significant straight-line regression" using an F test.

   (c) Can a test for lack of fit be performed with this data set? Explain.

   (d) Compare the value of the test statistic F obtained in part (b) with the value of T^2, the square of the test statistic for testing $H_0: \beta_1 = 0$ required in Problem 3 of Chapter 5.

   (e) Determine the value of r^2 using the ANOVA table obtained in part (a); check your answer with the r^2 value obtained in Problem 7 of Chapter 6.

6. (a) Determine the ANOVA table for the straight-line regression of DI (Y) on IQ (X) in Problem 4 of Chapter 5 (excluding the outlier).

   (b) Perform the test for lack of fit of the fitted straight-line model.

   (c) Determine the value of r^2 using the ANOVA table.

7–11. Answer the same questions as in parts (a)–(c) of Problem 6 for each of the data sets in Problems 5–8 and 10 of Chapter 5.

12. A biologist wished to study the effects of the temperature of a certain medium on the growth of human amniotic cells in a tissue culture. Using the same parent batch, she conducted an experiment in which five cell lines were cultured at each of four temperatures. The procedure involved initially inoculating a fixed number of (0.25 million) cells into a fresh culture flask and then, after 7 days, removing a small sample from the growing surface to be used for estimating the total number of cells in the flask. The results are given in the table, together with a computer printout for straight-line regression.

NUMBER OF CELLS × 10^{-6} AFTER 7 DAYS (Y)	TEMPERATURE (X)	NUMBER OF CELLS × 10^{-6} AFTER 7 DAYS (Y)	TEMPERATURE (X)
1.13	40	2.30	80
1.20	40	2.15	80
1.00	40	2.25	80
0.91	40	2.40	80
1.05	40	2.49	80
1.75	60	3.18	100
1.45	60	3.10	100
1.55	60	3.28	100
1.64	60	3.35	100
1.60	60	3.12	100

Computer printout for Problem 12

```
MEANS AND STANDARD DEVIATIONS

X: 70.0000 +- 22.9416    Y: 2.0450 +- 0.8334

CORRELATIONS

PEARSON: .9860

LINEAR REGRESSION OF Y ON X

SLOPE:        0.03582 +- 0.00143
INTERCEPT:   -0.46240 +- 0.10481
S(Y|X):       0.14263

ANALYSIS OF VARIANCE: F(1,18) = 630.716
```

(a) Construct the scatter diagram for this data set and comment on whether you think a straight line gives an adequate fit.

(b) Draw the least-squares straight line on the scatter diagram.

(c) Determine the ANOVA table for the straight-line regression of Y on X.

(d) Test for significance of the straight-line regression using an F test.

(e) Perform the test for lack of fit of the straight-line model.

CHAPTER 8 COMPARING TWO STRAIGHT-LINE REGRESSION MODELS

8.1 PREVIEW

Throughout the previous three chapters, an example involving observations (Table 5.1) on age and SBP for 30 individuals was used to illustrate the main principles and methods of straight-line regression analysis. This example, although hypothetical, was shown upon analysis to support the commonly found epidemiological observation that blood pressure increases with age.

Another such commonly found observation is that males tend to have higher blood pressure than females of similar age. This observation may also be demonstrated using the principles of regression analysis. However, the problem in this regard is one of comparison; that is, the regression of SBP on age for females must be compared with the corresponding regression for males (Figure 8.1). Such a comparison requires methodology that we have not discussed so far.

In this chapter our attention is focused on the comparison of two *straight-line* regression models. However, the methods we shall describe here can be alternatively done and even improved upon in some instances by appealing to more general regression techniques involving several independent variables. Such refinements will be discussed in Chapter 13.

[In Chapter 13 we shall describe how to compare two or more regression *curves*, which may not necessarily be straight lines. Such a comparison requires knowledge of the general method of regression analysis involving several independent variables.]

EXAMPLE To illustrate the straight-line comparison problem, it is convenient to continue with our previous hypothetical example using age and SBP. However, as mentioned at the end of Chapter 7, the outlier (47; 220) is to be discarded in any further analyses. We shall therefore assume that the 29 observations on age and SBP considered previously were made on females and that a second sample of observations on age and SBP was collected on 40 males. The entire data set is presented in Table 8.1. In addition, Table 8.1

FIGURE 8.1   *Comparison by sex of straight-line regressions of SBP on age*

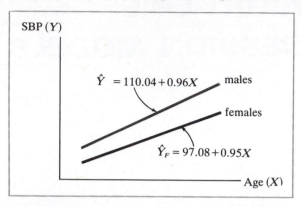

TABLE 8.1   *Data on SBP and age for 40 males and 29 females and associated information needed to compare two straight-line regression equations*

MALE (*i*)	SBP (*Y*)	AGE (*X*)	MALE (*i*)	SBP (*Y*)	AGE (*X*)	MALE (*i*)	SBP (*Y*)	AGE (*X*)
1	158	41	15	142	44	28	144	33
2	185	60	16	144	50	29	139	23
3	152	41	17	149	47	30	180	70
4	159	47	18	128	19	31	165	56
5	176	66	19	130	22	32	172	62
6	156	47	20	138	21	33	160	51
7	184	68	21	150	38	34	157	48
8	138	43	22	156	52	35	170	59
9	172	68	23	134	41	36	153	40
10	168	57	24	134	18	37	148	35
11	176	65	25	174	51	38	140	33
12	164	57	26	174	55	39	132	26
13	154	61	27	158	65	40	169	61
14	124	36						

FEMALE (*i*)	SBP (*Y*)	AGE (*X*)	FEMALE (*i*)	SBP (*Y*)	AGE (*X*)
1	144	39	16	135	45
2	138	45	17	114	17
3	145	47	18	116	20
4	162	65	19	124	19
5	142	46	20	136	36
6	170	67	21	142	50
7	124	42	22	120	39
8	158	67	23	120	21
9	154	56	24	160	44
10	162	64	25	158	53
11	150	56	26	144	63
12	140	59	27	130	29
13	110	34	28	125	25
14	128	42	29	175	69
15	130	48			

also provides the essential information needed to make a comparison of the two straight lines; for each data set, this information consists of the sample size (n), intercept ($\hat{\beta}_0$), slope ($\hat{\beta}_1$), sample mean $\bar{X}$ of the X's, sample mean $\bar{Y}$ of the Y's, sample variance S_X^2 of the X's, and residual mean-square error ($S_{Y|X}^2$). To distinguish between male and female data, we have used the subscripts M and F, respectively. Thus, n_M, $\hat{\beta}_{0M}$, $\hat{\beta}_{1M}$, and $S_{Y|X_M}^2$ denote the sample size, intercept, slope, and mean-square error for the male data, whereas n_F, $\hat{\beta}_{0F}$, $\hat{\beta}_{1F}$, and $S_{Y|X_F}^2$ denote the corresponding information for the female data. The least-squares straight lines[1] as as follows:

	n	$\hat{\beta}_0$	$\hat{\beta}_1$	$\bar{X}$	$\bar{Y}$	S_X^2	$S_{Y|X}^2$
Males	$n_M = 40$	$\hat{\beta}_{0M} = 110.04$	$\hat{\beta}_{1M} = 0.96$	$\bar{X}_M = 46.93$	$\bar{Y}_M = 155.15$	$S_{X_M}^2 = 221.15$	$S_{Y|X_M}^2 = 71.90$
Females	$n_F = 29$	$\hat{\beta}_{0F} = 97.08$	$\hat{\beta}_{1F} = 0.95$	$\bar{X}_F = 45.07$	$\bar{Y}_F = 139.86$	$S_{X_F}^2 = 242.14$	$S_{Y|X_F}^2 = 91.46$

Males: $\quad \hat{Y}_M = 110.04 + 0.96X$

Females: $\quad \hat{Y}_F = 97.08 + 0.95X$

These lines are sketched in Figure 8.1. It can be seen from the figure that the male line lies completely above the female line. This, in itself, lends support to the contention that males have higher blood pressure than females over the age range being considered. Nevertheless, it is necessary to explore statistically whether or not the observed differences between regression lines could have occurred by chance. In other words, to be statistically precise in the comparison of the two regression lines, it is necessary to take into consideration the sampling variability of the data through the use of statistical test(s) and/or confidence interval(s). In the sections below, a number of statistical procedures for dealing with this comparison problem are described.

8.2 COMPARING TWO STRAIGHT LINES

There are three basic questions to consider when making a comparison of two straight-line regression equations. These are:

1. Are the two slopes the same or different[2] (regardless of whether or not the intercepts are different)?
2. Are the two intercepts the same or different (regardless of whether or not the slopes are different)?
3. Are the two lines coincident (i.e., the same) or do they differ in slope and/or intercept?

Situations pertaining to these three questions are illustrated in Figure 8.2.

[1] It can be shown that the straight-line model for males, like that for females, is appropriate based on a lack-of-fit test.
[2] If the two slopes are not different, we say that the two lines are *parallel*.

FIGURE 8.2   *Straight-line comparison possibilities*

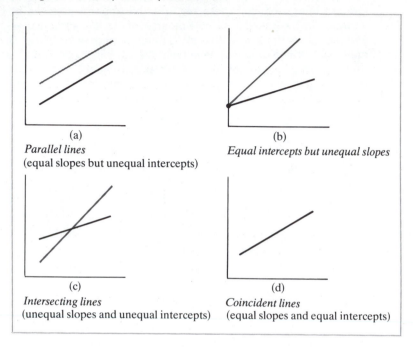

(a)
Parallel lines
(equal slopes but unequal intercepts)

(b)
Equal intercepts but unequal slopes

(c)
Intersecting lines
(unequal slopes and unequal intercepts)

(d)
Coincident lines
(equal slopes and equal intercepts)

For our particular age–SBP example, a conclusion that the lines are parallel (Figure 8.2a) can be interpreted to mean that one sex has consistently higher SBP than the other at all ages, but that the rate of change in SBP with respect to age is the same for both sexes. If it is concluded that the two lines have a common intercept but different slopes (Figure 8.2b), this means that the two sexes begin at an early age with the same average SBP, but that SBP changes with respect to age at a different rate for one sex than for the other. If the two lines have different slopes *and* different intercepts (Figure 8.2c[3]), this means that the relationship between age and SBP is different for different sexes in the sense that there are both different origins and rates of change. Furthermore, if the lines intersect in the range of X values of interest, this indicates that at early ages one sex has higher average SBP than the other, whereas at later ages the pattern is reversed.

There are two general approaches that can be used to answer the three questions given above concerning the comparison of two straight lines. One of these involves *fitting separately the two regression equations*

$$Y_M = \beta_{0M} + \beta_{1M}X + E \tag{8.1}$$

and

$$Y_F = \beta_{0F} + \beta_{1F}X + E \tag{8.2}$$

[3] It is possible for two lines to have unequal slopes and unequal intercepts and yet not intersect within the range of X values of interest. This is illustrated by our example in Figure 8.1.

for males and females, respectively, where Y represents SBP and X represents age. This fitting is then followed by the use of some standard statistical procedures for comparing corresponding regression coefficients, mainly β_{0M} versus β_{0F} and β_{1M} versus β_{1F}.

The second approach entails a pooling of the data on both males and females to allow consideration of a *single* regression equation involving an additional independent variable used to identify the sex in the pooled data. This additional variable is called a *dummy variable*. The single equation is a *multiple* regression model of the form

$$Y = \beta_0 + \beta_1 X + \beta_2 Z + \beta_3 ZX + E$$

where Z represents the additional dummy variable.

Unfortunately, we cannot go into the second approach at this point in the text, because the model involved obviously is not a straight-line model. We shall deal with this particular model and problem again in Chapter 13. For now, we shall restrict our attention to the first approach, which involves estimating the regression lines separately.

8.3 COMPARING TWO SLOPES—TEST FOR PARALLELISM

From (8.1) and (8.2), the appropriate null hypothesis for a comparison of slopes (i.e., for a test of parallelism) is given by

$$H_0: \beta_{1M} = \beta_{1F}$$

[When the null hypothesis $H_0: \beta_{1M} = \beta_{1F}$ is true, then two regression lines simplify to $Y_M = \beta_{0M} + \beta_1 X + E$ for males and $Y_F = \beta_{0F} + \beta_1 X + E$ for females, where β_1 $(= \beta_{1M} = \beta_{1F})$ is the common slope. An estimate of this common slope β_1 is given by the following formula, which is a weighted average of the two separate slope estimates:

$$\hat{\beta}_1 = \frac{(n_M - 1)S_{X_M}^2 \hat{\beta}_{1M} + (n_F - 1)S_{X_F}^2 \hat{\beta}_{1F}}{(n_M - 1)S_{X_M}^2 + (n_F - 1)S_{X_F}^2}$$

Note that $\hat{\beta}_1$ equals the slope computed from the pooled data.]

Any of the following three alternative hypotheses can be used:

$$H_a: \begin{cases} \beta_{1M} > \beta_{1F} & \text{(one-sided)} \\ \beta_{1M} < \beta_{1F} & \text{(one-sided)} \\ \beta_{1M} \neq \beta_{1F} & \text{(two-sided)} \end{cases}$$

Two methods for testing H_0 against any of these alternatives are presented below, followed by an example illustrating each method. The first method should be used if either of the sample sizes is reasonably small (e.g., less than 25). The second method can be used if both sample sizes are moderately large (e.g., greater than 25).

8.3.1 METHOD 1: A SMALL-SAMPLE t TEST FOR PARALLELISM

This method involves computing the test statistic

$$T = \frac{\hat{\beta}_{1M} - \hat{\beta}_{1F}}{S_{\hat{\beta}_{1M} - \hat{\beta}_{1F}}} \tag{8.3}$$

where

$\hat{\beta}_{1M}$ = least-squares estimate of the slope β_{1M} using the n_M observations on males

$\hat{\beta}_{1F}$ = least-squares estimate of the slope β_{1F} using the n_F observations on females

$S_{\hat{\beta}_{1M} - \hat{\beta}_{1F}}$ = estimate of the standard deviation of the estimated difference between slopes $(\hat{\beta}_{1M} - \hat{\beta}_{1F})$

This standard deviation involves a pooling and a summing of the estimated variances of the slopes of the fitted regression lines.[4] It is equal to the square root of the following variance:

$$S^2_{\hat{\beta}_{1M} - \hat{\beta}_{1F}} = S^2_{P,Y|X} \left[\frac{1}{(n_M - 1)S^2_{X_M}} + \frac{1}{(n_F - 1)S^2_{X_F}} \right] \tag{8.4}$$

where

$$S^2_{P,Y|X} = \frac{(n_M - 2)S^2_{Y|X_M} + (n_F - 2)S^2_{Y|X_F}}{n_M + n_F - 4} \tag{8.5}$$

is a pooled estimate of σ^2 (see footnote 4) based on combining residual mean-square errors for males and females,

and

$S^2_{Y|X_M}$ = residual mean-square error for the male data

$S^2_{Y|X_F}$ = residual mean-square error for the female data

[4] This pooling is valid only if the variance of Y_M, say σ^2_M, is equal to the variance of Y_F, say σ^2_F (i.e., only if the assumption of homogeneity of error variance holds). The assumption $\sigma^2_M = \sigma^2_F$ $(= \sigma^2,$ say) is not required for the second method.

$S_{X_M}^2 =$ variance of the X's for the male data

$S_{X_F}^2 =$ variance of the X's for the female data

$n_M =$ sample size for males

$n_F =$ sample size for females

The test statistic given by (8.3) will, under the usual regression assumptions, be distributed as a Student's t with $n_M + n_F - 4$ degrees of freedom when H_0 is true. We then have the following critical regions for different alternative hypotheses and significance level α:

$$\begin{cases} T \geq t_{n_M+n_F-4,1-\alpha} & \text{for } H_A: \beta_{1M} > \beta_{1F} \\ T \leq -t_{n_M+n_F-4,1-\alpha} & \text{for } H_A: \beta_{1M} < \beta_{1F} \\ |T| > t_{n_M+n_F-4,1-\alpha/2} & \text{for } H_A: \beta_{1M} \neq \beta_{1F} \end{cases}$$

The associated $100(1-\alpha)\%$ confidence interval for $\beta_{1M} - \beta_{1F}$ is of the form

$$(\hat{\beta}_{1M} - \hat{\beta}_{1F}) \pm t_{n_M+n_F-4,1-\alpha/2} \, S_{\hat{\beta}_{1M}-\hat{\beta}_{1F}}$$

8.3.2 METHOD 2: A LARGE-SAMPLE Z TEST FOR PARALLELISM

The second method involves computing the following test statistic:

$$\boxed{Z = \frac{\hat{\beta}_{1M} - \hat{\beta}_{1F}}{\sqrt{S_{\hat{\beta}_{1M}}^2 + S_{\hat{\beta}_{1F}}^2}}} \tag{8.6}$$

where

$S_{\hat{\beta}_{1M}}^2 = S_{Y|X_M}^2 / (n_M - 1) S_{X_M}^2$ estimates the variance of the estimated slope $\hat{\beta}_{1M}$ for males

$S_{\hat{\beta}_{1F}}^2 = S_{Y|X_F}^2 / (n_F - 1) S_{X_F}^2$ estimates the variance of the estimated slope $\hat{\beta}_{1F}$ for females

The distribution of Z will be approximately normal with zero mean and unit variance for moderately large sample sizes (i.e., $n_M > 25$ and $n_F > 25$) when $H_0: \beta_{1M} = \beta_{1F}$ is true. The appropriate critical regions are thus given by

$$\begin{cases} Z \geq z_{1-\alpha} & \text{for } H_A: \beta_{1M} > \beta_{1F} \\ Z \leq -z_{1-\alpha} & \text{for } H_A: \beta_{1M} < \beta_{1F} \\ |Z| \geq z_{1-\alpha/2} & \text{for } H_A: \beta_{1M} \neq \beta_{1F} \end{cases}$$

where, as before, $z_{1-\alpha}$ stands for the upper $100(1-\alpha)\%$ point of the standard normal distribution. The associated $100(1-\alpha)\%$ confidence interval for $\beta_{1M} - \beta_{1F}$ is

$$(\hat{\beta}_{1M} - \hat{\beta}_{1F}) \pm z_{1-\alpha/2} \sqrt{S^2_{\hat{\beta}_{1M}} + S^2_{\hat{\beta}_{1F}}}$$

EXAMPLES Using the data given in Table 8.1, the estimates $S^2_{P,Y|X}$, $S^2_{\hat{\beta}_{1M}-\hat{\beta}_{1F}}$, $S^2_{\hat{\beta}_{1M}}$, and $S^2_{\hat{\beta}_{1F}}$ are computed as follows:

$$S^2_{P,Y|X} = \frac{(n_M-2)S^2_{Y|X_M} + (n_F-2)S^2_{Y|X_F}}{n_M + n_F - 4} = \frac{38(71.90) + 27(91.46)}{40 + 29 - 4}$$

$$= \frac{5,201.62}{65} = 80.02$$

$$S^2_{\hat{\beta}_{1M}-\hat{\beta}_{1F}} = S^2_{P,Y|X} \left[\frac{1}{(n_M-1)S^2_{X_M}} + \frac{1}{(n_F-1)S^2_{X_F}} \right]$$

$$= 80.02 \left[\frac{1}{39(221.15)} + \frac{1}{28(242.14)} \right]$$

$$= 0.021$$

$$S^2_{\hat{\beta}_{1M}} = \frac{S^2_{Y|X_M}}{(n_M-1)S^2_{X_M}} = \frac{71.90}{39(221.15)} = 0.0083$$

$$S^2_{\hat{\beta}_{1F}} = \frac{S^2_{Y|X_F}}{(n_F-1)S^2_{X_F}} = \frac{91.46}{28(242.14)} = 0.0135$$

The test statistic (8.3) is then computed as

$$T = \frac{\hat{\beta}_{1M} - \hat{\beta}_{1F}}{S_{\hat{\beta}_{1M}-\hat{\beta}_{1F}}} = \frac{0.96 - 0.95}{\sqrt{0.021}} = \frac{0.01}{0.145} = 0.069$$

For this test statistic, the critical value for a two-sided test (i.e., H_A: $\beta_{1M} \neq \beta_{1F}$) with $\alpha = 0.05$ is given by

$$t_{65,0.975} = 1.9964$$

Since $|T| = 0.069$ does not exceed 1.9964, we would not reject H_0.
Using the estimates $S^2_{\hat{\beta}_{1M}}$ and $S^2_{\hat{\beta}_{1F}}$, we compute the test statistic (8.6) as

$$Z = \frac{\hat{\beta}_{1M} - \hat{\beta}_{1F}}{\sqrt{S^2_{\hat{\beta}_{1M}} + S^2_{\hat{\beta}_{1F}}}} = \frac{0.96 - 0.95}{\sqrt{0.0083 + 0.0135}} = \frac{0.01}{0.148} = 0.068$$

This value for Z turns out to be essentially the same as that obtained for T, although in general this may not happen, especially when the sample sizes are smaller. In using Z to perform a two-sided test for equal slopes, we then find that $|Z| = 0.068$ does not exceed $z_{0.975} = 1.96$. We would therefore not reject H_0 at $\alpha = 0.05$.
Thus, we may conclude from consideration of both tests that there is not sufficient evidence to reject the hypothesis of parallelism (i.e., the lines for males and females have the same slope).

8.4 COMPARING TWO INTERCEPTS

The procedures to be described in this section are appropriate for determining whether or not both straight lines originate at the same point, regardless of how the two slopes compare. The null hypothesis in this case is given by

$$H_0: \beta_{0M} = \beta_{0F}$$

[If $H_0: \beta_{0M} = \beta_{0F}$ is true, the two regression lines simplify to $Y_M = \beta_0 + \beta_{1M}X + E$ and $Y_F = \beta_0 + \beta_{1F}X + E$, where $\beta_0 (= \beta_{0M} = \beta_{0F})$ is the common intercept. An estimate of this common intercept β_0 is given by the following weighted average of the two separate intercept estimates:

$$\hat{\beta}_0 = \frac{n_M \hat{\beta}_{0M} + n_F \hat{\beta}_{0F}}{n_M + n_F}]$$

As with the test for parallelism, there are two methods for evaluating this hypothesis, one utilizing a t test and the other, a large-sample Z test.

8.4.1 METHOD 1: SMALL-SAMPLE t TEST FOR COMMON INTERCEPT

Compute the following test statistic:

$$T = \frac{\hat{\beta}_{0M} - \hat{\beta}_{0F}}{S_{\hat{\beta}_{0M} - \hat{\beta}_{0F}}} \tag{8.7}$$

where $\hat{\beta}_{0M}$ and $\hat{\beta}_{0F}$ are the intercept estimates for males and females, respectively, and where $S^2_{\hat{\beta}_{0M} - \hat{\beta}_{0F}}$ estimates the variance of the estimated difference between the intercepts by means of the formula

$$S^2_{\hat{\beta}_{0M} - \hat{\beta}_{0F}} = S^2_{P,Y|X} \left[\frac{1}{n_M} + \frac{1}{n_F} + \frac{\bar{X}_M^2}{(n_M - 1)S^2_{X_M}} + \frac{\bar{X}_F^2}{(n_F - 1)S^2_{X_F}} \right] \tag{8.8}$$

The statistic T given in (8.7) will have the t distribution with $n_M + n_F - 4$ degrees of freedom when $H_0: \beta_{0M} = \beta_{0F}$ is true. We therefore have the following critical regions for different alternative hypotheses and significance level α:

$$\begin{cases} T \geq t_{n_M + n_F - 4, 1 - \alpha} & \text{for } H_A: \beta_{0M} > \beta_{0F} \\ T \leq -t_{n_M + n_F - 4, 1 - \alpha} & \text{for } H_A: \beta_{0M} < \beta_{0F} \\ |T| \geq t_{n_M + n_F - 4, 1 - \alpha/2} & \text{for } H_A: \beta_{0M} \neq \beta_{0F} \end{cases}$$

The associated $100(1 - \alpha)\%$ confidence interval for $\beta_{0M} - \beta_{0F}$ is

$$(\hat{\beta}_{0M} - \hat{\beta}_{0F}) \pm t_{n_M + n_F - 4, 1 - \alpha/2} S_{\hat{\beta}_{0M} - \hat{\beta}_{0F}}$$

8.4.2 METHOD 2: LARGE-SAMPLE Z TEST FOR COMMON INTERCEPT

Compute the test statistic:

$$Z = \frac{\hat{\beta}_{0M} - \hat{\beta}_{0F}}{\sqrt{S_{\hat{\beta}_{0M}}^2 + S_{\hat{\beta}_{0F}}^2}}$$

(8.9)

where

$$S_{\hat{\beta}_{0M}}^2 = S_{Y|X_M}^2 \left[\frac{1}{n_M} + \frac{\bar{X}_M^2}{(n_M - 1)S_{X_M}^2} \right]$$

estimates the variance of the estimated intercept $\hat{\beta}_{0M}$ for males and

$$S_{\hat{\beta}_{0F}}^2 = S_{Y|X_F}^2 \left[\frac{1}{n_F} + \frac{\bar{X}_F^2}{(n_F - 1)S_{X_F}^2} \right]$$

estimates the variance of the estimated intercept $\hat{\beta}_{0F}$ for females.

The distribution of Z will be approximately standard normal for moderately large sample sizes when $H_0: \beta_{0M} = \beta_{0F}$ is true. The appropriate critical regions are thus given by

$$\begin{cases} Z \geq z_{1-\alpha} & \text{for } H_A: \beta_{0M} > \beta_{0F} \\ Z \leq -z_{1-\alpha} & \text{for } H_A: \beta_{0M} < \beta_{0F} \\ |Z| \geq z_{1-\alpha/2} & \text{for } H_A: \beta_{0M} \neq \beta_{0F} \end{cases}$$

The associated $100(1-\alpha)\%$ confidence interval for $\beta_{0M} - \beta_{0F}$ is

$$(\hat{\beta}_{0M} - \hat{\beta}_{0F}) \pm z_{1-\alpha/2} \sqrt{S_{\hat{\beta}_{0M}}^2 + S_{\hat{\beta}_{0F}}^2}$$

EXAMPLES For the data of Table 8.1 and the value of $S_{P,Y|X}^2$ obtained in Section 8.4, the estimates $S_{\hat{\beta}_{0M}-\hat{\beta}_{0F}}^2$, $S_{\hat{\beta}_{0M}}^2$, and $S_{\hat{\beta}_{0F}}^2$ are computed as follows:

$$S_{\hat{\beta}_{0M}-\hat{\beta}_{0F}}^2 = S_{P,Y|X}^2 \left[\frac{1}{n_M} + \frac{1}{n_F} + \frac{\bar{X}_M^2}{(n_M-1)S_{X_M}^2} + \frac{\bar{X}_F^2}{(n_F-1)S_{X_F}^2} \right]$$

$$= 80.02 \left[\frac{1}{40} + \frac{1}{29} + \frac{(46.93)^2}{39(221.15)} + \frac{(45.07)^2}{28(242.14)} \right]$$

$$= 80.02(0.0250 + 0.0345 + 0.2554 + 0.2996)$$

$$= 49.17$$

$$S^2_{\hat{\beta}_{0M}} = S^2_{Y|X_M}\left[\frac{1}{n_M} + \frac{\bar{X}^2_M}{(n_M-1)S^2_{X_M}}\right] = 71.90\left[\frac{1}{40} + \frac{(46.93)^2}{39(221.15)}\right]$$

$$= 71.40(0.0250 + 0.2554) = 20.16$$

$$S^2_{\hat{\beta}_{0F}} = S^2_{Y|X_F}\left[\frac{1}{n_F} + \frac{\bar{X}^2_F}{(n_F-1)S^2_{X_F}}\right] = 91.46\left[\frac{1}{29} + \frac{(45.07)^2}{28(242.14)}\right]$$

$$= 91.46(0.0345 + 0.2996) = 30.56$$

From these results, the T statistic of (8.7) is computed as follows:

$$T = \frac{\hat{\beta}_{0M} - \hat{\beta}_{0F}}{S_{\hat{\beta}_{0M} - \hat{\beta}_{0F}}} = \frac{110.04 - 97.08}{\sqrt{49.17}} = \frac{12.96}{7.01} = 1.85$$

For a two-sided test ($H_A: \beta_{0M} \neq \beta_{0F}$) with $\alpha = 0.05$, we find that $|T| = 1.85$ does not exceed $t_{65,0.975} = 1.9964$ (i.e., $0.05 < P < 0.10$). Thus, the null hypothesis of common intercepts is not rejected at $\alpha = 0.05$ but is rejected at $\alpha = 0.10$.

For the large-sample test, Z is computed as

$$Z = \frac{\hat{\beta}_{0M} - \hat{\beta}_{0F}}{\sqrt{S^2_{\hat{\beta}_{0M}} + S^2_{\hat{\beta}_{0F}}}} = \frac{110.04 - 97.08}{\sqrt{20.16 + 30.56}}$$

$$= \frac{12.96}{7.12} = 1.82$$

Since $|Z| = 1.82$ does not exceed $z_{0.975} = 1.96$, we would again have borderline rejection of the hypothesis of common intercepts with $0.05 < P < 0.10$.

8.5 TESTING FOR COINCIDENCE FROM SEPARATE STRAIGHT-LINE REGRESSION FITS

Two straight lines are *coincident* if their slopes are equal *and* their intercepts are equal. In considering the male–female regression equations given in (8.1) and (8.2), the hypothesis of coincidence is therefore equivalent to testing whether $\beta_{0M} = \beta_{0F}$ and $\beta_{1M} = \beta_{1F}$ simultaneously. If so, the two regression models both reduce to the general form

$$Y = \beta_0 + \beta_1 X + E$$

where $\beta_0 \ (= \beta_{0M} = \beta_{0F})$ and $\beta_1 \ (= \beta_{1M} = \beta_{1F})$ are the common intercept and slope, respectively.

[The estimates of the common slope β_1 and common intercept β_0 are obtained by simply pooling all observations on males and females together and determining the usual least-squares slope and intercept estimates using the pooled data set.]

As mentioned earlier, a preferred way to test for coincident lines is to employ a multiple regression model involving dummy variables. Nevertheless, another procedure that is generally not quite as efficient (e.g., in terms of the power of the test) is often found useful. This procedure has usually been found in practice to yield the same conclusion as that obtained from the dummy-variable-model approach.

8.5.1 THE PROCEDURE AND ITS DRAWBACKS

The obvious way to test for coincidence based on two separate straight-line fits is to perform both the test for H_0: $\beta_{0M} = \beta_{0F}$ of equal intercepts *and* the test of H_0: $\beta_{1M} = \beta_{1F}$ of equal slopes. If either one or both of these null hypotheses is rejected, one can conclude that the two lines are not coincident. If both are *not* rejected, one can conclude that the two lines are coincident.

A valid criticism of this testing procedure, which reflects on its power, is that it involves two separate tests rather than a single test. There are two difficulties in this regard:

1. The procedure does not "precisely" test for coincidence.
2. If α is the significance level of each separate test, the overall significance level for the two tests combined is greater than α; that is, there is more chance of rejecting a true H_0.

One way to get around difficulty 2 is to use $\alpha/2$ for each separate test to guarantee an overall significance level of no more than α. Nevertheless, using $\alpha/2$ for each test is conservative and may make it difficult to reject coincidence.

With regard to difficulty 1, it is important to point out that even if both tests are *not* rejected, it is still possible (although unlikely) for the two lines not to be coincident. This is so because of the fact that each separate test (e.g., the test of H_0: $\beta_{1M} = \beta_{1F}$) allows the remaining parameters (e.g., β_{0M} and β_{0F}) to be unequal. In other words, the test for equal slopes does not assume equal intercepts, nor does the test for equal intercepts assume equal slopes. The multiple regression procedure, which involves the use of the dummy-variable model, does not exhibit this drawback but rather permits testing for common slope and common intercept *simultaneously*.

EXAMPLE We saw in Sections 8.3 and 8.4 that the null hypothesis of equal slopes (regardless of the intercepts) was not rejected ($P > 0.40$), and the null hypothesis of equal intercepts (regardless of the slopes) was associated with a P value of between 0.05 and 0.10. Putting these two facts together, we would be inclined to support the position that there is no strong evidence in favor of noncoincidence.

8.6 COMPARING TWO CORRELATION COEFFICIENTS

In addition to the preceding tests dealing with the comparison of two straight lines, it is also of interest to determine whether or not the strength of the straight-line relationship is the same for one group (e.g., males) as for another (e.g., females). This amounts to testing for the equality of correlation coefficients.

For our example the null hypothesis may therefore be given by $H_0: \rho_M = \rho_F$, where ρ_M and ρ_F denote the population correlation coefficients for males and females, respectively.

To test this hypothesis it is necessary to use Fisher's Z transformation, discussed in Section 6.6. This transformation is given in (6.8) as $\frac{1}{2}\log_e (1+r)/(1-r)$. The distribution of this quantity is approximately normal with mean $\frac{1}{2}\log_e (1+\rho)/(1-\rho)$ and variance $1/(n-3)$.

It can be seen from simple algebra that a test of the hypothesis $H_0: \rho_M = \rho_F$ is equivalent to a test of the hypothesis

$$H_0^*: \tfrac{1}{2}\log_e \frac{1+\rho_M}{1-\rho_M} = \tfrac{1}{2}\log_e \frac{1+\rho_F}{1-\rho_F}$$

and a test of H_0^* is simply a test comparing two normally distributed statistics with known variances $1/(n_M-3)$ and $1/(n_F-3)$. The test statistic for such a comparison is shown in any elementary statistical text to have the general form of a ratio with the difference in sample values of the statistics in the numerator and the square root of the sum of their variances in the denominator. This yields the following test statistic for the test for equal correlations:

$$Z = \frac{\tfrac{1}{2}\log_e \dfrac{1+r_M}{1-r_M} - \tfrac{1}{2}\log_e \dfrac{1+r_F}{1-r_F}}{\sqrt{\dfrac{1}{n_M-3} + \dfrac{1}{n_F-3}}} \tag{8.10}$$

The test statistic given by (8.10) has an approximate standard normal distribution when the null hypothesis $H_0: \rho_M = \rho_F$ is true. Thus, we have the following critical regions involving percentage points of the standard normal distribution for significance level α:

$$\begin{cases} Z \geq z_{1-\alpha} & \text{for the one-sided alternative } H_A: \rho_M > \rho_F \\ Z \leq -z_{1-\alpha} & \text{for the one-sided alternative } H_A: \rho_M < \rho_F \\ |Z| \geq z_{1-\alpha/2} & \text{for the two-sided alternative } H_A: \rho_M \neq \rho_F \end{cases}$$

EXAMPLE   For the age–SBP data, the correlation coefficients are computed to be $r_M = 0.863$ and $r_F = 0.844$. Using Table A-5, these correlations are changed to the following Fisher Z values:

1.3050 for males   and   1.2349 for females

Substituting these values into (8.9), we obtain the following computed test statistic:

$$Z = \frac{1.3050 - 1.2349}{\sqrt{\dfrac{1}{40-3} + \dfrac{1}{29-3}}} = \frac{0.0701}{\sqrt{1.0655}} = 0.07$$

For a two-sided test with $\alpha = 0.05$, we compare $|Z| = 0.07$ with the critical point $z_{0.975} = 1.96$. Since $|Z|$ does not exceed 1.96, we therefore do not reject $H_0: \rho_M = \rho_F$.

PROBLEMS

1. Using the data in Problem 2 of Chapter 5 and/or the printout given here, answer the following questions concerning the separate straight-line regressions of SBP on QUET for smokers (SMK = 1) and nonsmokers (SMK = 0), respectively.
   (a) Determine the least-squares line of SBP (Y) on QUET (X) separately for smokers and nonsmokers.
   (b) Test "H_0: the slopes are the same for the populations of smokers and nonsmokers being sampled" versus "H_A: nonsmokers have a more positive slope."
   (c) Test "H_0: the intercepts are the same for the populations of smokers and nonsmokers being sampled" versus "H_A: the intercepts are different."
   (d) Test "H_0: the straight lines are coincident for the populations of smokers and nonsmokers being sampled" versus "H_A: the straight lines are not coincident."

Computer printout for Problem 1

SMOKERS

MEANS AND STANDARD DEVIATIONS

X: 3.40829 +- 0.56785   Y: 147.82353 +- 15.21198

CORRELATIONS

PEARSON: 0.750995   SPEARMAN: 0.718137   KENDALL: 0.588235

LINEAR REGRESSION OF Y ON X

SLOPE: 20.11804 +- 4.56719
INTERCEPT: 79.25533 +- 15.76837
S(Y|X):   10.37401

ANALYSIS OF VARIANCE: F(1,15) = 19.40317

NONSMOKERS

MEANS AND STANDARD DEVIATIONS

X: 3.47827 +- 0.41930   Y: 140.80000 +- 12.90183

CORRELATIONS

PEARSON: 0.854814   SPEARMAN: 0.919501   KENDALL: 0.778882

LINEAR REGRESSION OF Y ON X

SLOPE: 26.30283 +- 4.42865
INTERCEPT: 49.31176 +- 15.50814
S(Y|X):   6.94794

ANALYSIS OF VARIANCE: F(1,13) = 35.27456

2. A topic of major concern to demographers and economists alike is the effect of a high fertility rate on per capita income. The first two accompanying tables display

values of per capita income (PCI) and population percentage under age 15 (YNG) for a hypothetical sample of developing countries in Latin America and Africa, respectively. The third table summarizes the results of straight-line regressions of PCI (*Y*) on YNG (*X*) for each group of countries.

Latin American countries

YNG (*X*)	PCI (*Y*)	YNG (*X*)	PCI (*Y*)	YNG (*X*)	PCI (*Y*)
32.2	788	44.0	292	35.0	685
47.0	202	44.0	321	47.4	220
34.0	825	43.0	300	48.0	195
36.0	675	43.0	323	37.0	605
38.7	590	40.0	484	38.4	530
40.9	408	37.0	625	40.6	480
45.0	324	39.0	525	35.8	690
45.4	235	44.6	340	36.0	685
42.2	338	33.0	765		

African countries

YNG (*X*)	PCI (*Y*)	YNG (*X*)	PCI (*Y*)	YNG (*X*)	PCI (*Y*)
34.0	317	41.0	188	39.0	225
36.0	270	42.0	166	39.0	232
38.2	208	45.0	132	37.0	260
43.0	150	36.0	290	37.0	250
44.0	105	42.6	160	46.0	92
44.0	128	33.0	300	45.6	110
46.0	85	33.0	320	42.0	180
48.0	75	47.0	85	38.8	235
40.0	210	47.0	75		

Summary of separate straight-line fits

	n	$\hat{\beta}_0$	$\hat{\beta}_1$	$\bar{X}$	$\bar{Y}$	S_X^2	$S_{Y|X}^2$	r
Latin American	26	2170.67	−42.0	40.277	478.846	21.633	1391.756	−0.983
African	26	893.571	−17.28	40.931	186.462	20.611	157.238	−0.988

(a)–(d)   Do the same as in parts (a)–(d) of Problem 1 for the straight-line regressions of PCI (*Y*) on YNG (*X*) for Latin American and African countries, respectively.

(e)   Test "H_0: the population correlation coefficients are equal for the two groups of countries under study." (Use $\alpha = 0.05$.) Does your conclusion here clash with your findings regarding the equality of slopes?

(*Note*: For each test above, assume that the alternative hypothesis is two-sided.)

3.   A team of anthropologists and nutrition experts were investigating the influence of protein content in diet on the relationship between age and height for New Guinean children. The two accompanying tables display values of height (HT) in

centimeters and age (AGE) for a hypothetical sample of children with protein-rich and protein-poor diets, respectively.

Protein-rich diet

AGE (X)	0.2	0.5	0.8	1.0	1.0	1.4	1.8	2.0	2.0	2.5	2.5	3.0	2.7
HT (Y)	54	54.3	63	66	69	73	82	83	80.3	91	93.2	94	94

Protein-poor diet

AGE (X)	0.4	0.7	1.0	1.0	1.5	2.0	2.0	2.4	2.8	3.0	1.3	1.8	0.2	3.0
HT (Y)	52	55	61	63.4	66	68.5	67.9	72	76	74	65	69	51	77

The information for comparing the straight-line regression equations is as follows:

	n	$\hat{\beta}_0$	$\hat{\beta}_1$	$\bar{X}$	$\bar{Y}$	s_X^2	$s_{Y|X}^2$	r
Protein-rich	13	50.324	16.009	1.646	76.677	0.808	5.841	0.937
Protein-poor	14	51.225	3.686	1.650	65.557	0.873	4.598	0.969

(a)–(d) Do the same as in Problem 1(a)–(d) for the straight-line regressions of HT (Y) on AGE (X) for the two diets. (Consider a two-sided alternative in each case.)

(e) Test whether the population correlation coefficient for children with a protein-rich diet is significantly different from that for children with a protein-poor diet. (Consider a two-sided alternative.)

4. For the data of DI (Y) on IQ (X) in Problem 4 of Chapter 5, assume that this sample of 17 observations (*outlier removed*) consisted of males only. Now suppose that another sample of observations on DI (Y) and IQ (X) has been obtained for 14 females. The information needed to compare the straight-line regression equations for males and females is given in the table.

	n	$\hat{\beta}_0$	$\hat{\beta}_1$	$\bar{X}$	$\bar{Y}$	s_X^2	$s_{Y|X}^2$	r
Males	17	70.846	−0.444	101.411	25.812	215.882	24.335	−0.807
Females	19	61.871	−0.438	101.053	17.579	175.497	16.692	−0.825

(a)–(d) Do the same as in Problem 1(a)–(d) for the straight-line regressions of DI (Y) on IQ (X) for the two sexes. (Consider two-sided alternatives.)

(e) Perform a two-sided test as to whether the population correlation coefficients for males and females are equal or not.

5. For the data of VOTE (Y) on TVEXP (X) in Problem 5 of Chapter 5, assume that this sample of 20 observations came from congressional districts in New York. Now, suppose that a second sample of 17 congressional districts in California was then selected and the same information recorded. The table provides the information needed for comparing the straight-line regression equations for New York and California.

	n	$\hat{\beta}_0$	$\hat{\beta}_1$	$\bar{X}$	$\bar{Y}$	s_X^2	$s_{Y|X}^2$	r
New York	20	2.174	1.177	36.99	45.71	76.870	11.101	0.954
California	17	8.030	1.036	36.371	45.706	97.335	13.492	0.945

(a)–(d) Do the same as in Problem 1(a)–(d) for the straight-line regression of VOTE (Y) on TVEXP (X) for the two states. Consider the one-sided alternative $H_A: \beta_{1(CAL)} < \beta_{1(NY)}$ for the test for slope and the one-sided alternative $H_A: \beta_{0(CAL)} < \beta_{0(NY)}$ for the test for intercept.

(e) Perform a two-sided test regarding whether the correlation coefficients for New York and California are equal or not.

6. The data in the first table represent four-week growth rates for depleted chicks at different dosage levels of vitamin B, by sex.[5]

MALES		FEMALES	
GROWTH RATE (Y)	LOG$_{10}$ DOSE (X)	GROWTH RATE (Y)	LOG$_{10}$ DOSE (X)
17.1	0.301	18.5	0.301
14.3	0.301	22.1	0.301
21.6	0.301	15.3	0.301
24.5	0.602	23.6	0.602
20.6	0.602	26.9	0.602
23.8	0.602	20.2	0.602
27.7	0.903	24.3	0.903
31.0	0.903	27.1	0.903
29.4	0.903	30.1	0.903
30.1	1.204	28.1	0.903
28.6	1.204	30.3	1.204
34.2	1.204	33.0	1.204
37.3	1.204	35.8	1.204
33.3	1.505	32.6	1.505
31.8	1.505	36.1	1.505
40.2	1.505	30.5	1.505

Using the information provided in the next table:
(a) Determine the dose–response straight lines separately for each sex and plot them on the same graph.
(b) Test whether the slopes for males and females are different.
(c) Find a 99% confidence interval for the true difference between the male and female slopes.
(d) Test for coincidence of the two straight lines.

[5] Adapted from a study by Clark et al. (1940).

	n	$\hat{\beta}_0$	$\hat{\beta}_1$	$\bar{X}$	$\bar{Y}$	S_X^2	$S_{Y|X}^2$	r
Males	16	13.767	14.966	0.941	27.84	0.1797	11.4403	0.889
Females	16	15.656	12.735	0.903	27.16	0.1812	8.4719	0.888

7. The results in the first table[6] were obtained in a study concerning the amount of energy metabolized by two similar species of birds under constant-temperature conditions. Information based on separate straight-line fits to each data set is summarized in the second table.

SPECIES A		SPECIES B	
CALORIES (Y)	TEMPERATURE (X) (°C)	CALORIES (Y)	TEMPERATURE (X) (°C)
36.9	0	41.1	0
35.8	2	40.6	2
34.6	4	38.9	4
34.3	6	37.9	6
32.8	8	37.0	8
31.7	10	36.1	10
31.0	12	36.3	12
29.8	14	34.2	14
29.1	16	33.4	16
28.2	18	32.8	18
27.4	20	32.0	20
27.8	22	31.9	22
25.5	24	30.7	24
24.9	26	29.5	26
23.7	28	28.5	28
23.1	30	27.7	30

	n	$\hat{\beta}_0$	$\hat{\beta}_1$	$\bar{X}$	$\bar{Y}$	S_X^2	$S_{Y|X}^2$	r
Species A	16	36.579	−0.4528	15.00	29.79	90.6666	0.1662	0.9959
Species B	16	40.839	−0.4368	15.00	34.29	90.6666	0.1757	0.9953

(a) Plot the least-squares straight lines for each species on the same graph.
(b) Test whether the two lines are parallel.
(c) Test whether the two lines have the same intercept.
(d) At 15°C, give a 95% confidence interval for the true difference between the mean amounts of energy metabolized by each species.
(*Hint*: Use a confidence interval of the form

$$(\hat{Y}_{15}^A - \hat{Y}_{15}^B) \pm t_{n_A+n_B-4,1-\alpha/2}\sqrt{S_{\hat{Y}_{15}^A}^2 + S_{\hat{Y}_{15}^B}^2}$$

where $\hat{Y}_{15}^i$ is the predicted value at 15°C for species i ($i = A, B$) and $S_{\hat{Y}_{15}^i}^2$ is the estimated variance of $\hat{Y}_{15}^i$ given by the general formula (for $X = X_0$):

$$S_{\hat{Y}_{X_0}}^2 = S_{P,Y|X}^2\left[\frac{1}{n_i} + (X_0 - \bar{X})^2/(n_i-1)S_{X_i}^2\right]$$

[6] Adapted from a study by Davis (1955).

CHAPTER 9 POLYNOMIAL REGRESSION

9.1 PREVIEW

To this point we have restricted our discussion of regression analysis to the problem of fitting a straight-line model and of determining whether such a model is appropriate. We have seen for the age–SBP data of Table 5.1 and can show for the data on males in the previous chapter that the test for lack of fit of the straight-line model is not rejected. We have therefore concluded for both data sets that there is no need to consider a model of higher order (e.g., a parabola is of second order).

In this chapter we shall discuss what to do when it is of interest to consider a model of second order or higher. For the data sets above, in particular, we may ask whether or not we could significantly improve our prediction by increasing the complexity of the fitted model (regardless of the lack-of-fit test results). It is important to point out that the lack-of-fit test does not help in suggesting any specific alternative-model form which would be more appropriate than the one under consideration. Rather, the alternative hypothesis for this test is nonspecific in that it allows for the possibility that *any* model form other than the straight-line model might be preferable. Thus, it is conceivable that a specific change in the model structure may significantly improve the fit even though, in general, the straight-line fit is found to be adequate.

In our consideration of more complicated models, we shall again focus first on the data of Table 5.1, with the outlier corresponding to the second individual ($X = 47$; $Y = 220$) being dropped. For this data set, with $n = 29$ observations, we shall describe how to fit a higher-order model and how to interpret the results obtained. We will see that the methods for dealing with higher-order models are directly analogous to those used for the straight-line model. Furthermore, with regard to the age–SBP data of Table 5.1, we will see, as expected, that adding more complexity to the straight-line model does not significantly improve prediction.

9.2 POLYNOMIAL MODELS

The most general kind of curve usually considered for describing the relationship between a single independent variable and a single dependent variable is called a *polynomial*. Mathematically, a polynomial of order k in x is an expression of the form

$$y = c_0 + c_1 x + c_2 x^2 + c_3 x^3 + \cdots + c_k x^k$$

where the c's and k (which must be a nonnegative whole number) are constants. We have already considered the simple polynomial corresponding to $k = 1$ (i.e., the straight line), which has the form $y = c_0 + c_1 x$. We now focus primarily on the second-order polynomial corresponding to $k = 2$ (i.e., the parabola), which has the form $y = c_0 + c_1 x + c_2 x^2$.

In going from a *mathematical* model to a *statistical* model, as we did in the straight-line case, we may write a parabolic model in either of the following forms:

$$\mu_{Y|X} = \beta_0 + \beta_1 X + \beta_2 X^2 \tag{9.1}$$

or

$$Y = \beta_0 + \beta_1 X + \beta_2 X^2 + E \tag{9.2}$$

where we have used capital Y's and X's to denote statistical variables; β_0, β_1, and β_2 are unknown parameters called regression coefficients; $\mu_{Y|X}$ denotes the mean of Y at a given X; and E denotes the error component, which represents the difference between an individual's response and the true average response at X.

If we tentatively assume that a parabolic model as given either by (9.1) or (9.2) is appropriate for describing the relationship between X and Y, we must then determine a specific estimated parabola which best fits the data. As in the straight-line case, this best-fitting parabola may be determined by extending the least-squares method as described in the next section.

9.3 LEAST-SQUARES PROCEDURE FOR FITTING A PARABOLA

The least-squares estimates of a parabolic model are chosen so as to minimize the sum of squares of deviations of observed points from corresponding points on the fitted parabola (Figure 9.1). Letting $\hat{\beta}_0$, $\hat{\beta}_1$, and $\hat{\beta}_2$ denote the least-squares estimates of the unknown regression coefficients of the parabolic model (9.1) and letting $\hat{Y}_i$ denote the value of the predicted response at X_i, we then write the estimated parabola as follows:

$$\hat{Y} = \hat{\beta}_0 + \hat{\beta}_1 X + \hat{\beta}_2 X^2 \tag{9.3}$$

and the minimum sum of squares obtained using this least-squares parabola is

$$SSE = \sum_{i=1}^{n} (Y_i - \hat{Y}_i)^2 = \sum_{i=1}^{n} (Y_i - \hat{\beta}_0 - \hat{\beta}_1 X_i - \hat{\beta}_2 X_i^2)^2$$

FIGURE 9.1   *Deviations of observed points from the least-squares parabola*

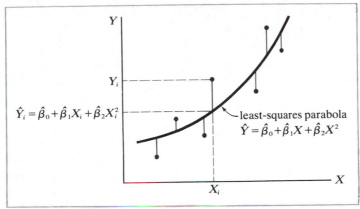

It is not necessary to present in this text the precise formulas for calculating the least-square estimates $\hat{\beta}_0$, $\hat{\beta}_1$, and $\hat{\beta}_2$. These formulas are quite complex and become even more so for polynomials of order higher than two. In addition, the researcher is not likely to use such regression analyses without having the availability of a computer, which can perform the necessary calculations internally and print out the numerical results directly. Even so, we have provided in Appendix C a discussion of matrices and their relationship to regression analysis; the use of matrix mathematics permits a compact representation of the general regression model and the associated least-squares methodology.

EXAMPLE   For the age–SBP data of Table 5.1 with the outlier removed, the least-squares estimates for the parabolic regression coefficients are computed to be

$$\hat{\beta}_0 = 113.41$$

$$\hat{\beta}_1 = 0.088$$

$$\hat{\beta}_2 = 0.010$$

The fitted model given by (9.3) then becomes

$$\hat{Y} = 113.41 + 0.088X + 0.010X^2 \tag{9.4}$$

which may be compared to the straight-line equation previously obtained (Section 5.5) for these data with the outlier removed,

$$\hat{Y} = 97.08 + 0.95X \tag{9.5}$$

It is important to notice on comparison of (9.4) and (9.5) that the constant $\hat{\beta}_0$ is different as well as the coefficient $\hat{\beta}_1$. This should not be surprising since the two models are different and therefore can result in different estimated coefficients.

9.4 ANOVA TABLE FOR SECOND-ORDER POLYNOMIAL REGRESSION IN ONE VARIABLE

As for the straight-line case, the essential results based on fitting a second (or higher)-order polynomial model can be summarized by an ANOVA table. The complete ANOVA table for a parabola fit to the age–SBP data of Table 5.1 (with the outlier removed) is given in Table 9.1.

[Some authors advocate comparing regression mean squares to the pure error mean square when testing for the significance of regression effects. We feel this is inappropriate because (1) if the test for lack of fit is *not* rejected, then the regression F tests can be made more powerful (due to increasing the denominator df) by pooling the lack of fit and pure error sums of squares together and then using the residual mean square in the denominator of such F statistics, and (2) if the test for lack of fit is rejected, then consideration of an alternative model is warranted, and so tests concerning regression effects in the model under consideration are of dubious value. We have adopted this philosophy regarding the F tests in Table 9.1, so that, for example, the entry 66.81 is an $F_{1,26}$ value calculated as

$$6{,}110.10 \left/ \left[\frac{380.50 + 1{,}925.55 + 163.30}{5 + 21 + 1} \right] \right.$$

For now we shall restrict our attention to Table 9.1.1, in which the first two rows of Table 9.1 are combined into a *regression* source and the last two rows into a *residual* source, so as to provide overall information about the fit of the second-order model.] As with the straight-line example, the total variation SSY can be partitioned into the residual sum of squares SSE about the regression curve and the regression sum of squares (SSY − SSE) due to the regression of Y on X and X^2. This partition is expressed by the equation

$$\sum_{i=1}^{n} (Y_i - \bar{Y})^2 = \sum_{i=1}^{n} (\hat{Y}_i - \bar{Y})^2 + \sum_{i=1}^{n} (Y_i - \hat{Y}_i)^2$$

$$(\text{SSY} = \text{SSY} - \text{SSE} + \text{SSE})$$

TABLE 9.1 *Complete ANOVA table for a parabola fit to age–SBP data (with outlier removed)*

SOURCE		df	SS	MS	F
Regression	X	1	6,110.10	6,110.10	66.81
	$X^2 \mid X$	1	163.30	163.30	1.84
Residual	lack of fit	21	1,925.55	91.69	1.20
	pure error	5	380.50	76.10	
Total (corrected) ($R^2 = 0.731$)		28	8,579.45		

TABLE 9.1.1 *Overall ANOVA table for the second-order model*

SOURCE	df	SS	MS	F
Regression	2	SSY − SSE = 6,273.40	$\dfrac{SSY - SSE}{2} = 3{,}136.70$	35.37 ($P < 0.001$)
Residual	26	SSE = 2,306.05	$\dfrac{SSE}{26} = 88.69$	
Total (corrected)	28	SSY = 8,579.45		

which we have referred to in Section 7.2 as the *fundamental equation of regression analysis*.

Any mean-square term in the table is, as usual, obtained by dividing the corresponding sum of squares by its degrees of freedom (df). In particular, the residual sum of squares SSE is divided by its degrees of freedom to yield an estimate of σ^2. This estimated variance is generally denoted as either MS residual or $S^2_{Y|X}$ and is given by the general formula

$$S^2_{Y|X} = \frac{1}{n-3} \sum_{i=1}^{n} (Y_i - \hat{Y}_i)^2 \tag{9.6}$$

The number of degrees of freedom associated with the estimate (9.6) is $(n-3)$, which follows from the general rule that the number of df associated with SSE is given by "n minus the number of estimated regression coefficients."

The F statistic in Table 9.1.1 is obtained by dividing the mean-square regression term by the mean-square residual term. We shall discuss in the next section the use of this statistic for inference purposes, as well as the use of the other F statistics in the complete ANOVA table (Table 9.1).

9.5 INFERENCES ASSOCIATED WITH SECOND-ORDER POLYNOMIAL REGRESSION

There are three basic inferential questions associated with second-order polynomial regression. These are:

1. Is the overall regression significant; that is, is the amount of variation in Y explained by the second-order model *significantly* more than that explained by ignoring X completely (and just using $\bar{Y}$)?

2. Does the second-order model provide significantly more predictive power than that provided by the straight-line model?

3. Is the second-order model appropriate, or should we add further terms to the model?

9.5.1 TEST FOR OVERALL REGRESSION AND MEASURE OF THE STRENGTH OF THE OVERALL PARABOLIC RELATIONSHIP

Answering question 1 above involves testing the following null hypothesis:

H_0: there is no significant overall regression using X and X^2 (9.7)

The testing procedure used for this null hypothesis is analogous to that used for straight-line regression. We simply compute the statistic

$$F = \frac{MS \text{ regression}}{MS \text{ residual}} \tag{9.8}$$

and compare this computed value with an appropriate critical point of the F distribution, which, in our example, has 2 and 26 df in numerator and denominator, respectively. For $\alpha = 0.001$, we find that $F = 35.37 > F_{2,26,0.999} = 9.12$, and so we reject the null hypothesis of no significant overall regression ($P < 0.001$).

> [The F statistic used for testing for *overall* regression significance using a polynomial model of order 2 is not equivalent to any t test. This can perhaps be best seen from the fact that the numerator df in the F statistic is now 2 instead of 1 (as it was in the straight-line regression case).]

To obtain a quantitative measure of how well the second-order model predicts the dependent variable, we can use the "squared multiple correlation coefficient" (i.e., the multiple R^2). As with r^2 in the straight-line regression case, R^2 represents the proportionate reduction in the error sum of squares obtained by using X and X^2 instead of the naive predictor $\bar{Y}$. The formula for calculating R^2 is given by

$$R^2 \text{ (second-order model)} = \frac{SSY - SSE \text{ (second-order model)}}{SSY} \tag{9.9}$$

For our example, $R^2 = 0.731$. The F test provides us with a decision as to whether or not this R^2 is significantly different from zero; our answer is as might be expected, since the straight-line regression was significant.

> [It is possible, although not likely, that the overall F test for the second-order model may not lead to rejection of H_0 even if the t test (or the equivalent F test) for significant regression of the straight-line model leads to rejection. This is possible mathematically because the loss of 1 degree of freedom in SSE in going from the straight-line model to the second-order model may result in a smaller computed F along with an altered critical point of the F distribution. In our example, the computed F is reduced from 66.81 for the straight-line case to 35.37 for the second-order case, and the critical point of the F distribution for $\alpha = 0.001$ is reduced from 13.6 to 9.12.]

9.5.2 TEST FOR THE ADDITION OF THE X^2 TERM TO THE MODEL

In order to answer the second question above, we must perform a test of the following null hypothesis:

H_0: the addition of the X^2 term to the model does not significantly improve the prediction of Y over and above that achieved by the straight-line model ... (9.10)

In dealing with this problem, we should first notice that the R^2 for the second-order model is 0.731, while the $R^2 = r^2$ for the first-order model was 0.712. Thus, we are certainly explaining more variation in Y by adding the X^2 term to the model (e.g., our improvement in R^2 is $0.731 - 0.712 = 0.019$). Nevertheless, it turns out, in general, that more variation will *always* be explained by adding extra terms to the model. The important question here is whether or not there has been a *significant* increase in the variation explained by the additional term, that is, whether the 0.019 increase is large enough to warrant adding the X^2 term to the model.

To answer this question we must compute what we call the *extra sum of squares due to the addition of X^2*, which we place in our ANOVA Table 9.1 under the source heading "Regression $X^2|X$." This sum of squares is computed simply from the formula

$$\begin{array}{ccc} \text{extra SS due} \\ \text{to adding } X^2 \end{array} = \begin{array}{c} \text{SS regression} \\ \text{(second-order model)} \end{array} - \begin{array}{c} \text{SS regression} \\ \text{(first-order model)} \end{array} \qquad (9.11)$$

For our example, recalling that the SS regression for the straight-line model was 6,110.10 (see Table 9.1), we obtain the extra sum of squares due to adding the X^2 term to the model as

$$6{,}273.40 - 6{,}110.10 = 163.30$$

To test the null hypothesis given by (9.10), we compute

$$F = \frac{(\text{extra SS due to adding } X^2)/1}{\text{mean-square residual for the second-order model}} \qquad (9.12)$$

and then compare this F value to an appropriate F percentage point (which is an $F_{1,26}$ value in our example).

[An alternative way to perform this test is to divide the estimated coefficient $\hat{\beta}_2$ by its standard deviation to form a statistic that has a t distribution under H_0 with 26 degrees of freedom. We shall discuss this testing procedure later in the context of another example.]

Again, for our example, the F statistic in (9.12) is computed to be

$$F = \frac{163.30}{88.69} = 1.84$$

Since $F_{1,26,0.90} = 2.91$, we would not reject H_0 at $\alpha = 0.10$. Furthermore, the P value for this test satisfies $0.10 < P < 0.25$, which is high. It is important to note here that we could have expected this result all along, since we previously found the straight-line model to be appropriate.

[We remind the reader that the test for lack of fit of the straight-line model is not equivalent to the test for the addition of the X^2 term. The lack-of-fit test is aimed at detecting any (nonspecific) kind of lack of fit. The test for addition of the X^2 term, on the other hand, tests whether a *specific* change in the model is helpful in improving the fit. Also, in the example just considered, the lack-of-fit test for deviations from linearity was based on an $F_{22,5}$ statistic (see Table 7.6) while the test for addition of the X^2 term was based on an $F_{1,26}$ statistic.]

9.5.3 TESTING FOR LACK OF FIT OF THE SECOND-ORDER MODEL

Because neither the test for lack of fit of the straight-line model nor the test for the addition of the X^2 term was significant, it would be superfluous to go still further and test for lack of fit of the second-order model. It should nevertheless be pointed out that, when appropriate, this test for lack of fit is exactly analogous to that used for the straight-line model. The null hypothesis is again given as

$$H_0: \text{no significant lack of fit of the assumed model} \tag{9.13}$$

The pure error sum of squares SS_{pe} (see Table 9.1) is computed exactly the same for the second-order model as for the straight-line model. The lack-of-fit sum of squares is computed by subtracting SS_{pe} from the residual sum of squares SSE. Finally, the same form of test statistic is used,

$$F = \frac{SS_{lof}/(n-3-df_{pe})}{SS_{pe}/df_{pe}} = \frac{MS_{lof}}{MS_{pe}} \tag{9.14}$$

where SS_{lof} and MS_{lof} denote the sum of squares and mean square for lack of fit, SS_{pe} and MS_{pe} are the corresponding pure error terms, and df_{pe} is the degrees of freedom for pure error. The F statistic in (9.14) has the F distribution with $n-3-df_{pe}$ and df_{pe} degrees of freedom when the H_0 of (9.13) is true. Thus, the test is performed in the usual way by comparing the computed value of the test statistic with a selected percentage point of the F distribution.

9.6 ANOTHER EXAMPLE REQUIRING A SECOND-ORDER MODEL

It should be apparent by now that we have squeezed out of the age–SBP data about as much information as possible in illustrating the principles and methods of straight-line and second-order regression analysis. We now turn to another hypothetical example which will again illustrate these methods, but which will lead us to a different conclusion with regard to the appropriateness of the second-order model.

Suppose that a laboratory study is undertaken to determine the relationship between the dose (X) of a certain drug and weight gain (Y). Eight laboratory animals of the same sex, age, and size are selected and randomly assigned to one of eight dosage levels of the given drug.

TABLE 9.2   *Laboratory data on weight gain after 2 weeks as a function of dosage level*

DOSAGE LEVEL (X)	1	2	3	4	5	6	7	8
WEIGHT GAIN (Y) (decagrams)	1	1.2	1.8	2.5	3.6	4.7	6.6	9.1

[The study design here can certainly be criticized for not having more than one animal receive the same dose. Replication at each dose would provide a much more reliable estimate of the variability in the data. However, it is often a characteristic of laboratory studies that sufficient numbers of animals are not easily available. It should also be noted that the data for this example has been contrived to simplify the presentation of the analysis and to bring out with clarity a relationship that is second-order in nature.]

The gain in weight, in decagrams, is measured for each animal after 2 weeks, during which time all animals have been subject to the same dietary regimen and general laboratory conditions. The data are given in Table 9.2 and a scatter diagram of these data appears as Figure 9.2.

By simply "eyeballing" the data in the scatter diagram, it is apparent that a parabolic curve is a more appropriate model than a straight line. We shall now proceed to quantify this visual impression.

The complete ANOVA table based on fitting a parabola to the data in Table 9.2 is given in Table 9.3. The equation for the least-squares parabola on which this ANOVA table is based has the form

$$\hat{Y} = 1.13 - 0.41X + 0.17X^2$$

Let us go into some detail concerning the information contained in the above ANOVA table. First, combining the "regression $X^2|X$" sum of squares with the residual

FIGURE 9.2   *Scatter diagram of the data in Table 9.2*

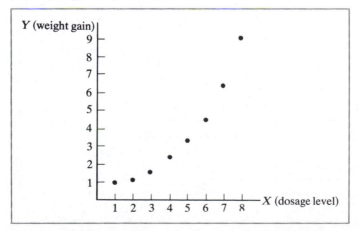

TABLE 9.3   *ANOVA table for data of Table 9.2*

SOURCE		df	SS	MS	F
Regression $\begin{cases} X \\ X^2\mid X \end{cases}$	X	1	52.04	52.04	61.95
	$X^2\mid X$	1	4.83	4.83	120.75
Residual		5	0.20	0.04	
Total (corrected) ($R^2 = 0.997$)		7	57.07		

sum of squares permits us to test whether there is a significant straight-line regression effect *before* adding the X^2 term to the model; that is, the ANOVA table for straight-line regression derived from Table 9.3 is as given in Table 9.4. The least-squares line is $\hat{Y} = -1.20 + 1.11X$. The null hypothesis of no significant "linear" regression is clearly rejected since an F of 61.95 exceeds $F_{1,6,0.999} = 35.51$ (i.e., $P < 0.001$). Because there are not repeated observations at any one X value, we cannot test for lack of fit. Our only recourse, then, is to examine the complete ANOVA table and decide whether or not the addition of the X^2 term significantly improves the prediction of Y over and above that achieved via a straight line. In other words, we ask whether the increase of 0.085 ($= 0.997 - 0.912$) in R^2 obtained by including the X^2 term in the model represents a significant improvement in the fit. The appropriate test statistic to be used in answering this question is as follows:

$$F = \frac{(\text{extra sum of squares due to adding } X^2)/1}{\text{mean-square residual for the second-order model}} = \frac{4.83}{0.04} = 120.75$$

which exceeds $F_{1,5,0.999} = 47.18$ (i.e., $P < 0.001$). Therefore, the addition of the X^2 term to the model significantly improves the prediction.

As might be expected, a test for overall significant regression of the second-order model yields a highly significant F:

$$F = \frac{\text{MS regression (second-order model)}}{\text{MS residual (second-order model)}} = \frac{(52.04 + 4.83)/2}{0.04} = 710.88$$

What have we shown up to this point? We have concluded so far that a first-order (straight-line) model is not as good as a second-order model. We now need to determine whether the second-order model can be improved upon. Again, we cannot test for lack of fit directly because there are no replicate observations. Instead, we can add an X^3 term to the model and test whether this addition significantly improves the

TABLE 9.4   *ANOVA table derived from Table 9.3*

SOURCE	df	SS	MS	F
Regression (X)	1	52.04	52.04	61.95
Residual	6	5.03	0.84	
Total (corrected) ($R^2 = 0.912$)	7	57.07		

TABLE 9.5   *ANOVA table for the third-order model*

SOURCE		df	SS	MS	F
Regression $\begin{cases} X \\ X^2 \mid X \\ X^3 \mid X, X^2 \end{cases}$		1 1 1	52.040 4.830 0.140	52.04 4.83 0.14	10.00 $(0.025 < P < 0.05)$
Residual		4	0.056	0.014	
Total (corrected) $(R^2 = 0.999)$		7	57.066		

prediction. Fitting the *third-order model* results in the ANOVA table given as Table 9.5. To test whether the addition of the third-order term significantly improves the fit, the following statistic is calculated:

$$F = \frac{\text{(extra sum of squares due to adding } X^3)/1}{\text{mean-square residual for the third-order model}} = \frac{0.14}{0.014} = 10.00$$

This F statistic has an F distribution with 1 and 4 degrees of freedom under H_0: the addition of the X^3 term is not worthwhile. Since $F_{1,4,0.95} = 7.71$ and $F_{1,4,0.975} = 12.22$, we have $0.025 < P < 0.05$. This P value would thus reject H_0 for $\alpha = 0.05$ and would not reject for $\alpha = 0.025$. This makes the decision as to whether or not to include the X^3 term a little difficult. However, some other factors should be taken into consideration: (1) The R^2 value for the parabola was very high, 0.997; (2) the R^2 value only increased from 0.997 to 0.999 in going from a second-order model to a third-order model; (3) the scatter diagram clearly suggests a second-order curve; and (4) when in doubt, the simplest model should be preferred, to promote ease in interpretation. All things considered, then, the most sensible decision is to conclude that the second-order model is the one to be preferred.

In summary, for the data in Table 9.2, the best-fitting model is

$$\hat{Y} = 1.13 - 0.41X + 0.17X^2$$

with an R^2 of 0.997.

Finally, it is also valuable to have the standard deviations (or standard errors) of the estimated regression coefficients. These are difficult to compute by hand whenever the number of independent variables in the model exceeds one. However, all commonly used regression programs print out the values of the coefficients and their standard errors. For the second-order model fit to the data in Table 9.2, we would obtain $S_{\hat{\beta}_1} = 0.141$ and $S_{\hat{\beta}_2} = 0.015$. For example, then, a $100(1 - \alpha)\%$ confidence interval for β_2 would be

$$\hat{\beta}_2 \pm t_{5, 1-\alpha/2} S_{\hat{\beta}_2}$$

where the df for the appropriate critical t value are the df associated with the residual sum of squares (5). In particular, a 95% confidence interval for β_2 in our example is

$$0.17 \pm (2.571)(0.015)$$

or (0.13, 0.21). Note that this interval does not include zero, which agrees with our ANOVA-table conclusion concerning the importance of the X^2 term in the model.

PROBLEMS   *1.*   In an environmental engineering study concerning a certain chemical reaction, the concentrations of 18 (separately prepared) solutions were recorded at different times (three measurements at each of six times). The *natural logarithms* of the concentrations were computed also. The data are as given in the table.

SOLUTION NUMBER (i)	TIME (X_i) (hours)	CONCENTRATION (Y_i) (mg/ml)	LOG$_e$ CONCENTRATION (ln Y_i)
1	6	0.029	−3.540
2	6	0.032	−3.442
3	6	0.027	−3.612
4	8	0.079	−2.538
5	8	0.072	−2.631
6	8	0.088	−2.430
7	10	0.181	−1.709
8	10	0.165	−1.802
9	10	0.201	−1.604
10	12	0.425	−0.856
11	12	0.384	−0.957
12	12	0.472	−0.751
13	14	1.130	0.122
14	14	1.020	0.020
15	14	1.249	0.222
16	16	2.812	1.034
17	16	2.465	0.902
18	16	3.099	1.131

(a)   Plot on separate sheets of graph paper:
(1)   Concentration (Y) versus time (X).
(2)   Natural logarithm of concentration (ln Y) versus time (X).
(b)   Using the computer printout, obtain the following:
(1)   The estimated equation of the straight-line (degree 1) regression of Y on X.
(2)   The estimated equation of the quadratic (degree 2) regression of Y on X.
(3)   The estimated equation of the straight-line (degree 1) regression of ln Y on X.
(4)   Plots of each of these fitted equations on their respective scatter diagrams.
(c)   Based on the computer printout, complete the table for the straight-line regression of Y on X.

SOURCE		df	SS	MS	F
Regression		1			
Residual $\begin{cases} \text{lack of fit} \\ \text{pure error} \end{cases}$		4 12			
Total		17			

Computer printout for Problem 1

```
Concentration (Y) on Time (X):
```

```
FITTING DEGREE 1                      FITTING DEGREE 2
MULTIPLE R-SQUARED = .732             MULTIPLE R-SQUARED = .957
```

```
REGRESSION COEFFICIENTS               REGRESSION COEFFICIENTS

0       -1.9318                       0            3.1721
1        .24597                       1           -.78102
                                      2            .46682E-01
```

ANALYSIS OF VARIANCE

SOURCE	DF	SS	MS
DEGREE 1	1	12.705	12.705
DEGREE 2	1	3.9051	3.9051
LACK OF FIT	3	.51446	.17149
PURE ERROR	12	.23248	.19374E-01
TOTAL	17	17.357	1.0210

```
Log  Concentration (ln Y) on Time (X):
   e
```

```
FITTING DEGREE 1                ANALYSIS OF VARIANCE
MULTIPLE R-SQUARED = .996
```

SOURCE	DF	SS	MS	
REGRESSION COEFFICIENTS				
DEGREE 1	1	42.746	42.746	
0 -6.2096	LACK OF FIT	4	.27836E-01	.69591E-02
1 .45117	PURE ERROR	12	.12247	.10206E-01
	TOTAL	17	42.896	2.5233

(d) Based on the computer output, complete the ANOVA table.

SOURCE		df	SS	MS	F
Regression {	degree 1 (X)	1			
	degree 2 ($X^2|X$)	1			
Residual {	lack of fit	3			
	pure error	12			
Total		17			

(e) Determine and compare the proportions of the total variation in Y explained by the straight-line regression on X and by the quadratic regression on X.

(f) Carry out F tests for the significance of the straight-line regression of Y on X and for the adequacy of fit of the estimated regression line.

(g) Carry out an overall F test for the significance of the quadratic regression of Y on X, a test for the significance of the addition of X^2 to the model, and an F test for the adequacy of fit of the estimated quadratic model.

(h) For the straight-line regression of ln Y on X carry out F tests for the significance of the overall regression and for the adequacy of fit of the straight-line model.

(i) What proportion of the variation in ln Y is explained by the straight-line regression of ln Y on X? Compare this result with that obtained in part (e) for the quadratic regression of Y on X.

(j) A fundamental assumption in regression analysis is variance homoscedasticity.

   (1) By examining the scatter diagrams constructed in part (a), state why taking natural logarithms of the concentrations helps with regard to the assumption of variance homogeneity.

   (2) Do you think the straight-line regression of ln Y on X is better for describing this set of data than the quadratic regression of Y on X? Explain.

(k) What key assumption about the data would be in question if, instead of 18 different solutions, there were only 3 different solutions, each of which was analyzed at the six different time points?

2. With the addition of five pairs of observations [(18,000; 39.2), (22,400; 27.9), (24,210; 22.3), (5,400; 11.7), and (9,340; 32.5)] to the data in Problem 3 of Chapter 5, the accompanying printout is obtained for the regression of TIME (Y) on INC (X).

(a) Using the computer output, complete the ANOVA table for the straight-line regression of TIME (Y) on INC (X).

SOURCE		df	SS	MS	F
Regression		1			
Residual {	lack of fit	18			
	pure error	5			
Total		24			

(b) Using the computer results, complete the ANOVA table for the quadratic regression of TIME (Y) on INC (X).

SOURCE		df	SS	MS	F
Regression {	degree 1 (X)	1			
	degree 2 ($X^2|X$)	1			
Residual {	lack of fit	17			
	pure error	5			
Total		24			

(c) Plot on the scatter diagram of the data for this problem the fitted straight-line (degree 1) equation and the fitted quadratic (degree 2) equation.

(d) Calculate and compare the R^2 values obtained for the straight-line, quadratic, and cubic fits.

Computer printout for Problem 2

```
FITTING DEGREE 1              FITTING DEGREE 2              FITTING DEGREE 3
MULTIPLE R-SQUARED = 0.153    MULTIPLE R-SQUARED = 0.880    MULTIPLE R-SQUARED = 0.901


REGRESSION COEFFICIENTS       REGRESSION COEFFICIENTS       REGRESSION COEFFICIENTS

0        20.17655             0        -19.86602            0        -35.29278
1         0.00061             1          0.00787            1          0.01223
                              2         -0.00000025         2         -0.0000006
                                                            3          0.0000000
```

```
ANALYSIS OF VARIANCE

        SOURCE        DF         SS             MS

        DEGREE 1       1     442.91410       442.91410
        DEGREE 2       1    2100.08113      2100.08113
        DEGREE 3       1      61.10018        61.10018
        LACK OF FIT   16     271.74872        16.98430
        PURE ERROR     5      15.20121         3.04024

        TOTAL         24    2891.04534       120.46022
```

(e)   Carry out F tests for the significance of the straight-line regression and for the adequacy of fit of the straight-line model.

(f)   Carry out F tests for the significance of the quadratic regression, of the addition of the quadratic term to the model, and of the adequacy of fit of the quadratic model.

(g)   Which model is most appropriate: straight-line, quadratic, or cubic?

3.   For the data on DIST (Y) and MPH (X) in Problem 7 of Chapter 5, use the information provided to answer the same questions as in parts (a)–(g) of Problem 2 above.

Degree 1 fit: $\hat{Y} = -122.345 + 6.227X$

Degree 2 fit: $\hat{Y} = 32.901 - 3.051X + 0.1176X^2$

Degree 3 fit: $\hat{Y} = 114.621 - 10.620X + 0.3247X^2 - 0.00173X^3$

SOURCE		df	SS	MS
Regression $\begin{cases} X \\ X^2\|X \\ X^3\|X, X^2 \end{cases}$		1 1 1	173,473.96 10,515.44 415.19	173,473.96 10,515.44 415.19
Residual $\begin{cases} \text{lack of fit} \\ \text{pure error} \end{cases}$		11 4	2,664.15 3,433.93	242.20 858.48
Total		18	190,502.67	

4.   For the data on VOTE (Y) and TVEXP (X) in Problem 5 of Chapter 5, it was found that the straight-line model was adequate. Using the computer printout for quadratic regression, do the following:

(a)   Plot the fitted straight-line model and the fitted quadratic model on the scatter diagram for the data of this problem.

(b) Determine the change in R^2 in going from a degree 1 to a degree 2 model.

(c) Test for the significance of the addition of the X^2 term to the model.

(d) Do the results in parts (a)–(c) contradict your earlier conclusion concerning the adequacy of fit of the straight-line model?

Computer printout for Problem 4

```
FITTING DEGREE 1                    FITTING DEGREE 2
MULTIPLE R-SQUARED = 0.910          MULTIPLE R-SQUARED = 0.911


REGRESSION COEFFICIENTS             REGRESSION COEFFICIENTS

0            2.17407                0            9.65678
1            1.17697                1            0.75077
                                    2            0.00575
```

SOURCE	df	SS	MS
Degree 1	1	2,023.20500	2,023.20500
Degree 2	1	2.45015	2.45015
Lack of fit	12	166.80279	13.90023
Pure error	5	30.55988	6.11198
Total	19	2,223.01782	117.00094

5. For the regression of PCI (Y) on YNG (X) for African countries considered in Problem 2 of Chapter 8, use the information provided to do the following:

(a) Plot both the estimated straight-line and quadratic models on the scatter diagram for the data of the African countries.

(b) Test for the significance of the straight-line regression and for the adequacy of fit of the straight-line model.

(c) Test for the significance of the addition of the X^2 term to the model.

(d) Which model is more appropriate, the straight-line model or the quadratic model?

Degree 1 fit: $\hat{Y} = 893.57 - 17.276X$

Degree 2 fit: $\hat{Y} = 732.05 - 9.203X - 0.0996X^2$

SOURCE		df	SS	MS
Regression $\begin{cases} X \\ X^2 \vert X \end{cases}$		1	153,784.8	153,784.8
		1	88.3	88.3
Residual $\begin{cases} \text{lack of fit} \\ \text{pure error} \end{cases}$		15	2,773.9	184.9
		8	911.5	113.9
Total		25	157,558.5	

6. For the data given in Problem 12 of Chapter 7, which concerns the relationship between the temperature (X) of a certain medium and the growth (Y) of human amniotic cells in a tissue culture, it is of interest to evaluate whether a parabolic

Computer printout for Problem 6 (from SPSS package)

DEPENDENT VARIABLE.. Y

VARIABLE(S) ENTERED ON STEP NUMBER 1.. X

		ANALYSIS OF VARIANCE	DF	SUM OF SQUARES	MEAN SQUARE	F
MULTIPLE R	0.98603	REGRESSION	1.	12.83072	12.83072	630.71608
R SQUARE	0.97225	RESIDUAL	18.	0.36618	0.02034	
ADJUSTED R SQUARE	0.97071					
STANDARD ERROR	0.14263					

——————— VARIABLES IN THE EQUATION ——————— ——————— VARIABLES NOT IN THE EQUATION ———————

VARIABLE	B	BETA	STD ERROR B	F	VARIABLE	BETA IN	PARTIAL	TOLERANCE	F
X	0.03582	0.98603	0.00143	630.716	XX	0.84502	0.64297	0.01606	11.981
(CONSTANT)	-0.46240								

VARIABLE(S) ENTERED ON STEP NUMBER 2.. XX

		ANALYSIS OF VARIANCE	DF	SUM OF SQUARES	MEAN SQUARE	F
MULTIPLE R	0.99183	REGRESSION	2.	12.98210	6.49105	513.73362
R SQUARE	0.98372	RESIDUAL	17.	0.21480	0.01264	
ADJUSTED R SQUARE	0.98181					
STANDARD ERROR	0.11241					

——————— VARIABLES IN THE EQUATION ——————— ——————— VARIABLES NOT IN THE EQUATION ———————

VARIABLE	B	BETA	STD ERROR B	F	VARIABLE	BETA IN	PARTIAL	TOLERANCE	F
X	0.00537	0.14782	0.00887	0.367					
XX	0.00022	0.84502	0.00006	11.981					
(CONSTANT)	0.49460								

MAXIMUM STEP REACHED

Note: B stands for the regression coefficient $\hat{\beta}$, XX stands for X^2, and you can ignore for now the terms "BETA," "PARTIAL," and "TOLERANCE." Also,

$$\text{adjusted } R^2 = R^2 - \left(\frac{k}{n-k-1}\right)(1 - R^2).$$

From *Statistical Package for the Social Sciences* by Nie et al. Copyright © 1975 by McGraw-Hill, Inc. Used with permission of McGraw-Hill Book Company and Dr. Norman Nie, President, SPSS Inc.

model is perhaps more appropriate than a straight-line model. Use the computer results given to answer the following questions:

(a) Plot the fitted straight-line model and the fitted quadratic model on the same scatter diagram.
(b) Test for the significance of the addition of the X^2 term to the model.
(c) Determine the change in R^2 in going from a straight line to a parabolic model.
(d) How do the results in parts (b) and (c) compare with your results in Problem 12 of Chapter 7 for the earlier test of adequacy of fit of the straight-line model?
(e) Which model is more appropriate, straight-line or parabolic?

7. The skin response in rats to different concentrations of a newly developed vaccine was measured in an experiment, resulting in the data given in the table and corresponding computer results.

CONCENTRATION (X) (ml/liter)	0.5	0.5	1.0	1.0	1.5	1.5	2.0	2.0	2.5	2.5	3.0	3.0
SKIN RESPONSE (Y) (mm)	13.90	13.81	14.08	13.99	13.75	13.60	13.32	13.39	13.45	13.53	13.59	13.64

Computer results for Problem 7

$$\text{Degree 1:} \quad \hat{Y} = 13.986 - 0.1802X$$

$$\text{Degree 2:} \quad \hat{Y} = 14.270 - 0.6065X + 0.1218X^2$$

$$\text{Degree 3:} \quad \hat{Y} = 13.362 + 1.6800X - 1.3929X^2 + 0.2885X^3$$

ANOVA tables

DEGREE 1			DEGREE 2			DEGREE 3		
SOURCE	df	SS	SOURCE	df	SS	SOURCE	df	SS
Regression	1	0.2844	Regression	2	0.3536	Regression	3	0.5222
Residual	10	0.3461	Residual	9	0.2769	Residual	8	0.1083

(a) Plot the straight-line, quadratic, and cubic equations on the scatter diagram for this data set.
(b) Test sequentially for significant straight-line fit, for significant addition of X^2, and for significant addition of X^3 to the model.
(c) Which of the three models do you recommend and why? (*Note*: You might also wish to consider R^2 for each model.)

CHAPTER 10 MULTIPLE REGRESSION ANALYSIS

10.1 PREVIEW

Multiple regression analysis can be looked upon as an extension of straight-line regression analysis (involving only one independent variable) to the situation where there are any number of independent variables to be considered. Several general applications of multiple regression analysis[1] were described in Chapter 4 and specific examples given in Chapter 1. In this chapter we shall describe the multiple regression method in detail, stating the required assumptions, describing the procedures for estimating important parameters, explaining how to make and interpret inferences about these parameters, and providing examples illustrating the use of the techniques of multiple regression analysis. Before proceeding to these details, however, it is important to mention that dealing with several independent variables simultaneously in a regression analysis is considerably more difficult than dealing with a single independent variable, for the following reasons:

1. It will be more difficult to determine the best choice of model, since there will sometimes be several reasonable candidates to choose from.

2. It will be more difficult to visualize what the fitted model looks like (especially if there are more than two independent variables involved), since it is not possible to plot directly in more than three dimensions either the data or the fitted model.

3. It will sometimes be more difficult to interpret what the best-fitting model means in real-life terms.

4. Computations will be virtually impossible without access to a high-speed computer and a reliable packaged computer program.

---

[1] We shall generally refer to multiple regression analysis simply as "regression analysis" throughout the remainder of the text.

10.2 MULTIPLE REGRESSION MODELS

In Chapter 9 we studied one example of a multiple regression model; such a model is given by any second- or higher-order polynomial. The addition of higher-order terms (e.g., an X^2 or X^3 term) to the model can be considered as equivalent to the addition of new independent variables. Thus, if we had renamed X as X_1 and renamed X^2 as X_2, the second-order model

$$Y = \beta_0 + \beta_1 X + \beta_2 X^2 + E$$

could be written as

$$Y = \beta_0 + \beta_1 X_1 + \beta_2 X_2 + E$$

Of course, in polynomial regression we really have only one basic independent variable, with the others being simple mathematical functions of this basic variable. In more general multiple regression problems, however, the number of basic independent variables may be greater than one. The general form of a regression model for k independent variables is given by

$$Y = \beta_0 + \beta_1 X_1 + \beta_2 X_2 + \beta_3 X_3 + \cdots + \beta_k X_k + E$$

where $\beta_0, \beta_1, \beta_2, \ldots, \beta_k$ are the *regression coefficients* that need to be estimated. The *independent variables* $X_1, X_2, \ldots, X_k$ may all be separate basic variables, or some of them may be functions of a few basic variables.

EXAMPLE
Suppose that we want to investigate how weight varies with height and age for children with a particular kind of nutritional deficiency. The dependent variable here is $Y =$ WGT and our two basic independent variables are $X_1 =$ HGT and $X_2 =$ AGE.

[Perhaps the main question associated with this type of study is whether the relationship for nutritionally deficient children is the same as that for "normal" children. To answer this question would require additional data on normal children and some kind of comparison of the models obtained for each group. Although we will learn how to deal with this kind of question in Chapter 13, we focus our attention here on the methods needed to describe the relationship of weight to height and age for this single group of nutritionally deficient children.]

To continue, then, suppose that a random sample consists of 12 children who attend a certain clinic. The weight, height, and age data obtained for each child are given in Table 10.1.

TABLE 10.1
Data on weight, height, and age of a random sample of 12 nutritionally deficient children

CHILD NO.	1	2	3	4	5	6	7	8	9	10	11	12
Y (WGT)	64	71	53	67	55	58	77	57	56	51	76	68
X_1 (HGT)	57	59	49	62	51	50	55	48	42	42	61	57
X_2 (AGE)	8	10	6	11	8	7	10	9	10	6	12	9

In describing the relationship of weight to height and age, we may want to consider the model

$$Y = \beta_0 + \beta_1 X_1 + \beta_2 X_2 + E$$

if we are interested only in first-order terms. If we want to consider, in addition, the higher-order term X_1^2, our model would be given by

$$Y = \beta_0 + \beta_1 X_1 + \beta_2 X_2 + \beta_3 X_3 + E$$

where $X_3 = X_1^2$. If we want to consider all possible first- and second-order terms, we would look at the model

$$Y = \beta_0 + \beta_1 X_1 + \beta_2 X_2 + \beta_3 X_3 + \beta_4 X_4 + \beta_5 X_5 + E$$

where $X_3 = X_1^2$, $X_4 = X_2^2$, and $X_5 = X_1 X_2$, or, equivalently,

$$Y = \beta_0 + \beta_1 X_1 + \beta_2 X_2 + \beta_3 X_1^2 + \beta_4 X_2^2 + \beta_5 X_1 X_2 + E$$

If we want to find the best predictive model, we might consider all of the models above (as well as some others) and then choose the best model according to some reasonable criterion (e.g., R^2).

We shall discuss the question of model selection in Chapter 15; the interpretation of product terms such as $X_1 X_2$ as "interaction" effects is explained in Chapter 12. For now we shall focus on the methods used and the interpretations that can be made when the choice of independent variables to be used in the model is not at issue.

10.3 GRAPHICAL LOOK AT THE PROBLEM

When we are dealing with only one independent variable, our problem can be easily described graphically as that of finding the curve that best fits the scatter of points (X_1, Y_1), (X_2, Y_2), ..., (X_n, Y_n) obtained on n individuals. Thus, we have a *two-dimensional* representation involving a plot of the form shown in Figure 10.1.

FIGURE 10.1   *Scatter plot for a single independent variable*

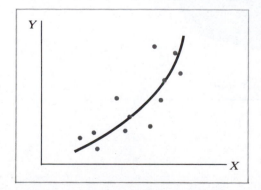

Furthermore, as we know, the *regression equation* for this problem is defined as the path described by the mean values of the distribution of Y when X is allowed to vary.

When the number of (basic) independent variables is two or more, the (graphical) dimension of the problem increases. The regression equation will no longer be a curve in two-dimensional space but will be a *hypersurface in* $(k+1)$-*dimensional space*, where k is the number of basic independent variables. Obviously, we will not be able to represent in a single plot either the scatter of data points or the regression equation if there are more than two basic independent variables. In the special case $k = 2$, as in the example just given where $X_1 = $ height, $X_2 = $ age, and $Y = $ weight, the problem is to find the *surface* in three-dimensional space that best fits the scatter of points (X_{11}, X_{21}, Y_1), $(X_{12}, X_{22}, Y_2), \dots, (X_{1n}, X_{2n}, Y_n)$, where (X_{1i}, X_{2i}, Y_i) denotes the X_1, X_2, and Y values for the ith individual in the sample. The *regression equation* in this case is therefore the surface described by the mean values of Y at various combinations of values of X_1 and X_2; that is, corresponding to *each* distinct pair of values of X_1 and X_2 is a distribution of Y values with mean $\mu_{Y|X_1,X_2}$ and variance $\sigma^2_{Y|X_1,X_2}$.

Just as the simplest curve in two-dimensional space is a straight line, the simplest surface in three-dimensional space is a *plane*, which has the statistical model form $Y = \beta_0 + \beta_1 X_1 + \beta_2 X_2 + E$. Thus, finding the best-fitting plane is frequently the first step in determining the best-fitting surface in three-dimensional space when there are two independent variables, just as fitting the best straight line is the first step when one independent variable is involved. A graphical representation of a planar fit to data in the three-dimensional situation is given in Figure 10.2.

For the three-dimensional case, the least-squares solution giving the best-fitting plane is determined by minimizing the sum of squares of the distances between the observed values Y_i and the corresponding predicted values $\hat{Y}_i = \hat{\beta}_0 + \hat{\beta}_1 X_{1i} + \hat{\beta}_2 X_{2i}$

FIGURE 10.2   *Best-fitting plane for three-dimensional data*

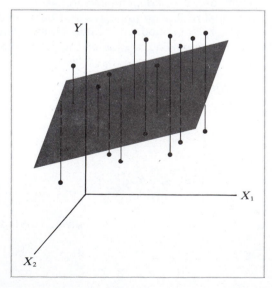

FIGURE 10.3 *Separate scatter diagrams of WGT versus HGT, WGT versus AGE, and AGE versus HGT*

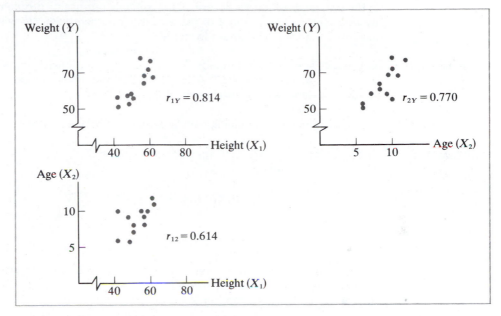

based on the fitted plane; that is, the quantity

$$\sum_{i=1}^{n} (Y_i - \hat{Y}_i)^2 = \sum_{i=1}^{n} (Y_i - \hat{\beta}_0 - \hat{\beta}_1 X_{1i} - \hat{\beta}_2 X_{2i})^2$$

is minimized to find least-squares estimates $\hat{\beta}_0$ of β_0, $\hat{\beta}_1$ of β_1, and $\hat{\beta}_2$ of β_2.

It is natural at this point to wonder how much can be learned by considering the independent variables in the multivariable problem separately. The best answer is probably that you can learn something about what is going on but that you will have too many separate (univariable) pieces of information to be able to complete the (multivariable) puzzle. For example, consider the data previously given for $Y = $ WGT, $X_1 = $ HGT, and $X_2 = $ AGE. If we plot separate scatter diagrams of WGT on HGT, WGT on AGE, and AGE on HGT, we get the results given in Figure 10.3.

First, we note that HGT is highly positively correlated with WGT ($r_{1Y} = 0.814$) and that AGE is also highly positively correlated with WGT ($r_{2Y} = 0.770$). Thus, if we used each of these independent variables separately, we would likely find two separate significant straight-line regressions. Does this mean that the best-fitting plane with both variables in the model together will also have significant predictive ability? The answer is probably yes. But what will the plane look like? This is difficult to say. We can get some idea of the difficulty if we consider the plot of HGT versus AGE, which reflects a positive correlation ($r_{12} = 0.614$). If, instead, these two variables were negatively correlated, we would expect a different orientation of the plane, although we could not clearly quantify either orientation. Thus, treating each independent variable separately does not help very much because the relationships between the independent variables themselves are not directly taken into account. The techniques of multiple regression, however, account for all these intercorrelations with regard to both estimation and inference making.

10.4 ASSUMPTIONS OF MULTIPLE REGRESSION

In the previous section we described the multiple regression problem in some general-ity and also hinted at some of the assumptions involved. We will now state these assumptions somewhat more formally.

ASSUMPTION 1　For each specific combination of values of the (basic) independent variables $X_1, X_2, \ldots, X_k$ (e.g., HGT $= 57$, AGE $= 8$ for the data given above), Y is a (univariate) random variable with a certain probability distribution.

ASSUMPTION 2　The Y observations are statistically independent of one another.

ASSUMPTION 3　The mean value of Y for each specific combination of $X_1, X_2, \ldots, X_k$ is a linear function of $X_1, \ldots, X_k$; that is,

$$\mu_{Y|X_1,X_2,X_3,\ldots,X_k} = \beta_0 + \beta_1 X_1 + \beta_2 X_2 + \cdots + \beta_k X_k \tag{10.1}$$

or

$$Y = \beta_0 + \beta_1 X_1 + \beta_2 X_2 + \cdots + \beta_k X_k + E \tag{10.2}$$

where E is the error component reflecting the difference between an individual's observed response Y and the true average response $\mu_{Y|X_1,X_2,\ldots,X_k}$. Some comments are in order regarding Assumption 3:

1. The surface described by (10.1) is called the *regression equation* (or *response surface* or *regression surface*).

2. If some of the independent variables are higher-order functions of a few basic independent variables (e.g., $X_3 = X_1^2$, $X_5 = X_1 X_2$), the expression $\beta_0 + \beta_1 X_1 + \cdots + \beta_k X_k$ is really nonlinear in the basic variables (hence the term "surface" rather than plane).

    [The techniques of multiple regression that we will be describing are applicable as long as the model under consideration is *inherently linear* in the regression coefficients (regardless of how the independent variables are defined). For example, a model of the form $\mu_{Y|X} = \beta_0 e^{\beta_1 X}$ is inherently linear because it can be transformed to the equivalent form $\mu_{Y|X}^* = \beta_0^* + \beta_1 X$, where $\mu_{Y|X}^* = \log_e \mu_{Y|X}$ and $\beta_0^* = \log_e \beta_0$. However, the model $\mu_{Y|X_1,X_2} = e^{\beta_1 X_1} + e^{\beta_2 X_2}$ cannot be transformed directly to a form that is linear in β_1 and β_2, and so estimation of β_1 and β_2 would require the use of *nonlinear regression* procedures [e.g., see Marquardt (1963)]. A discussion of these procedures is beyond the scope of this text.]

3. As with straight-line regression, E is the differential amount by which any indi-vidual's observed response deviates from the response surface. Thus, E is the *error component* in the model.

ASSUMPTION 4　The variance of Y is the same for any fixed combination of $X_1, X_2, \ldots, X_k$; that is,

$$\sigma_{Y|X_1,X_2,\ldots,X_k}^2 = \text{Var}\,(Y|X_1, X_2, \ldots, X_k) \equiv \sigma^2 \tag{10.3}$$

As before, this is called the assumption of homoscedasticity.

This assumption perhaps seems very restrictive. However, as mentioned earlier in the discussion of the assumptions for straight-line regression analysis, variance heteroscedasticity only has to be worried about when the data show very obvious and significant departures from homogeneity. In general, mild departures will not have too adverse an effect on the results.

ASSUMPTION 5   For any fixed combination of $X_1, X_2, \ldots, X_k$, Y is normally distributed. In other words,

$$Y \frown N(\mu_{Y|X_1, X_2, \ldots, X_k}, \sigma^2) \tag{10.4}$$

This assumption (or any of the others) is not necessary for the least-squares fitting of the regression model, but is required, in general, for inference-making purposes. In this regard, the usual parametric tests of hypotheses and confidence intervals used in a regression analysis are "robust" in the sense that only extreme departures of the distribution of Y from normality can yield spurious results. (This statement is based on both theoretical and experimental evidence.)

10.5 DETERMINING THE BEST ESTIMATE OF THE MULTIPLE REGRESSION EQUATION

As with straight-line regression, there are two basic approaches for determining the best estimate of a multiple regression equation: the least-squares approach and the minimum-variance approach. As for the straight-line case, both of these approaches yield the same solution.

[We are assuming, as previously noted, that we already know the best form of regression model to use; that is, we have already settled on a fixed set of k independent variables $X_1, X_2, \ldots, X_k$. The problem of determining the best model form via algorithms for choosing the most important independent variables will be discussed in detail in Chapter 15.]

10.5.1 LEAST-SQUARES APPROACH

In general, the least-squares method chooses the best-fitting model to be that model which minimizes the sum of squares of the distances between the observed responses and those predicted by the fitted model. Again, the idea is that the better the fit, the smaller will be the deviations of observed from predicted values. Thus, if we let

$$\hat{Y} = \hat{\beta}_0 + \hat{\beta}_1 X_1 + \hat{\beta}_2 X_2 + \cdots + \hat{\beta}_k X_k$$

denote the fitted regression model, the sum of squares of deviations of observed Y values from corresponding predicted values using the fitted regression model is given by

$$\sum_{i=1}^{n} (Y_i - \hat{Y}_i)^2 = \sum_{i=1}^{n} (Y_i - \hat{\beta}_0 - \hat{\beta}_1 X_{1i} - \cdots - \hat{\beta}_k X_{ki})^2 \tag{10.5}$$

The least-squares solution then consists of those values $\hat{\beta}_0, \hat{\beta}_1, \hat{\beta}_2, \ldots, \hat{\beta}_k$ (called the "least-squares estimates") for which the sum in (10.5) is a minimum. This minimum sum of squares is generally called the *residual sum of squares* (or, equivalently, the *error sum of squares* or the *sum of squares about regression*) and, as with polynomial regression, is referred to as SSE.

10.5.2 MINIMUM-VARIANCE APPROACH

As with the straight-line case, the minimum-variance approach determines the best-fitting surface to be the one that utilizes the minimum-variance (linear) unbiased estimates $\hat{\beta}_0, \hat{\beta}_1, \ldots, \hat{\beta}_k$ of $\beta_0, \beta_1, \ldots, \beta_k$, respectively.

10.5.3 COMMENTS ON THE LEAST-SQUARES SOLUTIONS

As previously noted for polynomial regression, it is not worthwhile in the general regression situation to write down explicitly the precise algebraic formulas for the least-squares solutions $\hat{\beta}_0, \hat{\beta}_1, \ldots, \hat{\beta}_k$.[2] This is because standard regression computer algorithms are readily available to perform the necessary calculations, making it unnecessary for the user to be concerned with such messy computational details.

It is of value, however, to mention some important properties of the least-squares solutions:

1. Each of the estimates $\hat{\beta}_0, \hat{\beta}_1, \ldots, \hat{\beta}_k$ is a linear function of the Y values. This linearity property makes it fairly straightforward to determine the statistical properties of these estimators. In particular, since the Y values are assumed to be normally distributed, each of the estimators $\hat{\beta}_0, \hat{\beta}_1, \hat{\beta}_2, \ldots, \hat{\beta}_k$ will be normally distributed, with easily computable standard deviations.

2. The least-squares regression equation $\hat{Y} = \hat{\beta}_0 + \hat{\beta}_1 X_1 + \hat{\beta}_2 X_2 + \cdots + \hat{\beta}_k X_k$ is that unique linear combination of the independent variables $X_1, \ldots, X_k$ which has maximum possible correlation with the dependent variable. In other words, of all possible choices of linear combinations of the form $b_0 + b_1 X_1 + b_2 X_2 + \cdots + b_k X_k$, the linear combination $\hat{Y}$ is such that the correlation

$$r_{Y,\hat{Y}} = \frac{\sum\limits_{i=1}^{n}(Y_i - \bar{Y})(\hat{Y}_i - \bar{\hat{Y}})}{\sqrt{\sum\limits_{i=1}^{n}(Y_i - \bar{Y})^2 \sum\limits_{i=1}^{n}(\hat{Y}_i - \bar{\hat{Y}})^2}} \qquad (10.6)$$

is a maximum, where $\hat{Y}_i$ is the predicted value of Y for the ith individual and $\bar{\hat{Y}}$ is the mean of the $\hat{Y}_i$'s. Incidentally, it is always true that $\bar{\hat{Y}} = \bar{Y}$; that is, the mean of the predicted values is equal to the mean of the observed values.

---

[2] However, the computational formula for $\hat{\beta}_0$ is easily expressible as a function of the other estimated coefficients as

$$\hat{\beta}_0 = \bar{Y} - \hat{\beta}_1 \bar{X}_1 - \hat{\beta}_2 \bar{X}_2 - \cdots - \hat{\beta}_k \bar{X}_k$$

Also, a compactly written expression involving all of the least-squares coefficients can be given in matrix notation, as described in Appendix B.

3. Just as straight-line regression is related to the bivariate normal distribution, multiple regression can be related to the multivariate normal distribution. We shall return to this point later.

EXAMPLE For the data given in Table 10.1 on the variables $Y = $ WGT, $X_1 = $ HGT, and $X_2 = $ AGE, the least-squares algorithm applied to the model

$$WGT = \beta_0 + \beta_1 HGT + \beta_2 AGE + \beta_3 AGE^2 + E$$

produces the estimated equation

$$\widehat{WGT} = 3.438 + 0.724 HGT + 2.777 AGE - 0.042 AGE^2$$

so that

$$\hat{\beta}_0 = 3.438, \quad \hat{\beta}_1 = 0.724, \quad \hat{\beta}_2 = 2.777, \quad \hat{\beta}_3 = -0.042$$

10.6 ANALYSIS-OF-VARIANCE TABLE, OVERALL *F* TEST, AND PARTIAL *F* TEST

As with straight-line and polynomial regression, an analysis-of-variance table can be used to provide an overall summary of a multiple regression analysis. For example, consider ANOVA Table 10.2, based on the use of HGT, AGE, and AGE^2 as independent variables for the data of Table 10.1.

As before, the term $SSY = \sum_{i=1}^{n} (Y_i - \bar{Y})^2 = 888.25$ is called the *total sum of squares*, and this figure represents the total variability in the Y observations before accounting for the joint effect of using the independent variables HGT, AGE, and AGE^2. The term $SSE = \sum_{i=1}^{n} (Y_i - \hat{Y}_i)^2 = 195.19$ is the *residual sum of squares* (or the *sum of squares due to error*), and this represents the amount of Y variation left unexplained after the independent variables have been used in the regression equation to predict Y. Finally, $SSY - SSE = \sum_{i=1}^{n} (\hat{Y}_i - \bar{Y})^2 = 693.06$ is called the *regression sum of squares* and measures the reduction in variation (or the variation explained) due to the independent variables in the regression equation. We thus have the familiar partition:

total sum of squares = regression sum of squares + residual sum of squares

or
$$\sum_{i=1}^{n} (Y_i - \bar{Y})^2 \quad = \quad \sum_{i=1}^{n} (\hat{Y}_i - \bar{Y})^2 \quad + \quad \sum_{i=1}^{n} (Y_i - \hat{Y}_i)^2$$

TABLE 10.2 *ANOVA table for WGT regressed on HGT, AGE, and AGE^2*

SOURCE	df	SS	MS	F	R^2
Regression	3	SSY − SSE = 693.06	231.02	9.47**	0.7802
Residual	8	SSE = 195.19	24.40		
Total	11	SSY = 888.25			

See Section 10.6.1 for an explanation of **.

10.6.1 TEST FOR SIGNIFICANT OVERALL REGRESSION

We can test the general null hypothesis "H_0: all k independent variables considered together do not explain a significant amount of the variation in Y"[3] by calculating the F statistic

$$F = \frac{\text{mean-square regression}}{\text{mean-square residual}} \tag{10.7}$$

and comparing it with the critical point $F_{k,n-k-1,1-\alpha}$, where k is the number of independent variables, n is the sample size, and α is a preselected significance level. We then reject H_0 if the computed F exceeds $F_{k,n-k-1,1-\alpha}$ in value. It can be shown that an equivalent expression for (10.7) is

$$F = \frac{R^2}{1-R^2} \frac{n-k-1}{k}$$

For the example considered, this F is computed to be

$$\frac{231.02}{24.40} = \frac{0.7802}{1-0.7802} \frac{12-3-1}{3} = 9.47$$

and the critical point for $\alpha = 0.01$ is $F_{3,8,0.99} = 7.59$. We would thus reject H_0 at $\alpha = 0.01$; that is, the P value is less than 0.01. [When $P < 0.01$, we usually denote this by ** next to the computed F (see Table 10.2). When $0.01 < P < 0.05$, we usually use only one *.]

[The *mean-square residual* term, which is the denominator of the F in (10.7) and is given by the formula

$$\frac{1}{n-k-1} \text{SSE} = \frac{1}{n-k-1} \sum_{i=1}^{n} (Y_i - \hat{Y}_i)^2$$

provides an estimate of σ^2 under the assumed model. The *mean-square regression* term, which is the numerator of the F in (10.7) and is given by $\sum_{i=1}^{n} (\hat{Y}_i - \bar{Y})^2 / k$, provides an independent estimate of σ^2 only if the null hypothesis H_0 of no significant overall regression is true. Otherwise, it overestimates σ^2 in proportion to the magnitude of the regression coefficients $\beta_1, \beta_2, \ldots, \beta_k$, and this is why an F value that is "too large" favors rejection of H_0. Thus, the F statistic (10.7) is the ratio of two independent estimates of the same variance only if the null hypothesis H_0 is true.]

The R^2 in the ANOVA table (with a value of 0.7802) provides a quantitative measure of how well the combination of independent variables predicts the dependent variable, and its formula is, as given earlier,

$$R^2 = \frac{\text{SSY} - \text{SSE}}{\text{SSY}} \tag{10.8}$$

---

[3] Alternatively, this null hypothesis can be stated "H_0: no significant overall regression using all k independent variables in the model," or "H_0: $\beta_1 = \beta_2 = \cdots = \beta_k = 0$."

10.6.2 PARTIAL F TEST

Some important additional information concerning the fitted regression model can be obtained by presenting the ANOVA table as in Table 10.3. What we have done in this representation is to partition the regression sum of squares into three components:

1. $SS(X_1)$: the sum of squares explained by using just $X_1 = HGT$ alone in predicting Y.
2. $SS(X_2|X_1)$: the extra sum of squares explained by using $X_2 = AGE$ in addition to X_1 in predicting Y.
3. $SS(X_3|X_1, X_2)$: the extra sum of squares explained by using $X_3 = AGE^2$ in addition to X_1 and X_2 in predicting Y.

We can use the extra information in the table to answer the following questions:

1. Does $X_1 = HGT$ alone significantly aid in predicting Y?
2. Does the addition of $X_2 = AGE$ significantly contribute to the prediction of Y after accounting (or controlling) for the contribution of X_1?
3. Does the addition of $X_3 = AGE^2$ significantly contribute to the prediction of Y after accounting for the contribution of X_1 and X_2?

We already know how to answer question 1. This simply involves fitting the straight-line regression model using $X_1 = HGT$ as the single independent variable. The value 588.92 therefore is the regression sum of squares for this straight-line regression model. The SSE for this model can be obtained from Table 10.3 by adding 195.19, 103.90, and 0.24 together, which yields the sum of squares 299.33, having 10 degrees of freedom (i.e., $10 = 8 + 1 + 1$). The F statistic for testing whether there is significant straight-line regression when using $X_1 = HGT$ only is then given by $F = (588.92/1)/(299.33/10) = 19.67$, which has a P value less than 0.01 (i.e., X_1 contributes significantly to the linear prediction of Y).

To answer questions 2 and 3, we must use what is called a *partial F test*. We have actually already performed such tests for polynomial regression situations when we were interested in adding higher-order terms to the model. For regression problems, the partial F test assesses whether the addition of any specific independent variable, given others already in the model, significantly contributes to the prediction of Y. The test, therefore, allows for the elimination of variables which are of no help in predicting Y and thus enables one to reduce the set of possible independent variables to an economical set of "important" predictors.

TABLE 10.3 *ANOVA table for WGT regressed on HGT, AGE, and AGE^2 containing components of the regression sum of squares*

SOURCE		df	SS	MS	F	R^2
Regression	X_1	1	588.92	588.92	19.67**	0.7802
	$X_2\|X_1$	1	103.90	103.90	4.78 $(0.05 < P < 0.10)$	
	$X_3\|X_1, X_2$	1	0.24	0.24	0.01	
Residual		8	195.19	24.40		
Total		11	888.25			

To perform a partial F test concerning a variable X^*, say, given that variables $X_1, X_2, \ldots, X_p$ are already in the model, we must first compute the "extra sum of squares from adding X^*, given $X_1, X_2, \ldots, X_p$," which we place in our ANOVA table under the source heading "Regression $X^*|X_1, X_2, \ldots, X_p$." This sum of squares is computed by the formula

$$\begin{array}{ccc} \text{extra sum of squares} & \text{regression sum of squares} & \text{regression sum of squares} \\ \text{from adding } X^*, \text{ given } = & \text{when } X_1, X_2, \ldots, X_p & - \quad \text{when } X_1, X_2, \ldots, X_p \\ X_1, X_2, \ldots, X_p & \text{and } X^* \text{ are } all & \text{(and } not\ X^*) \text{ are} \\ & \text{in the model} & \text{in the model} \end{array}$$

(10.9)

or, more compactly,

$$\underset{(X^*|X_1, X_2, \ldots, X_p)}{\text{SS}} = \underset{(X_1, X_2, \ldots, X_p, X^*)}{\text{regression SS}} - \underset{(X_1, X_2, \ldots, X_p)}{\text{regression SS}}$$

Thus, for our example,

$$SS(X_2|X_1) = \text{regression } SS(X_1, X_2) - \text{regression } SS(X_1)$$
$$= 692.82 - 588.92$$
$$= 103.90$$

and

$$SS(X_3|X_1, X_2) = \text{regression } SS(X_1, X_2, X_3) - \text{regression } SS(X_1, X_2)$$
$$= 693.06 - 692.82$$
$$= 0.24$$

To test the null hypothesis "H_0: the addition of X^* to a model already containing $X_1, X_2, \ldots, X_p$ does not significantly improve the prediction of Y," we compute

$$F(X^*|X_1, X_2, \ldots, X_p)$$
$$= \frac{\text{extra sum of squares from adding } X^*, \text{ given } X_1, X_2, \ldots, X_p}{\text{mean-square residual for the model containing all the variables } X_1, X_2, \ldots, X_p, X^*}$$

or, more compactly,

$$F(X^*|X_1, X_2, \ldots, X_p) = \frac{SS(X^*|X_1, X_2, \ldots, X_p)}{\text{MS residual } (X_1, X_2, \ldots, X_p, X^*)}$$

(10.10)

This F statistic has an F distribution with 1 and $n - p - 2$ degrees of freedom under H_0, so we would reject H_0 if the computed F exceeds $F_{1,n-p-2,1-\alpha}$. For our example, the partial F statistics are (from Table 10.3)

$$F(X_2|X_1) = \frac{SS(X_2|X_1)}{MS \text{ residual}(X_1, X_2)} = \frac{103.90}{(195.19 + 0.24)/9} = 4.78$$

and

$$F(X_3|X_1, X_2) = \frac{SS(X_3|X_1, X_2)}{MS \text{ residual}(X_1, X_2, X_3)} = \frac{0.24}{24.40} = 0.01$$

The quantity "MS residual(X_1, X_2)" can either be obtained directly from the ANOVA table for X_1 and X_2 only, or indirectly from the partitioned ANOVA table for X_1, X_2, and X_3 by using the formula

$$MS \text{ residual}(X_1, X_2) = \frac{SS \text{ residual}(X_1, X_2, X_3) + SS(X_3|X_1, X_2)}{8 + 1}$$

The statistic $F(X_2|X_1) = 4.78$ has a P value satisfying $0.05 < P < 0.10$, since $F_{1,9,0.90} = 3.36$ and $F_{1,9,0.95} = 5.12$. Thus, we would reject H_0 at $\alpha = 0.10$ and conclude that the addition of X_2 after accounting for X_1 significantly adds to the prediction of Y at the $\alpha = 0.10$ level. At $\alpha = 0.05$, however, we would not reject H_0.

The statistic $F(X_3|X_1, X_2) = 0.01$, and so obviously H_0 would not be rejected regardless of the significance level; we would therefore conclude that once $X_1 = $ height and $X_2 = $ age are in the model, the addition of $X_3 = $ age^2 is superfluous.

Actually, it should be pointed out that there is an alternative (but equivalent) way to perform the partial F test that is described above. This would involve a test of the null hypothesis $H_0: \beta^* = 0$, where β^* is the coefficient of X^* in the regression equation $Y = \beta_0 + \beta_1 X_1 + \cdots + \beta_p X_p + \beta^* X^* + E$. The equivalent test statistic for testing this null hypothesis is

$$T = \frac{\hat{\beta}^*}{S_{\hat{\beta}^*}} \tag{10.11}$$

where $\hat{\beta}^*$ is the corresponding estimated coefficient and $S_{\hat{\beta}^*}$ is the estimate of the standard error of $\hat{\beta}^*$, both of which are printed out by standard regression programs.

In performing this test, we reject $H_0: \beta^* = 0$ if

$$\begin{cases} |T| > t_{n-p-2,1-\alpha/2} & \text{(two-sided test; } H_A: \beta^* \neq 0) \\ T > t_{n-p-2,1-\alpha} & \text{(upper one-sided test; } H_A: \beta^* > 0) \\ T < -t_{n-p-2,1-\alpha} & \text{(lower one-sided test; } H_A: \beta^* < 0) \end{cases}$$

It can be shown that the two-sided t test is equivalent to the partial F test described above. For example, when testing $H_0: \beta_2 = 0$ in the model $Y = \beta_0 + \beta_1 X_1 + \beta_2 X_2 + E$ fit to the data in Table 10.1, we compute

$$T = \frac{\hat{\beta}_2}{S_{\hat{\beta}_2}} = \frac{2.050}{0.937} = 2.188$$

$$t_{9,0.95} = 1.833 \quad \text{and} \quad t_{9,0.975} = 2.2622$$

Squaring, we get

$$T^2 = 4.79 = \text{partial } F(X_2|X_1) \text{ from Table 10.3}$$

and

$$t^2_{9,1-\alpha/2} = F_{1,9,1-\alpha}$$

Similarly, when testing H_0: $\beta_3 = 0$ in the model $Y = \beta_0 + \beta_1 X_1 + \beta_2 X_2 + \beta_3 X_3 + E$, we compute

$$T = \frac{\hat{\beta}_3}{S_{\hat{\beta}_3}} = \frac{-0.042}{0.422} = -0.0995$$

which yields

$$T^2 = 0.01 = \text{partial } F(X_3|X_1, X_2) \qquad \text{from Table 10.3}$$

10.7 NUMERICAL EXAMPLES

We conclude this chapter with some examples of the type of output to be expected from a typical regression computer algorithm. This output generally consists of the values of the estimated regression coefficients, their estimated standard errors, the associated partial F (or T^2) statistics, and an ANOVA table. For the data of Table 10.1, the six models we have chosen to consider are, by no means, the only ones that could have been considered (e.g., no "interaction" terms were included). The results are as follows:

MODEL 1 $WGT = \beta_0 + \beta_1 HGT + E$

COEFFICIENT ($\hat{\beta}$)	STANDARD ERROR ($S_{\hat{\beta}}$)	PARTIAL F ($\hat{\beta}^2/S^2_{\hat{\beta}}$)
$\hat{\beta}_0 = 6.190$		
$\hat{\beta}_1 = 1.073$	$S_{\hat{\beta}_1} = 0.242$	19.66**

Estimated model: $\widehat{WGT} = 6.190 + 1.073 HGT$.

ANOVA table

SOURCE	df	SS	MS	F	R^2
Regression	1	588.92	588.92	19.67**	0.6630
Residual	10	299.33	29.93		
Total	11	888.25			

MODEL 2 $\quad \text{WGT} = \beta_0 + \beta_2\text{AGE} + E$

COEFFICIENT $(\hat{\beta})$	STANDARD ERROR $(S_{\hat{\beta}})$	PARTIAL F $(\hat{\beta}^2/S_{\hat{\beta}}^2)$
$\hat{\beta}_0 = 30.571$		
$\hat{\beta}_2 = 3.643$	$S_{\hat{\beta}_2} = 0.955$	14.55**

Estimated model: $\widehat{\text{WGT}} = 30.571 + 3.643\text{AGE}$.

ANOVA table

SOURCE	df	SS	MS	F	R^2
Regression	1	526.39	526.39	14.55**	0.5926
Residual	10	361.86	36.19		
Total	11	888.25			

MODEL 3 $\quad \text{WGT} = \beta_0 + \beta_3\text{AGE}^2 + E$

COEFFICIENT $(\hat{\beta})$	STANDARD ERROR $(S_{\hat{\beta}})$	PARTIAL F $(\hat{\beta}^2/S_{\hat{\beta}}^2)$
$\hat{\beta}_0 = 45.998$		
$\hat{\beta}_3 = 0.206$	$S_{\hat{\beta}_3} = 0.055$	14.03**

Estimated model: $\widehat{\text{WGT}} = 45.998 + 0.206\text{AGE}^2$.

ANOVA table

SOURCE	df	SS	MS	F	R^2
Regression	1	521.93	521.93	14.25**	0.5876
Residual	10	366.32	36.63		
Total	11	888.25			

(The F values disagree numerically due to rounding-off the values of $\hat{\beta}_3$ and $S_{\hat{\beta}_3}$.)

MODEL 4 $\quad \text{WGT} = \beta_0 + \beta_1\text{HGT} + \beta_2\text{AGE} + E$

COEFFICIENT $(\hat{\beta})$	STANDARD ERROR $(S_{\hat{\beta}})$	PARTIAL F $(\hat{\beta}^2/S_{\hat{\beta}}^2)$
$\hat{\beta}_0 = 6.553$		
$\hat{\beta}_1 = 0.722$	$S_{\hat{\beta}_1} = 0.261$	7.65*
$\hat{\beta}_2 = 2.050$	$S_{\hat{\beta}_2} = 0.937$	4.79 $(0.05 < P < 0.10)$

Estimated model: $\widehat{\text{WGT}} = 6.553 + 0.722\text{HGT} + 2.050\text{AGE}$.

ANOVA table

SOURCE	df	SS	MS	F	R^2
Regression	2	692.82	346.41	15.95**	0.7800
Residual	9	195.43	21.71		
Total	11	888.25			

MODEL 5 $WGT = \beta_0 + \beta_1 HGT + \beta_3 AGE^2 + E$

COEFFICIENT ($\hat{\beta}$)	STANDARD ERROR ($S_{\hat{\beta}}$)	PARTIAL F ($\hat{\beta}^2 / S_{\hat{\beta}}^2$)
$\hat{\beta}_0 = 15.118$		
$\hat{\beta}_1 = 0.726$	$S_{\hat{\beta}_1} = 0.263$	7.62*
$\hat{\beta}_3 = 0.115$	$S_{\hat{\beta}_3} = 0.054$	4.54 ($0.05 < P < 0.10$)

Estimated model: $W\hat{G}T = 15.118 + 0.726 HGT + 0.115 AGE^2$.

ANOVA table

SOURCE	df	SS	MS	F	R^2
Regression	2	689.65	344.82	15.63**	0.7764
Residual	9	198.60	22.07		
Total	11	888.25			

MODEL 6 $WGT = \beta_0 + \beta_1 HGT + \beta_2 AGE + \beta_3 AGE^2 + E$

COEFFICIENT ($\hat{\beta}$)	STANDARD ERROR ($S_{\hat{\beta}}$)	PARTIAL F ($\hat{\beta}^2 / S_{\hat{\beta}}^2$)
$\hat{\beta}_0 = 3.438$		
$\hat{\beta}_1 = 0.724$	$S_{\hat{\beta}_1} = 0.277$	6.83*
$\hat{\beta}_2 = 2.777$	$S_{\hat{\beta}_2} = 7.427$	0.14
$\hat{\beta}_3 = -0.042$	$S_{\hat{\beta}_3} = 0.422$	0.01

Estimated model: $W\hat{G}T = 3.438 + 0.724 HGT + 2.777 AGE - 0.042 AGE^2$.

ANOVA table

SOURCE	df	SS	MS	F	R^2
Regression	3	693.06	231.02	9.47**	0.7802
Residual	8	195.19	24.40		
Total	11	888.25			

Although we will deal with the problem of model selection more specifically in a later chapter, it is worthwhile to mention at this time that model 4, involving height and age, is the best of the lot if we use R^2 as our criterion in conjunction with a desire for model simplicity. This is because the R^2 value of 0.7800 achieved using this model is, for all practical purposes, the same as the maximum R^2 value obtained using all three variables.

PROBLEMS

1. The multiple regression relationship of SBP (Y) to AGE (X_1), SMK (X_2), and QUET (X_3) was studied using the data in Problem 2 of Chapter 5. Three regression models were considered, yielding least-squares estimates and ANOVA tables as shown.

MODEL	INDEPENDENT VARIABLES IN MODEL	$\hat{\beta}_0$	$\hat{\beta}_1$	$\hat{\beta}_2$	$\hat{\beta}_3$	$S_{\hat{\beta}_1}$	$S_{\hat{\beta}_2}$	$S_{\hat{\beta}_3}$
1	AGE (X_1)	59.092	1.605			0.2387		
2	AGE (X_1), SMK (X_2)	48.050	1.709	10.294		0.2018	2.7681	
3	AGE (X_1), SMK (X_2), QUET (X_3)	45.103	1.213	9.946	8.592	0.3238	2.6561	4.4987

ANOVA results

MODEL 1			MODEL 2		
SOURCE	df	SS	SOURCE	df	SS
Regression (X_1)	1	3,861.630	Regression (X_1, X_2)	2	4,689.684
Residual	30	2,564.338	Residual	29	1,736.285

MODEL 3		
SOURCE	df	SS
Regression (X_1, X_2, X_3)	3	4,889.826
Residual	28	1,536.143

(a) Using model 3: (1) What is the predicted SBP for a 50-year-old smoker with a quetelet index of 3.5? (2) What is the predicted SBP for a 50-year-old non-smoker with a quetelet index of 3.5? (3) For 50-year-old smokers, give an estimate of the change in SBP corresponding to an increase in quetelet index from 3.0 to 3.5.

(b) Using the ANOVA tables, compute and compare the R^2 values for models 1, 2, and 3.

(c) Carry out (separately) the overall F tests for significant regression under models 1, 2, and 3. Be sure to state your null hypothesis for each model in terms of regression coefficients.

(d) Carry out the partial F tests for (1) the addition of SMK to the model given AGE, and (2) the addition of QUET given AGE and SMK.

(e) Carry out the t tests equivalent to the partial F tests of part (d) and in each case verify this equivalence numerically.

(f) Several widely available regression computer programs have what is called a *stepwise* feature, which allows the computer to select independent variables to enter the regression model in the order of their ability to explain the residual variation in the dependent variable which has not been explained by the independent variables already included in the model. This type of algorithm selects the first variable to enter the model to be that independent variable with the highest value of F for straight-line regression (equivalently, this is the variable having the highest correlation with the dependent variable); the next variable to enter the model has the highest partial F among all the remaining candidate variables, and so on. This stepwise algorithm is often used to determine a best model by adding variables sequentially until a step is reached at which the addition of more independent variables will *not significantly improve* the prediction of the dependent variable, the significance of entering variables being assessed by partial F tests. Given that the order in which the independent variables of this study entered the model is AGE, SMK, and QUET, what would you recommend to be the best regression model based on the results of the partial F tests obtained in part (d)?[4]

2. A psychiatrist wanted to know whether the level of pathology (Y) in psychotic patients 6 months after treatment can be predicted with reasonable accuracy from knowledge of pretreatment symptom ratings of thinking disturbance (X_1) and hostile-suspiciousness (X_2). The table gives the data that were collected on 53 patients.

PT. NO.	Y	X_1	X_2	PT. NO.	Y	X_1	X_2	PT. NO.	Y	X_1	X_2	PT. NO.	Y	X_1	X_2
1	44	2.80	6.1	15	26	3.24	6.0	28	8	2.63	6.9	41	23	2.18	6.1
2	25	3.10	5.1	16	27	2.65	6.0	29	11	2.65	5.8	42	31	2.88	5.8
3	10	2.59	6.0	17	4	3.41	7.6	30	7	3.26	7.2	43	20	3.04	6.8
4	28	3.36	6.9	18	14	2.58	6.2	31	23	3.15	6.5	44	65	3.32	7.3
5	25	2.80	7.0	19	21	2.81	6.0	32	16	2.60	6.3	45	9	2.80	5.9
6	72	3.35	5.6	20	22	2.80	6.4	33	26	2.74	6.8	46	12	3.29	6.8
7	45	2.99	6.3	21	60	3.62	6.8	34	8	2.72	5.9	47	21	3.56	8.8
8	25	2.99	7.2	22	10	2.74	8.4	35	11	3.11	6.8	48	13	2.74	7.1
9	12	2.92	6.9	23	60	3.27	6.7	36	12	2.79	6.7	49	10	3.06	6.9
10	24	3.23	6.5	24	12	3.78	8.3	37	50	2.90	6.7	50	4	2.54	6.7
11	46	3.37	6.8	25	28	2.90	5.6	38	9	2.74	5.5	51	18	2.78	7.2
12	8	2.72	6.6	26	39	3.70	7.3	39	13	2.70	6.9	52	10	2.81	5.2
13	15	3.47	8.4	27	14	3.40	7.0	40	22	3.08	6.3	53	7	3.26	6.6
14	28	2.70	5.9												

(a) The least-squares equation involving both independent variables is given by $\hat{Y} = -0.628 + 23.639X_1 - 7.147X_2$. Using this equation, determine the predicted level of pathology ($\hat{Y}$) for a patient with pretreatment scores of 2.80 on thinking disturbance and 7.0 on hostile suspiciousness. How does the predicted value obtained compare with the value actually obtained for patient 5?

[4] We shall discuss the topic of model selection more formally in Chapter 15.

(b) Using the ANOVA tables, carry out the overall regression F tests for models containing both X_1 and X_2, X_1 alone, and X_2 alone.

(c) Carry out the partial F test for X_2 given X_1 in the model and the partial F test for X_1 given X_2 in the model.

(d) Based on your results in parts (b) and (c), how would you rate the two variables in order of their importance in predicting Y?

(e) What are the R^2 values for the three regressions referred to in part (b)?

(f) What is the best model involving either one or both of the independent variables?

ANOVA tables

SOURCE	df	SS	SOURCE	df	SS
X_1	1	1,546	X_2	1	160
$X_2 \| X_1$	1	1,238	$X_1 \| X_2$	1	2,623
Residual	50	11,008	Residual	50	11,008

3. The accompanying table presents the weight (X_1), age (X_2), and plasma lipid levels of total cholesterol (Y) for a hypothetical sample of 25 patients with hyperlipoproteinemia before being subject to drug therapy.

PATIENT	TOTAL CHOLESTEROL (Y) (mg/100 ml)	WEIGHT (X_1) (kg)	AGE (X_2) (years)
1	354	84	46
2	190	73	20
3	405	65	52
4	263	70	30
5	451	76	57
6	302	69	25
7	288	63	28
8	385	72	36
9	402	79	57
10	365	75	44
11	209	27	24
12	290	89	31
13	346	65	52
14	254	57	23
15	395	59	60
16	434	69	48
17	220	60	34
18	374	79	51
19	308	75	50
20	220	82	34
21	311	59	46
22	181	67	23
23	274	85	37
24	303	55	40
25	244	63	30

(a) Given the accompanying ANOVA tables for the separate straight-line regressions of Y on X_1 and Y on X_2, which the two independent variables would you say is the more important predictor of Y?

ANOVA tables

Y on X_1			Y on X_2		
SOURCE	df	SS	SOURCE	df	SS
Regression (X_1) Residual	1 23	10,231.7 135,144.3	Regression (X_2) Residual	1 23	101,932.7 43,444.3

(b) The estimated regression models resulting from the separate fits of Y on both X_1 and X_2, Y on X_1 alone, and Y on X_2 alone are given as follows:

$\hat{Y} = 77.983 + 0.417X_1 + 5.217X_2$

$\hat{Y} = 199.2975 + 1.622X_1$

$\hat{Y} = 102.5751 + 5.321X_2$

For each of these models, determine the predicted cholesterol level (Y) for patient 4 (with $Y = 263$, $X_1 = 70$, and $X_2 = 30$) and compare this predicted cholesterol level with the observed value. Comment on your findings.

(c) Given the ANOVA table based on the regression involving both X_1 and X_2, carry out the overall F test for this two-variable model, and the partial F test for the addition of X_1 to the model given X_2 is already in the model.

SOURCE	df	SS
Regression (X_1, X_2) Residual	2 22	102,570.8 42,806.2

(d) Compute and compare the R^2 values for each of the three models considered in part (b).

(e) Based on the results obtained in parts (a)–(d), what do you consider to be the best predictive model involving either one or both of the independent variables considered?

4. A sociologist investigating the recent upward shift in homicide trends throughout the United States studied the extent to which the homicide rate per 100,000 population (Y) is associated with population size (X_1), the percent of families with yearly incomes less than \$5,000 (X_2), and the rate of unemployment (X_3). Data are provided in the table for a hypothetical sample of 20 cities.

CITY	Y	X_1 (thousands)	X_2	X_3	CITY	Y	X_1 (thousands)	X_2	X_3
1	11.2	587	16.5	6.2	11	14.5	7,895	18.1	6.0
2	13.4	643	20.5	6.4	12	26.9	762	23.1	7.4
3	40.7	635	26.3	9.3	13	15.7	2,793	19.1	5.8
4	5.3	692	16.5	5.3	14	36.2	741	24.7	8.6
5	24.8	1,248	19.2	7.3	15	18.1	625	18.6	6.5
6	12.7	643	16.5	5.9	16	28.9	854	24.9	8.3
7	20.9	1,964	20.2	6.4	17	14.9	716	17.9	6.7
8	35.7	1,531	21.3	7.6	18	25.8	921	22.4	8.6
9	8.7	713	17.2	4.9	19	21.7	595	20.2	8.4
10	9.6	749	14.3	6.4	20	25.7	3,353	16.9	6.7

(a) Given the ANOVA tables shown, based on regressions involving each two-variable combination of the three independent variables above, carry out the overall regression F test for each two-variable regression model and the partial F test concerning the addition of the second variable given that the first variable is already in the model.

Y ON X_1 AND X_2			Y ON X_1 AND X_3		
SOURCE	df	SS	SOURCE	df	SS
X_2	1	1,308.34	X_3	1	1,360.14
$X_1\|X_2$	1	9.46	$X_1\|X_3$	1	35.35
Residual	17	537.40	Residual	17	459.71

Y ON X_2 AND X_3		
SOURCE	df	SS
X_3	1	1,360.14
$X_2\|X_3$	1	117.23
Residual	17	377.82

(b) Based on the results obtained in part (a), which two-variable model would you recommend?

(c) Compute the R^2 values for each two-variable model above and relate the results to the conclusion reached in part (b).

(d) Given the ANOVA table shown, involving all three independent variables, test whether the addition of X_1 to a model with X_2 and X_3 already included significantly improves the prediction of Y.

SOURCE	df	SS
Regression (X_1, X_2, X_3)	3	1,507.18
Residual	16	348.03

(e) Determine and comment on the increase in R^2 in going from a model with just X_2 and X_3 to a model that includes all three independent variables.

(f) The ANOVA table resulting from fitting a model with independent variables X_2, X_3, and $X_4 = X_2X_3$ is as given.[5]

SOURCE	df	SS	MS
X_3	1	1,360.14	1,360.14
$X_2\|X_3$	1	117.23	117.23
$X_4\|X_2, X_3$	1	0.09	0.09
Residual	16	377.73	23.61

[5] The coefficient of the product term X_4 measures what is generally called an *interaction effect* associated with the variables X_2 and X_3, which concerns whether or not the relationship between Y and one of these two variables depends upon the levels of the other variable. A more detailed discussion of the concept of interaction is given in Chapter 12.

Use this table to test whether X_4 significantly improves the prediction of Y given that X_2 and X_3 are already in the model.

(g) Based on the results for all preceding parts, which of the variables X_1, X_2, X_3, and X_4 would you consider to be important predictors of Y, and what is their order of importance? Based on R^2 values, what variables would you consider to be important?

5. A panel of educators in a large urban community was interested in evaluating the effects of educational resources on student performance. They examined the relationship between twelfth-grade mean verbal SAT scores (Y) and the following independent variables for a random sample of 25 high schools:

X_1: per pupil expenditure (in dollars)
X_2: percent of teachers with a master's degree or higher
X_3: pupil–teacher ratio

The ANOVA table given summarizes the key results obtained from the regression of Y on X_1, X_2, and X_3.

SOURCE	df	SS
X_1	1	18,953.04
$X_3 \mid X_1$	1	7,010.03
$X_2 \mid X_1, X_3$	1	10.93
Residual	21	2,248.23
Total	24	28,222.23

(a) Does X_1 alone significantly help in predicting Y? What is the R^2 value for a regression involving X_1 alone?

(b) Does the addition of X_3 significantly improve the prediction of Y after controlling for the contribution of X_1?

(c) Carry out the overall F test for the model containing X_1 and X_3, and then determine the increase in R^2 in going from a model with X_1 alone to a model containing both X_1 and X_3.

(d) Carry out the partial F test for the inclusion of the variable X_2 after controlling for X_1 and X_3. Also, determine the increase in R^2 due to the addition of X_2.

(e) Based on the preceding results, what do you consider to be the most appropriate model?

6. A team of environmental epidemiologists used data from 23 counties to investigate the relationship between respiratory cancer mortality rates (Y) for a given year and the following three independent variables:

X_1: air pollution index for the county
X_2: mean age (over 21) for the county
X_3: percent of working force in the county employed in a certain industry

The ANOVA table given summarizes the key results of the regression of Y on X_1, X_2, and X_3.

SOURCE	df	SS
X_1	1	1,523.658
$X_2 \mid X_1$	1	181.743
$X_3 \mid X_1, X_2$	1	130.529
Residual	19	551.723
Total	22	2,387.653

(a) Carry out the appropriate test procedure to determine whether variable X_3 significantly adds to the prediction of respiratory cancer mortality rate controlling for the contribution of X_1 and X_2. State your null hypothesis for this test in terms of an appropriate regression coefficient.

(b) Carry out the overall F test for the regression model which includes all three independent variables.

(c) Determine the R^2 for the model containing all three independent variables and for the model involving only X_1 and X_2. Comment on the difference in R^2 values obtained.

(d) Based on the ANOVA table, what independent variables should be retained as important predictors of Y?

7. In an experiment to describe the toxic action of a certain chemical on silkworm larvae,[6] the relationship of $\log_{10}$ dose and $\log_{10}$ larva weight to $\log_{10}$ survival time was sought. The data, obtained by feeding each larva a precisely measured dose of the chemical in an aqueous solution and then recording the survival time (i.e., time until death), are given in the table. Also given are relevant computer results.

LARVA NO.:	1	2	3	4	5	6	7	8
$\log_{10}$ survival time (Y)	2.836	2.966	2.687	2.679	2.827	2.442	2.421	2.602
$\log_{10}$ dose (X_1)	0.150	0.214	0.487	0.509	0.570	0.593	0.640	0.781
$\log_{10}$ weight (X_2)	0.425	0.439	0.301	0.325	0.371	0.093	0.140	0.406

LARVA NO.:	9	10	11	12	13	14	15
$\log_{10}$ survival time (Y)	2.556	2.441	2.420	2.439	2.385	2.452	2.351
$\log_{10}$ dose (X_1)	0.739	0.832	0.865	0.904	0.942	1.090	1.194
$\log_{10}$ weight (X_2)	0.364	0.156	0.247	0.278	0.141	0.289	0.193

Computer results for Problem 7

$$\hat{Y} = 2.952 - 0.550X_1$$

$$\hat{Y} = 2.187 + 1.370X_2$$

$$\hat{Y} = 2.593 - 0.381X_1 + 0.871X_2$$

[6] Adapted from a study by Bliss (1936).

ANOVA tables

SOURCE	df	SS	SOURCE	df	SS
Regression (X_1, X_2)	2	0.4637	Regression (X_1)	1	0.3633
Residual	12	0.0476	Residual	13	0.1480

SOURCE	df	SS
Regression (X_2)	1	0.3332
Residual	13	0.1780

(a) Test for the significance of overall regression involving both independent variables X_1 and X_2.

(b) Test to see whether using X_1 alone significantly helps in predicting survival time.

(c) Test to see whether using X_2 alone significantly aids in predicting survival time.

(d) Test whether the addition of X_2 significantly contributes to the prediction of Y over and above the contribution of X_1.

(e) Which independent variable do you consider to be the best single predictor of survival time?

(f) Which model involving one or both of the independent variables do you prefer and why?

(g) Using the fitted model containing both X_1 and X_2, what $\log_{10}$ dose (X_1) is required to kill a larva for which $X_2 = 0.200$ in the same time that a larva for which $X_2 = 0.400$ and $X_1 = 0.500$ is killed? [*Hint:* For the heavier larva, the estimated survival time $\hat{Y}_H$ is given by $\hat{Y}_H = \hat{\beta}_0 + \hat{\beta}_1(0.500) + \hat{\beta}_2(0.400)$, where $\hat{\beta}_0, \hat{\beta}_1$, and $\hat{\beta}_2$ are the least-squares coefficients for the model containing both X_1 and X_2. The estimated survival time for the lighter larva weighing $X_2 = 0.200$ at a $\log_{10}$ dose of X_1 is $\hat{Y}_L = \hat{\beta}_0 + \hat{\beta}_1 X_1 + \hat{\beta}_2(0.200)$. The problem is to find that value of X_1 which makes $\hat{Y}_L = \hat{Y}_H$.]

(h) Based on the hint in part (g), give a general formula for the $\log_{10}$ dose X_1 necessary to kill a larva weighing (in $\log_{10}$ units) X_2^0 in the same time as a larva weighing twice this $\log_{10}$ weight and subject to a $\log_{10}$ dose equal to X_1^0; that is, what is X_1 in terms of X_2^0 and X_1^0?

8. An experiment to evaluate the effects of certain variables on soil erosion was performed on 10-foot-square plots of sloped farmland subjected to 2 inches of artificial rain applied for a 20-minute period.[7] The data and related computer results are as given.

(a) The fitted model involving all three independent variables is given by $\hat{Y} = -1.879 + 77.326X_1 + 1.559X_2 - 23.904X_3$. Compute and compare observed and predicted values of Y for plots 1, 5, and 7.

(b) Test for significant overall regression of the model containing all three independent variables.

(c) Using the stepwise-regression results, test for the significance of each variable as it enters the model.

[7] Adapted from a study by Packer (1951).

PLOT NO.	1	2	3	4	5	6	7	8	9	10	11
SL (Y)	27.1	35.6	31.4	37.8	40.2	39.8	55.5	43.6	52.1	43.8	35.7
SG (X_1)	0.43	0.47	0.44	0.48	0.48	0.49	0.53	0.50	0.55	0.51	0.48
LOBS (X_2)	1.95	5.13	3.98	6.25	7.12	6.50	10.67	7.08	9.88	8.72	4.96
PGC (X_3)	0.34	0.32	0.29	0.30	0.25	0.26	0.10	0.16	0.19	0.18	0.28

SL denotes soil lost (in pounds/acre), SG denotes slope gradient of plot, LOBS denotes length (in inches) of the largest opening of bare soil on any boundary, and PGC denotes percent of ground cover.

Computer results for Problem 8

STEPWISE (i.e., in order of "relative importance")			FITTING X_1 FIRST, THEN LETTING X_2 AND X_3 ENTER STEPWISE		
SOURCE	df	SS	SOURCE	df	SS
X_2	1	667.7280	X_1	1	640.4250
$X_3\|X_2$	1	5.8228	$X_2\|X_1$	1	32.7819
$X_1\|X_3, X_2$	1	6.9405	$X_3\|X_1, X_2$	1	7.2844
Residual	7	16.0943	Residual	7	16.0943

(d) Which of the variables X_2 and X_3 seems to be a better predictor of Y given that X_1 is already in the model? Explain.

(e) Is the contribution of either X_2 or X_3 or both significant after taking into account the contribution of X_1 to the prediction of Y? Explain.

(f) Why might the investigator be interested in forcing X_1 into the model first rather than using the stepwise results for all three independent variables?

(g) What model for predicting soil erosion appears best based on a decision to force X_1 into the model first? Based on the stepwise results?

9. In a study by Yoshida (1961), the oxygen consumption of wireworm larvae groups was measured at five temperatures. The dependent variable giving the rate of oxygen consumption per larvae group in milliliters per hour was transformed to 0.5 less than the common log. Another independent variable (other than temperature) of importance was larvae group weight, which was also transformed to common logarithms. The data are given in the table as are relevant computer results.

(a) The fitted multiple regression model containing both X_1 and X_2 is given by

$$\hat{Y} = -0.6835 + 0.5917X_1 + 0.0393X_2$$

Using this fitted model, how much of a change in oxygen consumption would be predicted for a larvae group with fixed weight X_1 if the temperature was increased from $X_2 = 20$ to $X_2 = 25$?

(b) For a temperature of 20°C, compute and compare the predicted values of $\hat{Y}$ for weights of 0.250 and 0.500.

(c) Is temperature a better predictor of Y than weight is, or vice versa? Explain.

(d) What model involving either or both X_1 and X_2 is to be preferred? Explain.

(e) What is the change in R^2 in going from a model involving only X_2 to a model involving both X_1 and X_2?

OXYGEN CONSUMPTION (Y) (log ml/hr -0.5)	LARVAE GROUP WEIGHT (X_1) (log cg)	TEMPERATURE (X_2) (°C)
0.054	0.130	15.5
0.154	0.215	15.5
0.073	0.250	15.5
0.182	0.267	15.5
0.241	0.389	15.5
0.316	0.490	15.5
0.290	0.491	15.5
0.061	0.004	20.0
0.143	0.164	20.0
0.188	0.225	20.0
0.176	0.314	20.0
0.248	0.447	20.0
0.357	0.477	20.0
0.403	0.505	20.0
0.342	0.537	20.0
0.335	−0.046	25.0
0.408	0.176	25.0
0.366	0.199	25.0
0.482	0.292	25.0
0.545	0.380	25.0
0.596	0.483	25.0
0.590	0.491	25.0
0.631	0.491	25.0
0.610	0.519	25.0
0.482	0.053	30.0
0.477	0.114	30.0
0.551	0.137	30.0
0.516	0.190	30.0
0.561	0.210	30.0
0.588	0.230	30.0
0.561	0.240	30.0
0.580	0.260	30.0
0.674	0.389	30.0
0.718	0.470	30.0
0.754	0.521	30.0
0.800	0.544	30.0
0.654	−0.004	35.0
0.744	0.033	35.0
0.711	0.049	35.0
0.855	0.140	35.0
0.932	0.204	35.0
0.927	0.210	35.0
0.914	0.215	35.0
0.914	0.265	35.0
0.973	0.346	35.0
1.000	0.462	35.0
0.998	0.468	35.0

Computer results for Problem 9

Y ON X_1			Y ON X_2		
SOURCE	df	SS	SOURCE	df	SS
Regression	1	0.0662	Regression	1	2.7843
Residual	45	3.3398	Residual	45	0.6323

Y ON X_1 AND X_2		
SOURCE	df	SS
Regression	2	3.2104
Residual	44	0.1956

CHAPTER 11

CORRELATIONS: MULTIPLE, PARTIAL, AND MULTIPLE–PARTIAL

11.1 PREVIEW

We saw in Chapter 5 that the essential features of straight-line regression, apart from the quantitative prediction formula provided by the fitted regression equation, can be alternatively described in terms of the correlation coefficient r. These features are summarized as follows:

1. The squared correlation coefficient r^2 measures the strength of the linear relationship between the dependent variable Y and the independent variable X. The closer r^2 is to 1, the stronger the linear relationship; the closer r^2 is to 0, the weaker the linear relationship.

2. $r^2 = (\text{SSY} - \text{SSE})/\text{SSY}$ is the proportionate reduction in the total sum of squares achieved by using a straight-line model in X to predict Y.

3. $r = \hat{\beta}_1(S_X/S_Y)$, where $\hat{\beta}_1$ is the estimated slope of the regression line.

4. r (or r_{XY}) is an estimate of a population parameter ρ (or ρ_{XY}), which describes the correlation between X and Y, both considered as random variables.

5. Assuming that X and Y have a bivariate normal distribution with parameters μ_X, μ_Y, σ_X^2, σ_Y^2, and ρ_{XY}, the conditional distribution of Y given X is $N(\mu_{Y|X}, \sigma_{Y|X}^2)$, where

$$\mu_{Y|X} = \mu_Y + \rho\frac{\sigma_Y}{\sigma_X}(X - \mu_X) \quad \text{and} \quad \sigma_{Y|X}^2 = \sigma_Y^2(1 - \rho^2)$$

Here r^2 estimates ρ^2, which can be expressed as

$$\rho^2 = \frac{\sigma_Y^2 - \sigma_{Y|X}^2}{\sigma_Y^2}$$

6. r can be used as a general index of linear association between two random variables in the following sense:
   (a) The more highly positive r is, the more "positive" is the linear association; that is, an individual with a high value of one variable will likely have a high

value of the other, and an individual with a low value of one variable will probably have a low value of the other.

(b) The more highly negative r is, the more "negative" is the linear association; that is, an individual with a high value of one variable will likely have a low value of the other.

(c) If r is close to zero, there is little evidence of linear association, which indicates either that there is nonlinear association or no association at all.

This connection between regression and correlation can also be extended to the multiple regression case, as we will discuss in this chapter. However, when several independent variables are involved, the essential features of regression are alternatively described, not by a single correlation coefficient as in the straight-line case, but by several correlations. These include a set of (zero-order)[1] correlations such as r plus a whole group of additional (higher-order) indices called multiple correlations, partial correlations, and multiple–partial correlations. Examination of these higher-order correlations allows us to answer many of the same questions that can be answered by fitting a multiple regression model. In addition, the correlation analog has been found particularly useful in uncovering spurious relationships among variables, identifying intervening variables, and in making certain types of causal inferences.[2]

11.2 CORRELATION MATRIX

When dealing with more than one independent variable, the collection of all zero-order correlation coefficients (i.e., the r's between all possible pairs of variables) can be represented most compactly in *correlation-matrix form*. For example, when there are $k = 3$ independent variables X_1, X_2, and X_3, and one dependent variable Y, there are $C_2^4 = 6$ zero-order correlations, and the correlation matrix has the general form

$$
\begin{array}{c c}
 & \begin{array}{cccc} Y & X_1 & X_2 & X_3 \end{array} \\
\begin{array}{c} Y \\ X_1 \\ X_2 \\ X_3 \end{array} &
\left[\begin{array}{cccc}
1 & r_{Y1} & r_{Y2} & r_{Y3} \\
 & 1 & r_{12} & r_{13} \\
 & & 1 & r_{23} \\
 & & & 1
\end{array} \right]
\end{array}
$$

Here r_{Yj} ($j = 1, 2,$ or 3) is the correlation between Y and X_j, and r_{ij} ($i, j = 1, 2, 3$) is the correlation between X_i and X_j.

For the data involving weight, height, and age given in Table 10.1, this matrix takes the form

$$
\begin{array}{c c}
 & \begin{array}{cccc} \text{WGT} & \text{HGT} & \text{AGE} & \text{AGE}^2 \end{array} \\
\begin{array}{c} \text{WGT} \\ \text{HGT} \\ \text{AGE} \\ \text{AGE}^2 \end{array} &
\left[\begin{array}{cccc}
1 & 0.814 & 0.770 & 0.767 \\
 & 1 & 0.614 & 0.615 \\
 & & 1 & 0.994 \\
 & & & 1
\end{array} \right]
\end{array}
$$

---

[1] The "order" of a correlation coefficient, as referred to here, is the number of variables being controlled or adjusted for (see Section 11.5).

[2] See Blalock (1971) for a discussion of causal-inference techniques.

Each of these correlations taken separately describes the strength of the linear relationship between the two variables involved. In particular, the correlations $r_{Y1} = 0.814$, $r_{Y2} = 0.770$, and $r_{Y3} = 0.767$ measure the strength of the linear association with the dependent variable WGT for each of the independent variables taken separately. As we can see, HGT ($r_{Y1} = 0.814$) is the independent variable with the strongest linear relationship to WGT, next comes AGE, and last comes AGE^2.

Nevertheless, use of all these zero-order correlations describes neither (1) the overall relationship of the dependent variable WGT to the independent variables HGT, AGE, and AGE^2 considered together, nor (2) the relationship between WGT and AGE after controlling[3] for HGT, nor (3) the relationship between WGT and the combined effects of AGE and AGE^2 after controlling for HGT. The measure that describes (1) is called the multiple correlation coefficient of WGT on HGT, AGE, and AGE^2. The measure that describes (2) is called the partial correlation coefficient between WGT and AGE controlling for HGT. And the measure that describes (3) is called the multiple–partial correlation coefficient between WGT and the combined effects of AGE and AGE^2 controlling for HGT.

It should be pointed out here that even though AGE and AGE^2 are very highly correlated in our example, it is possible, if the general relationship of AGE to WGT is nonlinear, that AGE^2 will be significantly correlated with WGT even after AGE has been controlled for. In fact, this is what happens in general when the addition of a second-order term in polynomial regression significantly improves the prediction of the dependent variable.

11.3 MULTIPLE CORRELATION COEFFICIENT

The *multiple correlation coefficient*, denoted as $R_{Y|X_1,X_2,...,X_k}$, is a measure of the overall *linear association* of one (dependent) variable Y with several other (independent) variables $X_1, X_2, \ldots, X_k$.

[By linear association we mean that $R_{Y|X_1,X_2,...,X_k}$ measures the strength of the association between Y and the best-fitting linear combination of the X's, which is the least-squares solution $\hat{Y} = \hat{\beta}_0 + \hat{\beta}_1 X_1 + \hat{\beta}_2 X_2 + \cdots + \hat{\beta}_k X_k$. In fact, no other linear combination of the X's will have as great a correlation with Y. It can also be shown that $R_{Y|X_1,X_2,...,X_k}$ is always nonnegative.]

Thus, the multiple correlation coefficient is a direct generalization of the simple correlation coefficient r to the case of several independent variables. We have actually dealt with this measure up to now under the name R^2, which is the square of the multiple correlation coefficient.

There are two computational formulas which provide useful interpretations of the multiple correlation coefficient $R_{Y|X_1,X_2,...,X_k}$ and its square. These are

$$(1) \quad R_{Y|X_1,X_2,...,X_k} = \frac{\sum_{i=1}^{n} (Y_i - \bar{Y})(\hat{Y}_i - \bar{\hat{Y}})}{\left[\sum_{i=1}^{n} (Y_i - \bar{Y})^2 \sum_{i=1}^{n} (\hat{Y}_i - \bar{\hat{Y}})^2 \right]^{1/2}}$$

---

[3] In this case the term "controlling for" pertains to determining to what extent the variables WGT and AGE are related after removing the effect of HGT on the variables WGT and AGE.

and

$$(2) \quad R^2_{Y|X_1,X_2,\ldots,X_k} = \frac{\sum_{i=1}^{n} (Y_i - \bar{Y})^2 - \sum_{i=1}^{n} (Y_i - \hat{Y}_i)^2}{\sum_{i=1}^{n} (Y_i - \bar{Y})^2} = \frac{\text{SSY} - \text{SSE}}{\text{SSY}}$$

where $\hat{Y}_i = \hat{\beta}_0 + \hat{\beta}_1 X_{1i} + \cdots + \hat{\beta}_k X_{ki}$ (the predicted value for the ith individual) and $\bar{Y} = \sum_{i=1}^{n} \hat{Y}_i/n$. Formula (2), which we have seen several times before, is most useful for assessing the fit of the regression model. Also, it can be seen from formula (1) that $R_{Y|X_1,X_2,\ldots,X_k} = r_{Y,\hat{Y}}$, the *simple linear correlation between the observed values Y and the predicted values* $\hat{Y}$.

We shall illustrate the computation of $R_{Y|X_1,X_2,\ldots,X_k}$ by applying formula (1) to the data of Table 10.1, which considers the variables $X_1 = $ HGT, $X_2 = $ AGE, and $Y = $ WGT. Using only X_1 and X_2 in the model, the fitted regression equation is $\hat{Y} = 6.553 + 0.722 X_1 + 2.050 X_2$, and the observed and predicted values are given in Table 11.1.

One can check that $\bar{Y} = \bar{\hat{Y}} = 62.75$. And, as we mentioned in Chapter 10, this is no coincidence, since it is a mathematical fact that $\bar{Y} = \bar{\hat{Y}}$. With this result in mind, the best computational formula for $R_{Y|X_1,X_2,\ldots,X_k}$ takes the form

$$R_{Y|X_1,X_2,\ldots,X_k} = \frac{\sum_{i=1}^{n} Y_i \hat{Y}_i - n\bar{Y}^2}{\sqrt{\left(\sum_{i=1}^{n} Y_i^2 - n\bar{Y}^2\right)\left(\sum_{i=1}^{n} \hat{Y}_i^2 - n\bar{Y}^2\right)}} \tag{11.1}$$

so that, for example,

$$R_{\text{WGT}|\text{HGT,AGE}} = \frac{47{,}943.60 - 12(62.75)^2}{\sqrt{[48{,}139 - 12(62.75)^2][47{,}943.544 - 12(62.75)^2]}} = 0.8832$$

Two other multiple correlations calculated for this data set (which correspond to different regression models) are

$$R_{\text{WGT}|\text{HGT,AGE,AGE}^2} = 0.8833$$

and

$$R_{\text{WGT}|\text{HGT,AGE}^2} = 0.8811$$

TABLE 11.1 *Observed and predicted values for the regression of WGT on HGT and AGE*

CHILD No.:	1	2	3	4	5	6	7	8	9	10	11	12
(Observed) Y	64	71	53	67	55	58	77	57	56	51	76	68
(Predicted) $\hat{Y}$	64.11	69.65	54.23	73.87	59.78	57.01	66.77	59.66	57.38	49.18	75.20	66.16

From the three values above, we support our earlier finding that the use of HGT and AGE or HGT and AGE^2 does as well in predicting WGT as does the use of all three independent variables.

11.4 RELATIONSHIP OF $R_{Y|X_1, X_2, \ldots, X_k}$ TO THE MULTIVARIATE NORMAL DISTRIBUTION[4]

An informative way of looking at the sample multiple correlation coefficient $R_{Y|X_1, X_2, \ldots, X_k}$ is to consider it as an estimator of a population parameter characterizing the joint distribution of all the variables $Y, X_1, X_2, \ldots, X_k$ taken together. When we had two variables X and Y and we had assumed that their joint distribution was bivariate normal $N_2(\mu_Y, \mu_X, \sigma_Y^2, \sigma_X^2, \rho_{XY})$, we saw that r_{XY} estimated ρ_{XY}, which satisfied the formula $\rho_{XY}^2 = (\sigma_Y^2 - \sigma_{Y|X}^2)/\sigma_Y^2$, where $\sigma_{Y|X}^2$ was the variance of the conditional distribution of Y given X. Now, when we have k independent variables and one dependent variable, we get an analogous result if we assume that their joint distribution is *multivariate normal*. Let us now consider what happens with just *two* independent variables. Then, the *trivariate normal distribution* of Y, X_1, and X_2 can be described as

$$N_3(\mu_Y, \mu_{X_1}, \mu_{X_2}, \sigma_Y^2, \sigma_{X_1}^2, \sigma_{X_2}^2, \rho_{Y1}, \rho_{Y2}, \rho_{12})$$

where μ_Y, μ_{X_1}, and μ_{X_2} are the three (unconditional) means, σ_Y^2, $\sigma_{X_1}^2$, and $\sigma_{X_2}^2$ are the three (unconditional) variances, and ρ_{Y1}, ρ_{Y2}, and ρ_{12} are the three correlation coefficients. The *conditional distribution of Y given X_1 and X_2* is then a univariate normal distribution with a (conditional) mean denoted by $\mu_{Y|X_1, X_2}$ and a (conditional) variance denoted by $\sigma_{Y|X_1, X_2}^2$; we usually write this compactly as

$$Y|X_1, X_2 \frown N(\mu_{Y|X_1, X_2}, \sigma_{Y|X_1, X_2}^2)$$

In fact, it turns out that

$$\mu_{Y|X_1, X_2} = \mu_Y + \rho_{Y1} \frac{\sigma_{Y|X_2}}{\sigma_{X_1|X_2}}(X_1 - \mu_{X_1}) + \rho_{Y2} \frac{\sigma_{Y|X_1}}{\sigma_{X_2|X_1}}(X_2 - \mu_{X_2})$$

and

$$\sigma_{Y|X_1, X_2}^2 = (1 - \rho_{Y, \mu_{Y|X_1, X_2}}^2)\sigma_Y^2$$

where $\rho_{Y, \mu_{Y|X_1, X_2}}$ is the population correlation coefficient between the random variables Y and $\mu_{Y|X_1, X_2} = \beta_0 + \beta_1 X_1 + \beta_2 X_2$ (where we are considering X_1 and X_2 as random variables). Also, $\sigma_{Y|X_2}^2$, $\sigma_{Y|X_1}^2$, $\sigma_{X_1|X_2}^2$, and $\sigma_{X_2|X_1}^2$ are the conditional variances, respectively, of Y given X_2, Y given X_1, X_1 given X_2, and X_2 given X_1.

It is this parameter $\rho_{Y, \mu_{Y|X_1, X_2}}$ which is the *population analog of the sample multiple correlation coefficient* $R_{Y|X_1, X_2}$; and we write $\rho_{Y, \mu_{Y|X_1, X_2}}$ simply as $\rho_{Y|X_1, X_2}$. Furthermore, from the formula for $\sigma_{Y|X_1, X_2}^2$, it can be seen (with a little algebra) that

$$\rho_{Y|X_1, X_2}^2 \quad (\text{or} \quad \rho_{Y, \mu_{Y|X_1, X_2}}^2) = \frac{\sigma_Y^2 - \sigma_{Y|X_1, X_2}^2}{\sigma_Y^2}$$

[4] This section is not essential for the application-oriented reader.

which is the *proportionate reduction in the unconditional variance of Y due to conditioning on X_1 and X_2.*

Generalizing these findings to the case of k independent variables, we may summarize the characteristics of the multiple correlation coefficient $R_{Y|X_1, X_2,...,X_k}$ as follows:

1. $R^2_{Y|X_1, X_2,...,X_k}$ measures the proportionate reduction in the total sum of squares $\sum (Y_i - \bar{Y})^2$ to $\sum (Y_i - \hat{Y}_i)^2$ due to the multiple linear regression of Y on $X_1, X_2, \ldots, X_k$.

2. $R_{Y|X_1, X_2,...,X_k}$ is the correlation $r_{Y,\hat{Y}}$ of the observed values (Y) with the predicted values $(\hat{Y})$.

3. $R_{Y|X_1, X_2,...,X_k}$ is an estimate of $\rho_{Y|X_1, X_2,...,X_2}$, which is the correlation of Y with the mean of the conditional distribution of Y given $X_1, X_2, \ldots, X_k$ (i.e., the correlation of Y with the true regression equation $\beta_0 + \beta_1 X_1 + \beta_2 X_2 + \cdots + \beta_k X_k$, where the X's are considered to be random).

4. $R^2_{Y|X_1, X_2,...,X_k}$ is an estimate of the proportionate reduction in the unconditional variance of Y due to conditioning on $X_1, X_2, \ldots, X_k$; that is, it estimates

$$\rho^2_{Y|X_1, X_2,...,X_k} = \frac{\sigma^2_Y - \sigma^2_{Y|X_1,...,X_k}}{\sigma^2_Y}$$

11.5 PARTIAL CORRELATION COEFFICIENT

The *partial correlation coefficient* is a measure of the strength of the linear relationship between two variables after controlling for the effects of other variables. If the two variables of interest are Y and X, and the control variables are $Z_1, Z_2, \ldots, Z_p$, then we denote the corresponding partial correlation coefficient by $r_{YX|Z_1, Z_2,...,Z_p}$. The order of the partial correlation depends on the number of variables that are being controlled for. Thus,

first-order partials have the form $r_{YX|Z}$

second-order partials have the form $r_{YX|Z_1, Z_2}$

and, in general,

pth-*order partials* have the form $r_{YX|Z_1, Z_2,...,Z_p}$

For the three independent variables HGT, AGE, and AGE^2 using the data of Table 10.1, the highest-order partial possible is second-order. The values of most of the partial correlations that can possibly be computed from this data set are given in Table 11.2.

The easiest way to obtain a partial correlation is to use a standard computer program. A computing formula for use with a desk calculator, which helps to highlight the structure of the partial correlation coefficient, will be given a little later. First, however, let us see how we can use the information in Table 11.2 to describe our data.

If we look back at our (zero-order) correlation matrix, we see that the variable most highly correlated with WGT is HGT ($r_{Y1} = 0.814$). Thus, of the three independent variables we are considering, HGT is the "most important" variable based on the strength of its linear relationship with WGT.

TABLE 11.2   *Partial correlations for the WGT, HGT, and AGE data of Table 10.1*

ORDER	CONTROLLING VARIABLES	FORM OF CORRELATION	COMPUTED VALUE
1	HGT	$r_{\text{WGT,AGE}|\text{HGT}}$	0.589
1	HGT	$r_{\text{WGT,AGE}^2|\text{HGT}}$	0.580
1	HGT	$r_{\text{AGE,AGE}^2|\text{HGT}}$	0.988
1	AGE	$r_{\text{WGT,HGT}|\text{AGE}}$	0.678
1	AGE	$r_{\text{WGT,AGE}^2|\text{AGE}}$	0.015
1	AGE	$r_{\text{HGT,AGE}^2|\text{AGE}}$	0.060
1	AGE^2	$r_{\text{WGT,HGT}|\text{AGE}^2}$	0.677
1	AGE^2	$r_{\text{WGT,AGE}|\text{AGE}^2}$	0.111
1	AGE^2	$r_{\text{HGT,AGE}|\text{AGE}^2}$	0.022
2	HGT, AGE	$r_{\text{WGT,AGE}^2|\text{HGT,AGE}}$	−0.015
2	HGT, AGE^2	$r_{\text{WGT,AGE}|\text{HGT,AGE}^2}$	0.131
2	AGE, AGE^2	$r_{\text{WGT,HGT}|\text{AGE,AGE}^2}$	0.679

We might now ask the following question: After HGT, what is the "next most important" variable in terms of contributing to the linear preduction of WGT? Since the first-order partial $r_{\text{WGT,AGE}|\text{HGT}} = 0.589$ is larger than $r_{\text{WGT,AGE}^2|\text{HGT}} = 0.580$, it makes sense to conclude that AGE is next in importance, once we have accounted for HGT. (If we want to test the significance of this partial correlation coefficient, we would use a partial F test as described in Chapter 10. We shall return to this point shortly.)

The only variable left to consider is AGE^2. We might now ask: Once we have accounted for HGT and AGE, does AGE^2 add anything to our knowledge of WGT? To answer this, we can look at the second-order partial correlation $r_{\text{WGT,AGE}^2|\text{HGT,AGE}} = -0.015$. Notice that the magnitude of this partial correlation is very small. Thus, we would be inclined to conclude that AGE^2 provides essentially no additional information about WGT once HGT and AGE have been used together as predictors.

The kind of variable-selection procedure just described, which starts with the most important variable and continues step by step to add variables in order of importance while controlling for variables already selected, is called a *forward selection procedure*. Alternatively, we could have handled the variable selection problem by working backward. By this, we mean that we start with all the variables and delete (step by step) unimportant variables which do not contribute much to the description of the dependent variable. We shall discuss variable-selection procedures further in Chapter 15.

11.5.1 TESTS OF SIGNIFICANCE FOR PARTIAL CORRELATIONS

Regardless of the procedure chosen for variable selection, it will be necessary to decide at each step whether a particular partial correlation coefficient under consideration is significantly different from zero or not. Actually, we have already described how to test for such significance in a slightly different context when we were considering the various uses of the ANOVA table in regression analysis. When we wanted to test whether the addition of a variable to the regression model was worthwhile, given that certain other variables were already in the model, we used a partial F test. It can be shown that this partial F test is exactly equivalent to a test of significance for the corresponding partial correlation coefficient. Thus, to test whether $r_{YX|z_1,\ldots,z_p}$ is significantly different from zero, we compute the corresponding partial $F(X|Z_1, \ldots, Z_p)$ and reject the null hypothesis if this F statistic exceeds an appropriate critical value of the $F_{1,n-p-2}$ distribution. For example, in testing whether $r_{WGT,AGE^2|HGT,AGE}$ is significant or not, we find that the partial $F(AGE^2|HGT, AGE) = 0.010$ does not exceed $F_{1,12-2-2,0.90} = F_{1,8,0.90} = 3.46$. Therefore, we conclude that this partial correlation is not significantly different from zero, so that AGE^2 does not contribute to the prediction of WGT once we have accounted for HGT and AGE.

The null hypothesis for this test can be stated more formally by considering the population analog of the sample partial correlation coefficient $r_{YX|Z_1,\ldots,Z_p}$. This corresponding population parameter is usually denoted by $\rho_{YX|Z_1,\ldots,Z_p}$ and is called the *population partial correlation coefficient*. The null hypothesis can then be stated as $H_0: \rho_{YX|Z_1,\ldots,Z_p} = 0$ and the associated alternative hypothesis as $H_A: \rho_{YX|Z_1,\ldots,Z_p} \neq 0$.

11.5.2 RELATING THE TEST FOR PARTIAL CORRELATION TO THE PARTIAL F TEST

Consideration of the structure of the population partial correlation helps us to relate this form of higher-order correlation to regression. Let us, for simplicity, consider this relationship for the special case of two independent variables. The formula for the square of $\rho_{YX_1|X_2}$ can be written

$$\rho^2_{YX_1|X_2} = \frac{\sigma^2_{Y|X_2} - \sigma^2_{Y|X_1,X_2}}{\sigma^2_{Y|X_2}}$$

Thus, the square of the sample partial correlation $r_{YX_1|X_2}$ is an estimate of the proportionate reduction in the conditional variance of Y given X_2 due to conditioning on both X_1 and X_2.

[The partial correlation $\rho_{YX_1|X_2}$ can also be described as a zero-order correlation for a conditional bivariate distribution. If the joint distribution of Y, X_1, and X_2 is trivariate normal, the conditional joint distribution of Y and X_1 given X_2 is bivariate normal. The zero-order correlation between Y and X_1 for this conditional distribution is what we call $\rho_{YX_1|X_2}$; this is exactly the partial correlation between X_1 and Y controlling for X_2.]

It then follows that an analogous formula for the squared sample partial correlation coefficient is

$$r^2_{YX_1|X_2} = \frac{\text{SS residual (using only } X_2 \text{ in the model)} - \text{SS residual (using } X_1 \text{ and } X_2 \text{ in the model)}}{\text{SS residual (using only } X_2 \text{ in the model)}}$$

$$= \frac{\text{extra sum of squares due to adding } X_1 \text{ to the model given } X_2 \text{ is already in the model}}{\text{SS residual (using only } X_2 \text{ in the model)}} \tag{11.2}$$

It should be clear from the structure of (11.2) and from the discussion on partial F statistics given in Chapter 10 why the test of $H_0: \rho_{YX_1|X_2} = 0$ is performed using $F(X_1|X_2)$ as the test statistic.

11.5.3 ANOTHER WAY OF DESCRIBING PARTIAL CORRELATIONS

Another way to compute a first-order partial correlation is to use the formula

$$r_{YX|Z} = \frac{r_{YX} - r_{YZ}r_{XZ}}{\sqrt{(1 - r_{YZ}^2)(1 - r_{XZ}^2)}} \tag{11.3}$$

For example, to compute $r_{WGT,AGE|HGT}$, we calculate

$$\frac{r_{WGT,AGE} - r_{WGT,HGT}r_{AGE,HGT}}{\sqrt{(1 - r_{WGT,HGT}^2)(1 - r_{AGE,HGT}^2)}} = \frac{0.770 - 0.814(0.614)}{\sqrt{[1 - (0.814)^2][1 - (0.614)^2]}}$$

$$= \frac{0.770 - 0.500}{\sqrt{0.338(0.623)}} = 0.589$$

Notice that the first correlation in the numerator is the simple zero-order correlation between WGT and AGE. The *control variable* HGT appears in the second expression in the numerator (where it is correlated separately with each of the variables WGT and AGE) and in both terms in the denominator. *It is by use of (11.3) that we can interpret the partial correlation coefficient as an adjustment of the simple correlation coefficient to take into account the effect of the control variable.* In particular, if r_{YZ} and r_{XZ} have the same sign, then controlling for *Z* reduces (i.e., makes less positive or more negative, as the case may be) the zero-order correlation r_{YX} between *Y* and *X*. On the other hand, if r_{YZ} and r_{XZ} have opposite signs, controlling for *Z increases* r_{YX}.

Now, to compute higher-order partial correlations, we simply reapply this formula using the appropriate next-lower-order partials. For example, the second-order partial correlation is an adjustment of the first-order partial, which, in turn, is an adjustment of the simple zero-order correlation. In particular, we have the following general formula for a second-order partial correlation:

$$r_{YX|Z,W} = \frac{r_{YX|Z} - r_{YW|Z}r_{XW|Z}}{\sqrt{(1 - r_{YW|Z}^2)(1 - r_{XW|Z}^2)}} = \frac{r_{YX|W} - r_{YZ|W}r_{XZ|W}}{\sqrt{(1 - r_{YZ|W}^2)(1 - r_{XZ|W}^2)}} \tag{11.4}$$

To compute $r_{WGT,AGE^2|HGT,AGE}$, for example, we have

$$\frac{r_{WGT,AGE^2|HGT} - r_{WGT,AGE|HGT}r_{AGE^2,AGE|HGT}}{\sqrt{(1 - r_{WGT,AGE|HGT}^2)(1 - r_{AGE^2,AGE|HGT}^2)}} = \frac{0.580 - (0.589)(0.988)}{\sqrt{[1 - (0.589)^2][1 - (0.988)^2]}} = -0.015$$

11.5.4 PARTIAL CORRELATION AS A CORRELATION OF RESIDUALS OF REGRESSION

There is still one additional important interpretation that can be made concerning partial correlations. For the variables *Y*, *X*, and *Z*, suppose that we fit the two straight-line regression equations $Y = \beta_0 + \beta_1 Z + E$ and $X = \beta_0^* + \beta_1^* Z + E$. Let $\hat{Y} = \hat{\beta}_0 + \hat{\beta}_1 Z$ be the fitted line of *Y* on *Z*, and let $\hat{X} = \hat{\beta}_0^* + \hat{\beta}_1^* Z$ be the fitted line of *X* on *Z*. Then, the deviations or *residuals* $(\hat{Y} - Y)$ and $(\hat{X} - X)$ represent what remains after the variable *Z* has explained all the variation it can in the variables *Y* and *X* separately.

If we now correlate these residuals (i.e., find $r_{\hat{Y}-Y,\hat{X}-X}$), we obtain a measure that is "independent" of the effects of Z. It can be shown that *the partial correlation between Y and X controlling for Z can be defined as the correlation of the residuals of the straight-line regressions of Y on Z and X on Z*; that is, $r_{YX|Z} = r_{\hat{Y}-Y,\hat{X}-X}$.

11.5.5 SUMMARY OF THE FEATURES OF THE PARTIAL CORRELATION COEFFICIENT

1. The partial correlation $r_{YX|Z_1,Z_2,\ldots,Z_p}$ measures the strength of the linear relationship between two variables X and Y controlling for other variables $Z_1, Z_2, \ldots, Z_p$.

2. The square of the partial correlation $r_{YX|Z_1,\ldots,Z_p}$ measures the proportion of the residual sum of squares that is accounted for by the addition of X to a regression model already involving $Z_1, Z_2, \ldots, Z_p$; that is,

$$r^2_{YX|Z_1,\ldots,Z_p} = \frac{\text{extra sum of squares due to adding } X \text{ to a model already containing } Z_1, Z_2, \ldots, Z_p}{\text{SS residual (using only } Z_1, Z_2, \ldots, Z_p \text{ in the model)}}$$

3. The partial correlation coefficient $r_{YX|Z_1,\ldots,Z_p}$ is an estimate of the population parameter $\rho_{YX|Z_1,\ldots,Z_p}$, which is the correlation between Y and X in the conditional joint distribution of Y and X given $Z_1, \ldots, Z_p$. Also, the square of this population partial correlation coefficient is given by the equivalent formula

$$\rho^2_{YX|Z_1,\ldots,Z_p} = \frac{\sigma^2_{Y|Z_1,\ldots,Z_p} - \sigma^2_{Y|X,Z_1,\ldots,Z_p}}{\sigma^2_{Y|Z_1,\ldots,Z_p}}$$

where $\sigma^2_{Y|Z_1,\ldots,Z_p}$ is the variance of the conditional distribution of Y given $Z_1, \ldots, Z_p$ (and where $\sigma^2_{Y|X,Z_1,\ldots,Z_p}$ is similarly defined).

4. The partial F statistic $F(X|Z_1, Z_2, \ldots, Z_p)$ is used to test $H_0: \rho_{YX|Z_1,\ldots,Z_p} = 0$.

5. The (first-order) partial correlation coefficient $r_{YX|Z}$ is an adjustment of the (zero-order) correlation r_{YX} which takes into account the effect of the control variable Z. This is seen from the formula

$$r_{YX|Z} = \frac{r_{YX} - r_{YZ}r_{XZ}}{\sqrt{(1-r^2_{YZ})(1-r^2_{XZ})}}$$

Higher-order partial correlations are computed by reapplying this formula using the next-lower-order partials.

6. The partial correlation $r_{YX|Z}$ can be defined as the correlation of the residuals of the straight-line regressions of Y on Z and of X on Z; that is, $r_{YX|Z} = r_{\hat{Y}-Y,\hat{X}-X}$.

11.6 ALTERNATIVE REPRESENTATION OF THE REGRESSION MODEL

The correlation analog to multiple regression allows us to express the regression model $\mu_{Y|X_1,X_2,\ldots,X_k} = \beta_0 + \beta_1 X_1 + \beta_2 X_2 + \cdots + \beta_k X_k$ in terms of partial correlation coefficients and conditional variances. When $k = 3$, this representation takes the form

$$\mu_{Y|X_1, X_2, X_3} = \mu_Y + \rho_{Y1|23}\frac{\sigma_{Y|23}}{\sigma_{1|23}}(X_1 - \mu_{X_1}) + \rho_{Y2|13}\frac{\sigma_{Y|13}}{\sigma_{2|13}}(X_2 - \mu_{X_2}) + \rho_{Y3|12}\frac{\sigma_{Y|12}}{\sigma_{3|12}}(X_3 - \mu_{X_3})$$

$$(11.5)$$

where, for example, $\rho_{Y1|23} = \rho_{YX_1|X_2, X_3}$, and where we define

$$\beta_1 = \rho_{Y1|23}\frac{\sigma_{Y|23}}{\sigma_{1|23}}, \quad \beta_2 = \rho_{Y2|13}\frac{\sigma_{Y|13}}{\sigma_{2|13}}, \quad \beta_3 = \rho_{Y3|12}\frac{\sigma_{Y|12}}{\sigma_{3|12}}$$

Note the similarity between this representation and that for the straight-line case, where β_1 is equal to $\rho(\sigma_Y/\sigma_X)$. Also, we can see here that

$$\beta_0 = \mu_Y - \beta_1\mu_{X_1} - \beta_2\mu_{X_2} - \beta_3\mu_{X_3}$$

An equivalent method to least squares for estimating the coefficients β_0, β_1, β_2, and β_3 is to substitute for the population parameters in the above formulas the corresponding estimates

$$\hat{\mu}_Y = \bar{Y}, \qquad \hat{\beta}_1 = r_{Y1|23}\frac{S_{Y|23}}{S_{1|23}}, \qquad \hat{\mu}_{X_1} = \bar{X}_1, \qquad \hat{\beta}_2 = r_{Y2|13}\frac{S_{Y|13}}{S_{2|13}}, \qquad \hat{\mu}_{X_2} = \bar{X}_2,$$

$$\hat{\beta}_3 = r_{Y3|12}\frac{S_{Y|12}}{S_{3|12}}, \qquad \hat{\mu}_{X_3} = \bar{X}_3$$

11.7 MULTIPLE–PARTIAL CORRELATION COEFFICIENT AND ITS ASSOCIATED F TEST

The *multiple–partial correlation coefficient* is used to describe the overall relationship between a dependent variable and *two or more* independent variables while controlling for still other variables. For example, suppose, in addition to the independent variables HGT (X_1), AGE (X_2), and AGE^2 (X_2^2), we also consider the variable $X_1^2 = $ HGT^2 and the product term $X_1 X_2 = $ HGT $\times$ AGE. Our complete regression model is then of the form

$$Y = \beta_0 + \beta_1 X_1 + \beta_2 X_2 + \beta_{11} X_1^2 + \beta_{22} X_2^2 + \beta_{12} X_1 X_2 + E$$

We call such a model a *complete second-order model* since it includes all possible variables up through second-order terms. For such a complete model, it is frequently important to ask whether any of the second-order terms is important, or, in other words, whether a first-order model involving only X_1 and X_2 (i.e., $Y = \beta_0 + \beta_1 X_1 + \beta_2 X_2 + E$) is adequate. There are two equivalent ways to represent this question as a hypothesis-testing problem. One way is to test H_0: $\beta_{11} = \beta_{22} = \beta_{12} = 0$ (i.e., all second-order coefficients are zero). The other is to test the hypothesis H_0: $\rho_{Y(X_1^2, X_2^2, X_1 X_2)|X_1, X_2} = 0$, where $\rho_{Y(X_1^2, X_2^2 X_1 X_2)|X_1, X_2}$ is the population multiple–partial correlation of Y with the second-order variables, controlling for the effects of the first-order variables. [In general, we write the multiple partial as $\rho_{Y(X_1, X_2, ..., X_k)|Z_1, Z_2, ..., Z_p}$.] This parameter is estimated by the sample multiple-partial correlation $r_{Y(X_1^2, X_2^2, X_1 X_2)|X_1, X_2}$. This measure describes the overall multiple contribution of adding the second-order terms to the model after the effects of the first-order terms are partialed out or controlled for (hence the term "multiple–partial"). Two equivalent formulas for calculating $r_{Y(X_1^2, X_2^2, X_1 X_2)|X_1, X_2}^2$ are:

$r_{Y(X_1^2, X_2^2, X_1 X_2)|X_1, X_2}^2$

$$= \frac{\begin{array}{c} \text{SS residual (only } X_1 \text{ and } X_2 \text{ in the model)} \\ - \text{SS residual (all first- and second-order terms in the model)} \end{array}}{\text{SS residual (only } X_1 \text{ and } X_2 \text{ in the model)}}$$

$$= \frac{\begin{array}{c} \text{extra sum of squares due to the addition of the} \\ \text{second-order terms } X_1^2, X_2^2, \text{ and } X_1 X_2 \text{ to a model} \\ \text{containing only the first-order terms } X_1 \text{ and } X_2 \end{array}}{\text{SS residual (only } X_1 \text{ and } X_2 \text{ in the model)}} \qquad (11.6)$$

and

$$r_{Y(X_1^2, X_2^2, X_1 X_2)|X_1, X_2}^2 = \frac{R_{Y|X_1, X_2, X_1^2, X_2^2, X_1 X_2}^2 - R_{Y|X_1, X_2}^2}{1 - R_{Y|X_1, X_2}^2}$$

To test either H_0: $\rho_{Y(X_1^2, X_2^2, X_1 X_2)|X_1, X_2} = 0$ or, equivalently, H_0: $\beta_{11} = \beta_{22} = \beta_{12} = 0$, we calculate

$$F = \frac{[\text{SS residual (only } X_1 \text{ and } X_2 \text{ in the model)} - \text{SS residual (all first- and second-order terms in the model)}]/3}{\text{SS residual (all first- and second-order terms in the model)}/(n-6)}$$

and reject H_0 at the α significance level if $F \geq F_{3, n-6, 1-\alpha}$. In general, if we consider $r_{Y(X_1, ..., X_k)|Z_1, ..., Z_p}$, we compute

$$F = \frac{[\text{SS residual (only } Z_1, ..., Z_p \text{ in the model)} - \text{SS residual } (X_1, ..., X_k \text{ and } Z_1, ..., Z_p) \text{ in the model]}/k}{\text{SS residual } (X_1, ..., X_k \text{ and } Z_1, ..., Z_p \text{ in the model)}/(n-p-k-1)} \qquad (11.7)$$

and we reject H_0: $\rho_{Y(X_1, X_2, ..., X_k)|Z_1, Z_2, ..., Z_p} = 0$ if

$$F \geq F_{k, n-p-k-1, 1-\alpha}$$

For example, for the data of Table 10.1, the use of (11.6) gives

$$r_{WGT(AGE, AGE^2)|HGT}^2 = \frac{R_{WGT|HGT, AGE, AGE^2}^2 - R_{WGT|HGT}^2}{1 - R_{WGT|HGT}^2}$$

$$= \frac{0.7802 - 0.6630}{1 - 0.6630} = \frac{0.1172}{0.3370} = 0.348$$

which is not significantly different from zero since, from (11.7),

$$F = \frac{[SS \text{ residual (only HGT in the model)} - SS \text{ residual (HGT, AGE, AGE}^2 \text{ in the model)]}/2}{SS \text{ residual (HGT, AGE, AGE}^2 \text{ in the model)}/(12-1-2-1)}$$

$$= \frac{(299.33 - 195.19)/2}{195.19/8} = 2.13$$

PROBLEMS

1. The correlation matrix obtained for the variables SBP (Y), AGE (X_1), SMK (X_2), and QUET (X_3) using the data in Problem 2 of Chapter 5 is given by

	SBP	AGE	SMK	QUET
SBP	1	0.7752	0.2473	0.7420
AGE	0.7752	1	−0.1395	0.8028
SMK	0.2473	−0.1395	1	−0.0714
QUET	0.7420	0.8028	−0.0714	1

(a) Based on this matrix, which of the independent variables AGE, SMK, and QUET explains the largest proportion of the total variation in the dependent variable SBP? (*Note*: This variable is often referred to as the "most important" independent variable and is the first variable to enter the regression model using a stepwise regression program.)

(b) Using either an available computer program or an appropriate computational formula, determine the partial correlations $r_{SBP,SMK|AGE}$ and $r_{SBP,QUET|AGE}$. Based on these results, what independent variable should be the next variable after AGE to be considered for inclusion into the regression model? What must be done before this variable can be considered appropriate to add to the model?

(c) Test for the significance of $r_{SBP,SMK|AGE}$ using the ANOVA results given in Problem 1 of Chapter 10. Express the appropriate null hypothesis in terms of a population partial correlation coefficient.

(d) Determine the second-order partial $r_{SBP,QUET|AGE,SMK}$ and test for the significance of this partial correlation (again, using the results of Problem 1 in Chapter 10).

(e) Based on the results in parts (a)–(d), how would you rank the independent variables in terms of their importance in predicting Y? Which of these variables are relatively unimportant?

(f) Compute the squared multiple–partial correlation $r^2_{SBP(QUET,SMK)|AGE}$ using the ANOVA tables in Problem 1 of Chapter 10. Test for the significance of this correlation. Does this test result alter your previous choice in (e) of those variables to be included in the regression model?

(g) An alternative approach for ranking independent variables uses a "stepwise" program to determine a best model and then to order the variables in this model according to the absolute values of their *beta coefficients*. These quantities are simply adjusted regression coefficients which are all in the same units (and thus are comparable); the adjustment formula for the ith regression coefficient $\hat{\beta}_i$ is given by

$$\text{beta}(i) = \frac{S_{X_i}}{S_Y} \hat{\beta}_i$$

where S_Y and S_{X_i} are the sample standard deviations of Y and X_i, respectively. Given that $S_Y = 14.398$, $S_{X_1} = 6.956$, $S_{X_2} = 0.508$, $S_{X_3} = 0.497$, and that the fitted regression model is $\hat{Y} = 45.103 + 1.213X_1 + 9.946X_2 + 8.592X_3$, compute the beta coefficients, rank-order them from largest to smallest in absolute value, and compare this order to the previous stepwise order.

2. An equivalent way of performing a partial F test for the significance of the addition of a new variable to a model controlling for other variables already in the model is to perform a t test using the appropriate partial correlation coefficient. If the dependent variable is Y, the independent variable of interest is X, and the controlling variables are $Z_1, Z_2, \ldots, Z_p$, then the t test for $H_0: \rho_{YX|Z_1,Z_2,\ldots,Z_p} = 0$ against $H_A: \rho_{YX|Z_1,Z_2,\ldots,Z_p} \neq 0$ is given by the test statistic

$$T = r_{YX|Z_1,Z_2,\ldots,Z_p} \sqrt{n-p-2}/\sqrt{1-r^2_{YX|Z_1,Z_2,\ldots,Z_p}}$$

which has a t distribution under H_0 with $n-p-2$ degrees of freedom. The critical region for this test is therefore given by

$$|T| \geq t_{n-p-2,1-\alpha/2}$$

(a) In a study of the relationship of water hardness to sudden death rates in $n = 88$ North Carolina counties, the following partial correlations were obtained:

$$r_{YX_3|X_1} = 0.124 \quad \text{and} \quad r_{YX_3|X_1,X_2} = 0.121$$

where

$Y = $ sudden death rate in county

$X_1 = $ distance from county seat to main hospital center

$X_2 = $ population per physician

$X_3 = $ water hardness index for county

Test separately the following hypotheses:

$$\rho_{YX_3|X_1} = 0 \quad \text{and} \quad \rho_{YX_3|X_1,X_2} = 0$$

(b) An alternative way of forming the ANOVA table associated with a regression analysis is to use partial correlation coefficients. For example, if three independent variables X_1, X_2, and X_3 are used, the ANOVA table will look like:

SOURCE		df	SS
Regression	X_1	1	$r^2_{YX_1}SSY$
	$X_2|X_1$	1	$r^2_{YX_2|X_1}(1-r^2_{YX_1})SSY$
	$X_3|X_1, X_2$	1	$r^2_{YX_3|X_1,X_2}(1-r^2_{YX_2|X_1})(1-r^2_{YX_1})SSY$
Residual		$n-4$	$SSE = (1-r^2_{YX_3|X_1,X_2})(1-r^2_{YX_2|X_1})(1-r^2_{YX_1})SSY$
Total		$n-1$	$SSY = \sum\limits_{i=1}^{n} (Y_i - \bar{Y})^2$

Determine the ANOVA table for the three independent variables in the water hardness study using the following information:

$$r_{YX_1} = -0.196, \quad r_{YX_2} = 0.033, \quad r_{X_1X_2} = 0.038,$$

$$r_{YX_3|X_1,X_2} = 0.121, \quad SSY = 21.05, \quad n = 88$$

(c) In addition to the independent variables X_1, X_2, and X_3 considered above, the predictive ability of the following independent variables was also studied:

X_4 = per capita income

X_5 = coroner habit

$Z_1 = 1$ if county is in Piedmont area and 0 otherwise

$Z_2 = 1$ if county is in Coastal Plains area and 0 otherwise

$Z_3 = 1$ if county is in Tidewater area and 0 otherwise

Furthermore, 25 first-order product (i.e., "interaction") terms of the form X_1X_2, X_1Z_1, and so on, were also considered (excluding Z_1Z_2, Z_1Z_3, and Z_2Z_3). Three ANOVA tables that were obtained are given here.

Only X_1 used:

SOURCE	df	SS	MS
Regression	1	0.3846	0.3846
Residual	86	20.6703	0.2404

Only "main effects" used:

SOURCE	df	SS	MS
Regression	8	2.6853	0.3357
Residual	79	18.3696	0.2325

Main effects + first-order interactions used:

SOURCE	df	SS	MS
Regression	33	7.4143	0.2247
Residual	54	13.6406	0.2528

Test whether or not the addition of all the interaction terms to the model significantly aids in the prediction of the dependent variable after controlling for the main effects. State the null hypothesis for this test in terms of the appropriate multiple–partial correlation coefficient.

(d) Test whether there is significant overall prediction based on each of the three ANOVA tables. Determine the multiple R^2 values for each of the three tables.

(e) What can you conclude from these results about the relationship of water hardness to sudden death?

3. Two variables X and Y are said to have a *spurious correlation* if their correlation solely reflects each variable's relationship to a third (antecedent) variable Z (and to possibly other variables). For example, over a recent period of years, the correlation between U.S. Congressmen's total annual income from all sources (Y) and the number of persons owning color television sets (X) has been quite high. Yet, during this time, there has been a general upward trend in buying power (Z_1) and in wages of all types (Z_2), which, in turn, would be reflected in increased purchases of color TVs as well as in increased income of Congressmen. Thus, the high correlation between X and Y more than likely merely reflects the influence of inflation on each of these two variables. It is therefore a spurious correlation, since, without controlling for Z_1 and Z_2, say, this correlation misleadingly suggests a relationship between the color TV sales and Congressmen's income.

(a) How would you attempt to "statistically" detect whether or not a correlation between X and Y such as described is spurious?

(b) In a hypothetical study concerning socioecologic determinants of respiratory morbidity for a sample of 25 communities, the accompanying correlation matrix was obtained for four variables.

	UNEMPLOYMENT LEVEL (X_1)	AVERAGE TEMPERATURE (X_2)	AIR POLLUTION LEVEL (X_3)	RESPIRATORY MORBIDITY RATE (Y)
Unemployment level (X_1)	1	0.51	0.41	0.35
Average temperature (X_2)	—	1	0.29	0.65
Air pollution level (X_3)	—	—	1	0.50
Respiratory morbidity rate (Y)	—	—	—	1

(1) Determine the partial correlations $r_{YX_1|X_2}$, $r_{YX_1|X_3}$, and $r_{YX_1|X_2,X_3}$.

(2) Use the results in (1) to determine whether or not the correlation of 0.35 between unemployment level (X_1) and respiratory morbidity rate (Y) is spurious. (Use the testing formula given in Problem 2 to make the appropriate tests.)

(c) Describe a relevant example concerning spurious correlation in your field of interest. (Use only interval variables and define them carefully.)

4. (a) Using the information provided in Problem 2 of Chapter 10, determine the proportion of residual variation that is explained by the addition of X_2 to a model already containing X_1; that is, compute

$$Q = \frac{\text{SS regression } (X_1, X_2) - \text{SS regression } (X_1)}{\text{SS residual } (X_1)}$$

(b) How is the formula given in part (a) related to the partial correlation $r_{YX_2|X_1}$?

(c) Test the hypothesis H_0: $\rho_{YX_2|X_1} = 0$ using both an F test and a two-sided t test. Check to see that these tests are equivalent.

5. Refer to Problem 4 of Chapter 10 to answer the following questions concerning the relationship of homicide rate (Y) to population size (X_1), percentage of families with yearly incomes less than \$5,000 (X_2), and unemployment rate (X_3):
   (a) Determine the squared partial correlations $r^2_{YX_1|X_3}$ and $r^2_{YX_2|X_3}$ using the ANOVA tables. Check the computation of $r^2_{YX_2|X_3}$ by means of an alternative formula using the information that $r_{YX_2} = 0.8398$, $r_{YX_3} = 0.8562$, and $r_{X_2X_3} = 0.8074$.
   (b) Based on the results in part (a), which variable (if any) should next be considered for entry into the model given that X_3 is already in the model?
   (c) Test H_0: $\rho_{YX_2|X_3} = 0$ using the t test described in Problem 2.
   (d) Determine the squared partial correlation $r^2_{YX_1|X_2,X_3}$ from the ANOVA tables presented in Problem 4 and then test H_0: $\rho_{YX_1|X_2,X_3} = 0$.
   (e) Determine the squared multiple–partial correlation $r^2_{Y(X_1,X_2)|X_3}$ and test H_0: $\rho_{Y(X_1,X_2)|X_3} = 0$.
   (f) Based on the results in parts (a)–(e), what variables would you include in your final regression model? (Use $\alpha = 0.05$.)

6. Using the ANOVA table given in Problem 5 of Chapter 10, which considers the regression relationship of twelfth-grade mean verbal SAT scores (Y) to per pupil expenditures (X_1), percent of teachers with advanced degrees (X_2), and pupil–teacher ratio (X_3), test the following null hypotheses:
   (a) H_0: $\rho_{YX_3|X_1} = 0$.
   (b) H_0: $\rho_{YX_2|X_1,X_3} = 0$.
   (c) H_0: $\rho_{Y(X_2,X_3)|X_1} = 0$.
   (d) Based on these results, and assuming that X_1 is an important predictor of Y, what additional variables would you include in your regression model?

7. Using the ANOVA table given in Problem 6 of Chapter 10, which considers the regression relationship of respiratory cancer mortality rates (Y) to air pollution index (X_1), mean age (X_2), and percent of working force employed in a certain industry (X_3), test the following null hypotheses:
   (a) H_0: $\rho_{YX_2|X_1} = 0$.
   (b) H_0: $\rho_{YX_3|X_1,X_2} = 0$.
   (c) H_0: the addition of X_2 and X_3 to a model already containing X_1 does not significantly improve the prediction of Y.
   (d) Based on these results, which variables would you consider to be important predictors of Y? (Use $\alpha = 0.05$.)

8. Refer to the computer results given in Problem 8 of Chapter 10 to answer the following questions dealing with factors related to soil erosion:
   (a) Using the correlation matrix

$$
R = \begin{array}{c} Y \\ X_1 \\ X_2 \\ X_3 \end{array}
\begin{array}{cccc} Y & X_1 & X_2 & X_3 \end{array}
\left[\begin{array}{cccc}
1 & 0.959 & 0.979 & -0.904 \\
 & 1 & 0.951 & -0.819 \\
 & & 1 & -0.879 \\
 & & & 1
\end{array} \right]
$$

   compute $r_{YX_2|X_1}$ and $r_{YX_3|X_1}$.
   (b) Based on your results in part (a), which variable (if any) should next be entered into a regression model that already contains X_1?
   (c) Test H_0: $\rho_{YX_2|X_1} = 0$ using the t test described in Problem 2.

(d) Determine the squared multiple–partial correlation $r^2_{Y(X_2,X_3)|X_1}$ and test H_0: $\rho_{Y(X_2,X_3)|X_1} = 0$.

9. Using the computer results in Problem 9 in Chapter 10:
   (a) Test H_0: $\rho_{YX_1} = 0$ and H_0: $\rho_{YX_2} = 0$.
   (b) Test H_0: $\rho_{YX_1|X_2} = 0$ and H_0: $\rho_{YX_2|X_1} = 0$.
   (c) Based on your results in parts (a) and (b), which variables (if any) should be included in the regression model and what is their relative order of importance? Does your answer agree with that for Problem 9(d) of Chapter 10?

SUMMARY OF METHODS FOR TESTING HYPOTHESES IN MULTIPLE REGRESSION

12.1 PREVIEW

The purpose of this chapter is to illustrate the testing procedures described in Chapters 10 and 11 by means of an example involving seven independent variables and one dependent variable, and also to provide a brief discussion of the concept of statistical interaction.

EXAMPLE   The results presented below are based on data from a study by Gruber (1970) to determine the extent and manner in which changes in blood pressure over time depend on initial blood pressure (at the beginning of the study), the sex of the individual, and the relative weight of the individual. Data were collected on $n = 104$ persons, for which the $k = 7$ independent variables were examined by multiple regression. The variables used in the study are defined below:

$Y = $ SBPSL (estimated slope based on the straight-line regression of an individual's blood pressure over time)[1]
$X_1 = $ SBP1 (initial systolic blood pressure)
$X_2 = $ SEX (male $= 1$, female $= -1$)
$X_3 = $ RW (relative weight)
$X_4 = X_1 X_2$
$X_5 = X_1 X_3$ } A general discussion regarding the interpretation of such *product* (i.e., *interaction*) terms as these will be given in Section
$X_6 = X_2 X_3$ 12.2.
$X_7 = X_1 X_2 X_3$

---

[1] This choice of dependent variable can be criticized on the grounds that observations on an individual's blood pressure taken over time are *not* independent, thus violating Assumption 2 in Chapter 10. A preferred statistic to that used by Gruber could be obtained by weighted regression (see Chapter 16) or by growth-curve analysis [see Allen and Grizzle (1969)].

The complete model involving all seven independent variables is:

$$Y = \beta_0 + \beta_1 X_1 + \beta_2 X_2 + \beta_3 X_3 + \beta_4 X_4 + \beta_5 X_5 + \beta_6 X_6 + \beta_7 X_7 + E$$

and the estimated coefficients, their standard errors, and the associated partial F statistics (concerning the addition of each variable to a model already containing the other six variables)[2] are given in Table 12.1. The ANOVA table is given as Table 12.2.

TABLE 12.1 *Statistics for the model*

VARIABLE	$\hat{\beta}$	$S_{\hat{\beta}}$	PARTIAL F ($\hat{\beta}^2 / S_{\hat{\beta}}^2$)
X_1 (SBP1)	−0.045	0.00762	34.987
X_2 (SEX)	0.695	0.86644	0.643
X_3 (RW)	0.027	0.07049	0.149
X_4 ($X_1 X_2$)	−0.0029	0.00762	0.145
X_5 ($X_1 X_3$)	−0.00018	0.00062	0.084
X_6 ($X_2 X_3$)	−0.0092	0.07049	0.017
X_7 ($X_1 X_2 X_3$)	0.00022	0.00062	0.125
(constant)	4.667		

TABLE 12.2 *ANOVA table for the model*

SOURCE		df	SS	MS	F
Overall regression		7	37.148	5.307	$\dfrac{5.307}{76.246/96} = 6.68^{**}$
Regression	X_1	1	24.988	24.988	$\dfrac{24.988}{88.406/102} = 28.83^{**}$
	$X_2\|X_1$	1	7.886	7.886	$\dfrac{7.886}{80.520/101} = 9.89^{**}$
	$X_3\|X_1, X_2$	1	1.057	1.057	$\dfrac{1.057}{79.463/100} = 1.33$
	$X_4\|X_1, X_2, X_3$	1	0.020	0.020	$\dfrac{0.020}{79.443/99} = 0.025$
	$X_5\|X_1, X_2, X_3, X_4$	1	0.254	0.254	$\dfrac{0.254}{79.189/98} = 0.314$
	$X_6\|X_1, X_2, X_3, X_4, X_5$	1	2.844	2.844	$\dfrac{2.844}{76.345/97} = 3.613$
	$X_7\|X_1, X_2, X_3, X_4, X_5, X_6$	1	0.099	0.099	$\dfrac{0.099}{76.246/96} = 0.125$
Residual		96	76.246	0.794	
Total (corrected)		103	113.394		

Note: $F_{7,96,0.95} = 2.11$, $F_{1,96,0.95} = 3.95$, $F_{1,102,0.95} = 3.94$.

[2] The partial $F(X_i|\text{remaining six } X\text{'s})$ is computed either by using the "extra sum of squares" principle or by squaring a T statistic of the form $T = \hat{\beta}/S_{\hat{\beta}}$, as described in Chapter 10.

TEST I *Overall F test*

This test is used to determine whether or not all of the k independent variables taken together significantly contribute to the prediction of the dependent variable. This test statistic has the form

$$F = \frac{\text{MS regression}}{\text{MS residual}} = \frac{R^2}{1 - R^2} \frac{n - k - 1}{k}$$

where R^2 = squared multiple correlation coefficient (i.e., $R^2 = R^2_{Y|X_1, X_2, ..., X_k}$). The decision rule for this test is to reject H_0 if $F \geq F_{k, n-k-1, 1-\alpha}$, where H_0 is any one of the following equivalent hypotheses (for $k = 7$):

1. H_0: the overall regression is not significant (i.e., using all 7 independent variables together does not significantly help to predict Y).
2. H_0: $\beta_1 = \beta_2 = \beta_3 = \beta_4 = \beta_5 = \beta_6 = \beta_7 = 0$ (i.e., all the regression coefficients are simultaneously zero).
3. H_0: $\rho_{Y|X_1, X_2, X_3, X_4, X_5, X_6, X_7} = 0$, where $\rho_{Y|X_1, X_2, X_3, X_4, X_5, X_6, X_7}$ is the true multiple correlation between Y and all the independent variables.
4. H_0: the proportion of the total variability explained by the regression on the seven independent variables is not significant; that is,

$$R^2 = \sum_{i=1}^{n} (\hat{Y}_i - \bar{Y})^2 \bigg/ \sum_{i=1}^{n} (Y_i - \bar{Y})^2$$

is not significantly different from zero.

For our example, $n = 104$, $k = 7$, $R^2 = 37.148/113.394 = 0.328$, so that

$$F = \frac{5.307}{0.794} = \frac{0.328}{1 - 0.328} \frac{104 - 7 - 1}{7} = 6.68$$

which is significant at the $\alpha = 0.01$ level.

[It is possible that the overall regression for all seven independent variables is not significant even though the overall regression based on fewer variables (e.g., on one of the variables) is significant. This is because, as you add more variables, you are losing degrees of freedom for your estimate of error, even though you are explaining more variation.]

TEST II *Partial F test*

The partial F test is used to assess whether or not the addition of any specific independent variable to the model significantly improves the prediction of Y, given that other variables are already in the model. For example, the partial F test concerning β_7 given that all the other six variables are already in the model is

$$F = \frac{\text{SS residual} \left(\begin{array}{c} \text{all variables except } X_7 \\ \text{in model} \end{array} \right) - \text{SS residual} \left(\begin{array}{c} \text{all seven variables} \\ \text{in model} \end{array} \right)}{\text{MS residual (all seven variables in model)}}$$

$$= \frac{\text{SS regression} \left(\begin{array}{c} \text{all seven variables} \\ \text{in model} \end{array} \right) - \text{SS regression} \left(\begin{array}{c} \text{all variables except} \\ X_7 \text{ in model} \end{array} \right)}{\text{MS residual (all seven variables in model)}}$$

We would reject H_0 of no significant improvement in prediction by adding X_7 to a model already containing X_1, X_2, X_3, X_4, X_5, and X_6 if F exceeds $F_{1,96,1-\alpha}$. In this case we get

$$F = \frac{76.345 - 76.246}{76.246/96} = \frac{0.099}{0.794} = 0.125$$

which is clearly not significant.

The following are equivalent to the null hypothesis given above:

H_0: $\beta_7 = 0$ given that all seven variables are in the model

H_0: $\rho_{YX_7|X_1,X_2,X_3,X_4,X_5,X_6} = 0$, where $\rho_{YX_7|X_1,X_2,X_3,X_4,X_5,X_6}$ is the partial correlation of Y with X_7 controlling for X_1, X_2, X_3, X_4, X_5, and X_6

As another example, the partial F test concerning X_3 given that X_1 and X_2 are already in the model is

$$F = \frac{\text{SS residual} \begin{pmatrix} \text{only } X_1 \text{ and } X_2 \\ \text{in the model} \end{pmatrix} - \text{SS residual} \begin{pmatrix} X_1, X_2, \text{ and } X_3 \\ \text{in the model} \end{pmatrix}}{\text{SS residual } (X_1, X_2, \text{ and } X_3 \text{ in the model})/100}$$

$$= \frac{80.520 - 79.463}{79.463/100} = 1.33 \qquad \text{which is not significant when compared to } F_{1,100,0.90} = 2.76$$

TEST III *t test for a particular regression coefficient*
This test is equivalent to the partial F test described in II. To test H_0: $\beta = 0$ for a model involving k independent variables, compute $T = \hat{\beta}/S_{\hat{\beta}}$ and reject H_0 if $|T| \geq t_{n-k-1,1-\alpha/2}$ (two-sided test). For example, to test H_0: $\beta_7 = 0$ in the full model containing all seven independent variables, use

$$\hat{\beta}_7 = 0.00022 \quad \text{and} \quad S_{\hat{\beta}_7} = 0.00062$$

to calculate

$$T = \frac{\hat{\beta}_7}{S_{\hat{\beta}_7}} = 0.354 \qquad \text{which equals } \sqrt{F} = \sqrt{0.125}$$

[It would be incorrect to conclude from the use of a t test (or partial F test) concerning each of the seven variables separately (using the SSE for the full model) that two or more regression coefficients are zero. This is because once one variable out of the seven is dropped from the model, the model with six variables must be reestimated to provide the appropriate error term for testing for the significance of another coefficient. As an extreme example, it would be inappropriate to conclude that the overall regression is not significant if all seven t tests were not significant. However, if all partial F tests are done sequentially (as illustrated in the ANOVA table) and none are significant, the overall regression test will not be significant either.]

TEST IV *Test concerning any subset of regression coefficients*
This test is often used in regression problems to test whether higher-order interactions (product terms) are important. For example, we might ask whether the second- and

third-order interaction terms in the complete model considered earlier are significant, given that the first-order terms are already in the model; that is, we might want to test

$$H_0: \beta_4 = \beta_5 = \beta_6 = \beta_7 = 0$$

when the complete model has been fitted.

Equivalent hypotheses are:

H_0: the addition of second- and third-order terms to the model does not significantly improve prediction given that the first-order terms are already in the model

H_0: $\rho_{Y(X_4,X_5,X_6,X_7)|X_1,X_2,X_3} = 0$, where $\rho_{Y(X_4,X_5,X_6,X_7)|X_1,X_2,X_3}$ is the multiple correlation of Y on X_4, X_5, X_6, and X_7 controlling for X_1, X_2, and X_3 (i.e., a multiple-partial correlation).

The appropriate test statistic for our example is

$$F = \frac{\left[\text{SS residual } (X_1, X_2, X_3 \text{ in model}) - \text{SS residual} \left(\begin{array}{c} \text{all seven variables} \\ \text{in model} \end{array} \right) \right] \Big/ 4}{\text{SS residual (all seven variables in model)}/96}$$

$$= \frac{\left[\text{SS regression} \left(\begin{array}{c} \text{all seven variables} \\ \text{in model} \end{array} \right) - \text{SS regression } (X_1, X_2, X_3 \text{ in model}) \right] \Big/ 4}{\text{SS residual (all seven variables in model)}/96}$$

and we would reject H_0 if $F \geq F_{4,96,1-\alpha}$. Here we get

$$F = \frac{(79.463 - 76.246)/4}{76.246/96} = \frac{3.217/4}{0.794} = \frac{0.804}{0.794} = 1.01$$

which is not significant.

12.2 CONCEPT OF INTERACTION

It is our purpose in this section to provide a brief and general discussion of the concept of statistical "interaction." More specifically, we shall be concerned with describing the way in which two independent variables can "interact" to affect a dependent variable and with representing such an interaction phenomenon in terms of an appropriate regression model.

To help illustrate these concepts, we shall consider the following simple example. Suppose it is of interest to determine how two independent variables, temperature (T) and catalyst concentration (C), jointly affect the growth rate (Y) of organisms in a certain biological system. Further, suppose that two particular temperature levels (T_0 and T_1) and two particular catalyst concentration levels (C_0 and C_1) are to be examined, and that an experiment is performed in which an observation on Y is obtained for each of the four combinations of temperature–catalyst concentration levels, (T_0, C_0), (T_0, C_1), (T_1, C_0), and (T_1, C_1).

[Statistically speaking, this experiment is called a *complete factorial experiment*, complete in the sense that observations on Y are obtained for *all* combinations of settings for the independent variables (or factors). The advantage in employing a factorial experiment is that of being able to detect and measure interaction effects when they exist.]

Now, let us consider two graphs based on two different data sets which could conceivably have arisen via the experimentation scheme described above. Figure 12.1a suggests that the *rate of change* in Y as a function of temperature is the same regardless of the level of catalyst concentration; in other words, the relationship between Y and T does not in any way depend on C.

[For those readers having some familiarity with calculus, the phrase "rate of change" is related to the notion of a "derivative of a function." In particular, Figure 12.1a is meant to portray a situation where the partial derivative with respect to T of the response function relating the mean of Y to T and C is independent of C.]

It is important to point out that we are *not* saying that Y and C are unrelated, but that the relationship between Y and T is independent of the relationship between Y and C. When this is the case, we say that T and C do not interact, or, equivalently, that there is no T-by-C interaction effect. Practically speaking, this means that we can investigate the effects of T and C on Y independently of one another and that we can legitimately talk about the separate effects (sometimes called the "main effects") of T and C on Y.

One way to quantify the relationship depicted in Figure 12.1a is with a regression model of the form

$$\mu_{Y|T,C} = \beta_0 + \beta_1 T + \beta_2 C \tag{12.1}$$

Here the change in the mean of Y for a one-unit change in T is equal to β_1, regardless of the level of C. In fact, changing the level of C in (12.1) has only the effect of shifting the straight line relating $\mu_{Y|T,C}$ and T either up or down without affecting the value of the slope β_1, as is seen in Figure 12.1a. In particular, $\mu_{Y|T,C_0} = (\beta_0 + \beta_2 C_0) + \beta_1 T$ and $\mu_{Y|T,C_1} = (\beta_0 + \beta_2 C_1) + \beta_1 T$.

In general, then, one might say that "no interaction" is synonymous with "parallelism" in the sense that the response curves of Y versus T for various fixed

FIGURE 12.1 *"No interaction" versus "interaction"*

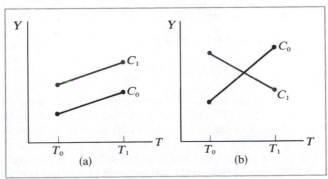

FIGURE 12.2   *Graphical illustration of "no interaction"*

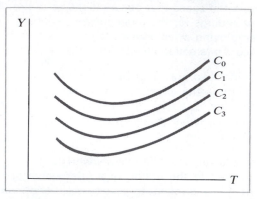

values of C are parallel; in other words, these response curves (which may be linear or nonlinear) all have the same general shape, differing from one another only by additive constants independent of T (e.g., see Figure 12.2).

In contrast, Figure 12.1b depicts the situation where the relationship between Y and T depends on C; in particular, Y appears to increase with increasing T with $C = C_0$ but appears to decrease with increasing T when $C = C_1$. In other words, the behavior of Y as a function of temperature cannot be considered independently of catalyst concentration. When this is the case, we say that T and C interact, or, equivalently, that there is a T-by-C interaction effect. Practically speaking, this means that it really does not make much sense to talk about the separate (or main) effects of T and C on Y, since T and C do not operate independently of one another with respect to their effects on Y.

One way to mathematically account for such an interaction effect is to consider a regression model of the form

$$\mu_{Y|T,C} = \beta_0 + \beta_1 T + \beta_2 C + \beta_{12} TC \tag{12.2}$$

Here, the change in the mean value of Y for a 1-unit change in T is equal to $\beta_1 + \beta_{12}C$, which clearly depends on the level of C. In other words, the introduction of a product term such as $\beta_{12}TC$ into a regression model of the type (12.2) is one way to account for the fact that two such factors as T and C do not operate independently of one another. For our particular example, when $C = C_0$, model (12.2) can be written as

$$\mu_{Y|T,C_0} = (\beta_0 + \beta_2 C_0) + (\beta_1 + \beta_{12} C_0) T$$

and, when $C = C_1$, model (12.2) becomes

$$\mu_{Y|T,C_1} = (\beta_0 + \beta_2 C_1) + (\beta_1 + \beta_{12} C_1) T$$

And, Figure 12.1b suggests that the interaction effect β_{12} would be negative, with the linear effect $(\beta_1 + \beta_{12}C_0)$ of T at C_0 being positive and the linear effect $(\beta_1 + \beta_{12}C_1)$ of T at C_1 being negative. A negative interaction effect is to be expected here, since Figure 12.1b suggests that the slope of the linear relationship between Y and T decreases as C changes from C_0 to C_1. Of course, it is possible for β_{12} to be positive, in which case the interaction effect would manifest itself in terms of a larger positive value for the slope when $C = C_1$ than when $C = C_0$.

PROBLEMS

1. A regression analysis using data on $n = 53$ males considered the following variables:

Y = SBPSL (estimated slope based on the straight-line regression of an individual's blood pressure over time)

X_1 = SBP1 (initial blood pressure)

X_2 = RW (relative weight)

$X_3 = X_1 X_2$ = SR (product of SBP1 and RW)

The computer printout was obtained using a standard stepwise-regression program (SPSS). Using this output, complete the following exercises:

(a) Fill in the ANOVA table for the fit of the model $Y = \beta_0 + \beta_1 X_1 + \beta_2 X_2 + \beta_3 X_3 + E$.

SOURCE	df	SS	MS
Regression $\begin{cases} X_1 \\ X_2\|X_1 \\ X_3\|X_1, X_2 \end{cases}$			
Residual			
Total	52		

(b) Test H_0: no significant overall regression using all three variables.

(c) Test H_0: no significant overall regression using only X_1 and X_2 in the model.

(d) Test H_0: $\rho_{YX_2\|X_1} = 0$.

(e) Test H_0: the addition of X_3 to the model given that X_1 and X_2 are already in the model is not significant.

(f) Test H_0: $\rho_{Y(X_2, X_3)\|X_1} = 0$.

(g) Based on the tests above, what would you consider to be the most appropriate regression model? (Use $\alpha = 0.05$.)

(h) Verify the values of the beta coefficients presented at step 3 of the computer output, given the values of the regression coefficients and the values $S_Y = 1.1636$, $S_{X_1} = 11.4450$, $S_{X_2} = 11.9711$, and $S_{X_3} = 1410.8213$. Does the rank order of the absolute values of the beta coefficients correspond to the order in which the variables entered the model?

2. A psychologist examined the regression relationship between anxiety level (Y), measured on a 1–50 scale as the average of an index determined at three time points in a 2-week period, and the following three independent variables:

X_1: systolic blood pressure

X_2: IQ

X_3: job satisfaction (measured on a 1–25 scale)

The ANOVA table summarizes results obtained from a stepwise regression analysis on data on 22 outpatients undergoing therapy at a certain clinic.

SOURCE	df	SS
Regression $\begin{cases} X_1 \\ X_2\|X_1 \\ X_3\|X_1, X_2 \end{cases}$	1 1 1	981.326 190.232 129.431
Residual	18	442.292

Computer printout for Problem 1 (SPSS)

DEPENDENT VARIABLE.. SBPSL

VARIABLE(S) ENTERED ON STEP NUMBER 1.. SBP1

		ANALYSIS OF VARIANCE	DF	SUM OF SQUARES	MEAN SQUARE	F
MULTIPLE R	0.45834	REGRESSION	1.	14.79083	14.79083	13.56308
R SQUARE	0.21007	RESIDUAL	51.	55.61661	1.09052	
STANDARD ERROR	1.04428					

————— VARIABLES IN THE EQUATION —————

VARIABLE	B	BETA	STD ERROR B	F
SBP1	-0.04660	-0.45834	0.01265	13.563
(CONSTANT)	5.10797			

————— VARIABLES NOT IN THE EQUATION —————

VARIABLE	BETA IN	PARTIAL	TOLERANCE	F
RW	0.23166	0.26007	0.99553	3.627
SR	0.23074	0.25953	0.99933	3.611

VARIABLE(S) ENTERED ON STEP NUMBER 2.. RW

		ANALYSIS OF VARIANCE	DF	SUM OF SQUARES	MEAN SQUARE	F
MULTIPLE R	0.51332	REGRESSION	2.	18.55240	9.27620	8.94435
R SQUARE	0.26350	RESIDUAL	50.	51.85504	1.03710	
STANDARD ERROR	1.01838					

————— VARIABLES IN THE EQUATION —————

VARIABLE	B	BETA	STD ERROR B	F
SBP1	-0.04817	-0.47382	0.01237	15.174
RW	0.02252	0.23166	0.01182	3.627
(CONSTANT)	5.38484			

————— VARIABLES NOT IN THE EQUATION —————

VARIABLE	BETA IN	PARTIAL	TOLERANCE	F
SR	0.04646	0.00450	0.00690	0.001

VARIABLE(S) ENTERED ON STEP NUMBER 3.. SR

		ANALYSIS OF VARIANCE	DF	SUM OF SQUARES	MEAN SQUARE	F
MULTIPLE R	0.51334	REGRESSION	3.	18.55345	6.18448	5.84409
R SQUARE	0.26352	RESIDUAL	49.	51.85399	1.05824	
STANDARD ERROR	1.02871					

————— VARIABLES IN THE EQUATION —————

VARIABLE	B	BETA	STD ERROR B	F
SBP1	-0.04798	-0.47193	0.01391	11.899
RW	0.01801	0.18527	0.14372	0.016
SR	0.00004	0.04646	0.00122	0.001
(CONSTANT)	5.36183			

————— VARIABLES NOT IN THE EQUATION —————

VARIABLE	BETA IN	PARTIAL	TOLERANCE	F

From *Statistical Package for the Social Sciences* by Nie et al. Copyright © 1975 by McGraw-Hill, Inc. Used with permission of McGraw-Hill Book Company and Dr. Norman Nie, President, SPSS, Inc.

(a) Test for the significance of each independent variable as it enters the model. State the null hypothesis for each test in terms of regression coefficients.

(b) Compute the R^2 values associated with each of the three steps in the variable-selection procedure and then comment on your results.

(c) Test for the significance of the addition of both X_2 and X_3 to a model already containing X_1. State the null hypothesis in terms of regression coefficients and in terms of (multiple-partial) correlation coefficients.

(d) Why can't we test the hypothesis $H_0: \rho_{Y(X_1,X_2)|X_3} = 0$ using the ANOVA table? Describe the appropriate test procedure.

(e) Based on the tests made, what would you recommend to be the most appropriate statistical model? (Use $\alpha = 0.05$.)

3. An educator examined (in persons over 60 years of age) the relationship between the number of hours devoted to reading each week (Y) and the independent variables social class (X_1), number of years of school completed (X_2), and reading speed measured by pages read per hour (X_3). The ANOVA table obtained from a stepwise regression analysis on data for a sample of 19 women over 60 is as shown.

SOURCE		df	SS
Regression $\begin{cases} X_3 \\ X_2 \vert X_3 \\ X_1 \vert X_2, X_3 \end{cases}$		1	1,058.628
		1	183.743
		1	37.982
Residual		15	363.300

(a) Assuming that all correlations are positive, determine the following:

$$r_{YX_3}, \quad r_{YX_2|X_3}, \quad r_{YX_1|X_2,X_3}, \quad r^2_{Y(X_1,X_2)|X_3}, \quad R^2_{Y|X_2,X_3}, \quad R^2_{Y|X_1,X_2,X_3}$$

(b) Test for the significance of each variable as it enters the model.

(c) Test $H_0: \beta_1 = \beta_2 = 0$ in the model $Y = \beta_0 + \beta_1 X_1 + \beta_2 X_2 + \beta_3 X_3 + E$.

(d) Why can't we test $H_0: \beta_1 = \beta_3 = 0$ using the ANOVA table given? What formula would you use for this test?

(e) What is your overall evaluation concerning the appropriate model to use given the results in parts (a)–(c)?

4. An experiment was conducted regarding a quantitative analysis of factors found in high-density lipoprotein (HDL) in a sample of human blood serum. Three variables thought to be predictive or associated with HDL measurement (Y) were the total cholesterol (X_1) and total triglyceride (X_2) concentrations in the sample, plus the presence or absence of a certain "sticky" component found in the serum called "sinking pre-beta or SPB" (X_3) and coded as 0 if absent and 1 if present. The data obtained are shown in the table and the computer results follow.

(a) Test whether X_1, X_2, or X_3 alone significantly helps in predicting Y.

(b) Test whether, taken together, X_1, X_2, and X_3 significantly help to predict Y.

(c) Test whether the true coefficients of the product terms $X_1 X_3$ and $X_2 X_3$ are simultaneously zero in the model containing X_1, X_2, and X_3 plus these product terms. State the null hypothesis in terms of a multiple-partial correlation coefficient. If this test is not rejected, what can you conclude about the

Y	X_1	X_2	X_3	Y	X_1	X_2	X_3
47	287	111	0	57	192	115	1
38	236	135	0	42	349	408	1
47	255	98	0	54	263	103	1
39	135	63	0	60	223	102	1
44	121	46	0	33	316	274	0
64	171	103	0	55	288	130	0
58	260	227	0	36	256	149	0
49	237	157	0	36	318	180	0
55	261	266	0	42	270	134	0
52	397	167	0	41	262	154	0
49	295	164	0	42	264	86	0
47	261	119	1	39	325	148	0
40	258	145	1	27	388	191	0
42	280	247	1	31	260	123	0
63	339	168	1	39	284	135	0
40	161	68	1	56	326	236	1
59	324	92	1	40	248	92	1
56	171	56	1	58	285	153	1
76	265	240	1	43	361	126	1
67	280	306	1	40	248	226	1
57	248	93	1	46	280	176	1

Computer results for Problem 4

SOURCE	df	SS	SOURCE	df	SS
Regression (X_1)	1	46.2356	Regression (X_2)	1	21.3397
Residual	40	4,567.3835	Residual	40	4,592.2793

SOURCE	df	SS	SOURCE	df	SS
Regression (X_3)	1	735.2054	Regression (X_1, X_2)	2	135.3820
Residual	40	3,878.4136	Residual	39	4,478.2369

SOURCE	df	SS	SOURCE	df	SS
Regression (X_1, X_3)	2	783.1691	Regression (X_2, X_3)	2	737.8069
Residual	39	3,830.4500	Residual	39	3,875.8122

SOURCE	df	SS	SOURCE	df	SS
Regression (X_1, X_2, X_3)	3	819.7473	Regression (X_1, X_3)	2	783.1691
$X_1X_3, X_2X_3 \mid X_1, X_2, X_3$	2	74.7443	$X_1X_3 \mid X_1, X_3$	1	62.4247
Residual	36	3,719.0517	Residual	38	3,768.0252

SOURCE	df	SS
Regression (X_2, X_3)	2	737.8069
$X_2X_3 \mid X_2, X_3$	1	1.5539
Residual	38	3,874.2583

relationship of Y to X_1 and X_2 when X_3 equals 1 as compared to when X_3 equals 0?

(d) Test (at $\alpha = 0.05$) whether X_3 is associated with Y after taking into account the combined contribution of X_1 and X_2. State the appropriate null hypothesis in terms of a partial correlation coefficient. What does your result together with your answer to part (c) tell you about the relationship of Y with X_1 and X_2 when SPB is present as compared to when it is absent?

(e) What overall conclusion can you draw about the association of Y with the three independent variables for this data set?

CHAPTER 13 DUMMY VARIABLES IN REGRESSION

13.1 PREVIEW

A *dummy* or *indicator* variable is any variable in a regression equation that takes on a finite number of values for the purpose of identifying different categories of a nominal variable. The term "dummy" simply relates to the property that the actual values taken on by such variables (usually values like 0, 1, and −1) describe no meaningful measurement level of the variable but rather act only to indicate (or designate) the categories of interest.

Examples of dummy variables include the following:

$$X_1 = \begin{cases} 1 & \text{if treatment A is used} \\ 0 & \text{otherwise} \end{cases}$$

$$X_2 = \begin{cases} 1 & \text{for females} \\ -1 & \text{for males} \end{cases}$$

$$Z_1 = \begin{cases} 1 & \text{if residence is in western U.S.} \\ 0 & \text{if residence is in central U.S.} \\ -1 & \text{if residence is in eastern U.S.} \end{cases}$$

$$Z_2 = \begin{cases} 0 & \text{if residence is in western U.S.} \\ 1 & \text{if residence is in central U.S.} \\ -1 & \text{if residence is in eastern U.S.} \end{cases}$$

The variable X_1 indicates a nominal variable describing "treatment group" (either treatment A or not treatment A); the variable X_2 indexes the levels of the nominal variable "sex"; and variables Z_1 and Z_2 work in tandem to describe the nominal variable "geographical residence." Note that in the latter case, the three categories of geographical residence are described by the following combination of the two variables Z_1 and Z_2:

Western U.S.: $Z_1 = 1$, $Z_2 = 0$

Central U.S.: $Z_1 = 0$, $Z_2 = 1$

Eastern U.S.: $Z_1 = -1$, $Z_2 = -1$

It is through the use of dummy variables that regression analysis takes on a broader range of application. In particular, the use of dummy variables allows one to employ regression analysis to produce the same information as is obtained by means of such seemingly distinct analytical procedures as analysis of variance, analysis of covariance, and discriminant analysis. Also, the use of dummy variables permits one to compare several regression equations by use of a single multiple regression model.

13.2 RULE FOR DEFINING DUMMY VARIABLES

The following simple rule should always be applied when defining a dummy variable: *If the nominal independent variable of interest has k categories, then one must define exactly $k - 1$ dummy variables to index these categories, provided that the regression model contains a constant term (i.e., an intercept). If the regression model does not contain a constant term, then k dummy variables are needed to index the k categories of interest.* For example, if there are $k = 3$ categories, the number of dummy variables should be $k - 1 = 2$ for a model containing an intercept.

[One exception needs to be pointed out. If a constant is not included in an overall regression model designed to compare several regression equations, it is possible to define the dummy variables in such a way as to allow for constant terms in each of these regression equations derived from the overall model. Thus, the use of a constant in such an overall model generally depends upon the preference of the investigator with regard to the coding of the dummy variables (see Section 13.4).]

In applying this rule, it should be noted that:

1. If an intercept is used in the regression equation, proper definition of the $k - 1$ dummy variables automatically indexes all the k categories.
2. If k dummy variables are used to describe a nominal variable with k categories in a model containing a constant term, all the coefficients in the model cannot be uniquely estimated.
3. There are many different ways to properly define $k - 1$ dummy variables to index the k categories of a given nominal variable. For example, two other alternative but equivalent ways to describe the nominal variable "geographical residence" (represented earlier by Z_1 and Z_2) are

$$Z_1^* = \begin{cases} 1 & \text{if residence is in western U.S.} \\ 0 & \text{otherwise} \end{cases}$$

$$Z_2^* = \begin{cases} 1 & \text{if residence is in central U.S.} \\ 0 & \text{otherwise} \end{cases}$$

and

$$Z_1' = \begin{cases} 1 & \text{if residence is in western U.S.} \\ 0 & \text{otherwise} \end{cases}$$

$$Z_2' = \begin{cases} 1 & \text{if residence is in eastern U.S.} \\ 0 & \text{otherwise} \end{cases}$$

In the remainder of this chapter we shall work through some examples concerning the use of dummy variables for the purpose of comparing several regression equations. In Chapter 14 we shall describe the connection between dummy variables and the analysis of covariance. In Chapters 17–20 we will show how dummy variables can be utilized in regression models for ANOVA problems, and, in particular, we will see the importance of such a regression formulation when there are unequal cell numbers in two-way or higher ANOVA problems.

13.3 COMPARING TWO STRAIGHT-LINE REGRESSION EQUATIONS

Suppose that we wish to compare males and females with regard to their separate straight-line regressions of SBP (Y) on AGE (X). We have, say, n_M pairs of (X, Y) observations on males and n_F such pairs on females. There are two methods for analyzing such data:

METHOD I Treat the male and female data separately by fitting the two separate regression equations $Y_M = \beta_{0M} + \beta_{1M}X + E$ and $Y_F = \beta_{0F} + \beta_{1F}X + E$, and then make appropriate two-sample t tests or (large-sample) Z tests as described earlier (Chapter 8).

METHOD II Define the dummy variable $Z = 0$ if male, 1 if female. Thus, for the n_M observations on males, $Z = 0$; and, for the n_F observations on females, $Z = 1$. Our data will then be of the form:

Males: $(X_{1M}, Y_{1M}, 0), (X_{2M}, Y_{2M}, 0), \ldots, (X_{n_M M}, Y_{n_M M}, 0)$

Females: $(X_{1F}, Y_{1F}, 1), (X_{2F}, Y_{2F}, 1), \ldots, (X_{n_F F}, Y_{n_F F}, 1)$

Then, fit to the combined data above the single multiple regression model

$$Y = \beta_0 + \beta_1 X + \beta_2 Z + \beta_3 XZ + E \tag{13.1}$$

which yields the following two models for the two values of Z:

$$\begin{cases} Z = 0: & Y_M = \beta_0 + \beta_1 X + E & \text{for males} \\ Z = 1: & Y_F = (\beta_0 + \beta_2) + (\beta_1 + \beta_3)X + E & \text{for females} \end{cases}$$

This allows us to write the regression coefficients from the separate models for method I in terms of the coefficients of model (13.1) as follows: $\beta_{0M} = \beta_0$, $\beta_{0F} = (\beta_0 + \beta_2)$, $\beta_{1M} = \beta_1$, and $\beta_{1F} = (\beta_1 + \beta_3)$. *Thus, model (13.1) incorporates the two separate regression equations within a single model* and allows for different slopes [β_1 for males and $(\beta_1 + \beta_3)$ for females] and different intercepts [β_0 for males and $(\beta_0 + \beta_2)$ for females].

13.3.1 TESTS CONCERNING SLOPES AND INTERCEPTS USING THE DUMMY-VARIABLE MODEL

For model (13.1), the following two hypotheses are especially relevant:

1. *The two regression lines are parallel*; that is, $H_0: \beta_3 = 0$. When $\beta_3 = 0$, the slope for females $\beta_{1F} = (\beta_1 + \beta_3)$ is then simply equal to β_1, which is the slope for males (i.e., the two lines are parallel). The test statistic concerning $H_0: \beta_3 = 0$ is the usual partial F test (or equivalent t test) for the significance of the addition of the variable XZ to a model already containing X and Z.[1]

2. *The two regression lines are coincident*; that is, $H_0: \beta_2 = \beta_3 = 0$. When both β_2 and β_3 are zero, the model $Y_F = (\beta_0 + \beta_2) + (\beta_1 + \beta_3)X + E$ reduces to the model $Y_M = \beta_0 + \beta_1 X + E$ for males (i.e., the two lines are coincident). The test of $H_0: \beta_2 = \beta_3 = 0$ is the usual test discussed earlier concerning a subset of regression coefficients.[2]

EXAMPLE For the age–SBP data described in the example following Section 8.1, $n_M = 40$, $n_F = 29$, and the fitted multiple regression model (13.1) turns out to have the form

$$\hat{Y} = 110.04 + 0.96X - 12.96Z - 0.012XZ$$

which yields the following separate straight-line equations:

$$Z = 0 \text{ (males):} \qquad \hat{Y}_M = 110.04 + 0.96X$$
$$Z = 1 \text{ (females):} \quad \hat{Y}_F = 97.08 + 0.95X$$

These two straight-line equations are exactly the same as obtained in Chapter 8 by fitting separate regressions (see Section 13.3.2 for further discussion on this point).

To test for *parallelism* or *coincidence*, we can appeal to a multiple regression printout for the appropriate F or t test required. The particular analysis of variance information needed (which is usually printed out by most standard regression programs) is as follows:

SOURCE	df	SS	MS	F
Regression (X)	1	14,951.25	14,951.25	121.27
Residual	67	8,260.51	123.29	
Regression (X, Z)	2	18,009.78	9,004.89	114.25
Residual	66	5,201.99	78.82	
Regression (X, Z, XZ)	3	18,010.33	6,003.44	75.02
Residual	65	5,201.44	80.02	

[1] If the test of $H_0: \beta_3 = 0$ results in not rejecting H_0, model (13.1) can be revised by eliminating the β_3 term; this revised (or reduced) model becomes $Y = \beta_0 + \beta_1 X + \beta_2 Z + E$, which is in the form of an analysis-of-covariance model (see Chapter 14).

[2] If the test for coincidence is not rejected, the complete model (13.1) can be reduced to the form $Y = \beta_0 + \beta_1 X + E$.

Using these results, we get the following tests:

H_0: $\beta_3 = 0$ (*parallelism*):

$$F(XZ|X, Z) = \frac{\text{SS regression } (X, Z, XZ) - \text{SS regression } (X, Z)}{\text{MS residual } (X, Z, XZ)}$$

$$= \frac{18,010.33 - 18,009.78}{80.02}$$

$$= 0.007$$

This F statistic, with 1 and 65 df, is extremely small, so we do not reject H_0 and therefore have no basis for believing that the two lines are not parallel. This was the same decision as made in Chapter 8 based on separate regression fits. In fact, the F computed here is the square of the corresponding T computed in Chapter 8 (see Section 13.3.2), although the numerical answers may not exactly agree due to round-off errors.

H_0: $\beta_2 = \beta_3 = 0$ (*coincidence*):

$$F(XZ, Z|X) = \frac{[\text{SS regression } (X, Z, XZ) - \text{SS regression } (X)]/2}{\text{MS residual } (X, Z, XZ)}$$

$$= \frac{[18,010.33 - 14,951.25]/2}{80.02}$$

$$= 19.1$$

Comparing this F with $F_{2,65,0.999} = 7.72$, we reject H_0 with $P < 0.001$ and conclude (as in Chapter 8) that there is very strong evidence that the two lines are not coincident.

13.3.2 COMPARISON OF METHODS I AND II

The question naturally arises at this point whether the method that uses dummy variables differs from the method that fits two separate regression equations and, if there is a difference, whether one of the methods is preferable to the other.

[The t tests described in Chapter 8 for separate regression fits assume that variance estimates for males and females can be pooled. The dummy-variable method described in this chapter also assumes that such pooling is appropriate. However, when pooled variance estimates are found to be inappropriate (i.e., when variance homogeneity does not hold), the dummy-variable approach should be avoided. Armitage (1971, p. 122) discusses some alternative procedures to pooling when there are unequal variances.]

To determine whether one method is preferable to the other, we first point out that the two methods will yield exactly the same estimated regression coefficients for the two straight-line models. That is, if we fit the model $Y = \beta_0 + \beta_1 X + \beta_2 Z + \beta_3 XZ + E$ by the least-squares method to obtain estimated coefficients $\hat{\beta}_0$, $\hat{\beta}_1$, $\hat{\beta}_2$, and $\hat{\beta}_3$, the straight-line equations obtained by setting $Z = 0$ and $Z = 1$ in this estimated model will be the same as would be obtained if two straight-line regressions were fit separately; that is, if $\hat{\beta}_{0M}$, $\hat{\beta}_{0F}$, $\hat{\beta}_{1M}$, and $\hat{\beta}_{1F}$ denote the estimated regression coefficients from separate regression fits, then $\hat{\beta}_{0M} = \hat{\beta}_0$, $\hat{\beta}_{0F} = \hat{\beta}_0 + \hat{\beta}_2$, $\hat{\beta}_{1M} = \hat{\beta}_1$, and $\hat{\beta}_{1F} = \hat{\beta}_1 + \hat{\beta}_3$. As for the statistical tests concerning the coefficients for the two methods, the following may be said:

1. *The tests for parallel lines are exactly equivalent*; that is, the T statistic with $n_1 + n_2 - 4$ df computed for testing H_0: $\beta_3 = 0$ in the dummy-variable model is exactly the same as the T statistic given in Chapter 8 for testing H_0: $\beta_{1M} = \beta_{1F}$ based on fitting two separate models.

2. *The tests for coincident lines are different*, and the one using the dummy-variable model is, in general, to be preferred. The separate regressions approach given in Chapter 8 tests H_0: $\beta_{1M} = \beta_{1F}$ and H_0: $\beta_{0M} = \beta_{0F}$ separately and rejects the hypothesis of coincident lines if either or both hypotheses are rejected. This is exactly equivalent to performing the two separate tests of H_0: $\beta_2 = 0$ and H_0: $\beta_3 = 0$ for the dummy-variable-model approach, but it is not equivalent to testing the single hypothesis H_0: $\beta_2 = \beta_3 = 0$ (i.e., testing whether β_2 and β_3 are both simultaneously zero).

13.3.3 OTHER DUMMY-VARIABLE MODELS

Two other dummy-variable models that could have been used instead of (13.1) are

$$Y = \beta_0^* + \beta_1^* X + \beta_2^* Z^* + \beta_3^* XZ^* + E \tag{13.2}$$

where

$$Z^* = \begin{cases} 1 & \text{if male} \\ -1 & \text{if female} \end{cases}$$

and

$$Y = \beta_1' X + \beta_2' Z_1' + \beta_3' Z_2' + \beta_4' XZ_1' + \beta_5' XZ_2' + E \tag{13.3}$$

where

$$Z_1' = \begin{cases} 1 & \text{if male} \\ 0 & \text{if female} \end{cases} \quad \text{and} \quad \begin{cases} Z_2' = 0 & \text{if male} \\ 1 & \text{if female} \end{cases}$$

For model (13.2), the separate regression equations are

$$Z^* = \ \ 1: \quad Y_M = (\beta_0^* + \beta_2^*) + (\beta_1^* + \beta_3^*)X + E$$

and

$$Z^* = -1: \quad Y_F = (\beta_0^* - \beta_2^*) + (\beta_1^* - \beta_3^*)X + E$$

The test for parallel lines is equivalent to testing $H_0: \beta_3^* = 0$ and the test for coincident lines is equivalent to testing $H_0: \beta_2^* = \beta_3^* = 0$.[3]

For model (13.3), the separate regression equations are

$$\begin{cases} Z_1' = 1, Z_2' = 0: & Y_M = \beta_2' + (\beta_1' + \beta_4')X + E \\ Z_1' = 0, Z_2' = 1: & Y_F = \beta_3' + (\beta_1' + \beta_5')X + E \end{cases}$$

The test for parallel lines is then equivalent to testing $H_0: \beta_4' = \beta_5'$ (not necessarily equal to zero) and the test for coincident lines is equivalent to testing $H_0: \beta_4' = \beta_5'$ and $\beta_2' = \beta_3'$ (simultaneously).[4]

[The test of $H_0: \beta_4' = \beta_5'$ and $\beta_2' = \beta_3'$ differs from previously discussed multiple-partial F tests because the coefficients in H_0 are not hypothesized to be zero. The general testing procedure in such a case is given as follows:

1. Reduce the model according to the specifications under the null hypothesis; e.g., under H_0, the full model (13.3) becomes

$$Y = \beta_1'X + \beta_2'(Z_1' + Z_2') + \beta_4'(XZ_1' + XZ_2') + E$$

2. Find the SS (residual) for this reduced model.
3. Compute the following F statistic:

$$F = \frac{[\text{SS residual (reduced model)} - \text{SS residual (full model)}]/\nu^*}{\text{MS residual (full model)}}$$

where $\nu^* = $ number of independent linear combinations specified to be zero under H_0 (e.g., in our case, $\nu^* = 2$ since H_0 above specifies that $\beta_4' - \beta_5' = 0$ and that $\beta_2' - \beta_3' = 0$).

4. Test H_0 using F tables with ν^* and $n - k - 1$ df, where k is the number of parameters in the full model (not counting the intercept).]

13.4 COMPARING FOUR REGRESSION EQUATIONS

Suppose that we wish to compare four social class (SC) groups with regard to their separate multiple regressions of SBP on age and weight. For each individual in each SC group, we observe values of the variables $Y = $ SBP, $X_1 = $ age, and $X_2 = $ weight; further, we shall suppose that there are n_i individuals in the ith SC group, $i = 1, 2, 3, 4$. We begin

---

[3] We can express the coefficients of model (13.2) in terms of the regression coefficients for the separate male and female models as follows:

$$\beta_0^* = \frac{\beta_{0M} + \beta_{0F}}{2}, \quad \beta_1^* = \frac{\beta_{1M} + \beta_{1F}}{2}, \quad \beta_2^* = \frac{\beta_{0M} - \beta_{0F}}{2}, \quad \beta_3^* = \frac{\beta_{1M} - \beta_{1F}}{2}$$

[4] We can write the coefficients of model (13.3) in terms of the regression coefficients for the separate male and female models as follows:

$$\beta_1' + \beta_4' = \beta_{1M}, \quad \beta_2' = \beta_{0M}, \quad \beta_3' = \beta_{0F}, \quad \beta_1' + \beta_5' = \beta_{1F}$$

by defining three dummy variables Z_1, Z_2, and Z_3 as follows:

$$Z_1 = \begin{cases} 1 & \text{if SC2} \\ 0 & \text{otherwise,} \end{cases} \quad Z_2 = \begin{cases} 1 & \text{if SC3} \\ 0 & \text{otherwise,} \end{cases} \quad Z_3 = \begin{cases} 1 & \text{if SC4} \\ 0 & \text{otherwise} \end{cases}$$

The complete model to be used (if no interaction between age and weight is considered) is then given as follows:

$$Y = \beta_0 + \beta_1 X_1 + \beta_2 X_2 + \beta_3 Z_1 + \beta_4 Z_2 + \beta_5 Z_3 + \beta_6 X_1 Z_1 + \beta_7 X_2 Z_1 + \beta_8 X_1 Z_2$$
$$+ \beta_9 X_2 Z_2 + \beta_{10} X_1 Z_3 + \beta_{11} X_2 Z_3 + E \tag{13.4}$$

For each particular social class, model (13.4) specializes as follows:

$$SC1(Z_1 = Z_2 = Z_3 = 0): \quad Y = \beta_0 + \beta_1 X_1 + \beta_2 X_2 + E$$
$$SC2(Z_1 = 1, Z_2 = Z_3 = 0): \quad Y = (\beta_0 + \beta_3) + (\beta_1 + \beta_6)X_1 + (\beta_2 + \beta_7)X_2 + E$$
$$SC3(Z_1 = Z_3 = 0, Z_2 = 1): \quad Y = (\beta_0 + \beta_4) + (\beta_1 + \beta_8)X_1 + (\beta_2 + \beta_9)X_2 + E$$
$$SC4(Z_1 = Z_2 = 0, Z_3 = 1): \quad Y = (\beta_0 + \beta_5) + (\beta_1 + \beta_{10})X_1 + (\beta_2 + \beta_{11})X_2 + E$$

13.4.1 TESTS OF HYPOTHESES

The following hypotheses concerning the parameters in model (13.4) are of interest.

1. *Are all four regression equations "parallel"?* (H_0: $\beta_6 = \beta_7 = \beta_8 = \beta_9 = \beta_{10} = \beta_{11} = 0$.) When H_0 is true, the models for each social class reduce to

    $$SC1: \quad Y = \beta_0 + \beta_1 X_1 + \beta_2 X_2 + E$$
    $$SC2: \quad Y = (\beta_0 + \beta_3) + \beta_1 X_1 + \beta_2 X_2 + E$$
    $$SC3: \quad Y = (\beta_0 + \beta_4) + \beta_1 X_1 + \beta_2 X_2 + E$$
    $$SC4: \quad Y = (\beta_0 + \beta_5) + \beta_1 X_1 + \beta_2 X_2 + E$$

    Thus, the coefficients of X_1 and X_2 are the same for each social class when H_0 is true (i.e., the four regression equations are said to be "parallel"). The test statistic for testing H_0 is given by the multiple-partial

    $$F(X_1 Z_1, X_2 Z_1, X_1 Z_2, X_2 Z_2, X_1 Z_3, X_2 Z_3 \,|\, X_1, X_2, Z_1, Z_2, Z_3)$$

    which has 6 and $(n_1 + n_2 + n_3 + n_4 - 12)$ df.

2. *Are all four regression equations coincident?* (H_0: $\beta_3 = \beta_4 = \beta_5 = \beta_6 = \beta_7 = \beta_8 = \beta_9 = \beta_{10} = \beta_{11} = 0$.) When H_0 is true, all four social class models reduce to the form

    $$Y = \beta_0 + \beta_1 X_1 + \beta_2 X_2 + E$$

    The test statistic is the multiple-partial

    $$F(Z_1, Z_2, Z_3, X_1 Z_1, X_1 Z_2, X_2 Z_1, X_2 Z_2, X_1 Z_3, X_2 Z_3 \,|\, X_1, X_2)$$

    which has 9 and $(n_1 + n_2 + n_3 + n_4 - 12)$ df.

13.4.2 OTHER DUMMY-VARIABLE MODELS

Two other dummy-variable coding schemes are as follows:

1. Define

$$Z_1^* = \begin{cases} -1 & \text{if SC1} \\ 1 & \text{if SC2} \\ 0 & \text{if SC3} \\ 0 & \text{if SC4} \end{cases}, \quad Z_2^* = \begin{cases} -1 & \text{if SC1} \\ 0 & \text{if SC2} \\ 1 & \text{if SC3} \\ 0 & \text{if SC4} \end{cases}, \quad Z_3^* = \begin{cases} -1 & \text{if SC1} \\ 0 & \text{if SC2} \\ 0 & \text{if SC3} \\ 1 & \text{if SC4} \end{cases}$$

and then use the model

$$Y = \beta_0^* + \beta_1^* X_1 + \beta_2^* X_2 + \beta_3^* Z_1^* + \beta_4^* Z_2^* + \beta_5^* Z_3^* + \beta_6^* X_1 Z_1^* + \beta_7^* X_2 Z_1^* + \beta_8^* X_1 Z_2^*$$
$$+ \beta_9^* X_2 Z_2^* + \beta_{10}^* X_1 Z_3^* + \beta_{11}^* X_2 Z_3^* + E \tag{13.5}$$

2. Define

$$Z_1' = \begin{cases} 1 & \text{if SC1} \\ 0 & \text{otherwise} \end{cases}, \quad Z_2' = \begin{cases} 1 & \text{if SC2} \\ 0 & \text{otherwise} \end{cases}, \quad Z_3' = \begin{cases} 1 & \text{if SC3} \\ 0 & \text{otherwise} \end{cases}$$

$$Z_4' = \begin{cases} 1 & \text{if SC4} \\ 0 & \text{otherwise} \end{cases}$$

and then use the model

$$Y = \beta_1' X_1 + \beta_2' X_2 + \beta_3' Z_1' + \beta_4' Z_2' + \beta_5' Z_3' + \beta_6' Z_4' + \beta_7' X_1 Z_1' + \beta_8' X_2 Z_1'$$
$$+ \beta_9' X_1 Z_1' + \beta_{10}' X_2 Z_2' + \beta_{11}' X_1 Z_3' + \beta_{12}' X_2 Z_3' + \beta_{13}' X_1 Z_4' + \beta_{14}' X_2 Z_4' + E \tag{13.6}$$

For the dummy-variable coding given by scheme 1, the four regression equations for the four social classes based on model (13.5) are

$$SC1(Z_1^* = Z_2^* = Z_3^* = -1): \quad Y = (\beta_0^* - \beta_3^* - \beta_4^* - \beta_5^*) + (\beta_1^* - \beta_6^* - \beta_8^* - \beta_{10}^*) X_1$$
$$+ (\beta_2^* - \beta_7^* - \beta_9^* - \beta_{11}^*) X_2 + E$$

$$SC2(Z_1^* = 1, Z_2^* = Z_3^* = 0): \quad Y = (\beta_0^* + \beta_3^*) + (\beta_1^* + \beta_6^*) X_1 + (\beta_2^* + \beta_7^*) X_2 + E$$

$$SC3(Z_1^* = 0, Z_2^* = 1, Z_3^* = 0): \quad Y = (\beta_0^* + \beta_4^*) + (\beta_1^* + \beta_8^*) X_1 + (\beta_2^* + \beta_9^*) X_2 + E$$

$$SC4(Z_1^* = Z_2^* = 0, Z_3^* = 1): \quad Y = (\beta_0^* + \beta_5^*) + (\beta_1^* + \beta_{10}^*) X_1 + (\beta_2^* + \beta_{11}^*) X_2 + E$$

So, the null hypotheses to be tested concerning parallelism and coincidence (using appropriate multiple-partial F tests) are:

Parallelism: $H_0: \beta_6^* = \beta_7^* = \beta_8^* = \beta_9^* = \beta_{10}^* = \beta_{11}^* = 0$

Coincidence: $H_0: \beta_3^* = \beta_4^* = \beta_5^* = \beta_6^* = \beta_7^* = \beta_8^* = \beta_9^* = \beta_{10}^* = \beta_{11}^* = 0$

For the dummy-variable coding given by scheme 2, the four regression equations based on model (13.6) are

$$SC1(Z_1' = 1, Z_2' = Z_3' = Z_4' = 0): \quad Y = \beta_3' + (\beta_1' + \beta_7')X_1 + (\beta_2' + \beta_8')X_2 + E$$

$$SC2(Z_2' = 1, Z_1' = Z_3' = Z_4' = 0): \quad Y = \beta_4' + (\beta_1' + \beta_9')X_1 + (\beta_2' + \beta_{10}')X_2 + E$$

$$SC3(Z_3' = 1, Z_1' = Z_2' = Z_4' = 0): \quad Y = \beta_5' + (\beta_1' + \beta_{11}')X_1 + (\beta_2' + \beta_{12}')X_2 + E$$

$$SC4(Z_4' = 1, Z_1' = Z_2' = Z_3' = 0): \quad Y = \beta_6' + (\beta_1' + \beta_{13}')X_1 + (\beta_2' + \beta_{14}')X_2 + E$$

So, the null hypotheses to be tested concerning parallelism and coincidence are:

Parallelism: $H_0: \beta_7' = \beta_9' = \beta_{11}' = \beta_{13}'$ and $\beta_8' = \beta_{10}' = \beta_{12}' = \beta_{14}'$

Coincidence: $H_0: \beta_3' = \beta_4' = \beta_5' = \beta_6'$ and $\beta_7' = \beta_9' = \beta_{11}' = \beta_{13}'$

               and $\beta_8' = \beta_{10}' = \beta_{12}' = \beta_{14}'$

[These tests for parallelism and coincidence concern the equality of sets of coefficients not necessarily equal to zero. Such a situation requires the use of the general testing procedure described on page 194. The ν^* values for the tests here are ν^* (parallelism) = 6 and ν^* (coincidence) = 9.]

13.5 COMPARING SEVERAL REGRESSION EQUATIONS INVOLVING TWO NOMINAL VARIABLES

Suppose that we want to compare eight regression equations of SBP (Y) on AGE (X_1) and WGT (X_2) corresponding to the eight combinations of SEX (Q) and social class (SC) groups. Then the following regression model can be used to make this comparison:

$$Y = \beta_0 + \beta_1 X_1 + \beta_2 X_2 + \beta_3 Z_1 + \beta_4 Z_2 + \beta_5 Z_3 + \beta_6 Q + \beta_7 Z_1 Q + \beta_8 Z_2 Q + \beta_9 Z_3 Q$$

$$+ \beta_{10} X_1 Z_1 + \beta_{11} X_2 Z_1 + \beta_{12} X_1 Z_2 + \beta_{13} X_2 Z_2 + \beta_{14} X_1 Z_3 + \beta_{15} X_2 Z_3$$

$$+ \beta_{16} X_1 Q + \beta_{17} X_2 Q + \beta_{18} X_1 Z_1 Q + \beta_{19} X_2 Z_1 Q + \beta_{20} X_1 Z_2 Q + \beta_{21} X_2 Z_2 Q$$

$$+ \beta_{22} X_1 Z_3 Q + \beta_{23} X_2 Z_3 Q + E \tag{13.7}$$

where the dummy variables are defined as

$$Z_1 = \begin{cases} 1 & \text{if SC2} \\ 0 & \text{otherwise} \end{cases}, \quad Z_2 = \begin{cases} 1 & \text{if SC3} \\ 0 & \text{otherwise} \end{cases}, \quad Z_3 = \begin{cases} 1 & \text{if SC4} \\ 0 & \text{otherwise} \end{cases}, \quad Q = \begin{cases} 1 & \text{if male} \\ 0 & \text{if female} \end{cases}$$

For each sex–SC combination, we have:

SC1–male: $Y = (\beta_0 + \beta_6) + (\beta_1 + \beta_{16})X_1 + (\beta_2 + \beta_{17})X_2 + E$

SC2–male: $Y = (\beta_0 + \beta_3 + \beta_6 + \beta_7) + (\beta_1 + \beta_{10} + \beta_{16} + \beta_{18})X_1$

                $+ (\beta_2 + \beta_{11} + \beta_{17} + \beta_{19})X_2 + E$

$$SC3\text{–male:} \quad Y = (\beta_0 + \beta_4 + \beta_6 + \beta_8) + (\beta_1 + \beta_{12} + \beta_{16} + \beta_{20})X_1$$
$$+ (\beta_2 + \beta_{13} + \beta_{17} + \beta_{21})X_2 + E$$

$$SC4\text{–male:} \quad Y = (\beta_0 + \beta_5 + \beta_6 + \beta_9) + (\beta_1 + \beta_{14} + \beta_{16} + \beta_{22})X_1$$
$$+ (\beta_2 + \beta_{15} + \beta_{17} + \beta_{23})X_2 + E$$

$$SC1\text{–female:} \quad Y = \beta_0 + \beta_1 X_1 + \beta_2 X_2 + E$$

$$SC2\text{–female:} \quad Y = (\beta_0 + \beta_3) + (\beta_1 + \beta_{10})X_1 + (\beta_2 + \beta_{11})X_2 + E$$

$$SC3\text{–female:} \quad Y = (\beta_0 + \beta_4) + (\beta_1 + \beta_{12})X_1 + (\beta_2 + \beta_{13})X_2 + E$$

$$SC4\text{–female:} \quad Y = (\beta_0 + \beta_5) + (\beta_1 + \beta_{14})X_1 + (\beta_2 + \beta_{15})X_2 + E$$

13.5.1 TESTS OF HYPOTHESES

The following hypotheses concerning the parameters in model (13.7) are of interest.

1. *Male and female regression equations are parallel* (*controlling for social class*). This is a test of

$$H_0: \beta_{16} = \beta_{17} = \beta_{18} = \beta_{19} = \beta_{20} = \beta_{21} = \beta_{22} = \beta_{23} = 0$$

When this null hypothesis is true, the eight equations above reduce to

$$SC1\text{–male:} \quad Y = (\beta_0 + \beta_6) + \beta_1 X_1 + \beta_2 X_2 + E$$
$$SC1\text{–female:} \quad Y = \beta_0 + \beta_1 X_1 + \beta_2 X_2 + E$$

$$SC2\text{–male:} \quad Y = (\beta_0 + \beta_3 + \beta_6 + \beta_7) + (\beta_1 + \beta_{10})X_1 + (\beta_2 + \beta_{11})X_2 + E$$
$$SC2\text{–female:} \quad Y = (\beta_0 + \beta_3) + (\beta_1 + \beta_{10})X_1 + (\beta_2 + \beta_{11})X_2 + E$$

$$SC3\text{–male:} \quad Y = (\beta_0 + \beta_4 + \beta_6 + \beta_8) + (\beta_1 + \beta_{12})X_1 + (\beta_2 + \beta_{13})X_2 + E$$
$$SC3\text{–female:} \quad Y = (\beta_0 + \beta_4) + (\beta_1 + \beta_{12})X_1 + (\beta_2 + \beta_{13})X_2 + E$$

$$SC4\text{–male:} \quad Y = (\beta_0 + \beta_5 + \beta_6 + \beta_9) + (\beta_1 + \beta_{14})X_1 + (\beta_2 + \beta_{15})X_2 + E$$
$$SC4\text{–female:} \quad Y = (\beta_0 + \beta_5) + (\beta_1 + \beta_{14})X_1 + (\beta_2 + \beta_{15})X_2 + E$$

Thus, within any specific social class group, the male and female regression equations are parallel (since they have the same X_1 and X_2 coefficients).

2. *Male and female regression equations are coincident* (*controlling for social class*). This is a test of

$$H_0: \beta_6 = \beta_7 = \beta_8 = \beta_9 = \beta_{16} = \beta_{17} = \beta_{18} = \beta_{19} = \beta_{20} = \beta_{21} = \beta_{22} = \beta_{23} = 0$$

3. *All four social class equations are parallel* (*controlling for sex*). This is a test of

$$H_0: \beta_{10} = \beta_{11} = \beta_{12} = \beta_{13} = \beta_{14} = \beta_{15} = \beta_{18} = \beta_{19} = \beta_{20} = \beta_{21} = \beta_{22} = \beta_{23} = 0$$

When this hypothesis is true, the eight equations reduce to

$$\begin{cases} \text{SC1–male:} & Y = (\beta_0+\beta_6)+(\beta_1+\beta_{16})X_1+(\beta_2+\beta_{17})X_2+E \\ \text{SC2–male:} & Y = (\beta_0+\beta_3+\beta_6+\beta_7)+(\beta_1+\beta_{16})X_1+(\beta_2+\beta_{17})X_2+E \\ \text{SC3–male:} & Y = (\beta_0+\beta_4+\beta_6+\beta_8)+(\beta_1+\beta_{16})X_1+(\beta_2+\beta_{17})X_2+E \\ \text{SC4–male:} & Y = (\beta_0+\beta_5+\beta_6+\beta_9)+(\beta_1+\beta_{16})X_1+(\beta_2+\beta_{17})X_2+E \end{cases}$$

$$\begin{cases} \text{SC1–female:} & Y = \beta_0+\beta_1 X_1+\beta_2 X_2+E \\ \text{SC2–female:} & Y = (\beta_0+\beta_3)+\beta_1 X_1+\beta_2 X_2+E \\ \text{SC3–female:} & Y = (\beta_0+\beta_4)+\beta_1 X_1+\beta_2 X_2+E \\ \text{SC4–female:} & Y = (\beta_0+\beta_5)+\beta_1 X_1+\beta_2 X_2+E \end{cases}$$

Thus, within any given sex group, all four regression equations have the same coefficients in X_1 and X_2.

4. *All four social class equations are coincident (controlling for sex)*. This is a test of

$$H_0: \beta_3=\beta_4=\beta_5=\beta_7=\beta_8=\beta_9=\beta_{10}=\beta_{11}=\beta_{12}=\beta_{13}=\beta_{14}$$
$$=\beta_{15}=\beta_{18}=\beta_{19}=\beta_{20}=\beta_{21}=\beta_{22}=\beta_{23}=0$$

5. *All eight regression equations are parallel*. This is a test of

$H_0: \beta_{10}$ through β_{23} are simultaneously zero

When this hypothesis is true, the eight equations are all of the form

$$Y = \beta_{0(i)}+\beta_1 X_1+\beta_2 X_2+E \qquad \text{for } i = 1, 2, \ldots, 8$$

(i.e., the eight models differ only in intercept).

6. *All eight regression equations are coincident*. This is a test of

$H_0: \beta_3$ through β_{23} are simultaneously zero

7. *There is no "interaction" effect between sex and social class*. This is a test of

$$H_0: \beta_7=\beta_8=\beta_9=\beta_{18}=\beta_{19}=\beta_{20}=\beta_{21}=\beta_{22}=\beta_{23}=0$$

From the form of the complete model (13.7), each of the coefficients in H_0 corresponds to a product term of the general form $Z_i Q$ or $X_j Z_i Q$, which involves the product of a social class variable and the sex variable.

PROBLEMS   1.   In Problem 1 of Chapter 8, separate straight-line regressions of SBP on QUET (using the data given in Problem 2 of Chapter 5) were compared for smokers (SMK = 1) and nonsmokers (SMK = 0).
(a)   Define a single multiple regression model which involves the data for both smokers and nonsmokers and which defines straight-line models for each

group with possibly differing intercepts and slopes. Define the intercept and slope for each straight-line model in terms of the regression coefficients of the single regression model.

(b) Using the computer results provided, determine and plot on graph paper the two fitted straight lines obtained from the fit of the regression model

(c) Test the following null hypotheses:

H_0: the two lines are parallel

H_0: the two lines are coincident

For each of these tests, state the appropriate null hypothesis in terms of the regression coefficients of the regression model.

(d) Compare your answers in parts (b) and (c) with those previously obtained from fitting separate regressions in Problem 1 of Chapter 8.

Computer results for Problem 1

VARIABLE ENTERED ON STEP 1: QUET				ANOVA (STEP 1)		
VARIABLE	$\hat{\beta}$	$S_{\hat{\beta}}$	PARTIAL F	SOURCE	df	SS
QUET	21.49167	3.54515	36.751	Regression	1	3,537.94538
(constant)	70.57643			Residual	30	2,888.02337

VARIABLE ENTERED ON STEP 2: SMK				ANOVA (STEP 2)		
VARIABLE	$\hat{\beta}$	$S_{\hat{\beta}}$	PARTIAL F	SOURCE	df	SS
QUET	22.11560	3.22996	46.882	Regression	2	4,120.36603
SMK	8.57101	3.16670	7.326	Residual	29	2,305.60272
(constant)	63.87606					

VARIABLE ENTERED ON STEP 3: QUET · SMK				ANOVA (STEP 3)		
VARIABLE	$\hat{\beta}$	$S_{\hat{\beta}}$	PARTIAL F	SOURCE	df	SS
QUET	26.30282	5.70349	21.268	Regression	3	4,184.10718
SMK	29.94357	24.16355	1.536	Residual	28	2,241.86157
QUET · SMK	−6.18479	6.93171	0.796			
(constant)	49.31178					

2. Use the computer results to compare the separate regressions of SBP on AGE and QUET for smokers and nonsmokers (using the data in Problem 2 of Chapter 5) as follows:

(a) State the appropriate regression model incorporating both equations for smokers and nonsmokers.

(b) Determine the fitted regression equations for smokers and nonsmokers separately using the fitted regression model.

(c) Test for parallelism of the two models, stating the null hypothesis in terms of appropriate regression coefficients.

(d) Test for coincidence of the two models, stating the null hypothesis in terms of regression coefficients.

Computer results for Problem 2

| VARIABLE(S) ENTERED ON STEP 1: AGE, QUET | | | | ANOVA (STEP 1) | | |
VARIABLE	$\hat{\beta}$	$S_{\hat{\beta}}$	PARTIAL F	SOURCE	df	SS
AGE	1.04516	0.38606	7.329	Regression	2	4,120.59219
QUET	9.75073	5.40245	3.258	Residual	29	2,305.37656
(constant)	55.32344					

| VARIABLE ENTERED ON STEP 2: SMK | | | | ANOVA (STEP 2) | | |
VARIABLE	$\hat{\beta}$	$S_{\hat{\beta}}$	PARTIAL F	SOURCE	df	SS
AGE	1.21271	0.32382	14.025	Regression	3	4,889.82563
QUET	8.59245	4.49868	3.648	Residual	28	1,536.14312
SMK	9.94557	2.65606	14.021			
(constant)	45.10320					

| VARIABLE(S) ENTERED ON STEP 3: AGE · SMK, QUET · SMK | | | | ANOVA (STEP 3) | | |
VARIABLE	$\hat{\beta}$	$S_{\hat{\beta}}$	PARTIAL F	SOURCE	df	SS
AGE	1.02892	0.50177	4.205	Regression	5	4,915.63033
QUET	10.45104	9.13014	1.310	Residual	26	1,510.33842
SMK	−0.53744	23.23004	0.001			
AGE · SMK	0.43733	0.71279	0.376			
QUET· SMK	−3.70682	10.76763	0.119			
(constant)	48.61271					

3. Using only the ANOVA table for the example discussed in Chapter 12 (based on Gruber's study of the determinants of blood pressure change), answer the following:
   (a) Determine the form of the separate fitted regression models of SBPSL (Y) on SBP1 (X_1), RW (X_3), and SBP1 · RW (X_5) for each sex in terms of the regression coefficients of the fitted regression model.
   (b) Using this ANOVA table, why can't one test whether these two regression equations are either parallel or coincident? Describe the appropriate testing procedure in each case.

4. For the data in Problem 2 of Chapter 8 concerning the relationship between per capita income (PCI) and population percentage under age 15 (YNG) for two groups of developing countries, address the following using the information provided:
   (a) State the regression model that incorporates the straight-line models for each group of countries.
   (b) Determine and plot the separate fitted straight lines based on the fitted regression model given in part (a).
   (c) Test for parallelism of the two straight lines.
   (d) Test for coincidence of the two straight lines.
   (e) Compare your results in parts (b)–(d) to those obtained in Problem 2 of Chapter 8.

Computer results for Problem 4

VARIABLE ENTERED ON STEP 1: YNG				ANOVA (STEP 1)		
VARIABLE	$\hat{\beta}$	$S_{\hat{\beta}}$	PARTIAL F	SOURCE	df	SS
YNG	−32.1236	4.6767	47.182	Regression (YNG)	1	1,095,554.6
(constant)	1,636.9963			Residual	50	1,160,995.1

VARIABLE ENTERED ON STEP 2: Z				ANOVA (STEP 2)		
VARIABLE	$\hat{\beta}$	$S_{\hat{\beta}}$	PARTIAL F	SOURCE	df	SS
YNG	−29.9394	1.9587	233.637	Regression (YNG, Z)	2	2,058,010.1
$Z\begin{cases}1 \text{ if Latin} \\ \quad\text{American} \\ 0 \text{ otherwise}\end{cases}$	272.8088	17.7008	237.536	Residual	49	198,539.6
(constant)	1,411.9048					

VARIABLE ENTERED ON STEP 3: YNG Z				ANOVA (STEP 3)		
VARIABLE	$\hat{\beta}$	$S_{\hat{\beta}}$	PARTIAL F	SOURCE	df	SS
YNG	−17.2758	1.2260	198.561	Regression (YNG, Z, YNG · Z)	3	2,219,373.9
Z	1,277.0988	70.0061	332.795	Residual	48	37,175.8
(YNG · Z)	−24.7291	1.7132	208.347			
(constant)	893.5713					

5. Using the information given, answer the same questions as in Problem 4 concerning the regression of height (Y) on age (X) for children in one of two diet categories (which is based on the data in Problem 3 of Chapter 8).

Computer results for Problem 5

VARIABLE ENTERED ON STEP 1: AGE				ANOVA (STEP 1)		
VARIABLE	$\hat{\beta}$	$S_{\hat{\beta}}$	PARTIAL F	SOURCE	df	SS
AGE	12.0445	1.5353	61.546	Regression (AGE)	1	3,053.35
(constant)	51.0600			Residual	25	1,240.28

VARIABLE ENTERED ON STEP 2: Z				ANOVA (STEP 2)		
VARIABLE	$\hat{\beta}$	$S_{\hat{\beta}}$	PARTIAL F	SOURCE	df	SS
AGE	12.0583	0.8897	183.699	Regression (AGE, Z)	2	3,893.80
$Z\begin{cases}1 \text{ if protein} \\ \quad\text{rich} \\ 0 \text{ otherwise}\end{cases}$	11.1662	1.5721	50.449	Residual	24	399.83
(constant)	45.6610					

VARIABLE ENTERED ON STEP 3: AGE · Z				ANOVA (STEP 3)		
VARIABLE	$\hat{\beta}$	$s_{\hat{\beta}}$	PARTIAL F	SOURCE	df	SS
AGE	8.6860	0.6762	164.999	Regression (AGE, Z, AGE · Z)	3	4,174.21
Z	−0.9015	1.8619	0.234	Residual	23	119.42
AGE · Z	7.3229	0.9965	54.005			
(constant)	51.2252					

6. In Gruber's (1970) study of $n = 104$ individuals (discussed in Chapter 12), the relationship between blood pressure change (SBPSL) and relative weight (RW) controlling for initial blood pressure (SBP1) was compared for three different geographical backgrounds and for three different psychosocial orientations using the following 15 variables:

Y = SBPSL
X_1 = SBP1 (initial blood pressure)
X_2 = R (1 if rural background, 0 if town, and −1 if urban)
X_3 = T (1 if town background, 0 if rural, and −1 if urban)
X_4 = TD (1 if traditional orientation, 0 if transitional, −1 if modern)
X_5 = TN (1 if transitional orientation, 0 if traditional, −1 if modern)
X_6 = RW (relative weight)
X_7 = T × TD
X_8 = T × TN
X_9 = R × TD
X_{10} = R × TN
X_{11} = R × TD × RW
X_{12} = R × TN × RW
X_{13} = T × TD × RW
X_{14} = T × TN × RW

A run of a standard stepwise-regression program using these data yielded the accompanying ANOVA table (variables were forced to enter in the order presented) based on the model

$$Y = \beta_0 + \beta_1 X_1 + \beta_2 X_2 + \cdots + \beta_{14} X_{14} + E$$

(a) Using the regression model given, determine the form of the nine fitted regression equations corresponding to the nine possible combinations of background with orientation (i.e., R = 1 and TD = 1, R = 0 and TD = 1, R = −1 and TD = 1, etc.). [*Note:* Each of the nine equations will be of the form $\hat{Y} = \hat{\beta}_0^* + \hat{\beta}_1^* (SBP1) + \hat{\beta}_2^* (RW)$.]

(b) Test the null hypothesis that the nine regression equations determined in part (a) are parallel. State the null hypothesis in terms of the regression coefficients of the original 14-variable regression model.

(c) Test the hypothesis "H_0: the three regression equations corresponding to the three backgrounds (rural, town, and urban) are parallel (but not necessarily coincident)" against the alternative "H_A: they are not parallel."

(d) Set up the formula for testing "H_0: the nine regression equations dealt with in part (a) are coincident" against "H_A: they are not coincident." State the null hypothesis in terms of the coefficients in the regression equation.

SOURCE		df	SS	F
	X_1	1	24.9878	28.830
	$X_2 \vert X_1$	1	0.5218	0.600
	$X_3 \vert X_1, X_2$	1	0.0057	0.006
	$X_4 \vert X_1, X_2, X_3$	1	1.0520	1.199
	$X_5 \vert X_1, X_2, X_3, X_4$	1	1.1116	1.271
	$X_6 \vert X_1$ through X_5	1	0.8321	0.951
Regression	$X_7 \vert X_1$ through X_6	1	0.2919	0.331
	$X_8 \vert X_1$ through X_7	1	1.6601	1.902
	$X_9 \vert X_1$ through X_8	1	0.5843	0.667
	$X_{10} \vert X_1$ through X_9	1	0.2266	0.257
	$X_{11} \vert X_1$ through X_{10}	1	1.1916	1.355
	$X_{12} \vert X_1$ through X_{11}	1	2.0853	2.407
	$X_{13} \vert X_1$ through X_{12}	1	1.5915	1.854
	$X_{14} \vert X_1$ through X_{13}	1	0.0208	0.024
Residual		89	77.2303	
Total		103	113.3934	

7. The Environmental Protection Agency conducted an experiment to assess the characteristics of sampling procedures designed to measure the suspended particulate concentration (X) in a particular city. At each of two distinct locations (designated location 1 and location 2), two identical sampling units were set up side by side and readings were taken on each of 10 days. The data are given here in tabular form, where X_{ij1} and X_{ij2} are, respectively, the measured concentration for samplers 1 and 2 at location i on day j ($i = 1, 2$; $j = 1, 2, \ldots, 10$). It was hypothesized that the inherent variation in the observations depends on the level of concentration being measured. To quantify this hypothesis, it is proposed to fit a model of the form

$$|d_{ij}| = \beta_0 + \beta_1 Z + \beta_2(X_{ij1} + X_{ij2}) + \beta_3(X_{ij1} + X_{ij2})Z + E$$

where Z is $+1$ if the observation pertains to location 1 and is zero otherwise, and where $|d_{ij}| = |X_{ij1} - X_{ij2}|$.

(a) Using the results provided, determine and interpret the fitted straight-line relationship between $|d_{ij}|$ and $(X_{ij1} + X_{ij2})$ at each location.

(b) Test for parallelism of the two lines.

	LOCATION 1			LOCATION 2		
DAY	X_{1j1}	X_{1j2}	$d_{1j} = X_{1j1} - X_{1j2}$	X_{2j1}	X_{2j2}	$d_{2j} = X_{2j1} - X_{2j2}$
1	4	3	+1	6	5	+1
2	8	6	+2	3	1	+2
3	12	16	−4	1	2	−1
4	1	1	0	10	12	−2
5	7	6	+1	17	17	0
6	11	8	+3	4	7	−3
7	14	10	+4	8	6	+2
8	10	12	−2	12	12	0
9	2	2	0	10	9	+1
10	15	20	−5	20	19	+1

(c) Test for coincidence of the two lines.
(d) How would one test whether the level of concentration is significantly related to the inherent variation in the observations for at least one of the two locations?

Computer results for Problem 7

VARIABLE ENTERED ON STEP 1: $X_{ij1}+X_{ij2}$				ANOVA (STEP 1)		
VARIABLE	$\hat{\beta}$	$S_{\hat{\beta}}$	PARTIAL F	SOURCE	df	SS
$(X_{ij1}+X_{ij2})$	0.0455	0.0288	2.493	Regression $(X_{ij1}+X_{ij2})$	1	4.8348
(constant)	0.9560			Residual	18	34.9152

VARIABLE ENTERED ON STEP 2: Z				ANOVA (STEP 2)		
VARIABLE	$\hat{\beta}$	$S_{\hat{\beta}}$	PARTIAL F	SOURCE	df	SS
$(X_{ij1}+X_{ij2})$	0.0482	0.0277	3.031	Regression $(X_{ij1}+X_{ij2}, Z)$	2	9.4514
$Z\begin{pmatrix}1 \text{ if loc. 1} \\ 0 \text{ otherwise}\end{pmatrix}$	0.9626	0.5981	2.590	Residual	17	30.2986
(constant)	0.4279					

VARIABLE ENTERED ON STEP 3: $(X_{ij1}+X_{ij2})\cdot Z$				ANOVA (STEP 3)		
VARIABLE	$\hat{\beta}$	$S_{\hat{\beta}}$	PARTIAL F	SOURCE	df	SS
$(X_{ij1}+X_{ij2})$	−0.0392	0.0202	3.746	Regression (all three variables)	3	31.3395
Z	−2.4279	0.6177	15.448	Residual	16	8.4105
$(X_{ij1}+X_{ij2})\cdot Z$	0.1951	0.0302	41.640			
(constant)	2.0086					

8. A biologist compared the effect of temperature of each of two mediums on the growth of human amniotic cells in a tissue culture. The data obtained are as shown.

(a) Assuming that a parabolic model is appropriate for describing the relationship between Y and X for each medium, provide a single regression model that will incorporate two separate parabolic models, one corresponding to each medium.
(b) Use the computer printout provided to determine and plot the separate fitted parabolas for each medium. (*Note:* $Z=0$ for medium A and $Z=1$ for medium B.)
(c) Test for "parallelism" of the two parabolas.
(d) Test for coincidence of the two parabolas.
(e) Using only the computer results given, is it possible to test whether a quadratic term should be included in the model for each medium? Explain.

MEDIUM A		MEDIUM B	
NO. OF CELLS $\times 10^{-6}$ (Y)	TEMPERATURE (X)	NO. OF CELLS $\times 10^{-6}$ (Y)	TEMPERATURE (X)
1.13	40	0.98	40
1.20	40	1.05	40
1.00	40	0.92	40
0.91	40	0.90	40
1.05	40	0.89	40
1.75	60	1.60	60
1.45	60	1.45	60
1.55	60	1.40	60
1.64	60	1.50	60
1.60	60	1.56	60
2.30	80	2.20	80
2.15	80	2.10	80
2.25	80	2.20	80
2.40	80	2.30	80
2.49	80	2.26	80
3.18	100	3.10	100
3.10	100	3.00	100
3.28	100	3.13	100
3.35	100	3.20	100
3.12	100	3.07	100

Computer results for Problem 8

SOURCE	df	SS	SOURCE	df	SS
Regression (X)	1	25.6686	Regression (X, X^2)	2	25.9593
Residual	38	0.7036	Residual	37	0.4129

SOURCE	df	SS	SOURCE	df	SS
Regression (X, X^2, Z)	3	26.0685	Regression (X, X^2, Z, ZX)	4	26.0685
Residual	36	0.3037	Residual	35	0.3037

SOURCE	df	SS
Regression (X, X^2, Z, ZX, ZX^2)	5	26.0686
Residual	34	0.3036

Regression coefficients for the fitted model containing all five independent variables: constant: 0.4946; X: 0.00537; X^2: 0.00022; Z: -0.1437; ZX: 0.00123; ZX^2: -0.00001.

9. Answer the following questions using the computer results given based on the data of Problem 6 of Chapter 8 concerning the growth rates (Y) of depleted chicks at different (log) dosage levels (X) of vitamin B for males and females:

Computer printout for Problem 9 (SPSS)

VARIABLE(S) ENTERED ON STEP NUMBER 1.. X

		ANALYSIS OF VARIANCE	DF	SUM OF SQUARES	MEAN SQUARE	F
MULTIPLE R	0.88583	REGRESSION	1.	1041.33728	1041.33728	109.33735
R SQUARE	0.78470	RESIDUAL	30.	285.72230	9.52408	
ADJUSTED R SQUARE	0.77752					
STANDARD ERROR	3.08611					

———————— VARIABLES IN THE EQUATION ————————

VARIABLE	B	BETA	STD ERROR B	F
X	13.85501	0.88583	1.32502	109.337
(CONSTANT)	14.72828			

———————— VARIABLES NOT IN THE EQUATION ————————

VARIABLE	BETA IN	PARTIAL	TOLERANCE	F
Z	0.01293	0.02784	0.99791	0.022
XZ	0.04475	0.08801	0.83267	0.226

VARIABLE(S) ENTERED ON STEP NUMBER 2.. Z (= 1 if MALE, = 0 if FEMALE)

		ANALYSIS OF VARIANCE	DF	SUM OF SQUARES	MEAN SQUARE	F
MULTIPLE R	0.88592	REGRESSION	2.	1041.55874	520.77937	52.89862
R SQUARE	0.78486	RESIDUAL	29.	285.50084	9.84486	
ADJUSTED R SQUARE	0.77002					
STANDARD ERROR	3.13765					

———————— VARIABLES IN THE EQUATION ————————

VARIABLE	B	BETA	STD ERROR B	F
X	13.84577	0.88524	1.34856	105.413
Z	0.16655	0.01293	1.11049	0.022
(CONSTANT)	14.65352			

———————— VARIABLES NOT IN THE EQUATION ————————

VARIABLE	BETA IN	PARTIAL	TOLERANCE	F
XZ	0.19139	0.15355	0.13847	0.676

VARIABLE(S) ENTERED ON STEP NUMBER 3.. XZ

		ANALYSIS OF VARIANCE	DF	SUM OF SQUARES	MEAN SQUARE	F
MULTIPLE R	0.88878	REGRESSION	3.	1048.28988	349.42996	35.09721
R SQUARE	0.78993	RESIDUAL	28.	278.76970	9.95606	
ADJUSTED R SQUARE	0.76743					
STANDARD ERROR	3.15532					

———————— VARIABLES IN THE EQUATION ————————

VARIABLE	B	BETA	STD ERROR B	F
X	12.73533	0.81424	1.91389	44.278
Z	-1.88944	-0.14670	2.73852	0.476
XZ	2.23020	0.19139	2.71233	0.676
(CONSTANT)	15.65624			

———————— VARIABLES NOT IN THE EQUATION ————————

VARIABLE	BETA IN	PARTIAL	TOLERANCE	F

From *Statistical Package for the Social Sciences* by Nie et al. Copyright © 1975 by McGraw-Hill Book Company and Dr. Norman Nie, President, SPSS Inc. Used with permission of McGraw-Hill Book Company and Dr. Norman Nie, President, SPSS Inc.

(a) Define a single multiple regression model that incorporates different straight-line models for males and females.

(b) Plot the fitted straight lines for each sex on graph paper.

(c) Test for parallelism.

(d) Test for coincidence.

(e) Compare your results regarding parts (b), (c), and (d) to those obtained for Problem 6 of Chapter 8.

CHAPTER 14

ANALYSIS OF COVARIANCE AND OTHER METHODS FOR ADJUSTING CONTINUOUS DATA

14.1 ADJUSTMENT PROBLEM

Suppose, as in Example 1 of Section 13.3 (which dealt with the data described in Chapter 8), that we have a sample of observations on the dependent variable SBP and the independent variable age for $n_F = 29$ women and $n_M = 40$ men.

Two questions often of interest when analyzing such data are as follows:

1. Is the true straight-line relationship between blood pressure and age (given that a straight-line model is adequate) the same for the male and female populations?
2. Do the mean blood pressure levels for the male and female groups differ significantly after taking into account (i.e., after adjusting for or controlling for) the possible confounding effect due to differing age distributions in the two groups?

Although these questions are not completely unrelated in terms of the statistical techniques required to answer them, they nevertheless differ in emphasis: question 1 focuses on a comparison of straight-line regression equations, whereas question 2 focuses on a comparison of the mean blood pressure levels in the two groups.

We have already considered question 1 (in Chapter 13) through use of the model

$$Y = \beta_0 + \beta_1 X + \beta_2 Z + \beta_3 XZ + E \tag{14.1}$$

where Z is a dummy variable identifying the sex group (i.e., $Z = 1$ if female and $Z = 0$ if male). Based on appropriate tests of hypotheses concerning the parameters in this model, three important conclusions could be reached regarding question 1. These are:

1. The lines are coincident (i.e., $\beta_2 = \beta_3 = 0$).
2. The lines are parallel, but not coincident (i.e., $\beta_3 = 0$, but $\beta_2 \neq 0$).
3. The lines are not parallel ($\beta_3 \neq 0$).

These conclusions have a great deal to do with the answer to question 2. If conclusion 1 is appropriate, we would say that the two sex groups do not differ in mean blood pressure level after controlling for the effect of age. If conclusion 2 holds, we would say that the sex group associated with the "higher" straight line has a higher mean blood pressure level at all ages. If conclusion 3 is valid, we would have to look closer at the orientation of the two straight lines relative to one another: if they do not intersect in the age range of interest, we would say that the sex group associated with the higher curve has a higher mean blood pressure level at each age; if they do intersect in the age range of interest, what we could say is that one sex group has a higher mean blood pressure level than the other group at lower ages and a lower mean blood pressure level at higher ages (i.e., there is an age–SBP interaction effect).

Thus, by considering question 1 as outlined above, we may draw reasonable inferences concerning the relationship between the true average blood pressure levels in the two groups as a function of age. Nevertheless, additional statistical considerations are necessary for estimating the true adjusted mean difference as well as for estimating the adjusted means for each group. In this regard question 2 is concerned with the problem of determining an appropriate method for adjusting the sample mean blood pressure levels to take into account the effect of age and the problem of providing a statistical test to compare these adjusted mean scores.

In the example we are considering, the need for adjustment stems from the fact that age is a factor known to be strongly associated with blood pressure and that the two groups, as sampled, may have widely different age distributions. Without adjusting the sample mean values to reflect any differences in the age distributions in the two groups, it would not be possible to determine (e.g., through the use of a two-sample t test based on the unadjusted sample means) whether or not a significant difference was due solely to the effect of age or to the effects of factors of interest other than age. With adjustment, however, it could be determined whether any findings were or were not solely attributable, for example, to the females in the sample being older than the males (or vice versa).

14.2 ANALYSIS OF COVARIANCE

The usual statistical technique for handling the adjustment problem described above is called the *analysis of covariance*. This approach involves fitting a regression model of the form

$$Y = \beta_0 + \beta_1 X + \beta_2 Z + E \tag{14.2}$$

where X (age) is referred to as the *covariate* and Z is a dummy variable that indexes the two groups to be compared ($Z = 1$ if female, $Z = 0$ if male). This model assumes, in contrast to model (14.1), which contains a $\beta_3 XZ$ term, that the regression lines for males and females are parallel. Under this model, *the adjusted mean scores for males and females are defined to be the predicted values obtained by evaluating the model at $Z = 0$ and $Z = 1$ when X is set equal to the overall mean age for the two groups. Furthermore, a partial F test of the hypothesis $H_0: \beta_2 = 0$ is used to determine whether or not these adjusted mean scores are significantly different from one another.*

In computing these adjusted scores, we need to consider the two straight lines obtained by fitting model (14.2):

Males   ($Z = 0$): $\hat{Y}_M = \hat{\beta}_0 + \hat{\beta}_1 X$

Females ($Z = 1$): $\hat{Y}_F = (\hat{\beta}_0 + \hat{\beta}_2) + \hat{\beta}_1 X$ (14.3)

Explicit formulas for the estimated coefficients in (14.3) are as follows:

$$\hat{\beta}_1 = \frac{(n_M - 1)S^2_{X_M}\hat{\beta}_{1M} + (n_F - 1)S^2_{X_F}\hat{\beta}_{1F}}{(n_M - 1)S^2_{X_M} + (n_F - 1)S^2_{X_F}}$$

$$\hat{\beta}_0 = \bar{Y}_M - \hat{\beta}_1\bar{X}_M \tag{14.4}$$

$$\hat{\beta}_0 + \hat{\beta}_2 = \bar{Y}_F - \hat{\beta}_1\bar{X}_F$$

where $\hat{\beta}_{1M}$ and $\hat{\beta}_{1F}$ are the estimated slopes based on separate straight-line fits for males and females, $S^2_{X_M}$ and $S^2_{X_F}$ are the sample variances (on X) for males and females, $\bar{Y}_M$ and $\bar{Y}_F$ are the mean blood pressures for the male and female samples, and $\bar{X}_M$ and $\bar{X}_F$ are the mean ages for the male and female samples, respectively. Note that $\hat{\beta}_1$ is a weighted average of the slopes $\hat{\beta}_{1M}$ and $\hat{\beta}_{1F}$, which are estimated separately from the male and female data sets.

Based on (14.3) and (14.4), two alternative formulas for computing adjusted mean scores are as follows:

SEX	Z	ADJUSTED SCORE	FORMULA 1	FORMULA 2
Male	0	$\bar{Y}_M$ (adj.)	$\hat{\beta}_0 + \hat{\beta}_1\bar{X}$	$\bar{Y}_M - \hat{\beta}_1(\bar{X}_M - \bar{X})$
Female	1	$\bar{Y}_F$ (adj.)	$(\hat{\beta}_0 + \hat{\beta}_2) + \hat{\beta}_1\bar{X}$	$\bar{Y}_F - \hat{\beta}_1(\bar{X}_F - \bar{X})$

$$\tag{14.5}$$

In this table, $\bar{X}$ is the overall mean age for the combined data on males and females,

$$\bar{X} = \frac{n_M\bar{X}_M + n_F\bar{X}_F}{n_M + n_F}$$

Formula 1 is useful when model (14.2) has been estimated directly using a standard multiple regression program; formula 2, on the other hand, can be used without resorting to multiple regression procedures, although separate straight-line fits to the male and female data sets are required.

Given that the parallel straight-line assumption of model (14.2) is appropriate,[1] the two formulas provide a comparison of the mean blood pressure levels for the two sex groups as if they both have the same age distribution. In this regard, the covariance approach described above attempts to "artificially equate" the age distributions by treating both sex groups as if they have the same mean age, the best estimate of which is $\bar{X}$. The adjusted scores, then, represent the predicted $\hat{Y}$ values for each fitted line at $\bar{X}$, the assumed common mean age. This is depicted graphically in Figure 14.1.

That the partial F test of H_0: $\beta_2 = 0$ addresses the question of whether there is a significant difference between the adjusted means follows because, from formula 1 in (14.5), the difference in the two adjusted mean scores is exactly equal to $\hat{\beta}_2$; that is,

$$\hat{\beta}_2 = \bar{Y}_F(\text{adj.}) - \bar{Y}_M(\text{adj.}) = [(\hat{\beta}_0 + \hat{\beta}_2) + \hat{\beta}_1\bar{X}] - (\hat{\beta}_0 + \hat{\beta}_1\bar{X})$$

EXAMPLE   The least-squares fitting of the model (14.2) for the male–female data we have been discussing yields the following estimated model:

$$\hat{Y} = 110.29 + 0.96X - 13.51Z$$

[1] We shall discuss this assumption in Section 14.3.

FIGURE 14.1   *Adjusted mean SBP scores for males and females controlling for age using analysis of covariance*

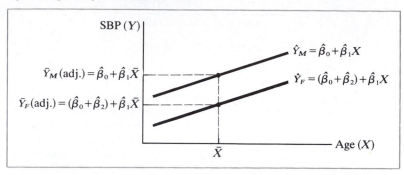

The separate fitted equations for males and females are

$$\text{Males} \quad (Z = 0): \quad \hat{Y}_M = 110.29 + 0.96X$$

$$\text{Females} (Z = 1): \quad \hat{Y}_F = 96.78 + 0.96X$$

The adjusted mean scores obtained from these equations using formula 1 of (14.5) with $\bar{X} = 46.14$ are

$$\bar{Y}_M(\text{adj.}) = 110.29 + 0.96(46.14) = 154.40$$

$$\bar{Y}_F(\text{adj.}) = 96.78 + 0.96(46.14) = 140.89$$

A comparison of these adjusted mean scores with the unadjusted means gives the following:

	UNADJUSTED $\bar{Y}$	$\bar{Y}(\text{adj.})$
Males	155.15	154.40
Females	139.86	140.89

Notice that the adjusted mean for males is slightly lower than the corresponding unadjusted mean for males, whereas the female adjusted mean is slightly higher than its unadjusted counterpart. The direction of these changes resulting from the covariance adjustment accurately reflects the fact that, in this sample, the males are somewhat older on the average ($\bar{X}_M = 46.93$) than the females ($\bar{X}_F = 45.07$). Use of the adjusted mean scores, in effect, removes the influence of age on the comparison of mean blood pressures by considering what the mean blood pressures in the two groups would be if both groups had the same mean age ($\bar{X} = 46.14$).

Nevertheless, whether adjusted or not, the mean blood pressure for males in this example appears to be considerably higher than that for females. In fact, the covariance adjustment has done little to change this impression, since the discrepancy between the male and female groups is 15.29 using unadjusted mean scores and is 13.51 based on adjusted mean scores. To test whether this difference in adjusted mean scores is significant, we use the partial F test of the hypothesis H_0: $\beta_2 = 0$ based on model (14.2), which can be computed from the following analysis-of-variance presentation:

SOURCE		SS	df	MS
Reduced model ($\beta_2 = 0$)	Regression (X)	14,951.25	1	14,951.25
	Residual	8,260.51	67	123.29
Complete model (14.2)	Regression (X, Z)	18,009.78	2	9,004.89
	Residual	5,201.99	66	78.82

From this presentation, the appropriate partial F statistic is obtained as follows:

$$F(Z|X) = \frac{\text{SS regression } (X, Z) - \text{SS regression } (X)}{\text{MS residual } (X, Z)}$$

$$= \frac{18,009.78 - 14,951.25}{78.82}$$

$$= 38.80$$

which has 1 and 66 df. The P value for this test satisfies $P < 0.001$, so we would reject H_0 and conclude that the two adjusted scores are significantly different from one another.

14.3 ASSUMPTION OF PARALLELISM: A POTENTIAL DRAWBACK

A potential problem with regard to the use of the analysis of covariance pertains to the assumption of parallelism of the regression lines being compared. It is certainly conceivable that, in certain applications, these regression lines may have different slopes. In such a case, the parallelism assumption would be invalid and the covariance method of adjustment as previously described should be avoided. To guard against the possibility of incorrectly applying the covariance method of adjustment, we recommend that a test for parallelism be made before proceeding with the analysis of covariance. This amounts to a test of H_0: $\beta_3 = 0$ for the complete model (14.1).

[We saw in Chapter 13 that the parallelism hypothesis (H_0: $\beta_3 = 0$) is not rejected for the age–SBP data being considered. This result supports the use of the analysis-of-covariance model for these data (given, of course, that the variance homogeneity assumption also holds).]

If this test supports the conclusion that the regression lines are not truly parallel, the question arises as to what should be done, if anything, about adjustment. It is usually recommended in this case that no adjustment at all be made to the sample means, since any such adjustment would be misleading; that is, a direct comparison of means is not appropriate when the true difference between the mean blood pressure levels in the two groups varies with age (i.e., when there is an age–sex interaction). In this case, since the main feature of the data would then be that the two regression lines describe very different relationships between age and blood pressure, an analysis that allows for the fitting of two separate regression lines (without assuming parallelism) and which quantifies how the lines differ would be sufficient.

14.4 ANALYSIS OF COVARIANCE: SEVERAL GROUPS AND SEVERAL COVARIATES

The example discussed in the previous sections considered the comparison of two groups with an adjustment for the single covariate, age. In general, analysis of covariance may be used to provide adjusted scores when there are several (say, k) groups and when it is necessary to adjust simultaneously for several (say, p) covariates. The regression model describing this general situation is written

$$Y = \beta_0 + \beta_1 X_1 + \cdots + \beta_p X_p + \beta_{p+1} Z_1 + \cdots + \beta_{p+k-1} Z_{k-1} + E \tag{14.6}$$

For this model, there are p covariates $X_1, \ldots, X_p$ and k groups, which are represented by the $k-1$ dummy variables $Z_1, \ldots, Z_{k-1}$. As discussed in Chapter 13, there is flexibility with regard to how these dummy variables are defined, but, for our purposes, we will assume that the $Z's$ are defined as follows:

$$Z_j = \begin{cases} 1 & \text{if group } j \\ 0 & \text{otherwise} \end{cases}, \quad j = 1, 2, \ldots, k-1$$

The fitted regression equations for the k different groups are then determined by specifying the appropriate combinations of values for the $Z's$. These are

Group 1 ($Z_1 = 1$, other $Z_j = 0$): $\quad \hat{Y}_1 = (\hat{\beta}_0 + \hat{\beta}_{p+1}) + \hat{\beta}_1 X_1 + \hat{\beta}_2 X_2 + \cdots + \hat{\beta}_p X_p$

Group 2 ($Z_2 = 1$, other $Z_j = 0$): $\quad \hat{Y}_2 = (\hat{\beta}_0 + \hat{\beta}_{p+2}) + \hat{\beta}_1 X_1 + \hat{\beta}_2 X_2 + \cdots + \hat{\beta}_p X_p$

$$\vdots \qquad\qquad\qquad \vdots \qquad\qquad\qquad \vdots \qquad\qquad \tag{14.7}$$

Group $k-1$ ($Z_{k-1} = 1$, other $Z_j = 0$): $\quad \hat{Y}_{k-1} = (\hat{\beta}_0 + \hat{\beta}_{p+k-1}) + \hat{\beta}_1 X_1 + \hat{\beta}_2 X_2 + \cdots + \hat{\beta}_p X_p$

Group k (all $Z_j = 0$): $\quad \hat{Y}_k = \hat{\beta}_0 + \hat{\beta}_1 X_1 + \hat{\beta}_2 X_2 + \cdots + \hat{\beta}_p X_p$

Notice from (14.7) that the corresponding coefficients of the covariates X_1, $X_2, \ldots, X_p$ in each of the k equations are exactly the same; thus, these regression equations represent "parallel" hypersurfaces in $(k+1)$-space, which is a natural generalization of the situation for the single covariate case. The adjusted mean score for a particular group is then computed as the predicted Y value obtained by evaluating the fitted equation for that group at the mean values $\bar{X}_1, \bar{X}_2, \ldots, \bar{X}_p$ of the p covariates based on the combined data for all k groups:

$$\bar{Y}_j(\text{adj.}) = (\hat{\beta}_0 + \hat{\beta}_{p+j}) + \hat{\beta}_1 \bar{X}_1 + \cdots + \hat{\beta}_p \bar{X}_p, \quad j = 1, 2, \ldots, k-1$$

$$\bar{Y}_k(\text{adj.}) = \hat{\beta}_0 + \hat{\beta}_1 \bar{X}_1 + \cdots + \hat{\beta}_p \bar{X}_p \tag{14.8}$$

To determine whether the k adjusted mean scores are significantly different from one another, we test the null hypothesis

$$H_0: \beta_{p+1} = \beta_{p+2} = \cdots = \beta_{p+k-1} = 0$$

using a multiple-partial F test with $(k-1)$ and $(n-p-k)$ df based on model (14.6). If H_0 is rejected, we would conclude that there are significant differences among the adjusted means (although we could not, without further inspection, determine where the major differences are).

EXAMPLE   In the Ponape study (Patrick et al., 1974) of the effect of rapid cultural change on health status, one research goal was to determine whether blood pressure was associated with a measure of the strength of a Ponapean's prestige in the modern (i.e., western) part of his culture relative to the traditional cultural framework. A measure of prestige discrepancy (PD) was developed and then measured by questionnaire for each of 550 Ponapean males. It was of particular interest to consider whether higher PD scores corresponded to higher blood pressures. It was also considered necessary to adjust or control for the effects of the covariates age and body size in considering these questions.

The analysis performed in this case first involved categorizing PD into three groups:

Group 1:   modern prestige much higher than traditional prestige

Group 2:   modern prestige not much different from traditional prestige

Group 3:   traditional prestige much higher than modern prestige

Then, an analysis of covariance was carried out with diastolic blood pressure (DBP) as the dependent variable, and with age and quetelet (QUET) index (a measure of body size) as the two covariates. A covariance model in this case is given by

$$DBP = \beta_0 + \beta_1(AGE) + \beta_2(QUET) + \beta_3 Z_1 + \beta_4 Z_2 + E \tag{14.9}$$

where

$$Z_1 = \begin{cases} 1 & \text{if group 1} \\ 0 & \text{otherwise} \end{cases} \quad Z_2 = \begin{cases} 1 & \text{if group 2} \\ 0 & \text{otherwise} \end{cases}$$

[A test of the parallelism assumption implicit in model (14.9) considers the null hypothesis $H_0: \beta_5 = \beta_6 = \beta_7 = \beta_8 = 0$ in the full model

$$DBP = \beta_0 + \beta_1(AGE) + \beta_2(QUET) + \beta_3 Z_1 + \beta_4 Z_2 + \beta_5 Z_1(AGE)$$
$$+ \beta_6 Z_2(AGE) + \beta_7 Z_1(QUET) + \beta_8 Z_2(QUET) + E$$

This multiple-partial F test (with 4 and 541 degrees of freedom) was not rejected using the Ponape data, thus supporting the use of model (14.9).]

The adjusted DBP mean scores for the three groups are then determined by substituting the values of the overall means $\overline{AGE}$ and $\overline{QUET}$ into the fitted equations for

TABLE 14.1   *Analysis-of-covariance example using Ponape study data*

	PD GROUP 1	PD GROUP 2	PD GROUP 3	P VALUE FOR F TEST OF H_0: $\beta_3 = \beta_4 = 0$ UNDER MODEL (14.9) (2 AND 545 df)
$\overline{DBP}$(unadj.) $\overline{DBP}$(adj.)	71.68 72.07	68.16 68.02	68.55 68.80	$0.001 < P < 0.005$
	$n_1 = 87$	$n_2 = 383$	$n_3 = 80$	

the three groups as follows:

$$\overline{DBP}_1(adj.) = (\hat{\beta}_0 + \hat{\beta}_3) + \hat{\beta}_1(\overline{AGE}) + \hat{\beta}_2(\overline{QUET})$$

$$\overline{DBP}_2(adj.) = (\hat{\beta}_0 + \hat{\beta}_4) + \hat{\beta}_1(\overline{AGE}) + \hat{\beta}_2(\overline{QUET})$$

$$\overline{DBP}_3(adj.) = \hat{\beta}_0 + \hat{\beta}_1(\overline{AGE}) + \hat{\beta}_2(\overline{QUET})$$

The test for the equality of these adjusted means is based on the use of a multiple-partial F test of the null hypothesis H_0: $\beta_3 = \beta_4 = 0$ under model (14.9). Table 14.1 summarizes these calculations. The results indicate the presence of highly significant differences among the adjusted mean blood pressure scores, with group 1 having a somewhat higher adjusted mean blood pressure than the other two groups. It should be noted, however, that, despite the statistical significance found, the adjusted mean blood pressure of 72.07 for group 1 is close enough (clinically speaking) to the adjusted means for the other groups to cast doubt on the clinical significance of these results.

14.5 *Z*-SCORE METHOD OF ADJUSTMENT: ALTERNATIVE TO ANALYSIS OF COVARIANCE

A method for adjusting continuous data due to Roberts, which is described in Hamilton et al. (1954), has often been used in epidemiologic and other health-related research. In fact, because this method, the *Z-score method*, is intuitively appealing and has no overwhelming disadvantages when compared with the analysis of covariance, it remains the method of choice of many epidemiologic investigators. The *Z*-score method does not involve the use of regression analysis but only requires the computation of means and standard deviations for different categories of the covariates of interest and the use of a (large-sample) *Z* test for comparing the adjusted scores for two groups. Furthermore, this method requires no assumption about variance homogeneity (as does the covariance approach), and, in fact, takes into account in the adjustment process differences in the variability of the dependent variable in the various covariate categories.

14.5.1 THE METHOD

In describing the *Z*-score method, we shall again consider the age–blood pressure data discussed earlier in the chapter. We therefore again seek an adjustment of the mean

TABLE 14.2   *Categorization of the response variable SBP by age and sex*

AGE CATEGORY (i)	AGE RANGE	MALES ($n_M = 40$)	FEMALES ($n_F = 29$)
1	15–24	128, 130, 138, 134, 139	114, 116, 124, 120
2	25–34	144, 140, 132	110, 125, 130
3	35–44	158, 152, 138, 124, 134, 142, 148, 150, 153	144, 124, 128, 136, 120, 160
4	45–54	159, 156, 144, 149, 156, 174, 160, 157	138, 145, 142, 130, 135, 142, 158
5	55–64	185, 168, 164, 154, 174, 165, 172, 170, 169	154, 162, 150, 140, 144
6	65+	176, 184, 172, 176, 158, 180	162, 170, 158, 175

systolic blood pressure levels in the male and female groups which corrects for age differences in the two groups and which permits a test of whether males and females have significantly different adjusted scores. To compute the adjusted scores, we first divide the age (X) variable into q (≥ 2) categories; the results of such a categorization when $q = 6$ are given in Table 14.2 for the data of Table 8.1.

By combining the data for males and females within each age category, it is possible to form a table of sample means and standard deviations as given in Table 14.3.

[Although the example being used involves the comparison of just two groups controlling for a single covariate, the *Z*-score method is nevertheless adaptable to the comparison of k (≥ 2) groups controlling for p (≥ 1) covariates. In this more general case, the categories used in the adjustment process are defined in terms of combinations of categories of each covariate.]

Then, for each individual in the combined sample of males and females, an adjusted score (i.e., the *Z* score) is computed using the formula

$$Z = \frac{Y - \bar{Y}_i}{S_i}, \qquad i = 1, 2, \ldots, q (= 6 \text{ in this example}) \qquad (14.10)$$

where *Y* is the observed value of SBP for the individual in question and *i* indexes the age category to which that individual belongs. For example, the male in the 35–44 age

TABLE 14.3   *Sample means and standard deviations for each age category (male and female data combined)*

AGE CATEGORY (i)	AGE RANGE	$\bar{Y}_i$	S_i
1	15–24	127.000	9.165
2	25–34	130.167	12.040
3	35–44	140.733	12.887
4	45–54	149.667	11.648
5	55–64	162.214	12.448
6	65+	171.100	9.049

category with a blood pressure of 158 would get a Z score of

$$Z = \frac{158 - 140.733}{12.887} = 1.34$$

and a female in the same age category with a systolic blood pressure of 124 would get a Z score of

$$Z = \frac{124 - 140.733}{12.887} = -1.30$$

Thus, an SBP value of 158 is 1.34 standard deviations above the norm for the age group 35–44, whereas an SBP value of 124 is 1.30 standard deviations below the norm for that age group. In general, the Z score adjusts for age by transforming each individual's SBP value into standard units based on the overall mean and standard-deviation values for the particular age group to which the individual belongs.

To determine whether males and females differ significantly in adjusted scores, the following (large-sample) statistic may be used:

$$Z = \frac{(\bar{Z}_M - \bar{Z}_F) - 0}{\sqrt{S_{Z_M}^2/n_M + S_{Z_F}^2/n_F}} \qquad (14.11)$$

where $\bar{Z}_M$ is the mean of the Z scores for all n_M males, $\bar{Z}_F$ is the mean of the Z scores for all n_F females, $S_{Z_M}^2$ is the sample variance for the male Z scores, and $S_{Z_F}^2$ is the sample variance for the female Z scores. When both n_M and n_F are large, the test statistic Z will have the standard normal distribution under H_0 of no difference between the true adjusted scores.

In our example, the test statistic of (14.11) is computed as follows:

$$\bar{Z}_M = 0.509, \qquad \bar{Z}_F = -0.701$$

$$S_{Z_M}^2 = 0.534, \qquad S_{Z_F}^2 = 0.627$$

$$n_M = 40, \qquad n_F = 29$$

so

$$Z = \frac{0.509 - (-0.701)}{\sqrt{0.534/40 + 0.627/29}} = \frac{1.21}{\sqrt{0.0350}} = 6.47$$

This computed Z exceeds $Z_{0.9999} = 3.80$, so H_0 would be rejected.

14.5.2 ADJUSTED SCORES BASED ON A REFERENCE GROUP

Because the Z score just discussed has the drawback of not being in the same units as the original dependent variable Y, it is of interest to consider an alternative adjusted score having the same units as Y; this score is formed using a standard (or reference) age category of interest. The choice of this reference category is up to the investigator, although we recommend either choosing that category bracketing the overall mean age or a category for which the variance is approximately the same for males and females. Given such a standard, this adjusted score for an individual with response Y is then defined as

$$Y(\text{adj.}) = \bar{Y}_s + \frac{Y - \bar{Y}_i}{S_i} \cdot S_s \qquad (14.12)$$

where, as before, i denotes the age category to which the individual belongs and $\bar{Y}_s$ and S_s are the mean and standard deviation for the reference category chosen. For example, with the 45–54 age group (which brackets the overall mean age of 46.14) as the standard, an observed blood pressure of 158 in the 35–44 age category would have an adjusted score of

$$149.667 + \frac{158 - 140.733}{12.887} \cdot 11.648 = 165.27$$

Similarly, an observed blood pressure of 124 in the same age category would have an adjusted score of

$$149.667 + \frac{124 - 140.733}{12.887} \cdot 11.648 = 134.52$$

Furthermore, the means of these adjusted scores for the male and female groups would be computed using (14.12) as follows:

$$\bar{Y}_M(\text{adj.}) = \bar{Y}_s + \bar{Z}_M S_s = 149.667 + (0.509)(11.648) = 155.89$$

$$\bar{Y}_F(\text{adj.}) = \bar{Y}_s + \bar{Z}_F S_s = 149.667 + (-0.701)(11.648) = 141.50$$

[Note that the difference between $\bar{Y}_M(\text{adj.})$ and $\bar{Y}_F(\text{adj.})$ is a constant multiple of the difference between $\bar{Z}_M$ and $\bar{Z}_F$, so the test for a significant difference between adjusted mean scores given by (14.11) is still applicable.]

These adjusted scores should be interpreted as estimating the mean blood pressure scores to be expected if all persons in both groups were in the 45–54 age category. Table 14.4 compares these mean scores with those obtained from the

TABLE 14.4 *Comparison of adjusted mean blood pressure scores using different methods of adjustment*

	UNADJUSTED MEAN SBP	ADJUSTED MEAN SBP USING ANALYSIS OF COVARIANCE (predicted Y at overall mean age of 46.14)	ADJUSTED MEAN SBP USING *Z*-SCORE METHOD (mean SBP score expected using 45–54 age group as standard)
Males	155.15	154.50	155.89
Females	139.86	140.88	141.50
Difference	15.29	13.52	14.39

analysis of covariance. From Table 14.4 it can be seen that both adjustment procedures have reduced the original difference between the unadjusted mean blood pressures. Thus, both methods properly reflect the fact that the sample of males was somewhat older, on the average, than the sample of females.

14.5.3 COMPARISON OF Z-SCORE METHOD WITH ANALYSIS OF COVARIANCE

One drawback of the Z-score method is that the choice of categories, as well as the choice of standard group, is a somewhat arbitrary process and certainly affects the values of the adjusted mean scores. Analysis of covariance, on the other hand, does not require a categorization scheme for the covariates, nor does it require specification of a reference group.

Analysis of covariance adjusts by treating the two comparison groups as if they had the same mean age. Alternatively, the Z-score method adjusts by removing the effect of age through standardization within each age category. The latter approach has the advantage of taking into account differences in variability of the response in various age categories. Also, the Z-score method does not require the specification of a particular model form.

Both methods are, nevertheless, inappropriate when the relationship between the covariates and the response is not the same in each group. Such "nonparallelism" might be reflected, for example, by males having higher blood pressures than females at older ages and females having higher blood pressures than males at younger ages. This could result in large positive and large negative individual Z scores canceling each other out when added up to form the mean Z scores for each group. Consequently, the adjusted scores for each group would be close to zero, giving the misleading impression that there was little difference between male and female blood pressures, when, in fact, there were large differences in certain age categories.

With all the considerations above, then, it is somewhat difficult to claim that one method is clearly superior to the other in all instances. We happen to find analysis of covariance somewhat more direct in its approach to adjustment, but the choice of method is still up to the investigator. The Z-score method perhaps is more appropriate when the investigator is concerned about the variance homogeneity assumption.

PROBLEMS 1. In Problem 1 of Chapter 13, straight-line regression fits of SBP on QUET were compared for smokers and nonsmokers, and it was demonstrated that these straight lines could be considered to be parallel. Use the results based on that problem (and the fact that the overall sample mean value $\overline{QUET} = 3.441$) to address the following issues relating to an analysis-of-covariance for these data:

(a) State the appropriate analysis-of-covariance regression model to be used for comparing the mean blood pressures in the two smoking categories controlling for QUET.

(b) Determine the adjusted SBP means for smokers and nonsmokers, respectively. Compare these values to the unadjusted mean values

$\overline{SBP}(\text{smokers}) = 147.823$ and $\overline{SBP}(\text{nonsmokers}) = 140.800$

(c) Perform a test of hypothesis concerning whether the true adjusted mean blood pressures in the two groups are equal or not. State the null hypothesis in terms of the regression coefficients in the analysis-of-covariance model given in part (a).

(d) Obtain a 95% confidence interval for the true difference in adjusted SBP means.

(e) Based on the grouping of SBP scores by categories of QUET as given in the table, carry out the Z-score method of adjustment to obtain:

    (1) Mean (adjusted) Z scores for smokers and nonsmokers, respectively.

    (2) Mean (adjusted) blood pressure scores for smokers and nonsmokers using QUET category 3 as the reference category.

    (3) A test of significance concerning the difference between mean (adjusted) Z scores for smokers and nonsmokers.

(f) Compare the results obtained in part (e) with the analysis-of-covariance results obtained earlier.

QUET CATEGORY NO.	QUET RANGE	SBP OBSERVATIONS	
		SMOKERS	NONSMOKERS
$i=1$	<3.110	134, 142, 132, 126, 146, 129, 144	135, 130, 120
$i=2$	3.110–3.440	145, 149	122, 135, 142, 132, 137
$i=3$	3.440–3.770	162, 160, 140, 150	148, 138, 144
$i=4$	>3.770	180, 166, 138, 170	152, 161, 152, 164

2. Answer the same questions as in parts (a)–(d) of Problem 1 regarding an analysis of covariance designed to control for *both* AGE and QUET. (*Note*: AGE = 53.250.) Use the results based on Problem 2 of Chapter 13.

    (e) Is it necessary to control for *both* AGE and QUET as opposed, say, to controlling for just one of the two covariates?

3. In an experiment conducted at the National Institute of Environmental Health Sciences, the absorption (or uptake) of a chemical in an organism of a rat for two different diets, I and II, was known to be affected by the weight (or size) of the rat. A completely randomized design utilizing four rats for each diet was employed in the experiment and the initial weight of each rat was recorded so that a comparison of diets could be made after adjusting for the effect of initial weight. The data for the experiment are given in the table.

INITIAL WEIGHT (X)	3	1	4	4	5	2	3	2
DIET (Z)	I	I	I	I	II	II	II	II
RESPONSE (Y)	14	13	14	15	16	15	15	14

(a) Using the initial weight as a covariate, state the analysis-of-covariance regression model for comparing the two diets (set $Z = -1$ if diet I is used and $Z = +1$ if diet II is used).

(b) Use the results in the table to determine the adjusted mean responses for each diet, controlling for initial weight.

(c) Test whether the two diets are significantly different using the analysis-of-covariance regression model defined in part (a).

(d) Test whether the two diets are significantly different, completely ignoring the covariate. How do the two testing procedures compare?

(e) Determine a 95% confidence interval for the true difference in the adjusted mean responses.

FITTED COVARIANCE MODEL			ANOVA		
VARIABLE	$\hat{\beta}$	$s_{\hat{\beta}}$	SOURCE	df	SS
X	0.5	0.1291	X	1	3.0
Z	0.5	0.1581	$Z\|X$	1	2.0
(constant)	13.0		Residual	5	1.0
			Total	7	6.0

4. A political scientist developed a questionnaire to determine political tolerance scores (Y) for a random sample of faculty members at her university. She wanted to compare mean scores adjusted for age (X) for each of three categories: full professors, associate professors, and assistant professors. The results are given in the tables (the higher the score, the more tolerant the individual).

Group 1: full professors

TOLERANCE (Y)	3.03	2.70	4.31	2.70	5.09	4.02	3.71	5.52	5.29	4.62
AGE (X)	65	61	47	52	49	45	41	41	40	39

Group 2: associate professors

TOLERANCE (Y)	4.62	5.22	4.85	4.51	5.12	4.47	4.50	4.88	5.17	5.21
AGE (X)	34	31	30	35	49	31	42	43	39	49

Group 3: assistant professors

TOLERANCE (Y)	5.20	5.86	4.61	4.55	4.47	5.71	4.77	5.82	3.67	5.29
AGE (X)	26	33	48	32	25	33	42	30	31	27

(a) State an analysis-of-covariance regression model that can be used to compare the three groups, controlling for age.

(b) What model should be used to check whether the analysis-of-covariance model in part (a) is appropriate? Carry out the appropriate test. (Use $\alpha = 0.01$.)

(c) Using analysis of covariance, determine adjusted mean tolerance scores for each group, and test whether these are significantly different from one another. Also, compare the adjusted means with the unadjusted means. [*Note*: $\bar{X}$(overall) $= 39.667$, $\bar{Y}$(group 1) $= 4.10$, $\bar{Y}$(group 2) $= 4.86$, $\bar{Y}$(group 3) $= 5.00$.]

FITTED COVARIANCE MODEL			ANOVA		
VARIABLE	$\hat{\beta}$	$S_{\hat{\beta}}$	SOURCE	df	SS
X	−0.0364	0.0174	X	1	6.2050
$Z_1 \left(\begin{array}{l} 1 \text{ if group 1} \\ 0 \text{ otherwise} \end{array} \right)$	−0.3398	0.4146	$Z_1, Z_2 \mid X$	2	0.6434
$Z_2 \left(\begin{array}{l} 1 \text{ if group 2} \\ 0 \text{ otherwise} \end{array} \right)$	0.0636	0.3323	$XZ_1, XZ_2 \mid X, Z_1, Z_2$	2	3.3610
(constant)	6.1837		Residual	24	9.7628

5. A psychological experiment was performed to determine whether in problem-solving dyads each containing one male and one female, "influencing" behavior depended on the sex of the experimenter. The problem for each dyad was a strategy game called "Rope a Steer," which required 20 separate decisions as to which way to proceed toward a defined goal in a game board. For each subject in the dyad, a verbal-influence activity score was derived as a function of the number of statements made by the subject to influence the dyad to move in a particular direction. The difference in verbal-influence activity scores within a dyad was denoted as the variable VIAD, which was then used as the dependent variable in an analysis of covariance designed to control for the effects of differing IQ scores of the male and female in each dyad. The data are given in the tables.

Group 1: male experimenter

VIAD	−10	−4	9	−15	−15	5	−8	−4	−1	13
IQM	115	112	106	123	125	105	115	122	138	110
IQF	100	110	108	135	115	112	121	132	135	126

Group 2: female experimenter

VIAD	8	−5	2	−7	15	−10	−3	10	2	4
IQM	120	130	110	113	102	141	120	113	114	102
IQF	141	128	104	98	106	130	128	105	107	111

(a) State an analysis-of-covariance model appropriate for these data.
(b) Determine adjusted mean VIAD scores for each group and compare these to the unadjusted means for each group.
(c) Test whether the adjusted mean scores are significantly different.
(d) Find a 95% confidence interval for the true difference in adjusted mean scores.
(e) Based on the grouping of VIAD scores by categories of IQ as given in the table, carry out the Z-score method of adjustment to obtain:
(1) Mean (adjusted) Z scores for group 1 and group 2.
(2) Mean (adjusted) VIAD scores using IQ category 2 as the reference category.

FITTED COVARIANCE MODEL			ANOVA		
VARIABLE	$\hat{\beta}$	$s_{\hat{\beta}}$	SOURCE	df	SS
IQM	−0.6930	0.1939	IQM, IQF	2	612.05
IQF	0.2811	0.1597	Z|IQM, IQF	1	131.47
$Z\left(\begin{array}{l}\text{1 if female experimenter}\\\text{0 if male experimenter}\end{array}\right)$	5.1961	3.1329	(IQM∗Z, IQF∗Z)|IQM, IQF, Z	2	3.22
(constant)	44.5900		Residual	14	761.46

    (3)   A test of significance concerning the difference between mean (adjusted) Z scores for groups 1 and 2.

  (f)   Compare the results for analysis of covariance with those for the Z-score method.

IQ CATEGORY NO.	IQ RANGE (female–male difference)	VIAD SCORES	
		GROUP 1	GROUP 2
$i = 1$	−15 to −8	−15	−7, −10, 10
$i = 2$	−7 to 0	−10, −4, −1	−5, 2, 2
$i = 3$	+1 to +8	9, 5, −8	15, −3
$i = 4$	+9 to +24	−15, −4, 13	8, 4

6. An experiment was conducted to compare the effects of four different drugs (A, B, C, D) in delaying the atrophy of denervated muscles. A certain leg muscle in each of 48 rats was deprived of its nerve supply by severing the appropriate nerves. The rats were then put randomly into four groups, and each group was assigned treatment by one of the drugs. After 12 days, four rats from each group were killed and the weight W (in grams) of the denervated muscle was obtained.

    Theoretically, atrophy should be measured by the *loss* in weight of the muscle, but the initial weight of the muscle could not have been obtained without killing the rat. Instead, the initial total body weight X (in grams) of the rat was measured. It was assumed that this is closely related to the initial weight of the muscle.

    Drugs A and C were large and small doses, respectively, of atropine sulfate. Drug B was quinidine sulfate. Drug D acted as a control; it was simply a saline solution, which could not have had any effect on atrophy. Compare the effects of the four drugs using analysis of covariance, controlling for initial total body weight (X). Use the results in the table to perform your analysis. (*Note*: $\bar{X} = 226.125$.)

DRUG A		DRUG B		DRUG C		DRUG D	
X	W	X	W	X	W	X	W
198	0.34	233	0.41	204	0.57	186	0.81
175	0.43	250	0.87	234	0.80	286	1.01
199	0.41	289	0.91	211	0.69	245	0.97
224	0.48	255	0.87	214	0.84	215	0.87

FITTED COVARIANCE MODEL			ANOVA		
VARIABLE	$\hat{\beta}$	$S_{\hat{\beta}}$	SOURCE	df	SS
X	0.0032	0.0013	Regression (X)	1	0.3202
$Z_1 \begin{pmatrix}1 \text{ if drug A} \\ 0 \text{ otherwise}\end{pmatrix}$	−0.3917	0.0954	Residual	14	0.4571
$Z_2 \begin{pmatrix}1 \text{ if drug B} \\ 0 \text{ otherwise}\end{pmatrix}$	−0.2257	0.0902	SOURCE	df	SS
$Z_3 \begin{pmatrix}1 \text{ if drug C} \\ 0 \text{ otherwise}\end{pmatrix}$	−0.1351	0.0878	Regression (X, Z_1, Z_2, Z_3)	4	0.6185
(constant)	0.1728		Residual	11	0.1587

7. Chemical analysis for sodium content was made on trough urine samples for each of two collection periods, one before and one after administration of Mercuhydrin, for each of 30 dogs. The experimenter used 7 dogs as a control group for the study,

EXPERIMENTAL GROUP ($Z=1$)			CONTROL GROUP ($Z=0$)		
ANIMAL NO.	FIRST COLLECTION PERIOD (X) ([Na], mM/liter)	SECOND COLLECTION PERIOD (Y) ([Na], mM/liter)	ANIMAL NO.	FIRST COLLECTION PERIOD (X) ([Na], mM/liter)	SECOND COLLECTION PERIOD (Y) ([Na], mM/liter)
1	17.5	22.1	1	11.1	9.4
2	9.4	12.0	2	5.1	5.9
3	10.0	15.2	3	6.5	14.8
4	7.4	23.1	4	17.2	15.5
5	8.8	9.8	5	11.8	23.4
6	18.9	26.9	6	6.6	7.3
7	10.8	11.1	7	4.1	8.2
8	8.8	13.6			
9	8.8	12.8	Mean	8.9	12.1
10	9.2	7.5	Std. dev.	4.7	6.4
11	8.1	8.1			
12	10.3	27.5			
13	10.1	11.2			
14	7.3	11.0			
15	11.1	15.3			
16	9.4	11.5			
17	8.2	8.4			
18	6.3	12.7			
19	9.7	17.1			
20	7.1	9.5			
21	7.2	11.0			
22	5.3	8.2			
23	14.3	15.8			
24	7.9	9.7			
25	14.1	14.7			
26	12.8	17.0			
27	12.8	20.2			
28	10.7	13.9			
29	5.9	11.8			
30	3.8	9.0			
Mean	9.7	13.9			
Std. dev.	3.3	5.3			

Computer results for Problem 7

MULTIPLE R-SQUARE = 0.406				ANOVA		
VARIABLE	$\hat{\beta}$	S.D. OF $\hat{\beta}$	$\hat{\beta}$/S.D.	SOURCE	df	MS
X	0.96155	0.20418	4.709	X	1	434.4861
Z	1.06435	1.83729	0.5793	Added by Z	1	6.3764
(constant)	3.49992			Lack of fit	30	21.0913
				Pure error	4	3.3179

with results being collected for two similar time periods not involving administration of the drug.

(a) Using analysis of covariance (computer results are given above) with the before measure as the covariate, find the adjusted mean sodium contents for both experimental and control groups.

(b) Test whether the two adjusted means are significantly different.

(c) What alternative testing approach (involving a t test) could be used for these data? Carry out this test. [*Hint*: Variances of before–after differences for each group are 16.78517 (experimental) and 25.915 (control).]

(d) Are the two testing methods (t test versus covariance analysis) equivalent in this problem? Explain in terms of a comparison of regression models appropriate for each method.

(e) What do the results of performing a lack-of-fit test using the computer output given indicate about the appropriateness of the covariance model used in parts (a) and (b)?

CHAPTER 15 SELECTING THE BEST REGRESSION EQUATION

15.1 PREVIEW

The general problem to be discussed in this chapter is as follows. We have one dependent variable Y and a set of k independent variables $X_1, X_2, \ldots, X_k$. We want to determine the *best* (i.e., the *most important*) subset of these k independent variables and the corresponding *best-fitting* regression model for describing the relationship between Y and the X's.

There are four basic statistical procedures currently in use for selecting this *best* regression equation. These are:

1. The all-possible-regressions procedure.
2. The backward elimination procedure.
3. The forward selection procedure.
4. The stepwise regression procedure.

(There are also some slight variations on these basic procedures and a few other algorithms that are sometimes used.)

Before proceeding to describe each of these procedures in detail, some general comments are in order:

1. Some of the k independent variables may consist of higher-order functions of a few basic variables (e.g., $X_4 = X_1 X_2$, $X_6 = X_2^2$, etc.). In practice, the use of any of the procedures above gives more acceptable results when there are not very many of such higher-order terms under consideration, one reason being that a model containing several such terms is invariably quite difficult to interpret.

    [Variable selection procedures like those discussed in this chapter sometimes lead to unstable subset selection when the candidate independent variables are highly correlated (e.g., when some independent variables are functions of other independent variables); see Marquardt and Snee (1975) for further discussion in this regard.]

2. It is possible to arrive at different solutions by using the four different methods. When this happens, it is necessary to weigh all the results and make a choice of best model based on practical considerations regarding the variables under study, the nature of the data, and the interpretations that can be made regarding the different candidate models.

3. Sometimes even a single procedure will provide a number of reasonably good candidate models, from which a choice will have to be made.

15.2 DESCRIPTION OF THE METHODS

We shall illustrate the use of the four procedures in the context of the following example considered in Chapter 10, which involves the three independent variables $X_1 = $ HGT, $X_2 = $ AGE, and $X_3 = $ AGE^2, and the dependent variable $Y = $ WGT:

OBSERVATION NO.:	1	2	3	4	5	6	7	8	9	10	11	12
Y (WGT)	64	71	53	67	55	58	77	57	56	51	76	68
X_1 (HGT)	57	59	49	62	51	50	55	48	42	42	61	57
X_2 (AGE)	8	10	6	11	8	7	10	9	10	6	12	9
X_3 (AGE^2)	64	100	36	121	64	49	100	81	100	36	144	81

15.2.1 ALL-POSSIBLE-REGRESSIONS PROCEDURE

The *all-possible-regressions procedure* requires that we fit every possible regression equation associated with every possible combination of the independent variables. In particular, for our example, fit the seven models corresponding to the following seven combinations of independent variables: (1) HGT; (2) AGE; (3) AGE^2; (4) HGT, AGE; (5) HGT, AGE^2; (6) AGE, AGE^2; and (7) HGT, AGE, AGE^2. (For k independent variables, the number of models to be fitted is $2^k - 1$.) Then partition the different equations obtained into sets involving 1, 2, 3, . . . , and k variables, and order the models within each set according to some criterion (e.g., R^2). The leaders in each set are then selected for further examination.

   For our data a summary of the results of this all-possible-regressions procedure is given in Table 15.1. From the table the leaders (in terms of R^2 values) in each of the sets involving one, two, and three variables are given by

   One-variable set: (1) HGT with $R^2 = 0.6630$
   Two-variable set: (4) HGT, AGE with $R^2 = 0.7800$
   Three-variable set: (7) HGT, AGE, AGE^2 with $R^2 = 0.7802$

Of the three models, clearly model (4), involving HGT and AGE, should be our choice since its R^2 value is essentially the same as the value for model (7) and is much higher than the value for model (1).

   Thus, our choice of best regression equation based on the all-possible-regressions procedure is given by

$$\widehat{\text{WGT}} = 6.553 + 0.722\text{HGT} + 2.050\text{AGE}.$$

TABLE 15.1 *Summary of results of all-possible-regressions procedure*

NUMBER OF VARIABLES IN MODEL	VARIABLES IN MODEL	ESTIMATED COEFFICIENTS				PARTIAL F STATISTICS*			OVERALL F STATISTICS	R^2 VALUES
		$\hat{\beta}_0$	$\hat{\beta}_1$	$\hat{\beta}_2$	$\hat{\beta}_3$	X_1	X_2	X_3		
1	(1) HGT (X_1)	6.190	1.073			19.67**			19.67**	0.6630
1	(2) AGE (X_2)	30.571		3.643			14.55**		14.55**	0.5926
1	(3) AGE^2 (X_3)	45.998			0.206			14.25**	14.25**	0.5876
2	(4) HGT, AGE	6.553	0.722	2.050		7.665*	4.785		15.95**	0.7800
2	(5) HGT, AGE^2	15.118	0.726		0.115	7.601*		4.565	15.63**	0.7764
2	(6) AGE, AGE^2	32.404		3.205	0.025		0.113	0.002	6.55*	0.5927
3	(7) HGT, AGE, AGE^2	3.438	0.724	2.777	−0.042	6.827*	0.140	0.010	9.47**	0.7802

The partial F statistic for a given variable in a given model assesses the additional contribution of that variable to the prediction of Y (=WGT) over and above what is contributed by the other variables in the given model. For example, for model (4), which involves X_1 and X_2, the partial F for X_2 is $F(X_2|X_1) = 4.785$; but for model (7), which includes X_1, X_2, and X_3, the partial F for X_2 is $F(X_2|X_1, X_3) = 0.140$.

15.2.2 BACKWARD ELIMINATION PROCEDURE

In the *backward elimination procedure*, we proceed as follows:

Step 1: Determine the fitted regression equation containing all independent variables. For our example we obtain

$$\widehat{WGT} = 3.438 + 0.724HGT + 2.777AGE - 0.042AGE^2$$

with ANOVA table

SOURCE	df	SS	MS	F	R^2
Regression	3	693.06	231.02	9.47**	0.7802
Residual	8	195.19	24.40		
Total	11	888.25			

Step 2: Calculate the partial F statistic for every variable in the model as though it were the last variable to enter (see Table 15.1):

VARIABLE	PARTIAL F (BASED ON 1 AND 8 df)
HGT	6.827**
AGE	0.140
AGE^2	0.010

(Recall that the partial F statistics above test whether the addition of the last variable to the model significantly helps in predicting the dependent variable given that the other variables are already in the model.)

Step 3: Focus on the lowest observed partial F test value (F_L, say). In our example, $F_L = 0.010$ for the variable AGE^2.

Step 4: Compare this value F_L with a preselected critical value (F_C, say) of the F distribution (i.e., test for the significance of the partial F_L). (a) If $F_L < F_C$, remove from the model the variable under consideration, recompute the regression equation for the remaining variables, and repeat steps 2, 3, and 4. (b) If $F_L > F_C$, adopt the complete regression equation as calculated. In our example, if we work at the 10% level, $F_L = 0.010 < F_{1,8,0.90} = F_C = 3.46$. Therefore, we remove AGE^2 from the model and recompute the equation using only HGT and AGE. We then obtain

$$\widehat{WGT} = 6.553 + 0.722HGT + 2.050AGE,$$

with ANOVA table

SOURCE	df	SS	MS	F	R^2
Regression	2	692.82	346.41	15.95**	0.7800
Residual	9	195.43	21.71		
Total	11	888.25			

With AGE^2 out of the picture, the partial F's become 7.665 for HGT and 4.785 for AGE (see Table 15.1). Our new $F_C = F_{1,9,0.90} = 3.36$, which is less than 4.785. Therefore, the partial F for AGE is significant and we stop here with this model, which is the same model that we arrived at using the all-possible-regressions procedure.

15.2.3 FORWARD SELECTION PROCEDURE

In the *forward selection procedure*, we proceed as follows:

Step 1: Select as the first variable to enter the model that variable most highly correlated with the dependent variable, and then fit the associated straight-line regression equation. For our example we have the following correlations: $r_{YX_1} = 0.814$, $r_{YX_2} = 0.770$, and $r_{YX_3} = 0.767$. Thus, the first variable to enter is $X_1 = HGT$.

The straight-line regression equation relating WGT and HGT is

$$\widehat{WGT} = 6.190 + 1.073HGT$$

with ANOVA table

SOURCE	df	SS	MS	F	R^2
Regression	1	588.92	588.92	19.67**	0.6630
Residual	10	299.33	29.93		
Total	11	888.25			

If the F statistic in the table is not significant, we stop and conclude that no independent variables are important predictors. If the F statistic is significant (as it is), we include this variable (in our case, height) in the model and proceed to step 2.

Step 2: Calculate the partial F statistic associated with each remaining variable based on a regression equation containing that variable and the variable initially selected. For our example,

partial $F(X_2|X_1) = 4.785$

partial $F(X_3|X_1) = 4.565$

Step 3: Focus on the variable with the largest partial F statistic. For our data, the variable AGE has the largest partial $F (= 4.785)$.

Step 4: Test for the significance of the partial F statistic associated with the variable selected in step 3. (a) If this test is significant, add the new variable to the regression equation. (b) If this test is not significant, use in the model only the variable added in step 1. For our example, since the partial F for AGE is significant at the 10% level ($F_{1,9,0.90} = 3.36$), we add AGE to get the two-variable model

$$\widehat{WGT} = 6.553 + 0.722HGT + 2.050AGE$$

Step 5: At each subsequent step, determine the partial F statistics for those variables not yet in the model and then add to the model that variable which has the largest partial F value if it is statistically significant. At any step, if the largest partial F is not significant, no more variables are included in the model and the process is terminated.

[The reader should be aware that the probability of finding at least one significant independent variable when actually there are none increases rapidly as the number k of candidate independent variables increases, an upper bound on this overall significance level being $1 - (1 - \alpha)^k$, where α is the significance level of any one test. To control this overall significance level so that it does not exceed α in value, a conservative but easily implemented approach is to conduct any one test at level α/k [see Pope and Webster (1972) and Kupper et al., (1976) for further discussion in this regard].]

For our example we have already added HGT and AGE to the model. We now see if we should add AGE^2. The partial F for AGE^2, controlling for HGT and AGE, is given by

partial $F(X_3|X_1, X_2) = 0.010$

and this value is not statistically significant, since $F_{1,8,0.90} = 3.46$. Again, we have arrived at the same two-variable model chosen via the previously discussed methods.

15.2.4 STEPWISE REGRESSION PROCEDURE

Stepwise regression is an improved version of forward regression which permits reexamination, at every step, of the variables incorporated in the model in previous steps. A variable that entered at an early stage may, at a later stage, become superfluous because of its relationship with other variables now in the model. To check on this possibility, at each step a partial F test for each variable presently in the model is made, treating it as though it were the most recent variable entered, irrespective of its actual entry point into the model. That variable with the smallest nonsignificant partial F statistic (if there is such a variable) is removed, the model is refitted with the remaining variables, the partial F's are obtained and similarly examined, and so on. The whole process continues until no more variables can be entered or removed.

For our example the first step would be, as with the forward selection procedure, to add the variable HGT to the model, since it has the highest significant correlation with WGT. Next, as before, we would add AGE to the model, since it has a higher significant partial correlation with WGT than does AGE^2, controlling for HGT. Now, before testing to see if AGE^2 should also be added to the model, we look at the partial F of HGT given that AGE is already in the model to see if we should now remove HGT. This partial is given by $F(X_1|X_2) = 7.665$ (see Table 15.1), which exceeds $F_{1,9,0.90} = 3.36$. Thus, we do not remove HGT from the model. Next, we check to see if we need to add AGE^2; of course, the answer is no, since we have dealt with this situation before.

The analysis-of-variance table that best summarizes the results obtained for our example is thus given by

SOURCE		df	SS	MS	F	R^2
Regression $\begin{cases} X_1 \\ X_2\|X_1 \end{cases}$		1	588.92	588.92	19.67**	0.7800
		1	103.90	103.90	4.79 ($P<0.10$)	
Residual		9	195.43	21.71		
Total		11	888.25			

The ANOVA table that considers all variables is

SOURCE		df	SS	MS	F	R^2
Regression $\begin{cases} X_1 \\ X_2\|X_1 \\ X_3\|X_1, X_2 \end{cases}$		1	588.92	588.92	19.67**	0.7802
		1	103.90	103.90	4.79 ($P<0.10$)	
		1	0.24	0.24	0.01	
Residual		8	195.19	24.40		
Total		11	888.25			

PROBLEMS

1. Add the variables HGT^2 and (AGE · HGT) to the data set on WGT, HGT, AGE, and AGE^2 given in Section 15.2. Then, using an available regression program, determine the best regression model relating WGT to the five independent variables using (a) the forward approach, (b) the backward approach, and (c) an approach that first determines the best model using HGT and AGE as the only candidate independent variables and then determines whether any second-order terms should be added to the model. (Use $\alpha = 0.10$.) Compare and discuss the models obtained for each of the three approaches used.

2. Using the data given in Problem 2 of Chapter 5 (with SBP as the dependent variable), find the best regression model using $\alpha = 0.05$ and the independent variables AGE, QUET, and SMK by the (a) forward approach, (b) backward approach, and (c) all-possible-regressions approach. (d) Based on the results in parts (a)–(c), select a model for further analysis to determine which, *if any*, of the

following interaction (i.e., product) terms should be added to the model: $AQ = AGE \cdot QUET$, $AS = AGE \cdot SMK$, $QS = QUET \cdot SMK$, and $AQS = AGE \cdot QUET \cdot SMK$.

3. For the same data set as considered in Problem 2, use the stepwise-regression approach to find the best regression models of SBP on QUET, AGE, and (QUET · AGE) for *smokers and nonsmokers separately*. (Use $\alpha = 0.05$.) Compare the results for each group with the answer to Problem 2(d).

4. For the data given in Problem 4 of Chapter 10, find (using $\alpha = 0.10$) the best regression model relating homicide rate (Y) to population size (X_1), percent of families with income less than \$5,000 (X_2), and unemployment rate (X_3). Use (a) the stepwise approach, (b) the backward approach, and (c) the all-possible-regressions approach.

5. The data set listed contains information on AGE, SEX (1 = male, 2 = female), work problems index (WP), marital conflict index (MC), and depression index (DEP) for a

ID NO.	AGE	SEX	WP	MC	DEP
1	45	2	90	70	69
2	35	1	90	75	75
3	32	2	70	32	35
4	32	2	80	30	73
5	39	2	85	55	86
6	25	2	85	6	161
7	22	1	75	20	202
8	30	2	70	63	91
9	49	2	75	4	113
10	47	1	84	12	68
11	48	1	64	11	109
12	49	2	85	7	92
13	45	2	80	8	80
14	41	2	80	15	82
15	45	2	82	6	156
16	59	2	72	5	198
17	42	2	70	17	170
18	35	1	70	29	188
19	31	2	70	80	82
20	45	1	70	126	37
21	28	1	85	30	194
22	37	1	90	9	294
23	29	1	80	14	94
24	29	1	70	24	126
25	31	1	80	21	192
26	29	1	60	11	232
27	29	1	70	10	184
28	23	2	80	10	238
29	44	2	78	19	112
30	28	1	70	22	141
31	32	2	70	21	108
32	36	2	74	77	87
33	22	2	78	67	33
34	46	2	70	25	73
35	21	1	70	14	168
36	34	1	80	17	218
37	27	2	80	18	175
38	31	2	80	42	126
39	19	2	75	36	135

sample of 39 new admissions to a psychiatric clinic at a large university hospital. For each sex *separately*, determine (using $\alpha = 0.10$) the best regression model relating DEP to MC and WP, controlling for age, using the following sequential procedure: (1) Force AGE into the model first; (2) Use the all-possible-regressions approach on the remaining two independent variables WP and MC; (3) Determine whether the interaction term (MC · WP) should be added to the model. Compare and discuss the results obtained for each sex.

CHAPTER 16 RESIDUAL ANALYSIS, TRANSFORMATIONS, AND WEIGHTED LEAST SQUARES

16.1 PREVIEW

The purpose of this chapter is to provide a general overview of the statistical technique known as residual analysis, along with brief discussions regarding the use of data transformations and the method of weighted least squares. These procedures are all concerned with issues regarding the assumptions of multiple regression analysis. Residual analysis represents an attempt to discern which assumptions (if any) are being violated, while data transformations and weighted-least-squares procedures are used to help correct for such violations.

16.2 RESIDUAL ANALYSIS

Given n data points $(Y_i, X_{1i}, X_{2i}, \ldots, X_{ki})$, $i = 1, 2, \ldots, n$, recall that the methodology of regression analysis is concerned with the least-squares fitting of a model describing the observed response Y_i as

$$Y_i = \beta_0 + \beta_1 X_{1i} + \beta_2 X_{2i} + \cdots + \beta_k X_{ki} + E_i \qquad i = 1, 2, \ldots, n$$

The fitted model we write as

$$\hat{Y} = \hat{\beta}_0 + \hat{\beta}_1 X_1 + \hat{\beta}_2 X_2 + \cdots + \hat{\beta}_k X_k$$

and the predicted response at the ith data point is then

$$\hat{Y}_i = \hat{\beta}_0 + \hat{\beta}_1 X_{1i} + \hat{\beta}_2 X_{2i} + \cdots + \hat{\beta}_k X_{ki}$$

With this framework in mind, we define the ith *residual* e_i to be the difference between the observed value Y_i and the predicted value $\hat{Y}_i$,

$$e_i = Y_i - \hat{Y}_i \qquad i = 1, 2, \ldots, n$$

In words, then, the $\{e_i\}$ reflect the amount of discrepancy between observed and predicted values which is still present after having fit the least-squares model. Now, the usual assumptions which are made about the errors when performing a regression analysis are that the $\{E_i\}$ are independent, have zero mean, have a common variance σ^2, and follow a normal distribution (the normality assumption is required for performing parametric tests of significance). If the model is indeed appropriate for the data under analysis, it is reasonable to expect that the observed residuals should exhibit properties not at odds with the stated assumptions. The basic strategy underlying the statistical procedure generally referred to as *residual analysis* is to assess the appropriateness of a model in terms of the behavior of the set of observed residuals, and it is the purpose here to discuss methods for making such an assessment. The methodology to be discussed will be applicable in any context where a model is fit and a set of residuals is produced (e.g., as in analysis of variance and nonlinear regression situations).

16.2.1 SOME PROPERTIES OF RESIDUALS

It is important to keep in mind the following characteristics of the n residuals $e_1, e_2, \ldots, e_n$:

1. The mean of the $\{e_i\}$ is zero:

$$\bar{e} = \frac{1}{n} \sum_{i=1}^{n} e_i = 0$$

2. The sample variance of the n residuals is defined to be

$$S^2 = \frac{1}{n-(k+1)} \sum_{i=1}^{n} e_i^2$$

which is exactly the residual mean square $SSE/(n-k-1)$ obtained by use of regression analysis and provides an unbiased estimate of σ^2 if the model is correct. The quantity e_i/S is called a *standardized residual* and is sometimes examined in a residual analysis instead of e_i.

3. The $\{e_i\}$ are *not* independent random variables. This is obvious from the fact that the $\{e_i\}$ sum to zero. This same lack of independence clearly is present for the standardized residuals. In general, if the number of residuals (n) is large relative to the number of independent variables (k), this dependency effect can, for all practical purposes, be ignored in any analysis of the residuals [see Anscombe and Tukey (1963) for a discussion of the effect of this dependency on graphical procedures involving residuals].

16.2.2 GRAPHICAL ANALYSIS OF RESIDUALS

Often, the most direct and most revealing way to examine a set of residuals is to make a series of plots of the $\{e_i\}$ versus a number of carefully selected variables. In this regard it is recommended practice to start first by looking at a scatter diagram of the $\{e_i\}$, followed by plots of the $\{e_i\}$ versus the X_{ji}'s for each j, versus the $\hat{Y}_i$'s, versus time, and

TABLE 16.1 *Residuals for the calibration experiment data*

X_i	Y_i	$\hat{Y}_i$	e_i	e_i/S
0	10.70	15.39	−4.69	−1.14
0.5	14.20	17.29	−3.09	−0.75
1	16.70	19.18	−2.48	−0.60
1.5	19.10	21.07	−1.97	−0.48
2	24.90	22.97	1.93	0.47
2.5	25.40	24.86	0.54	0.13
3	32.30	26.75	5.55	1.35
3.5	30.80	28.65	2.15	0.52
4	39.60	30.54	9.06	2.20
4.5	30.30	32.43	−2.13	−0.52
5	37.20	34.33	2.87	0.70
5.5	37.80	36.22	1.58	0.38
6	37.50	38.11	−0.61	−0.15
6.5	38.60	40.01	−1.41	−0.34
7	42.60	41.90	0.70	0.17
7.5	44.30	43.79	0.51	0.12
8	37.20	45.69	−8.49	−2.07

versus any other variables that might help to bring out revealing patterns among the residuals. In such a graphical analysis, a violation of a specific assumption (e.g., independence, model appropriateness, normality, homogeneity of variance) will sometimes be much more evident from one type of plot than from another, and it is our purpose here to discuss both available types of plots and the interpretations that can be drawn from them.

We shall illustrate these graphical techniques with a specific example. For an environmental engineering instrument calibration experiment, 17 paired values of the "concentration of a certain pollutant" (X) and the "instrument reading" (Y) are given in Table 16.1, along with the predicted values $\{\hat{Y}_i\}$, the residuals $\{e_i\}$, and the standardized residuals $\{e_i/S\}$ based on the least-squares straight line $\hat{Y} = 15.39 + 3.79X$ for which $S = 4.11$.

The first thing that we will do is to examine a scatter diagram of the standardized residuals; we would, of course, obtain the same shape of plot using the $\{e_i\}$, since only a change in scale is involved. This scatter plot is as shown in Figure 16.1.

Now, what should we look for in a plot of this type? Since we usually assume that $E_i \frown N(0, \sigma^2)$ so that $E_i/\sigma \frown N(0, 1)$, we would expect (if the normality assumption holds) that the scatter plot of the $\{e_i/S\}$ would look like one based on a random sample of 17 observations from a $N(0, 1)$ distribution. One can simulate what such a plot might be expected to look like by selecting a number of random samples each of size 17 from (widely available) tables of standard normal random deviates and then constructing a series of such scatter plots. In this way one can get a feel for the variation in the pattern of points to be anticipated under the $N(0, 1)$ assumption.

FIGURE 16.1 *Scatter plot of the $\{e_i/S\}$*

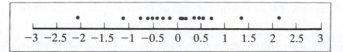

FIGURE 16.2   *Standardized residual plot on normal probability paper for the model* $\hat{Y} = 15.39 + 3.79X$

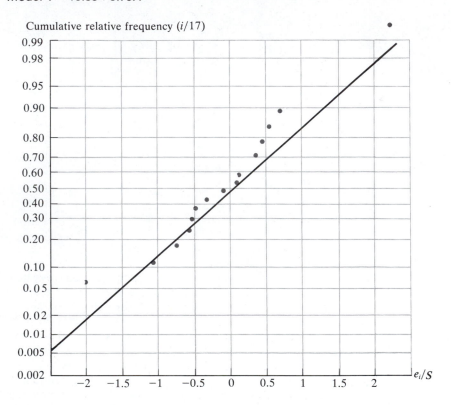

An alternative graphical approach is to construct a *normal* or *half-normal* plot of the residuals (or standardized residuals) on normal probability paper. Figure 16.2 represents such a plot for the data we are considering. The plotting procedure first entails ordering the standardized residuals from smallest to largest, followed by a marking off of the horizontal axis to include all the $\{e_i/S\}$. The cumulative relative frequencies up to each value are then plotted as ordinates versus the e_i/S values as abscissas.[1] For example, the standardized residual 1.35 is the next to largest of the 17 residuals, so we plot 16/17 (or 0.94) as the ordinate value corresponding to 1.35. An important property associated with the use of this particular type of graph paper is that the cumulative relative frequencies for a normal distribution plot as a straight line. The line for the $N(0, 1)$ distribution is drawn on the graph and can be used as a yardstick in assessing whether the scatter of points reflects any obvious deviation from normality. In our case, neither the scatter diagram nor the normal plot suggest any blatant departures from the normality assumption, although such a statement is necessarily qualitative in nature.

More quantitative criteria for assessing the validity of the normality assumption can, of course, be based on the use of standard statistical testing procedures (e.g., the

[1] To avoid having to deal with a cumulative relative frequency of 1 (see Figure 16.2), it is often recommended that $(i - 1/2)/n$ or $i/(n + 1)$ be plotted versus e_i/S, rather than i/n.

chi-square and Kolmogorov–Smirnov tests). In this regard, another way to examine the distribution of residuals is the following. Since the $\{e_i/S\}$ are assumed to represent a sample from a standard normal distribution, we would expect *approximately* 68% of these standardized residuals to lie in the interval $(-1.65, +1.65)$, about 95% to be contained in the interval $(-1.96, +1.96)$, etc. [If $n - (k+1)$ is small, 68% and 95% limits of the $t_{n-(k+1)}$ distribution can be used instead.] For these data, only 2 (i.e., 12%) of the 17 standardized residuals exceed 1.35 in absolute value. Also, the values of these two particular standardized residuals are large $(-2.07$ and $2.20)$, so the validity of the normality assumption might be questioned.

As a word of caution, the analysis of residuals with respect to departures from normality is generally difficult because the distribution of a set of residuals is affected by several factors. For example, the residuals may appear to exhibit a nonnormal pattern because an inappropriate regression model has been used, or because of variance nonhomogeneity, or even simply because the number of residuals is too small to provide a pattern of sufficient stability to permit one to make a valid statistical inference concerning the nature of the underlying probability distribution. Hence, it is good practice to gather evidence concerning each of the possible types of departures from the assumptions and then to examine this evidence *in toto* before passing out specific indictments. Following this strategy, we shall defer judgment concerning violation of the normality assumption (e.g., with respect to the two standardized residuals which are possible "outliers"), until we have examined the possibility of other violations.

The next step in our graphical analysis will be to construct a plot of the standardized residuals versus the values of the independent variable X. (In a multivariable situation, one could make such a plot for each independent variable in the model.) Possible types of patterns that can emerge from such a plot are portrayed in Figure 16.3. Figure 16.3a illustrates the type of pattern to be expected if all basic assumptions hold. A horizontal band of points should be obtained, with no hint of the presence of any systematic trends.

FIGURE 16.3   *Typical residual plots*

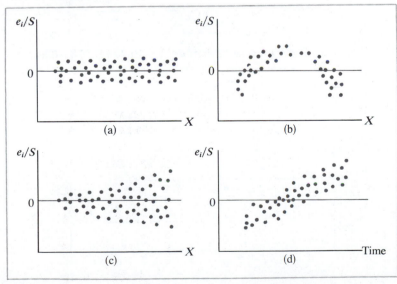

Figure 16.3b illustrates the systematic pattern to be expected from a departure from linearity, reflecting a need for curvilinear terms in the regression model. Naturally, different types of model inappropriateness result in different residual patterns.

Figure 16.3c represents the pattern to be expected when the error variance increases directly with X. It is possible, of course, to encounter situations where the error variance appears to be an even more complex function of X. In any case, transformations of the data are very often helpful in eliminating such variance heteroscedasticity (see Section 16.3). Also, it is a good idea when planning to examine residuals for evidence of error variance nonhomogeneity and model inappropriateness to take replicate observations at as many values of X as possible.

Figure 16.3d is a plot of the $\{e_i/S\}$ versus time, and a linear time-related effect is clearly present. Whenever there are variables not included in the regression model (e.g., the variable "time" when the data are collected in a time sequence) but which may have a significant effect (e.g., such as that reflected in Figure 16.3d by a strong positive correlation with the residuals), it is often informative to construct plots involving such variables.

FIGURE 16.4   *Plot of e_i/S versus X_i for the model $\hat{Y} = 15.39 + 3.79X$*

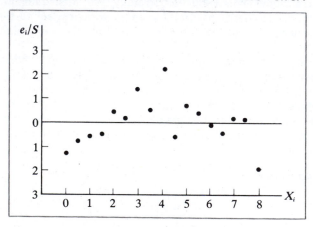

FIGURE 16.5   *Plot of e_i/S versus X_i for the model $\hat{Y} = 10.00 + 8.10X - 0.54X^2$*

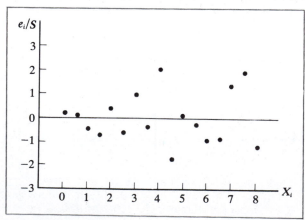

For our data a plot of the standardized residuals versus X for the model $\hat{Y} = 15.39 + 3.79X$ is presented in Figure 16.4. One can immediately recognize a systematic pattern similar to the prototype pattern given by Figure 16.3b, and this suggests introducing a quadratic term into the model. If we do this, the resulting model is $\hat{Y} = 10.00 + 8.10X - 0.54X^2$ with $S = 2.84$, and a plot of e_i/S versus X for this model is given in Figure 16.5. Although Figure 16.5 reflects considerable variation (due in part to a small sample size), the pattern of points mimics the horizontal band of Figure 16.3a more than the other prototypes discussed.

16.2.3 MISCELLANEOUS TOPICS CONCERNING RESIDUALS

Outliers

An outlier among a set of residuals is one that is much larger than the rest in absolute value, perhaps lying as many as three or more standard deviations from the mean of the residuals. Clearly, the presence of such an extreme value can significantly affect the least-squares fitting of a model, and so it is important to determine if the observation in question should be retained or discarded. Since the presence of an outlier in the data may be an indication of special circumstances warranting further investigation (e.g., such as the presence of an unanticipated interaction effect), an immediate discarding of the observation is not recommended unless there is strong evidence that it resulted from a mistake (e.g., an error in data recording) or from some other cause independent of the process under study (e.g., an obvious instrument malfunction). There are statistical testing procedures for evaluating outliers, and a discussion of some of these can be found in papers by Anscombe (1960) and Anscombe and Tukey (1963).

Tests involving residuals

As mentioned earlier, a graphical analysis of a set of residuals is necessarily somewhat subjective in nature. However, as demonstrated in the previous section, a careful graphical approach involving the simultaneous evaluation of several different types of residual plots is often sufficient to reveal whatever anomalies exist in the data. Nevertheless, there are occasions when it is desirable to utilize statistical testing procedures to answer specific questions. We have already alluded to the fact that standard goodness-of-fit tests are available for examining the normality assumption and that tests for outliers are discussed in the statistical literature. In fact, there are statistical tests available to examine each of the specific assumptions in question. For example, when a set of residuals is gathered in a time sequence, a nonparametric "runs test" [e.g., see Siegel (1956, pp. 52–60)] is frequently used to determine whether the time sequence of positive and negative residuals is unusual enough to be considered more than just a random occurrence. And, tests for variance nonhomogeneity could be based on a comparison, by means of F statistics, of sample variances calculated from replicate observations obtained at each of a series of values of the independent variable [e.g., see Bartlett (1947)]. Or, in assessing the amount of variance nonhomogeneity in terms of a suspected monotonic relationship between the error variance and an independent variable, one can test the significance of the Spearman rank correlation [again, see Siegel (1956, pp. 202–213)] between the absolute value of the residual and the value of the independent variable.

16.3 TRANSFORMATIONS

Two general options are open to an investigator when one or more of the underlying assumptions of regression analysis are not satisfied: *either* the method of analysis can be modified to deal directly with the actual characteristics of the data (e.g., by using weighted-least-squares analysis, nonlinear least-squares procedures, or growth-curve analysis), *or* the original variables (particularly the dependent variable) can be transformed to new variables for which the standard regression assumptions are more closely satisfied. In this section we shall focus briefly on the transformation approach, and in Section 16.4 we shall discuss one aspect of the modification approach: weighted-least-squares analysis.

The three primary reasons for using data transformations are (1) *to stabilize the variance* of the dependent variable when the homoscedasticity assumption is violated; (2) *to normalize* (i.e., to transform to the normal distribution) the dependent variable when the normality assumption is noticeably violated; and (3) *to linearize* the regression model when the original data suggests a model that is nonlinear in either the regression coefficients and/or the original variables (dependent or independent). It is fortunate that the same transformation often simultaneously helps to accomplish the first two goals and sometimes even the third, rather than to achieve one goal at the expense of either of the other two.

A more thorough discussion than given here concerning the properties of various transformations can be found in several sources, notably Armitage (1971), Draper and Smith (1966), and Neter and Wasserman (1974). Nevertheless, we consider it to be useful to describe a few of the more commonly used transformations:

1. *The log transformation* $(Y' = \log Y)$. Used (provided Y takes on only positive values) to stabilize the variance when the variance of Y increases markedly with increasing Y; to normalize the dependent variable when the distribution of the residuals for Y is positively skewed; and to linearize the regression model when the relationship of Y with some independent variable suggests a model with consistently increasing slope.

2. *The square-root transformation* $(Y' = \sqrt{Y})$. Used to stabilize the variance when it is proportional to the mean of Y. This is particularly appropriate when the dependent variable has the Poisson distribution.

3. *The reciprocal transformation* $(Y' = 1/Y)$. Used to stabilize the variance when it is proportional to the fourth power of the mean of Y, or, in other words, when there is a huge increase in variance above some threshold value of Y. The purpose of this transformation is to minimize the effect of large values of Y, since for these values the transformed Y' values will be close to zero, and large increases in Y will cause only trivial decreases in Y'.

4. *The square transformation* $(Y' = Y^2)$. Used to stabilize the variance when it decreases with the mean of Y; to normalize the dependent variable when the distribution of the residuals for Y is negatively skewed; and to linearize the model when the original relationship with some independent variable is curvilinear downward (i.e., the slope consistently decreases as the independent variable increases).

5. *The arcsin transformation* $(Y' = \arcsin \sqrt{Y} = \sin^{-1} \sqrt{Y})$. Used for variance stabilization when Y is a proportion or rate.

16.4 WEIGHTED-LEAST-SQUARES ANALYSIS

The *weighted-least-squares* method of analysis is a modification of standard regression analysis procedures and is designed for the situation in which a regression model is to be fit to a set of data for which the assumptions of variance homogeneity and/or independence do not hold. We shall briefly describe here the weighted-least-squares approach for dealing with variance heterogeneity. We refer the reader to other sources, such as Draper and Smith (1966) and Neter and Wasserman (1974), for a discussion of the general method of weighted regression, which incorporates the nonindependence situation.

When the variance of Y varies for different values of the independent variable(s), weighted-least-squares analysis can be used provided these variances (i.e., σ_i^2 for the ith observation on Y) are *known* or can be assumed to be of the form $\sigma_i^2 = \sigma^2/W_i$, where the weights $\{W_i\}$ are *known*. The methodology then involves determining the regression coefficients $\hat{\beta}_0', \hat{\beta}_1', \ldots, \hat{\beta}_k'$, which minimize the expression

$$\sum_{i=1}^{n} W_i(Y_i - \hat{\beta}_0' - \hat{\beta}_1'X_{1i} - \cdots - \hat{\beta}_k'X_{ki})^2$$

where the "weight" W_i is given by $1/\sigma_i^2$ when the $\{\sigma_i^2\}$ are known, or is exactly the W_i in the expression σ^2/W_i when this latter form applies.

The specific weighted-least-squares solution for the straight-line regression case (i.e., $Y = \beta_0 + \beta_1 X + E$) is given by the formulas

$$\hat{\beta}_1' = \frac{\sum_{i=1}^{n} W_i(X_i - \bar{X}')(Y_i - \bar{Y}')}{\sum_{i=1}^{n} W_i(X_i - \bar{X}')^2} = \frac{\sum_{i=1}^{n} W_i X_i Y_i - \left(\sum_{i=1}^{n} W_i X_i\right)\left(\sum_{i=1}^{n} W_i Y_i\right)\Big/\left(\sum_{i=1}^{n} W_i\right)}{\sum_{i=1}^{n} W_i X_i^2 - \left(\sum_{i=1}^{n} W_i X_i\right)^2\Big/\left(\sum_{i=1}^{n} W_i\right)}$$

and

$$\hat{\beta}_0' = \bar{Y}' - \hat{\beta}_1 \bar{X}'$$

where

$$\bar{Y}' = \sum_{i=1}^{n} W_i Y_i \Big/ \sum_{i=1}^{n} W_i \quad \text{and} \quad \bar{X}' = \sum_{i=1}^{n} W_i X_i \Big/ \sum_{i=1}^{n} W_i$$

Under the usual normality assumption on the Y variable, the same general procedures as used earlier in the unweighted case regarding t tests, F tests, and confidence intervals about the various regression parameters are applicable; e.g., to test $H_0: \beta_1 = 0$, the following test statistic may be used:

$$T = \frac{\hat{\beta}_1' - 0}{S_{Y|X}'/S_X'\sqrt{n-1}} \frown t_{n-2} \quad \text{under } H_0$$

where

$$S_{Y|X}^{2'} = \frac{1}{n-2} \sum_{i=1}^{n} W_i(Y_i - \hat{\beta}_0' - \hat{\beta}_1'X_i)^2$$

and

$$S_X'^2 = \frac{1}{n-1} \sum_{i=1}^{n} W_i(X_i - \bar{X}')^2$$

ONE-WAY ANALYSIS OF VARIANCE

17.1 PREVIEW

This chapter is the first of four chapters concerned with the subject of *analysis of variance*. Analysis of variance, abbreviated *ANOVA*, has been mentioned previously. Earlier, we generally described ANOVA as a technique for assessing how several *nominal* independent variables affect a *continuous* dependent variable. An example was given in Table 2.2 (the Ponape study); the problem involved describing the effects on blood pressure of two cultural incongruity indices, each dichotomized into the two categories "high" and "low" (see Example 1.5). In this case, the dependent variable (blood pressure) is continuous and the two independent variables (the cultural incongruity indices) are both nominal.

A second example (Daly's 1973 study) concerns whether or not four Turnkey housing developments with differing racial compositions in surrounding neighborhoods have significantly different mean CMI scores (Example 1.6). Here, the dependent variable (CMI score) is continuous, and the four Turnkey developments comprise the four categories of a single nominal variable.[1]

The very fact that ANOVA is generally restricted only to consideration of *nominal* independent variables allows for an interesting interpretation of the purpose of the technique. Loosely speaking, *ANOVA is usually concerned with comparisons involving several population means*. In fact, in the simplest special case involving a comparison of *two* population means, the ANOVA comparison procedure is equivalent to the usual two-sample *t* test (which, as we know, requires the assumption of equal population variances).

The population means to be compared are generally easily specified[2] by cross-classifying the nominal independent variables under consideration to form different

---

[1] Recall that a regression model that incorporates such a nominal variable with four categories will contain three dummy variables to represent these four categories.

[2] Such specification is not really possible if the categories of any nominal variable are considered as only a sample from a much larger population of categories of interest. We will consider such situations later when discussing what are called *random-effects models*.

combinations of categories. In the first example above (dealing with the Ponape study), we need only cross-classify the HI and LO categories of incongruity index 1 with the HI and LO categories of index 2. This yields the four population means corresponding to the four combinations HI–HI, HI–LO, LO–HI, and LO–LO, as indicated by the following configuration:

INDEX 2

	HI	LO
HI	μ_1	μ_2
LO	μ_3	μ_4

INDEX 1

Assessing whether the two indices have some *effect* on the dependent variable "blood pressure" is equivalent to determining what kind of differences, if any, exist among the four means.

For the second example above (Daly's study), there are four mean CMI scores to be compared corresponding to the four Turnkey developments. The main questions concern whether or not the four means are equal, and, if not, where the differences are.

17.1.1 WHY THE NAME ANOVA?

If the technique called ANOVA usually deals with a comparison of means, it seems somewhat inappropriate to call it analysis of "variance." Instead, why not use the acronym *ANOME*, where ME is short for "means"? Actually, however, the use of the designation ANOVA is quite justifiable, since, although means are usually being compared, the comparisons are made using estimates of variance. The ANOVA test statistics, as with regression analysis, are F statistics and are actually ratios of estimates of variance. In fact, it is even possible, and in some cases appropriate, to specify the null hypotheses of interest in terms of population variances.

17.1.2 ANOVA VERSUS REGRESSION

Another general distinction to be made concerns the difference between what may be called an "ANOVA problem" and what may be called a "regression problem." For ANOVA, *all* the independent variables are treated as being *nominal*; for regression analysis, any mixture of measurement scales (nominal, ordinal, or interval) is allowable for the independent variables. In fact, ANOVA is often conveniently looked upon as a special case of regression analysis, and almost any ANOVA model can be represented by a regression model whose parameters can be estimated and inferred about in the usual manner. The same may be said for certain other multivariable techniques, such as analysis of covariance and even discriminant analysis. In effect, then, we may view the various names given to these techniques as indicators of different (linear) models with the same general form:

$$Y = \beta_0 + \beta_1 X_1 + \cdots + \beta_p X_p + E$$

yet involving different types of variables and perhaps different assumptions regarding these variables. The choice of method, in this light, can then be regarded simply as being equivalent to the choice of an appropriate linear model.

17.1.3 FACTORS AND LEVELS

Some additional terms need to be introduced at this point. When considering the use of dummy variables in regression in Chapter 13, we saw that a nominal variable with k categories is generally incorporated into a regression model by defining $k-1$ dummy variables. These $k-1$ variables collectively describe the *basic* nominal variable being considered. It is usually convenient to be able to refer to such a basic variable without having to identify the specific dummy variables used to define it in the regression model. The approach generally adopted is to call the basic nominal variable a *factor*. The different categories of the factor are often referred to as its *levels*.

For example, if we wanted to compare the effects of several drugs on some human health response, we would consider the nominal variable "drugs" as a single factor and the specific drug categories as the levels. If k drugs are being compared, these would be incorporated into a regression model by defining $k-1$ dummy variables. If, in addition to comparing the drugs, we also wanted to consider whether males and females respond differently, we would consider the nominal variable "sex" as a second factor and the specific setting categories (male and female) as the levels of this dichotomous factor.

17.1.4 FIXED VERSUS RANDOM FACTORS

One final point to be discussed before proceeding with the methodology of one-way ANOVA is the need to distinguish between what are called "fixed" and "random" factors. A *random factor* is one whose levels may be regarded as a sample from some large population of levels.[3] A *fixed factor* is one whose levels are the only relevant levels of interest. The distinction is important in any analysis-of-variance situation, since different tests of significance are required for different configurations of random and fixed factors. We will see this more specifically when considering two-way analysis of variance. For now it is perhaps useful to give some examples of random and fixed factors. These are described briefly below and are summarized in Table 17.1.

TABLE 17.1 *Examples of random and fixed factors*

RANDOM	FIXED	EITHER (DEPENDS ON THE EXPERIMENTAL SITUATION)
Subjects	Sex	Locations
Litters	Age	Treatments
Observers	Marital status	Drugs
Days	Education	Tests
Weeks		

[3] Practically speaking, the experimental levels of a random factor need not be actually selected at random as long as they are reasonably representative of the large *population* of levels of interest.

1. "Subjects" or "groups of subjects" (e.g., litters of mice) is usually considered a random factor, since we ordinarily wish to infer from the subjects used to a large population of potential subjects.

2. "Observers" is a random factor often considered when examining the effect of interobserver variability on the response variable of interest.

3. "Days," "weeks," and so on, are usually considered as random factors when investigating the effect of time on a response variable observed during different time periods. Here we would usually consider such "temporal" factors to have levels representative of a large number of possible time periods.

4. "Sex" is always a fixed factor, since consideration of its two levels involves all possible relevant levels of interest.

5. "Locations" (e.g., cities, plants, states) may be fixed or random depending on whether the locations included are the only ones of interest or whether it is of interest to make inferences about a larger geographical universe.

6. "Age" is usually considered to be a fixed factor regardless of the way in which the different age groups are defined.

7. "Treatments," "drugs," "tests," and so on, are usually considered as fixed factors, but they may be considered random if their levels are representative of a much larger group of possible levels.

8. "Marital status" is considered to be a fixed factor.

9. "Education" is considered to be a fixed factor.

17.2 ONE-WAY ANOVA: THE PROBLEM, ASSUMPTIONS, AND DATA CONFIGURATION

One-way ANOVA deals with the effect of a *single factor* on a (single) response variable. When that one factor is a fixed factor, one-way ANOVA (often referred to as *fixed-effects one-way ANOVA*) involves a comparison of several (two or more) population means.[4] It can be said that the different populations correspond to the different levels of the single factor "populations."

17.2.1 THE PROBLEM

The main analysis problem in fixed-effects one-way ANOVA concerns the question of whether or not the population means are all equal. Thus, if there are k means (denoted as $\mu_1, \mu_2, \ldots, \mu_k$), the basic *null hypothesis* of interest is given by

$$H_0: \mu_1 = \mu_2 = \cdots = \mu_k \tag{17.1}$$

The *alternative hypothesis* is given by

$$H_A: \text{the } k \text{ population means are not all equal}$$

---

[4] We shall be focusing our attention throughout this section on situations involving only fixed factors. Random factors will be discussed in Section 17.6.

If the null hypothesis (17.1) is rejected, the next problem is to find out where the differences are. For example, if $k = 3$ and $H_0: \mu_1 = \mu_2 = \mu_3$ is rejected, we might wish to determine whether the main differences are between μ_1 and μ_2, or between μ_1 and μ_3, or between μ_1 and the average of the other two means, and so on. Such questions fall under the general statistical subject-matter heading referred to as "multiple comparison procedures," which will be discussed in Section 17.7.

17.2.2 THE ASSUMPTIONS

The assumptions needed for using fixed-effects one-way ANOVA may be simply stated as follows:

1. Random samples (e.g., individuals, animals, etc.) are selected from each of k populations or groups.
2. A value of a specified (dependent) variable is recorded for each experimental unit (e.g., individual, animal, etc.) sampled.
3. The dependent variable is normally distributed in each population.
4. The variance of the dependent variable is the same in each population (and this common variance is denoted as σ^2).

Although these assumptions provide the theoretical justification for applying this method, it is nevertheless sometimes necessary to compare several means using fixed-effects one-way ANOVA when the necessary assumptions are not clearly satisfied. Indeed, it is a rare instance when these assumptions hold exactly. It is therefore important to briefly consider the consequences of applying fixed-effects one-way ANOVA when the preceding assumptions are in question.

In general, it may be said that one can be reasonably confident in applying fixed-effects one-way ANOVA if none of the assumptions are very badly violated. This characteristic applies to more complex ANOVA situations as well as to fixed-effects one-way ANOVA. The term generally used to denote this property is called *robustness* (i.e., the procedure is robust with respect to moderate departures from the basic assumptions).

We must nevertheless be careful to avoid using robustness as an automatic justification for blindly applying the ANOVA method. In this regard, certain facts should be kept in mind when considering the use of ANOVA in a given situation. For example, the *normality assumption* does not have to be exactly satisfied as long as we are dealing with relatively large samples (e.g., 20 or more observations from each population), although the consequences of large deviations from normality are somewhat more severe for random factors than for fixed factors. The *assumption of variance homogeneity* can also be mildly violated without serious risk, provided that the numbers of observations selected from each population are more or less the same, although, again, the consequences are more severe for random factors. Violation of the *assumption of independence* of the observations, however, can lead to very serious errors in inference for both the fixed and random cases. In general, great care should be taken to ensure that the observations are independent. The necessity for worrying about dependence arises primarily in studies where repeated observations are recorded on the same experimental subjects, since very often the level of response of a subject on one occasion has a decided effect on subsequent responses.

What, then, does one do when one or more of these assumptions are in serious question? For one thing, the data might be transformed (e.g., by means of a log, square root, or other transformation) so that they more closely satisfy the assumptions. For another thing, a more appropriate method of analysis (e.g., nonparametric ANOVA or growth-curve analysis) may be utilized.[5]

17.2.3 DATA CONFIGURATION FOR ONE-WAY ANOVA

Computations necessary for one-way ANOVA can be easily performed even with an ordinary desk calculator when the data are conveniently arranged. Table 17.2 illustrates a useful way of presenting the data for the general one-way situation. It can be seen from this table that the number of observations selected from each population does *not* have to be the same; that is, there are n_i observations from the ith population, and n_i need not equal n_j if $i \neq j$. Notice also that the double-subscript notation (Y_{ij}) is used to distinguish one observation from another. The first subscript for a given observation denotes the population number and the second subscript distinguishes that observation from the others in that sample. Thus, Y_{23} denotes the third observation from the second population, Y_{62} denotes the second observation from the sixth population, and Y_{knk} denotes the last observation from the kth population. The totals for each sample (from each population) are denoted alternatively by T_i or $Y_i.$ for the ith sample, where the "." denotes that we are summing over all values of j (i.e., we are adding together all observations within the given sample); and the grand total over all samples is denoted as $G = Y_{..}$ The sample means are alternatively denoted by $\bar{Y}_i.$ or T_i/n_i for the ith sample; these statistics are particularly important because they are the estimates of the population means of interest. Finally, the overall (grand) mean over all samples is $\bar{Y} = G/n$.

EXAMPLE   In the study by Daly concerning the effects of neighborhood characteristics on health, a stratified random sample of 100 households was selected, 25 from each of the four Turnkey neighborhoods included in the study. The data configuration of CMI scores for women heads of household is given in Table 17.3. Such scores are measures (derived

TABLE 17.2   *General data configuration for one-way ANOVA*

POPULATION	SAMPLE SIZE	OBSERVATIONS	TOTALS	SAMPLE MEANS
1	n_1	$Y_{11}, Y_{12}, Y_{13}, \ldots, Y_{1n_1}$	$T_1 = Y_1.$	$\bar{Y}_1. = T_1/n_1$
2	n_2	$Y_{21}, Y_{22}, Y_{23}, \ldots, Y_{2n_2}$	$T_2 = Y_2.$	$\bar{Y}_2. = T_2/n_2$
3	n_3	$Y_{31}, Y_{32}, Y_{33}, \ldots, Y_{3n_3}$	$T_3 = Y_3.$	$\bar{Y}_3. = T_3/n_3$
⋮	⋮	⋮	⋮	⋮
k	n_k	$Y_{k1}, Y_{k2}, Y_{k3}, \ldots, Y_{knk}$	$T_k = Y_k.$	$\bar{Y}_k. = T_k/n_k$
	$n = \sum\limits_{i=1}^{k} n_i$		$G = Y_{..}$	$\bar{Y} = G/n$

[5] A description of nonparametric methods that can be used when these assumptions are clearly and strongly violated can be found in Siegel (1956) and Lehmann (1975). A discussion of growth-curve analysis can be found in Allen and Grizzle (1969).

TABLE 17.3   *CMI scores for a sample of women from different households in four Turnkey housing neighborhoods*

POPULATION	NO. HOUSEHOLDS IN NEIGHBORHOOD	% BLACKS IN SURROUNDING NEIGHBORHOOD	SAMPLE SIZE (n_i)	OBSERVATIONS (Y_{ij})	TOTALS (T_i)	SAMPLE MEAN $(\bar{Y}_{i.})$
1: Cherryview	98	17	25	49, 12, 28, 24, 16, 28, 21, 48, 30, 18, 10, 10, 15, 7, 6, 11, 13, 17, 43, 18, 7, 10, 9, 12, 12,	$T_1 = 473$	$\bar{Y}_{1.} = 18.92$
2: Morningside	211	100	25	5, 1, 44, 11, 4, 3, 14, 2, 13, 68, 34, 40, 36, 40, 22, 25, 14, 23, 26, 11, 20, 4, 16, 25, 17	$T_2 = 518$	$\bar{Y}_{2.} = 20.72$
3: Northhills	212	36	25	20, 31, 19, 9, 7, 16, 11, 17, 9, 14, 10, 5, 15, 19, 29, 23, 70, 25, 6, 62, 2, 14, 26, 7, 55	$T_3 = 521$	$\bar{Y}_{3.} = 20.84$
4: Easton	40	65	25	13, 10, 20, 20, 22, 14, 10, 8, 21, 35, 17, 23, 17, 23, 83, 21, 17, 41, 20, 25, 49, 41, 27, 37, 57	$T_4 = 671$	$\bar{Y}_{4.} = 26.84$
			$\sum_{i=1}^{4} n_i = 100$		$G = 2,183$	$\bar{Y} = 21.83$

from questionnaires) of the overall (self-perceived) health of individuals; the higher the score, the poorer the health. It is important to point out, in addition, that each of the Turnkey neighborhoods differed in the total number of residents and in the percentage of blacks in the surrounding neighborhoods. The racial composition of the Turnkey neighborhoods themselves was over 95% black. Daly's main thesis was that the health of persons living in similar federal or state housing projects varied according to the racial composition of the surrounding neighborhoods. The "friendlier" (in terms of racial composition) the surrounding neighborhood was, the better would be the health of the residents in the project. According to Daly, federal housing planners had never used information concerning overall neighborhood racial composition and other neighborhood characteristics and their relationship to health as criteria for selection of areas for such projects. This study, it was hoped, could provide some concrete recommendations for improved federal planning.

On first examination of the data in Table 17.3, it can be seen that the sample means vary as we go down the table. In order to determine whether or not the observed differences in the sample means can be attributed solely to chance, one may perform a one-way ANOVA. The possibility of violations of the assumptions underlying this methodology should not be of great concern for this data set, since the sample sizes are equal and reasonably large and since observations on women from different households may be considered to be independent.

17.3 METHODOLOGY FOR ONE-WAY ANOVA

The null hypothesis of equal population means ($H_0: \mu_1 = \mu_2 = \cdots = \mu_k$) is tested using an F test. The test statistic is calculated as follows:

$$F = \frac{\text{MST}}{\text{MSE}} \qquad (17.2)$$

where

$$\text{MST} = \left[\sum_{i=1}^{k} (T_i^2/n_i) - G^2/n \right] \bigg/ (k-1) \qquad (17.3)$$

and

$$\text{MSE} = \left[\sum_{i=1}^{k} \sum_{j=1}^{n_i} Y_{ij}^2 - \sum_{i=1}^{k} (T_i^2/n_i) \right] \bigg/ (n-k) \qquad (17.4)$$

When H_0 is true (i.e., when the population means are all equal), the F statistic of (17.2) has the F distribution with $(k-1)$ numerator and $(n-k)$ denominator degrees of freedom. Thus, for a given α we would reject H_0 and conclude that some (i.e., at least two) of the population means are different from another if

$$F \geq F_{k-1, n-k, 1-\alpha}$$

where $F_{k-1, n-k, 1-\alpha}$ is the $100(1-\alpha)\%$ point of the F distribution with $k-1$ and $n-k$ degrees of freedom. It should be pointed out that the critical region for this test involves only upper percentage points of the F distribution, since only large values of the F statistic (usually values much greater than 1) will provide evidence for rejection of H_0.

17.3.1 NUMERICAL ILLUSTRATION

For the data given in Table 17.3, the calculations needed to perform the F test proceed as follows:

$$\sum_{i=1}^{4} \sum_{j=1}^{25} Y_{ij}^2 = \underbrace{\frac{(49)^2 + (12)^2 + \cdots + (37)^2 + (57)^2}{\text{(a sum of 100 squared observations)}}} = 72{,}851.40$$

$$\sum_{i=1}^{4} (T_i^2/n_i) = \frac{(473)^2}{25} + \frac{(518)^2}{25} + \frac{(521)^2}{25} + \frac{(671)^2}{25} = 48{,}549.40$$

$$G^2/n = (2{,}183)^2/100 = 47{,}654.89$$

$$\text{MST} = \left[\sum_{i=1}^{4} (T_i^2/n_i) - G^2/n \right] \bigg/ (4-1)$$

$$= (48{,}549.40 - 47{,}654.89)/3$$

$$= 298.17$$

$$MSE = \left[\sum_{i=1}^{4} \sum_{j=1}^{25} Y_{ij}^2 - \sum_{i=1}^{4} (T_i^2/n_i) \right] \Big/ (100-4)$$

$$= (72{,}851.40 - 48{,}549.40)/96$$

$$= 253.14$$

$$F = MST/MSE$$

$$= 298.17/253.14$$

$$= 1.178$$

Using these calculations we may test "H_0: $\mu_1 = \mu_2 = \mu_3 = \mu_4$" (i.e., there are *no* differences among the true mean CMI scores for the four neighborhoods) against "H_A: there are differences among the true mean CMI scores." For example, if $\alpha = 0.10$, we would find from the F tables that $F_{3,96,0.90} = 2.15$, which is greater than the computed F of 1.178. Thus, we would *not reject* H_0 at $\alpha = 0.10$.

   To find the P value for this test, we note further that $F_{3,96,0.75} = 1.41$, which also exceeds the computed F. Thus, we know that $P > 0.25$. Hence, we must conclude (as does Daly) that there are no significant differences among the observed mean CMI scores for the four neighborhoods.

   It should be noted at this point that if a significant difference among the sample means had been found, it would still be up to the investigator to determine whether the actual magnitudes of these differences are meaningful in a practical sense and whether the patterns of these differences are as hypothesized. In our example, examination of the percentages of blacks in the surrounding neighborhoods (see Table 17.3) indicates that the observed differences among the sample means are clearly not in the pattern hypothesized. That is, under Daly's conjecture, it would be expected that Cherryview (with 17% black in the surrounding neighborhood) would have the highest observed mean CMI score, followed by Northhills (36%), Easton (65%), and finally Morningside (100%). This was not the order actually obtained. Daly also examined whether her conjecture was supported when controlling for other possibly relevant factors, such as "months lived in the neighborhood," "number of children," and "marital status." However, no significant results were obtained from these analyses either.

17.3.2 RATIONALE FOR THE F TEST IN ONE-WAY ANOVA

The use of the F test described above may be motivated in a number of ways, three of which are given below. Hopefully, these will not only provide the user of this F test with an intuitive theoretical appreciation of its purpose, but will also provide some insight into the rationale behind more complex ANOVA testing procedures.

   1. The F test in one-way ANOVA is a generalization of the two-sample t test. It may be easily shown, with the help of a little algebra, that the numerator and denominator components in the F statistic (17.2) for one-way ANOVA are simple generalizations of the corresponding components in the square of the ordinary two-sample t-test statistic. In fact, when $k = 2$ it can be shown that the F statistic for one-way ANOVA is exactly equal to the square of the corresponding t statistic. Such a result is intuitively reasonable since the numerator degrees of freedom of the F when $k = 2$ is 1, and we have previously noted that the square of a t statistic with ν degrees of freedom has the F distribution with 1 and ν degrees of freedom in numerator and denominator, respectively.

In particular, recall that the two-sample t-test statistic is given by the formula

$$T = \frac{(\bar{Y}_{1.} - \bar{Y}_{2.})/\sqrt{1/n_1 + 1/n_2}}{S_p}$$

where the pooled sample variance S_p^2 is given by

$$S_p^2 = \frac{1}{n_1 + n_2 - 2} \sum_{i=1}^{2} \sum_{j=1}^{n_i} (Y_{ij} - \bar{Y}_{i.})^2$$

or, equivalently, by

$$S_p^2 = \frac{(n_1 - 1)S_1^2 + (n_2 - 1)S_2^2}{n_1 + n_2 - 2}$$

where S_1^2 and S_2^2 are the sample variances for groups 1 and 2, respectively. Focusing first on the denominator (MSE) of the F statistic (17.2), it can be shown with some algebra that

$$\text{MSE} = \left[\sum_{i=1}^{k} \sum_{j=1}^{n_i} Y_{ij}^2 - \sum_{i=1}^{k} (T_i^2/n_i) \right] \Big/ (n - k)$$

$$= \frac{(n_1 - 1)S_1^2 + (n_2 - 1)S_2^2 + \cdots + (n_k - 1)S_k^2}{n_1 + n_2 + \cdots + n_k - k}$$

Thus, MSE is a pooled estimate of the common population variance σ^2, being a weighted sum of the k estimates of σ^2 obtained using the k different sets of observations. Furthermore, when $k = 2$, MSE is equal to S_p^2.

Looking at the numerator (MST) of the F statistic, it can be shown that

$$\text{MST} = \left[\sum_{i=1}^{k} (T_i^2/n_i) - G^2/n \right] \Big/ (k - 1)$$

$$= \frac{1}{k - 1} \sum_{i=1}^{k} n_i (\bar{Y}_{i.} - \bar{Y})^2$$

which simplifies to $(\bar{Y}_{1.} - \bar{Y}_{2.})^2/(1/n_1 + 1/n_2)$ when $k = 2$. Thus, the equivalence has been established.

2. *The F statistic is the ratio of two variance estimates.* We have already seen that MSE is a pooled estimate of the common population variance σ^2; that is, the true average (or mean) value (μ_{MSE}, say) of MSE is σ^2. It turns out, however, that MST estimates σ^2 *only* when H_0 is true, that is, *only* when the population means $\mu_1, \mu_2, \ldots, \mu_k$ are all equal. In fact, the true mean value (μ_{MST}, say) of MST has the general form

$$\mu_{\text{MST}} = \sigma^2 + \frac{1}{k - 1} \sum_{i=1}^{k} n_i (\mu_i - \bar{\mu})^2 \tag{17.5}$$

where $\bar{\mu} = \sum_{i=1}^{k} n_i \mu_i / n$. It can be seen by inspection of expression (17.5) that MST estimates σ^2 *only* when all the μ_i are equal, in which case $\mu_i = \bar{\mu}$ for every i and so $\sum_{i=1}^{k} n_i (\mu_i - \bar{\mu})^2 = 0$. Otherwise, both terms on the right-hand side of (17.5) are positive and MST estimates something greater in value than σ^2. In other words,

$$\mu_{\text{MST}} = \sigma^2 \quad \text{when } H_0 \text{ is true}$$

and

$$\mu_{\text{MST}} > \sigma^2 \quad \text{when } H_0 \text{ is not true}$$

Loosely speaking, then, the F statistic MST/MSE may be viewed as approximating in some sense the ratio of population means

$$\frac{\mu_{\text{MST}}}{\mu_{\text{MSE}}} = \frac{\sigma^2 + [1/(k-1)] \sum_{i=1}^{k} n_i (\mu_i - \bar{\mu})^2}{\sigma^2} \tag{17.6}$$

When H_0 is true, the numerator and denominator of (17.6) both equal σ^2, and the F statistic in this case is thus the ratio of two estimates of the same variance. Furthermore, F can be expected to give different values depending on whether or not H_0 is true; that is, F should take a value close to 1 if H_0 is true (since it would be approximating $\sigma^2/\sigma^2 = 1$), whereas F should be larger than 1 if H_0 is false [since the numerator of (17.6) would be greater than the denominator].

3. *The F statistic compares the variability between groups to the variability within groups.* As with regression analysis, the "total" variability in the observations in a one-way ANOVA situation is measured by a "total" sum of squares,

$$\text{TSS} = \sum_{i=1}^{k} \sum_{j=1}^{n_i} (Y_{ij} - \bar{Y})^2 \tag{17.7}$$

Furthermore, it can be shown that

$$\text{TSS} = \text{SST} + \text{SSE} \tag{17.8}$$

where

$$\text{SST} = (k-1)\text{MST} = \sum_{i=1}^{k} n_i (\bar{Y}_{i.} - \bar{Y})^2$$

and

$$\text{SSE} = (n-k)\text{MSE} = \sum_{i=1}^{k} \sum_{j=1}^{n_i} (Y_{ij} - \bar{Y}_{i.})^2$$

The term "SST" can be seen to be a measure of the variability *between* (or *across*) populations. (The designation "SST" is read "sum of squares due to treatments," since the populations often represent treatment groups.) It involves components of the general form $(\bar{Y}_{i.} - \bar{Y})$, which is the difference between the ith-group mean and the overall mean.

The term "SSE" is a measure of the variability *within* populations and gives no information concerning variability between populations. It involves components of the general form $(Y_{ij} - \bar{Y}_{i.})$, which is the difference between the jth observation in the ith group and the mean for the ith group.

If SST is quite large in comparison to SSE, we know that most of the total variability is due to differences *between* populations rather than to differences *within* populations. Thus, it is quite natural in such a case to suspect that the population means are not all equal.

By writing the F statistic (17.2) in the form

$$F = \frac{SST}{SSE} \frac{n-k}{k-1}$$

we can see that F will be large whenever SST accounts for a much larger proportion of the total sum of squares than does SSE.

17.3.3 ANOVA TABLE FOR ONE-WAY ANOVA

As with regression analysis, the results of any analysis-of-variance procedure can be summarized in a table appropriately named for the method. The ANOVA table for one-way ANOVA is given in general form in Table 17.4; and Table 17.5 is the ANOVA table specific to our example involving the CMI data. The source and sum-of-squares[6]

TABLE 17.4 *General ANOVA table for one-way ANOVA (k populations)*

SOURCE	df	SS	MS	F
Between	$k-1$	$SST = \sum_{i=1}^{k} (T_i^2/n_i) - G^2/n$	$MST = SST/(k-1)$	MST/MSE
Within	$n-k$	$SSE = TSS - SST$	$MSE = SSE/(n-k)$	
Total	$n-1$	$TSS = \sum_{i=1}^{k} \sum_{j=1}^{n_i} Y_{ij}^2 - G^2/n$		

TABLE 17.5 *ANOVA table for CMI data ($k = 4$)*

SOURCE	df	SS	MS	F
Between (neighborhoods)	3	894.51	298.17	1.178
Within (error)	96	24,302.00	253.14	
Total	99	25,196.51		

[6] The usual method for calculating sums of squares involves computing TSS and SST separately and then computing SSE by subtraction (i.e., SSE = TSS − SST).

columns in Table 17.4 display the components of the fundamental equation of one-way ANOVA:

$$TSS = SST + SSE$$

The mean-square column contains the sums of squares divided by their corresponding degrees of freedom; the two mean squares are then used to form the numerator and denominator for the F test.

17.4 REGRESSION MODEL FOR FIXED-EFFECTS ONE-WAY ANOVA

We have stated earlier that most ANOVA procedures can be alternatively considered in a regression-analysis setting; this can be done by defining appropriate dummy variables in a regression model.[7] The ANOVA F tests are then equivalently formulated in terms of hypotheses concerning the coefficients of the dummy variables in the regression model.[8]

EXAMPLE   For the example involving the CMI data of Daly's study (see Table 17.3), a number of alternative regression models could be used to describe the situation, depending on the coding schemes used for the dummy variables. One such model is

$$Y = \mu + \alpha_1 X_1 + \alpha_2 X_2 + \alpha_3 X_3 + E \tag{17.9}$$

where the regression coefficients are denoted as μ, α_1, α_2, and α_3 and the independent variables are defined as

$$X_1 = \begin{cases} 1 & \text{if neighborhood 1} \\ -1 & \text{if neighborhood 4,} \\ 0 & \text{if otherwise} \end{cases} \quad X_2 = \begin{cases} 1 & \text{if neighborhood 2} \\ -1 & \text{if neighborhood 4,} \\ 0 & \text{if otherwise} \end{cases}$$

$$X_3 = \begin{cases} 1 & \text{if neighborhood 3} \\ -1 & \text{if neighborhood 4} \\ 0 & \text{if otherwise} \end{cases}$$

[Although we have previously used the Greek letter β with subscripts to denote regression coefficients, we have changed the notation for our ANOVA regression model in order that these coefficients correspond to the parameters in the classical fixed-effects ANOVA model described in Section 17.5.

[7] As mentioned earlier, we are restricting our attention here entirely to models with *fixed* factors. Models involving random factors will be treated in Section 17.6.

[8] We will see later that a regression formulation often is desirable, if not mandatory, when dealing with certain *nonorthogonal* ANOVA problems involving two or more factors. We shall discuss such problems in Chapter 20.

Other coding schemes for the independent variables will yield exactly the same ANOVA table and F test as model (17.9), although the regression coefficients themselves will represent different parameters and will have different least-squares estimators. One other frequently used coding scheme defines the independent variables as

$$X_i = \begin{cases} 1 & \text{if neighborhood } i \\ 0 & \text{otherwise} \end{cases}, \quad i = 1, 2, 3]$$

The coefficients μ, α_1, α_2, and α_3 for this (dummy variable) model can each be shown to be expressible in terms of the underlying population means (μ_1, μ_2, μ_3, μ_4) as follows:

$$\begin{cases} \mu = \dfrac{\mu_1 + \mu_2 + \mu_3 + \mu_4}{4} \quad (= \bar{\mu}^*, \text{ say}) \\ \alpha_1 = \mu_1 - \bar{\mu}^* \\ \alpha_2 = \mu_2 - \bar{\mu}^* \\ \alpha_3 = \mu_3 - \bar{\mu}^* \end{cases} \qquad (17.10)$$

[We show the coefficients to be expressible as in (17.10) as follows: $\mu_{Y|X_1,X_2,X_3} = \mu + \alpha_1 X_1 + \alpha_2 X_2 + \alpha_3 X_3$. Thus,

$\mu_1 = \mu_{Y|1,0,0} = \mu + \alpha_1$ since $X_1 = 1, X_2 = 0, X_3 = 0$ for neighborhood 1

$\mu_2 = \mu_{Y|0,1,0} = \mu + \alpha_2$ since $X_1 = 0, X_2 = 1, X_3 = 0$ for neighborhood 2

$\mu_3 = \mu_{Y|0,0,1} = \mu + \alpha_3$ since $X_1 = 0, X_2 = 0, X_3 = 1$ for neighborhood 3

$\mu_4 = \mu_{Y|-1,-1,-1} = \mu - \alpha_1 - \alpha_2 - \alpha_3$ since $X_1 = X_2 = X_3 = -1$ for neighborhood 4

Adding the left-hand sides and right-hand sides of these equations yields

$$\mu_1 + \mu_2 + \mu_3 + \mu_4 = 4\mu$$

or

$$\mu = \frac{1}{4} \sum_{i=1}^{4} \mu_i \quad (= \bar{\mu}^*)$$

Solution (17.10) is then obtained by replacing μ by $\bar{\mu}^*$ in the equations above and then solving for the regression coefficients α_i in terms of μ_1, μ_2, μ_3, and μ_4.]

Model (17.9), then, involves coefficients that describe separate comparisons of the first three population means with the overall unweighted mean, $\bar{\mu}^*$.

[If the coding scheme

$$X_i = \begin{cases} 1 & \text{if neighborhood } i \\ 0 & \text{otherwise} \end{cases}, \quad i = 1, 2, 3 \text{ is used}$$

the regression coefficients will describe separate comparisons of the first three population means with μ_4:

$$\left\{\begin{array}{l} \mu = \mu_4 \\ \alpha_1 = \mu_1 - \mu_4 \\ \alpha_2 = \mu_2 - \mu_4 \\ \alpha_3 = \mu_3 - \mu_4] \end{array}\right.$$

Also, for model (17.9), $(\mu_4 - \bar{\mu}^*)$ is expressible as the negative sum of α_1, α_2, and α_3. Moreover, this regression model can be fitted to provide *exactly* the same F statistic as required in one-way ANOVA for the test of H_0: $\mu_1 = \mu_2 = \mu_3 = \mu_4$. The equivalent regression null hypothesis is H_0: $\alpha_1 = \alpha_2 = \alpha_3 = 0$,[9] the regression F statistic will have the same degrees of freedom (i.e., $k - 1 = 3$ and $n - k = 96$) as given previously, and the ANOVA table will be exactly the same as given in the last section (when we pool the dummy-variable effects into one source of variation with 3 df).

REGRESSION MODEL FOR FIXED-EFFECTS ONE-WAY ANOVA
17.4.1 INVOLVING k POPULATIONS

An analogous model to that of the previous section may be given for the general situation involving k populations as follows:

$$Y = \mu + \alpha_1 X_1 + \alpha_2 X_2 + \cdots + \alpha_{k-1} X_{k-1} + E \qquad (17.11)$$

where

$$X_i = \left\{\begin{array}{ll} 1 & \text{for population } i \\ -1 & \text{for population } k, \qquad i = 1, 2, \ldots, k-1 \\ 0 & \text{otherwise} \end{array}\right.$$

The coefficients of this model are expressible in terms of the k population means $\mu_1, \mu_2, \mu_3, \ldots, \mu_k$ as follows:

$$\left\{\begin{array}{l} \mu = (\mu_1 + \mu_2 + \cdots + \mu_k)/k = \bar{\mu}^* \\ \alpha_1 = \mu_1 - \bar{\mu}^* \\ \alpha_2 = \mu_2 - \bar{\mu}^* \\ \qquad \cdot \\ \qquad \cdot \\ \qquad \cdot \\ \alpha_{k-1} = \mu_{k-1} - \bar{\mu}^* \\ -(\alpha_1 + \alpha_2 + \cdots + \alpha_{k-1}) = \mu_k - \bar{\mu}^* \end{array}\right. \qquad (17.12)$$

[9] When $\alpha_1 = \alpha_2 = \alpha_3 = 0$, it follows from simple algebra using (17.10) that $\mu_1 = \mu_2 = \mu_3 = \mu_4$ (e.g., $\alpha_1 = \mu_1 - \bar{\mu}^* = 0$ implies that $\mu_1 = \bar{\mu}^*$, $\alpha_2 = \mu_2 - \bar{\mu}^* = 0$ implies that $\mu_2 = \bar{\mu}^* = \mu_1$, etc.).

for model (17.11). Also, the F statistic for one-way ANOVA with k populations can be equivalently obtained by testing the null hypothesis $H_0: \alpha_1 = \alpha_2 = \cdots = \alpha_{k-1} = 0$ in model (17.11).

17.5 FIXED-EFFECTS MODEL FOR ONE-WAY ANOVA

Many textbooks and articles dealing strictly with analysis-of-variance procedures use a more classical type of model than the regression model given above to describe the fixed-effects one-way ANOVA situation. This type of model is often referred to as a *fixed-effects ANOVA model*; in such a model, all the factors being considered are fixed (i.e., the levels of each factor are the only levels of interest). The "effects" referred to in such a model represent measures of the influence (i.e., the effect) that different levels of the factor[10] have on the dependent variable. Such measures are often expressed in the form of differences between a given mean and an overall mean. That is, the *effect* of the ith population is often measured by the amount that the ith population mean differs from an overall mean.

EXAMPLE　For the CMI data ($k = 4$), the fixed-effects ANOVA model is given as follows:

$$Y_{ij} = \mu + \alpha_i + E_{ij}, \qquad i = 1, 2, 3, 4; \qquad j = 1, 2, \ldots, 25 \tag{17.13}$$

where

$Y_{ij} = j$th observation from the ith population,

$\mu = (\mu_1 + \mu_2 + \mu_3 + \mu_4)/4$ ($= \bar{\mu}^*$, the overall *unweighted* mean since $n_i = 25$ for all i)

$\alpha_1 = \mu_1 - \mu = $ differential effect of neighborhood 1

$\alpha_2 = \mu_2 - \mu = $ differential effect of neighborhood 2

$\alpha_3 = \mu_3 - \mu = $ differential effect of neighborhood 3

$\alpha_4 = \mu_4 - \mu = $ differential effect of neighborhood 4

$E_{ij} = Y_{ij} - \mu - \alpha_i = $ error component associated with the jth observation from the ith population

One important property of this model is that the sum of the four α effects is zero; that is, $\alpha_1 + \alpha_2 + \alpha_3 + \alpha_4 = 0$. Thus, these effects represent differentials from the overall population mean μ which average out to zero. Nevertheless, the effect of one level (i.e., a neighborhood) may be considerably different from the effect of another. If this is so, we would more than likely find that our F test leads to rejection of the null hypothesis of equal population mean CMI scores for the four neighborhoods.

Another important property of this model is that the effects α_1, α_2, α_3, and α_4, which are actually population parameters defined in terms of population means, can be

---

[10] When dealing with models with two or more factors, "effects" can also refer to measures of the influence of combinations of levels of the different factors on the dependent variable.

simply estimated from the data. This is done for any given effect by appropriately substituting the usual estimates of the means into the expression for the effect. For our example, the estimated effects are given by

$$\begin{cases} \hat{\alpha}_1 = \bar{Y}_{1.} - \bar{Y} = \text{sample mean CMI score for neighborhood 1} \\ \qquad\qquad - \text{overall sample mean CMI score for all neighborhoods} \\ \hat{\alpha}_2 = \bar{Y}_{2.} - \bar{Y} = \text{sample mean CMI score for neighborhood 2} \\ \qquad\qquad - \text{overall sample mean CMI score for all neighborhoods} \\ \hat{\alpha}_3 = \bar{Y}_{3.} - \bar{Y} = \text{sample mean CMI score for neighborhood 3} \\ \qquad\qquad - \text{overall sample mean CMI score for all neighborhoods} \\ \hat{\alpha}_4 = \bar{Y}_{4.} - \bar{Y} = \text{sample mean CMI score for neighborhood 4} \\ \qquad\qquad - \text{overall sample mean CMI score for all neighborhoods} \end{cases}$$

The actual numerical values, using these formulas, are as follows:

$$\begin{cases} \hat{\alpha}_1 = 18.92 - 21.83 = -2.91 \\ \hat{\alpha}_2 = 20.72 - 21.83 = -1.11 \\ \hat{\alpha}_3 = 20.84 - 21.83 = -0.99 \\ \hat{\alpha}_4 = 26.84 - 21.83 = 5.01 \end{cases}$$

As with the population effects, it can be easily seen that the estimated effects also sum to zero, that is, $\sum_{i=1}^{4} \hat{\alpha}_i = 0$.

17.5.1 GENERAL FIXED-EFFECTS ONE-WAY ANOVA MODEL

If we consider the general one-way ANOVA situation (i.e., there are k populations and n_i observations from the ith population), then the fixed-effects one-way ANOVA model may be written as follows:

$$Y_{ij} = \mu + \alpha_i + E_{ij}, \qquad i = 1, 2, \ldots, k; \;\; j = 1, 2, \ldots, n_i \tag{17.14}$$

where[11]

$Y_{ij} = j$th observation from the ith population

$\mu = (\mu_1 + \mu_2 + \cdots + \mu_k)/k \quad (= \bar{\mu}^*)$

$\alpha_i = \mu_i - \mu = $ differential effect of population i

$E_{ij} = Y_{ij} - \mu - \alpha_i = $ error component associated with the jth observation from the ith population

Here, it is easy to show that the sum of the α effects is zero, that is, $\sum_{i=1}^{k} \alpha_i = 0$. Similarly, the estimated effects, $\hat{\alpha}_i = \bar{Y}_{i.} - \bar{Y}^*$, where $\bar{Y}^* = \sum_{i=1}^{k} \bar{Y}_{i.}/k$, satisfy the constraint $\sum_{i=1}^{k} \hat{\alpha}_i = 0$.

[11] An alternative definition of μ may be $\bar{\mu} = \sum_{i=1}^{k} n_i \mu_i / n$, the overall weighted mean of means.

[If μ is defined to be $\mu = \bar{\mu} = \sum_{i=1}^{k} n_i \mu_i / n$, it can be shown that the weighted sum $\sum_{i=1}^{k} n_i \alpha_i = 0$ and that, similarly, the weighted sum of the estimated effects $\hat{\alpha}_i = \bar{Y}_{i.} - \bar{Y}$, where $\bar{Y} = \sum_{i=1}^{k} n_i \bar{Y}_{i.} / n$, satisfies $\sum_{i=1}^{k} n_i \hat{\alpha}_i = 0$.]

Another property worth noting is that this model (17.14) corresponds in structure to the regression model given by (17.9); that is, the regression coefficients $\alpha_1, \ldots, \alpha_{k-1}$ are precisely the effects $\alpha_1 = \mu_1 - \mu^*, \ldots, \alpha_{k-1} = \mu_{k-1} - \mu^*$, the regression constant μ represents the overall (unweighted) mean μ^*, and the negative sum of the regression coefficients $(-\sum_{i=1}^{k-1} \alpha_i)$ represents the effect $\alpha_k = \mu_k - \mu^*$. This is why we have defined each of these models using the same notation for the unknown parameters:

$$
\begin{cases}
Y = \mu + \sum_{i=1}^{k-1} \alpha_i X_i + E & \text{(dummy-variable regression model)} \\
\qquad\quad \Updownarrow \quad \nearrow \\
Y_{ij} = \mu + \alpha_i + E_{ij} & \text{(fixed-effects ANOVA model)}
\end{cases}
$$

[Note that μ represents the unweighted average of the k population means, μ^*, rather than the weighted average, $\bar{\mu}$, even though the sample sizes are allowed to be different in the different populations. Correspondingly, the least-squares estimate of μ is $\sum_{i=1}^{k} \bar{Y}_{i.} / k$, the unweighted average of the k sample means, rather than $\bar{Y}$. Nevertheless, it is possible to redefine the dummy variables in the regression model to obtain a least-squares solution yielding $\bar{Y}$ as the estimate of μ. This requires the following dummy-variable definitions:

$$
X_i = \begin{cases}
-n_i & \text{if population } k \\
n_k & \text{if population } i, \qquad i = 1, 2, \ldots, k-1] \\
0 & \text{otherwise.}
\end{cases}
$$

17.6 RANDOM-EFFECTS MODEL

We have earlier (in Section 17.1) distinguished between factors which are *fixed* and those which are *random*. We have also stated that for ANOVA situations involving two or more factors, the F tests required for making inferences differ depending on whether (1) all factors are fixed, (2) all factors are random, or (3) there is some mixture of fixed and random factors. It turns out, furthermore, that the null hypotheses to be tested must be stated in different terms when random factors are involved than when only fixed factors are involved.

EXAMPLE   In order to get some insight into the structure of random-effects models, it is useful to consider Daly's study (see Table 17.3). It might be argued for this example that the four different Turnkey neighborhoods are merely a representative sample of a larger population of similar types of neighborhoods (some of which might even be predominantly white with differing percentages of blacks in the surrounding neighborhoods). If so, the neighborhood factor would have to be considered as random. The appropriate ANOVA model for this example would then be a *random-effects* one-way ANOVA model.[12] Its form would be essentially the same as that given in (17.13), except that the α-effect components would now be treated differently. That is, the random-effects model[13] would be of the form

$$Y_{ij} = \mu + A_i + E_{ij}, \qquad i = 1, 2, 3, 4; \quad j = 1, 2, \ldots, 25 \qquad (17.15)$$

In this model, the A_i's can be viewed as a random sample of random variables having a common distribution. This common distribution represents the entire population of possible effects (in our example, neighborhoods).

To perform the appropriate analysis, it is necessary to assume that the distribution of A_i is normal with zero mean:

$$A_i \frown N(0, \sigma_A^2), \qquad i = 1, 2, 3, 4 \qquad (17.16)$$

where σ_A^2 denotes the variance of A_i. We must also assume that the A_i's are independent of the E_{ij}'s and of each other.

The requirement of zero mean in (17.16) is similar in philosophy to the requirement that $\sum_{i=1}^{k} \alpha_i = 0$ for the fixed-effects model. When the random model (17.15) applies, we are assuming that the average (i.e., mean) effect of neighborhoods is zero over the entire population of neighborhoods. That is, we assume that $\mu_{A_i} = 0$, $i = 1, 2, 3, 4$.

How then do we state our null hypothesis? Because we have required the neighborhood effects to average out to zero over the entire population of possible effects, there is only one way to assess whether or not there are any significant neighborhood effects at all, and this involves consideration of σ_A^2. If there is no variability (i.e., $\sigma_A^2 = 0$), all neighborhood effects must be zero. If there is variability (i.e., $\sigma_A^2 > 0$), there are some nonzero effects in the population of neighborhood effects.

Thus, our null hypothesis of no neighborhood effects should be stated as follows:

$$H_0: \sigma_A^2 = 0 \qquad (17.17)$$

This hypothesis is, therefore, analogous to the null hypothesis (17.1) used in the fixed-effects case, although it happens to be stated in terms of a population variance rather than in terms of population means.

It still remains to explain why the F test given by (17.2) for the fixed-effects model should be exactly the same as that used for the random-effects model.[14] Such an explanation is best made by considering the properties of the mean squares MST and MSE. Recalling our previous argument in Section 17.3.2 for the fixed-effects model, we saw that the F statistic, MST/MSE, could be considered as a rough approximation to the ratio of the means of these mean squares,[15]

$$\frac{\mu_{MST}}{\mu_{MSE}} = \frac{\sigma^2 + [1/(k-1)] \sum_{i=1}^{k} n_i (\mu_i - \bar{\mu})^2}{\sigma^2}$$

---

[12] This type of model is also referred to as a *components-of-variance model*.

[13] Our usual convention up to now has been to use Latin letters (X, Y, Z, etc.) to denote random variables and Greek letters (β, μ, σ, τ) to denote parameters. This would require using A_i's rather than α_i's to denote random effects.

[14] Again, we point out that the F tests are computationally equivalent for fixed-effects and random-effects models only in one-way ANOVA. When dealing with two-way or higher-way ANOVA, the testing procedures may be different.

[15] The parameters μ_{MST} and μ_{MSE} are often called *expected mean squares*.

A similar argument can be made with regard to the F statistic for the random-effects model. In particular, it still holds, for the random- as well as for the fixed-effects model, that the denominator MSE estimates σ^2, that is,

$$\mu_{\text{MSE}} = \sigma^2$$

Furthermore, it can be shown for the random-effects model applied to our example ($k = 4$, $n_i = 25$) that MST estimates

$$\mu_{\text{MST(random)}} = \sigma^2 + 25\sigma_A^2 \tag{17.18}$$

Thus, for the random-effects model, F approximates the ratio

$$\frac{\mu_{\text{MST(random)}}}{\mu_{\text{MSE}}} = \frac{\sigma^2 + 25\sigma_A^2}{\sigma^2} \tag{17.19}$$

Since the null hypothesis in this case is $H_0: \sigma_A^2 = 0$, we can see that the ratio (17.19) simplifies to $\sigma^2/\sigma^2 = 1$ when H_0 is true. Thus, we see again that the F statistic under H_0 consists of the ratio of two estimates of the same variance σ^2. Furthermore, because $\sigma_A^2 > 0$ when H_0 is not true, the more the variability among neighborhood effects, the larger should be the observed value of F.

17.6.1 GENERAL RANDOM-EFFECTS MODEL FOR ONE-WAY ANOVA

In general, the random-effects model for one-way ANOVA is given by

$$Y_{ij} = \mu + A_i + E_{ij}, \qquad i = 1, 2, \ldots, k; \quad j = 1, 2, \ldots, n_i \tag{17.20}$$

where A_i and E_{ij} are independent random variables satisfying $A_i \frown N(0, \sigma_A^2)$ and $E_{ij} \frown N(0, \sigma^2)$.

For this model it can be shown that F approximates the following ratio of expected mean squares:

$$\frac{\mu_{\text{MST(random)}}}{\mu_{\text{MSE}}} = \frac{\sigma^2 + n_0 \sigma_A^2}{\sigma^2}$$

where

$$n_0 = \left[\sum_{i=1}^{k} n_i - \left(\sum_{i=1}^{k} n_i^2 \right) \bigg/ \left(\sum_{i=1}^{k} n_i \right) \right] \bigg/ (k-1)$$

is like an average of the n_i observations selected from each population.[16] The F statistic

[16] When all the n_i are equal as in the Daly example (i.e., $n_i = n^*$), then n_0 is equal to n^* since

$$n_0 = \frac{kn^* - kn^{*2}/kn^*}{k-1} = n^*$$

In the Daly example, $n^* = 25$.

TABLE 17.6   *Combined one-way ANOVA table for fixed- and random-effects models*

SOURCE	df	MS	F	EXPECTED MEAN SQUARES (EMS)	
				FIXED EFFECTS	RANDOM EFFECTS
Between	$k-1$	MST	MST/MSE	$\sigma^2 + \dfrac{1}{k-1} \sum\limits_{i=1}^{k} n_i(\mu_i - \bar{\mu})^2$	$\sigma^2 + n_0\sigma_A^2$
Within	$n-k$	MSE		σ^2	σ^2
Total	$n-1$				
				$H_0: \mu_1 = \mu_2 = \cdots = \mu_k$	$H_0: \sigma_A^2 = 0$

for the random-effects model is therefore the ratio of two estimates of σ^2 when $H_0: \sigma_A^2 = 0$ is true.

Table 17.6 summarizes both the similarities and differences between the fixed- and random-effects models. Tables with similar formats will be used in subsequent chapters to highlight distinctions for higher-way (i.e., two-or-more-factor) ANOVA situations.

17.7 MULTIPLE-COMPARISON PROCEDURES

Whenever an ANOVA *F* test for simultaneously comparing several population means is found to be statistically significant, it is then customarily of interest to determine what *specific* differences there are among the population means. For example, if four means are being compared (fixed-effects case)[17] and the null hypothesis $H_0: \mu_1 = \mu_2 = \mu_3 = \mu_4$ is rejected, one may wish to determine which subgroups of means are different by considering any number of more specific hypotheses like $H_{01}: \mu_1 = \mu_2$, $H_{02}: \mu_2 = \mu_3$, $H_{03}: \mu_3 = \mu_4$, or even one like $H_{04}: (\mu_1 + \mu_2)/2 = (\mu_3 + \mu_4)/2$, which compares the average effect of populations 1 and 2 with the average effect of populations 3 and 4. Such specific comparisons may have either been of interest to the investigator in advance (a priori) of data collection or may arise in completely exploratory studies only after (a posteriori) examination of the data. Regardless of whether such comparisons are decided upon a priori or a posteriori to data collection and examination, a seemingly reasonable first approach to inference-making regarding specific differences among the population means would be to make several *t* tests and to focus on all those tests found significant. For example, if all *pairwise* comparisons among the means were desired, this would require $_4C_2 = 6$ such tests when considering four means (or, in general, $_kC_2 = k(k-1)/2$ tests when dealing with *k* means). Thus, for testing $H_0: \mu_i = \mu_j$ at the α level of significance, we could reject this H_0 when

$$|T| \geq t_{n-k, 1-\alpha/2}$$

[17] This section deals only with fixed-effects ANOVA problems, since the random-effects model treats the observed factor levels as a sample from a larger population of levels of interest and is therefore not directed at inferences regarding just these sampled levels.

where

$$T = \frac{(\bar{Y}_i - \bar{Y}_j) - 0}{\sqrt{MSE(1/n_i + 1/n_j)}}$$

where n = the total number of observations, k = the number of means under consideration, n_i and n_j are the sizes of the samples selected from the ith and jth populations, respectively, $\bar{Y}_i$ and $\bar{Y}_j$ are the corresponding sample means, and MSE is the mean-square-error term with $(n - k)$ df which estimates the (homoscedastic) variance σ^2. Note that MSE is used instead of just a two-sample estimate of σ^2 based only on data from groups i and j; this is because MSE is a better estimate of σ^2 (in terms of degrees of freedom) under the assumption of variance homogeneity over all k populations.

Equivalently, one could reject H_0 if the $100(1 - \alpha)\%$ confidence interval

$$(\bar{Y}_i - \bar{Y}_j) \pm t_{n-k,1-\alpha/2}\sqrt{MSE(1/n_i + 1/n_j)}$$

does not include zero.

Unfortunately, there is a serious drawback to the approach of performing several such t tests; this drawback arises from the fact that the more null hypotheses there are to be tested, the more likely it is to reject one of them even if all null hypotheses are actually true. Another way of saying this is that if several such tests are made, each at the α level, then the probability of incorrectly rejecting *at least one* H_0 will be much larger than α and will increase with the number of tests made. Moreover, if in an exploratory study, the investigator decides to compare only those sample means which are most discrepant (e.g., like the largest versus the smallest), then the testing procedure would be biased in favor of rejecting H_0 because only those comparisons most likely to be significant would have been made, this bias being reflected in the fact that the true probability of falsely rejecting a given null hypothesis would exceed the α level specified for the test.

17.7.1 LSD APPROACH

An approximate way to circumvent the problem of distorted significance levels when making several tests involves reducing the significance level used for each individual test sufficiently so that the *overall significance level* (i.e., the probability of falsely rejecting at least one of the null hypotheses being tested) is fixed at some desired value (say, α). Now, it can be shown that if one makes l such tests, the maximum possible value for this overall significance level is $l\alpha$. Thus, one simple way to ensure an overall significance level of at most α would be to use α/l as the significance level for each separate test; this approach is often referred to as the *least significant difference* (*LSD*) method. For example, if all $l = {}_kC_2 = k(k-1)/2$ pairwise comparisons of k population means are to be made, each test could be performed at the $\alpha/{}_kC_2$ significance level to ensure that the overall significance level would not exceed α.

A disadvantage to the LSD method, however, is that the *true* overall significance level may be so much less than the maximum value α that none of the individual tests is very likely to be rejected (i.e., the overall power of the method is low). Consequently, several other more powerful procedures have been devised which can be used to provide an overall significance level of α, the whole collection of such procedures being

TABLE 17.7   *Potencies (dosages at death) for four cardiac substances*

SUBSTANCE	SAMPLE SIZE (n_i)	DOSAGE AT DEATH (Y_{ij})	TOTALS	SAMPLE MEANS $(\bar{Y}_i)$	SAMPLE VARIANCES (S_i^2)
1	10	29, 28, 23, 26, 26 19, 25, 29, 26, 28	259	25.9	9.4333
2	10	17, 25, 24, 19, 28 21, 20, 25, 19, 24	222	22.2	12.1778
3	10	17, 16, 21, 22, 23 18, 20, 17, 25, 21	200	20.0	8.6667
4	10	18, 20, 25, 24, 16 20, 20, 17, 19, 17	196	19.6	8.7111

grouped under the subject-matter heading "multiple-comparison procedures." We shall focus here on two such methods, one due to Tukey and the other due to Scheffé. Other discussions of these and other multiple-comparison methods can be found in Miller (1966), Guenther (1964), Lindman (1974), and Neter and Wasserman (1974).

EXAMPLE   As an illustration of a multiple-comparison type of problem, let us now consider the set of data given in Table 17.7, which was collected in a laboratory experiment designed to compare the relative potencies of four cardiac substances, the experimental method used being to infuse slowly a suitable dilution of one of the substances into an anesthetized guinea pig and then to record the dose at which death occurs. The experiment was designed so that 10 guinea pigs were used for each substance, and it was assumed that the laboratory environment and the measurement procedures were the same for each guinea pig used. The main research goal involves determining whether or not there are any differences among the potencies of the four substances and, if so, to quantify where those differences are. The overall ANOVA table for comparing the mean potencies of the four cardiac substances is given in Table 17.8.

The global F test for testing the null hypothesis of equality of the four population means strongly rejects ($P < 0.001$) this null hypothesis. So, the multiple-comparison question now arises as to what is the best way to account for the differences found. As a crude first step, we can descriptively examine the nature of the differences with the help of a schematic diagram of ordered sample means (Figure 17.1). In the diagram a line has been drawn connecting substances 3 and 4 to indicate that the sample mean potencies for these two substances are quite similar. On the other hand, no line has

TABLE 17.8   *ANOVA table for potency data of Table 17.7*

SOURCE	df	SS	MS	F
Substances	3	249.875	83.292	8.545 ($P < 0.001$)
Error	36	350.900	9.747	
Total	39	600.775		

FIGURE 17.1   *Crude comparison of sample means for potency data*

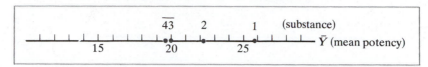

been drawn connecting 1 and 2 with each other or with 3 and 4, suggesting that 1 and 2 appear to be different from each other as well as from both 3 and 4.

Such an overall quantification of the nature of the differences among the population means is what is desired from a multiple-comparison analysis. Nevertheless, the purely descriptive approach above does not take into account the sampling variability associated with any estimated comparison of interest, so that, for example, two sample means that appear to be "practically" different may not, in fact, be "statistically" different. Since the only multiple-comparison method that we have discussed so far is the LSD method, let us now consider how we can apply this method to the data of Table 17.7, using an overall significance level of $\alpha = 0.05$ for all pairwise comparisons of the mean potencies of the four cardiac substances. This approach requires the computation of $_4C_2 = 6$ confidence intervals, each associated with a significance level of $\alpha/6 = 0.05/6 = 0.0083$, utilizing the formula

$$(\bar{Y}_i - \bar{Y}_j) \pm t_{36,1-0.0083/2}\sqrt{MSE(\tfrac{1}{10} + \tfrac{1}{10})}$$

The right-hand side of this expression is calculated as follows:

$$t_{36,0.99585}\sqrt{9.747(\tfrac{1}{5})} = 2.72(1.396) = 3.798$$

Thus, the pairwise confidence intervals are given as follows:

$\mu_1 - \mu_4$: 6.3 ± 3.798; i.e., $2.502 \le (\mu_1 - \mu_4) \le 10.098^*$

$\mu_1 - \mu_3$: 5.9 ± 3.798; i.e., $2.102 \le (\mu_1 - \mu_3) \le 9.698^*$

$\mu_1 - \mu_2$: 3.7 ± 3.798; i.e., $-0.098 \le (\mu_1 - \mu_2) \le 7.498$

$\mu_2 - \mu_4$: 2.6 ± 3.798; i.e., $-1.198 \le (\mu_2 - \mu_4) \le 6.398$

$\mu_2 - \mu_3$: 2.2 ± 3.798; i.e., $-1.598 \le (\mu_2 - \mu_3) \le 5.998$

$\mu_3 - \mu_4$: 0.4 ± 3.798; i.e., $-3.398 \le (\mu_3 - \mu_4) \le 4.198$

[With regard to this example, the term "least significant difference" refers to the fact that *any two* sample means will be said to be significantly different if their absolute difference exceeds 3.798. Of course, such a blanket statement cannot be made when the n_i's are not all equal.]

The intervals above indicate only two significant comparisons (starred), and translate into a diagrammatic overall ranking (Figure 17.2). These results are somewhat ambiguous since there are overlapping "sets of similarities," which indicate that substances 2, 3, and 4 have essentially the same potency, that 1 and 2 have about the same potency, but that 1 differs from both 3 and 4. In other words, one conclusion to be made here is that 2, 3, and 4 are to be grouped together and that 1 and 2 are to be

FIGURE 17.2   *LSD comparison of sample means for potency data*

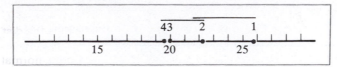

grouped together, which is difficult to reconcile since substance 2 is common to both groups. Having to confront this ambiguity is actually quite fortuitous from a pedagogical standpoint, since such ambiguous results do not occur infrequently in practice when carrying out multiple-comparison procedures. In our case the results indicate that the procedure used was not sensitive enough regarding substance 2. Repeating the analysis with a larger data set would help to clear up the ambiguity. Alternatively, since the LSD approach tends to be conservative (i.e., the confidence intervals tend to be wider than necessary to achieve the overall significance level desired), other multiple-comparison methods, such as those of Tukey or Scheffé, may provide more precise results (i.e., smaller width confidence intervals) and so may possibly reduce or eliminate any ambiguity with regard to the conclusions made.

17.7.2   TUKEY'S METHOD

Tukey's method is applicable when:

1.   The sizes of the samples selected from each population are equal; that is, $n_i \equiv n^*$, say, for all $i = 1, 2, \ldots, k$, where k is the number of means being compared.
2.   Pairwise comparisons of the means are of primary interest; that is, null hypotheses of the form H_0: $\mu_i = \mu_j$ are to be considered.

[Actually, a generalized version of Tukey's procedure is available for considering more complex comparisons than just simple pairwise differences between means. However, since it is most powerful in the situation when just simple differences between means are to be examined and not in the situation when more complex comparisons need to be studied, the general recommendation is that Tukey's method be used only in the former situation; the latter situation is best handled using Scheffé's method (to be discussed in Section 17.7.4).]

To use Tukey's method, the following confidence interval is computed for the pairwise comparison regarding population means μ_i and μ_j:

$$(\bar{Y}_i - \bar{Y}_j) \pm T\sqrt{\text{MSE}} \tag{17.21}$$

where

$$T = \frac{1}{\sqrt{n^*}} q_{k,n-k,1-\alpha}$$

and $q_{k,n-k,1-\alpha}$ is the $100(1-\alpha)\%$ point of the distribution of the *studentized range* with k and $n - k$ degrees of freedom (see Table A-6 in Appendix A), n^* is the common sample size (i.e., $n_i \equiv n^*$), k is the number of populations or groups, and n ($= kn^*$) is the total number of observations.

[The *studentized range distribution* with k and r df is the distribution of a statistic of the form R/S, where $R = \{\max_i (Y_i's) - \min_i (Y_i's)\}$ is the range of a set of k independent observations $Y_1, \ldots, Y_k$ from a normal distribution with mean μ and variance σ^2, and S^2 is an estimate of σ^2 based on r df which is independent of the Y's. In particular, when k means are being compared in fixed-effects one-way ANOVA, the statistic $\{\max_i (\bar{Y}_i's) - \min_i (\bar{Y}_i's)\}/\sqrt{MSE/n^*}$ has the studentized range distribution with k and $(n-k)$ df under $H_0: \mu_1 = \mu_2 = \cdots = \mu_k$, where $n_i = n^*$ for each i and $n = kn^*$.

An alternative to Tukey's procedure, which uses the studentized range distribution but with a modified numerator df, is called the *Student–Newman–Keuls (SNK)* method. The SNK procedure replaces the first k in $q_{k,n-k,1-\alpha}$ by $k^* =$ the number of means in the range of means being tested. Thus, for comparing the second largest with the smallest of four means, we would have $k^* = 3$, whereas $k^* = 2$ when comparing the second largest with the third largest.

For unequal sample sizes, a slight modification to the Tukey procedure has been recommended in Steele and Torrie (1960). For each comparison involving unequal sample sizes, the term $T\sqrt{MSE}$ in (17.21) is replaced by $q_{k,n-k,1-\alpha}\sqrt{(MSE/2)(1/n_i + 1/n_j)}$, where n_i and n_j are the sample sizes associated with the ith and jth groups, respectively.]

The set of all $_kC_2$ Tukey pairwise-type confidence intervals of the form (17.21) has the property that the probability is $(1-\alpha)$ that these intervals will simultaneously contain the associated population mean differences being estimated; that is, $(1-\alpha)$ is the *overall confidence coefficient* for all pairwise comparisons taken together. In particular, if each confidence interval is used to test the appropriate pairwise hypothesis of the general form $H_0: \mu_i = \mu_j$ by determining whether the value zero is contained in the calculated interval or not, the probability of falsely rejecting the null hypothesis for *at least one* of the $_kC_2$ tests is equal to α.

In applying Tukey's method to a set of data, the procedure is usually carried out stepwise as follows:

1. Rank-order the sample means $\bar{Y}_i$ from largest to smallest (e.g., in our example, the order is $\bar{Y}_1 > \bar{Y}_2 > \bar{Y}_3 > \bar{Y}_4$).

2. Compare the largest sample mean with the smallest using (17.21), then the largest with the next smallest and so on, until the largest has either been compared with the second largest or until a nonsignificant result has been obtained, whichever comes first (e.g., in our example, we would first look at 1 versus 4, then 1 versus 3, and finally 1 versus 2).

3. Continue by comparing the second largest mean with the smallest, the second largest with the next smallest, and so on, but make no further comparisons involving the second largest mean once any nonsignificant result is obtained.

4. Continue making such comparisons involving the third largest mean, then the fourth largest, and so on. At each stage, once a nonsignificant comparison is located, conclude that no difference exists between any means enclosed by the first nonsignificant pair.

5. Represent the overall conclusions about similarities and differences among the population means using a schematic diagram involving the ordered sample means, with lines drawn to connect those means which are not significantly different.

To illustrate this method using an overall significance level of $\alpha = 0.05$ for the potency data of Table 17.7, the ordering of sample means from largest to smallest indicates that the following sequence of pairwise comparisons is to be made:

$$1 \text{ vs. } 4, \quad 1 \text{ vs. } 3, \quad 1 \text{ vs. } 2, \quad 2 \text{ vs. } 4, \quad 2 \text{ vs. } 3, \quad 3 \text{ vs. } 4$$

Since the value of $T\sqrt{MSE}$ in (17.21) is needed for any such comparison, we compute this first as follows: for $n^* = 10$, $k = 4$, $n = 40$, MSE $= 9.747$,

$$T = \frac{1}{\sqrt{n^*}} q_{k,n-k,1-\alpha} = \frac{1}{\sqrt{10}} q_{4,36,0.95} = \frac{1}{\sqrt{10}} (3.84) = 1.206$$

(the value $q_{4,36,0.95} = 3.84$ was obtained from Table A-6 by interpolation), and so

$$T\sqrt{MSE} = 1.206\sqrt{9.747} = 3.765$$

Now, using (17.21), we compare 1 and 4 as follows:

$$(\bar{Y}_1 - \bar{Y}_4) \pm T\sqrt{MSE}$$

or

$$(25.9 - 19.6) \pm 3.765$$

or

$$6.3 \pm 3.765$$

or

$$2.535 \le (\mu_1 - \mu_4) \le 10.065$$

This confidence interval does not contain the value 0, so that we can conclude (based on an overall significance level of $\alpha = 0.05$) that $\mu_1 \ne \mu_4$.

Next, we look at 1 versus 3 as follows:

$$(\bar{Y}_1 - \bar{Y}_3) \pm T\sqrt{MSE}$$

or

$$(25.9 - 20.0) \pm 3.765$$

or

$$2.135 \le (\mu_1 - \mu_3) \le 9.665$$

FIGURE 17.3   *Tukey comparison of sample means for potency data*

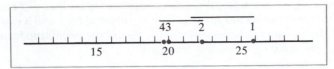

This confidence interval also does not contain the value 0, so that we can conclude (with overall significance level $\alpha = 0.05$) that $\mu_1 \neq \mu_3$.

Next, we compare 1 and 2, obtaining

$$3.7 \pm 3.765$$

or

$$-0.065 \leq (\mu_1 - \mu_2) \leq 7.465$$

which contains the value 0. We thus conclude that there is no evidence to reject $H_0: \mu_1 = \mu_2$ and also that all other remaining comparisons, which involve smaller (in absolute value) pairwise mean differences, are nonsignificant, which supports the hypothesis that $\mu_2 = \mu_3 = \mu_4$. In summary, we may schematically represent the results based on applying Tukey's method as shown in Figure 17.3.

[In general, it is possible that a pairwise difference between, say, the largest mean and the second largest may not be significant, even though there may be another pairwise difference (say, between the third and fourth largest means) which is significant. Thus, in general, one should not stop making *all* remaining comparisons on first encountering a nonsignificant pairwise difference, unless all remaining pairwise differences are smaller than that involved in the first nonsignificant pair. As an example, if it was found that $\bar{Y}_1 = 100$, $\bar{Y}_2 = 99$, $\bar{Y}_3 = 80$, $\bar{Y}_4 = 20$, then $(\bar{Y}_1 - \bar{Y}_2) = 1$ is quite small (and possibly nonsignificant) while $(\bar{Y}_3 - \bar{Y}_4) = 60$ is large.]

As with the LSD method, there remains some ambiguity here, since substance 2 has again been associated with substance 1 and also with substances 3 and 4. Again, it should be noted that the occurrence of such an ambiguity is not uncommon, and, in this instance, suggests that the amount of data collected was not sufficient to clearly categorize substance 2.

17.7.3   SCHEFFÉ'S METHOD

Scheffé's method is generally recommended when:

1. The sizes of the samples selected from the different populations are not all equal; and/or

2. Comparisons other than simple pairwise comparisons between two means are of interest; these more general types of comparisons are referred to as *contrasts*.

As an illustration of what we mean by a *contrast*, suppose the investigator who collected the potency data of Table 17.7 suspected that substances 1 and 3 were similar in potency, that substances 2 and 4 were also similar in potency, and that the potencies of 1 and 3, on the average, differed significantly from those of 2 and 4. Then it would be of interest to compare the average results obtained for 1 and 3 with the average results for 2 and 4 in order to assess whether $(\mu_1 + \mu_3)/2$ really differed from $(\mu_2 + \mu_4)/2$. In other words, one could consider the "contrast"

$$L_1 = \frac{\mu_1 + \mu_3}{2} - \frac{\mu_2 + \mu_4}{2}$$

which would be zero if the null hypothesis H_0: $(\mu_1 + \mu_3)/2 = (\mu_2 + \mu_4)/2$ were true. Notice further that we can rewrite L_1 as follows:

$$L_1 = \frac{\mu_1 + \mu_3}{2} - \frac{\mu_2 + \mu_4}{2} = \frac{1}{2}\mu_1 - \frac{1}{2}\mu_2 + \frac{1}{2}\mu_3 - \frac{1}{2}\mu_4$$

or

$$L_1 = \sum_{i=1}^{4} c_{1i}\mu_i$$

so that L_1 is a *linear* function of the population means, with $c_{11} = \frac{1}{2}$, $c_{12} = -\frac{1}{2}$, $c_{13} = \frac{1}{2}$, and $c_{14} = -\frac{1}{2}$. Also, notice that

$$c_{11} + c_{12} + c_{13} + c_{14} = \frac{1}{2} - \frac{1}{2} + \frac{1}{2} - \frac{1}{2} = 0$$

In general, a *contrast* is defined to be any linear function of the population means, say

$$L = \sum_{i=1}^{k} c_i\mu_i$$

such that

$$\sum_{i=1}^{k} c_i = 0$$

As another example, examination of the data in Table 17.7 suggests that the mean potency of substance 1 is definitely higher than the mean potencies of the other three substances. Such an observation suggests a comparison of the mean potency of substance 1 with the average potency of substances 2, 3, and 4; in this case the appropriate contrast to be considered is

$$L_2 = \mu_1 - \frac{\mu_2 + \mu_3 + \mu_4}{3}$$

or $L_2 = \sum_{i=1}^{4} c_{2i}\mu_i$, where $c_{21} = 1$, $c_{22} = c_{23} = c_{24} = -\frac{1}{3}$. (Notice again that $\sum_{i=1}^{4} c_{2i} = 0$.)

A third contrast possibly of interest would involve a comparison of the average results for substances 1 and 2 with those for 3 and 4. The appropriate contrast here would then be

$$L_3 = \frac{\mu_1 + \mu_2}{2} - \frac{\mu_3 + \mu_4}{2} = \sum_{i=1}^{4} c_{3i}\mu_i$$

where $c_{31} = c_{32} = \frac{1}{2}$ and $c_{33} = c_{34} = -\frac{1}{2}$.

Finally, any pairwise comparison is also a contrast, since, for example, a comparison of 1 with 4 takes the form

$$L_4 = \mu_1 - \mu_4 = \sum_{i=1}^{4} c_{4i}\mu_i$$

where $c_{41} = 1$, $c_{42} = c_{43} = 0$, and $c_{44} = -1$.

Scheffé's method provides a family of confidence intervals for evaluating *all possible contrasts* that can be defined given a fixed number (say, k) of population means, such that the overall confidence coefficient associated with the entire family is $(1 - \alpha)$, where α is to be specified by the investigator. In other words, the probability is $(1 - \alpha)$ that these confidence intervals will simultaneously contain the true values of all the contrasts being considered. In terms of significance levels, we can say that the overall significance level is α for testing hypotheses of the general form $H_0 : L = \sum_{i=1}^{k} c_i\mu_i = 0$ concerning all possible contrasts; that is, the probability is α that at least one such null hypothesis will falsely be rejected.

The general form of a Scheffé-type confidence interval is as follows. Let $L = \sum_{i=1}^{k} c_i\mu_i$ be some contrast of interest. Then the appropriate confidence interval concerning L is given by

$$\sum_{i=1}^{k} c_i \bar{Y}_i \pm S \sqrt{\text{MSE}\left(\sum_{i=1}^{k} c_i^2 / n_i \right)} \tag{17.22}$$

where $\hat{L} = \sum_{i=1}^{k} c_i \bar{Y}_i$ is the unbiased point estimator of L and where $S^2 = (k-1)F_{k-1, n-k, 1-\alpha}$ with $n = \sum_{i=1}^{k} n_i$.

As a special case, when one is concerned only with pairwise comparisons, then this formula simplifies to

$$(\bar{Y}_i - \bar{Y}_j) \pm S \sqrt{\text{MSE}(1/n_i + 1/n_j)} \tag{17.23}$$

when considering inferences regarding $(\mu_i - \mu_j)$.

It is important to point out that if the investigator is only interested in making pairwise comparisons, then Scheffé's method using (17.23) is not recommended because Tukey's method will always provide narrower confidence intervals (i.e., will give more precise estimates of the true pairwise differences). However, if the sample sizes are unequal and/or if contrasts other than pairwise comparisons are of interest, then Scheffé's method is to be preferred. Furthermore, Scheffé's method has the desirable property that whenever the overall F test regarding the null hypothesis that all k population means are equal is rejected, then at least one contrast will be found that differs significantly from zero. Tukey's method, on the other hand, may not turn up any significant pairwise comparisons even when the overall F statistic is significant.

To illustrate the use of Scheffé's method, let us first consider all pairwise comparisons for the data of Table 17.7 and follow the same procedure as used with Tukey's method; that is, we shall consider in order 1 vs. 4, 1 vs. 3, 1 vs. 2, 2 vs. 4, 2 vs. 3, and 3 vs. 4. We begin by first computing the quantity

$$S\sqrt{\text{MSE}(1/n_i + 1/n_j)}$$

which will have the same value for all pairwise comparisons since all the sample sizes are equal to 10. So

$$\sqrt{\text{MSE}(1/n_i + 1/n_j)} = \sqrt{9.747(\tfrac{1}{10} + \tfrac{1}{10})} = 1.3962$$

and with $k = 4$, $n = 40$, and $\alpha = 0.05$, we have

$$S = \sqrt{(k-1)F_{k-1,n-k,1-\alpha}} = \sqrt{3F_{3,36,0.95}} = \sqrt{3(2.886)} = 2.9424$$

Thus, $S\sqrt{\text{MSE}(1/n_i + 1/n_j)} = 2.9424(1.3962) = 4.1082$. Now, to compare substances 1 and 4, we obtain the following confidence interval:

$$(\bar{Y}_1 - \bar{Y}_4) \pm S\sqrt{\text{MSE}(1/n_1 + 1/n_4)}$$

or

$$(25.9 - 19.6) \pm 4.1082$$

or

$$2.192 \le (\mu_1 - \mu_4) \le 10.408$$

Since this interval does not contain the value zero (which was also the case for the corresponding Tukey interval), we would support the contention that $\mu_1 \ne \mu_4$ and so proceed with the 1 vs. 3 comparison:

$$(\bar{Y}_1 - \bar{Y}_3) \pm S\sqrt{\text{MSE}(1/n_1 + 1/n_3)}$$

or

$$(25.9 - 20.0) \pm 4.1082$$

or

$$1.792 \le (\mu_1 - \mu_3) \le 10.008$$

As before with Tukey's method, this interval does not contain the value zero, which supports the rejection of H_0: $\mu_1 = \mu_3$.

The next comparison, between substances 1 and 2, yields the interval

$$-0.408 \le (\mu_1 - \mu_2) \le 7.808$$

TABLE 17.9   *Comparison of some Tukey and Scheffé confidence intervals for the potency data of Table 17.7*

PAIRWISE COMPARISON	TUKEY		SCHEFFÉ	
	LOWER LIMIT	UPPER LIMIT	LOWER LIMIT	UPPER LIMIT
$\mu_1 - \mu_4$	2.535	10.065	2.192	10.408
$\mu_1 - \mu_3$	2.135	9.665	1.792	10.008
$\mu_1 - \mu_2$	−0.065	7.465	−0.408	7.808
$\mu_2 - \mu_4$	−1.165	6.365	−1.508	6.708
$\mu_2 - \mu_3$	−1.565	5.965	−1.908	6.308
$\mu_3 - \mu_4$	−3.365	4.165	−3.708	4.508

which contains the value zero. Thus, we cannot reject H_0: $\mu_1 = \mu_2$, and noting that the remaining pairwise comparisons would be nonsignificant, we would favor both the conclusion that $\mu_2 = \mu_3 = \mu_4$ and that $\mu_1 = \mu_2$. Thus, when considering only pairwise comparisons for this data set, Tukey's and Scheffé's methods yield the same general conclusions regarding the relative potencies of the four substances, including the ambiguity associated with substance 2. However, on closer inspection of the pairwise intervals derived (see Table 17.9), it can be seen that the Tukey intervals are narrower and so provide the more precise inferences.

Let us now illustrate how to use Scheffé's method to make inferences regarding two contrasts previously identified as being of interest. In particular, we shall consider

$$L_2 = \mu_1 - \frac{\mu_2 + \mu_3 + \mu_4}{3} = \sum_{i=1}^{4} c_{2i}\mu_i$$

where $c_{21} = 1$, $c_{22} = c_{23} = c_{24} = -\frac{1}{3}$; and

$$L_3 = \frac{\mu_1 + \mu_2}{2} - \frac{\mu_3 + \mu_4}{2} = \sum_{i=1}^{4} c_{3i}\mu_i$$

where $c_{31} = c_{32} = \frac{1}{2}$ and $c_{33} = c_{34} = -\frac{1}{2}$.

Using our previously computed value $S = \sqrt{(k-1)F_{k-1,n-k,1-\alpha}} = 2.9424$, we then calculate from (17.22) as follows:

$$\left(\bar{Y}_1 - \frac{\bar{Y}_2 + \bar{Y}_3 + \bar{Y}_4}{3}\right) \pm S\sqrt{MSE\left[\frac{(1)^2}{10} + \frac{(-1/3)^2}{10} + \frac{(-1/3)^2}{10} + \frac{(-1/3)^2}{10}\right]}$$

or

$$\left(25.9 - \frac{22.2 + 20.0 + 19.6}{3}\right) \pm 2.9424\sqrt{9.747\left(\tfrac{12}{90}\right)}$$

or

$$(25.9 - 20.6) \pm 2.9424(1.1400)$$

or

$$5.3 \pm 3.354$$

or

$$1.946 \leq L_2 \leq 8.654$$

Since this interval does not contain the value 0, we have evidence that the potency of substance 1 is different from the average potency of substances 2, 3, and 4.

Next, we calculate the following Scheffé interval regarding L_3:

$$\left(\frac{\bar{Y}_1 + \bar{Y}_2}{2} - \frac{\bar{Y}_3 + \bar{Y}_4}{2} \right) \pm S\sqrt{MSE\left[\frac{(1/2)^2}{10} + \frac{(1/2)^2}{10} + \frac{(-1/2)^2}{10} + \frac{(-1/2)^2}{10} \right]}$$

or

$$\left(\frac{25.9 + 22.2}{2} - \frac{20.0 + 19.6}{2} \right) \pm 2.9424\sqrt{9.747(\tfrac{1}{10})}$$

or

$$(24.05 - 19.80) \pm 2.9424(0.9873)$$

or

$$4.24 \pm 2.9049$$

or

$$1.345 \leq L_3 \leq 7.155$$

This interval also does not contain the value 0, so that we would be inclined to support the conclusion that the average potency of substances 1 and 2 is different from the average potency of substances 3 and 4.

How do the above results concerning the contrasts L_2 and L_3 help in clearing up the ambiguity created by considering all pairwise comparisons? First, it is obvious that uncertainty regarding substance 2 still remains. Nevertheless, if forced to make a clear-cut recommendation, we should consider the fact that the confidence interval for comparing 1 with the average of 2, 3, and 4 (i.e., $1.946 \leq L_2 \leq 8.654$) is "farther away" from 0 than is the confidence interval for comparing the average of 1 and 2 with the average of 3 and 4 (i.e., $1.345 \leq L_3 \leq 7.155$), suggesting that substance 2 is "closer to" 3 and 4 in potency than it is to substance 1. The pairwise comparisons also support this contention, since the confidence interval for $(\mu_1 - \mu_2)$ is farther from zero than is the interval for $(\mu_2 - \mu_3)$, indicating again that 2 is closer to 3 than to 1. Thus, if a definite decision regarding substance 2 is warranted for this set of data, the most logical thing to do would be to consider the potency of 1 as distinct from the potencies of 2, 3, and 4, which, as a threesome, are too similar to separate. Schematically, this conclusion is represented in Figure 17.4.

FIGURE 17.4   *Conclusion regarding sample means for potency data*

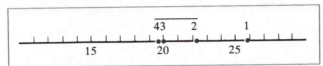

17.8 ORTHOGONAL CONTRASTS AND PARTITIONING AN ANOVA SUM OF SQUARES

From our previous discussions on multiple regression, we are familiar with the notion of a partitioned sum of squares in regression analysis, where the sum of squares due to regression (SSR) has been broken down into various components reflecting the relative contributions of various terms in the fitted model. In an ANOVA framework, it is also possible via the use of "orthogonal contrasts" to partition SST, the treatment sum of squares, into meaningful components associated with certain specific comparisons of particular interest. To illustrate how such a partitioning can be accomplished, we need to discuss two new concepts, *orthogonal contrasts* and the *sum of squares associated with a contrast*.

In the notation of Section 17.7, two estimated contrasts

$$\hat{L}_A = \sum_{i=1}^{k} c_{Ai}\bar{Y}_i \quad \text{and} \quad \hat{L}_B = \sum_{i=1}^{k} c_{Bi}\bar{Y}_i$$

are said to be *orthogonal* to one another (i.e., are orthogonal contrasts) if

$$\sum_{i=1}^{k} \frac{c_{Ai}c_{Bi}}{n_i} = 0 \tag{17.24}$$

In the special case when the n_i's are equal, then (17.24) reduces to the condition

$$\sum_{i=1}^{k} c_{Ai}c_{Bi} = 0 \tag{17.25}$$

As an illustration consider the three contrasts discussed earlier with regard to the potency data of Table 17.7:

$$\hat{L}_1 = \frac{\bar{Y}_1 + \bar{Y}_3}{2} - \frac{\bar{Y}_2 + \bar{Y}_4}{2} = \frac{1}{2}\bar{Y}_1 - \frac{1}{2}\bar{Y}_2 + \frac{1}{2}\bar{Y}_3 - \frac{1}{2}\bar{Y}_4$$

where $c_{11} = c_{13} = \frac{1}{2}$ and $c_{12} = c_{14} = -\frac{1}{2}$;

$$\hat{L}_2 = \bar{Y}_1 - \frac{1}{3}(\bar{Y}_2 + \bar{Y}_3 + \bar{Y}_4) = \bar{Y}_1 - \frac{1}{3}\bar{Y}_2 - \frac{1}{3}\bar{Y}_3 - \frac{1}{3}\bar{Y}_4$$

where $c_{21} = 1$ and $c_{22} = c_{23} = c_{24} = -\frac{1}{3}$;

$$\hat{L}_3 = \frac{\bar{Y}_1 + \bar{Y}_2}{2} - \frac{\bar{Y}_3 + \bar{Y}_4}{2} = \frac{1}{2}\bar{Y}_1 + \frac{1}{2}\bar{Y}_2 - \frac{1}{2}\bar{Y}_3 - \frac{1}{2}\bar{Y}_4$$

where $c_{31} = c_{32} = \frac{1}{2}$ and $c_{33} = c_{34} = -\frac{1}{2}$. Hence, since $n_i = 10$ for every i, we need only verify that condition (17.25) holds to demonstrate orthogonality. In particular, we have: for $\mathcal{L}_1$ and $\mathcal{L}_2$,

$$\sum_{i=1}^{4} c_{1i}c_{2i} = (\tfrac{1}{2})(1) + (-\tfrac{1}{2})(-\tfrac{1}{3}) + (\tfrac{1}{2})(-\tfrac{1}{3}) + (-\tfrac{1}{2})(-\tfrac{1}{3}) = \tfrac{2}{3} \neq 0$$

for $\mathcal{L}_1$ and $\mathcal{L}_3$,

$$\sum_{i=1}^{4} c_{1i}c_{3i} = (\tfrac{1}{2})(\tfrac{1}{2}) + (-\tfrac{1}{2})(\tfrac{1}{2}) + (\tfrac{1}{2})(-\tfrac{1}{2}) + (-\tfrac{1}{2})(-\tfrac{1}{2}) = 0$$

for $\mathcal{L}_2$ and $\mathcal{L}_3$,

$$\sum_{i=1}^{4} c_{2i}c_{3i} = (1)(\tfrac{1}{2}) + (-\tfrac{1}{3})(\tfrac{1}{2}) + (-\tfrac{1}{3})(-\tfrac{1}{2}) + (-\tfrac{1}{3})(-\tfrac{1}{2}) = \tfrac{2}{3} \neq 0$$

Thus, we conclude that $\mathcal{L}_1$ and $\mathcal{L}_3$ are orthogonal to one another but that neither is orthogonal to $\mathcal{L}_2$.

Orthogonality is a desirable property for the following reasons. Suppose that SST denotes the treatment sum of squares with $k-1$ degrees of freedom for a fixed-effects one-way ANOVA, and suppose that $\mathcal{L}_1, \mathcal{L}_2, \ldots, \mathcal{L}_t$ are a set of t ($\leq k-1$) mutually orthogonal contrasts of the k sample means (by mutually orthogonal, we mean that any two contrasts selected from the set of t contrasts are orthogonal to one another). Then it can be shown that SST can be partitioned into $t+1$ statistically independent sums of squares, t of these sums of squares having one degree of freedom each and being associated with the t orthogonal contrasts, and the remaining sum of squares having $k-1-t$ degrees of freedom and being associated with what is "left over" after accounting for the t orthogonal contrast sums of squares. In other words, we can write

$$\text{SST} = \text{SS}(\mathcal{L}_1) + \text{SS}(\mathcal{L}_2) + \cdots + \text{SS}(\mathcal{L}_t) + \text{SS (remainder)}$$

where $\text{SS}(\mathcal{L})$ is the notation for the sum of squares (with 1 df) associated with the contrast $\mathcal{L}$. In particular, it can be shown that

$$\text{SS}(\mathcal{L}) = \frac{(\mathcal{L})^2}{\sum_{i=1}^{k} (c_i^2/n_i)} \tag{17.26}$$

when

$$\mathcal{L} = \sum_{i=1}^{k} c_i \bar{Y}_i$$

that

$$\frac{\text{SS}(\mathcal{L})}{\text{MSE}} \sim F_{1,n-k} \quad \text{under} \quad H_0: L = \sum_{i=1}^{k} c_i \mu_i = 0$$

and in general that

$$\frac{[SS(\mathcal{L}_1) + \cdots + SS(\mathcal{L}_t)]/t}{MSE} \frown F_{t, n-k}$$

under $H_0: \mathcal{L}_1 = \mathcal{L}_2 = \cdots = \mathcal{L}_t = 0$ when $\mathcal{L}_1, \mathcal{L}_2, \ldots, \mathcal{L}_t$ are mutually orthogonal. Thus, the advantage in being able to partition SST as above is that it enables one to test hypotheses concerning sets of orthogonal contrasts which are of more specific interest than the global hypothesis $H_0: \mu_1 = \mu_2 = \cdots = \mu_k$.

For example, to test $H_0: \mathcal{L}_2 = \mu_1 - \frac{1}{3}(\mu_2 + \mu_3 + \mu_4) = 0$ for the potency data in Table 17.7, we first calculate, using (17.26),

$$SS(\mathcal{L}_2) = \frac{(5.30)^2}{[(1)^2 + (-\frac{1}{3})^2 + (-\frac{1}{3})^2 + (-\frac{1}{3})^2]/10} = 210.675$$

and then form the ratio

$$\frac{SS(\mathcal{L}_2)}{MSE} = \frac{210.675}{9.747} = 21.614$$

which is highly significant ($P < 0.001$ based on the $F_{1,36}$ distribution). A modification of Table 17.8 to reflect this partitioning of SST (the sum of squares for "substances") would give

SOURCE		df	SS	MS	F
Substances $\begin{cases} 1 \text{ vs. } (2, 3, 4) \\ \text{remainder} \end{cases}$		1	210.675	210.675	21.614 ($P < 0.001$)
		2	39.200	Irrelevant	—
Error		36	350.900	9.747	
Total		39	600.775		

Similarly, the partitioned ANOVA table for testing

$$H_0: \mathcal{L}_3 = \frac{\mu_1 + \mu_2}{2} - \frac{\mu_3 + \mu_4}{2} = 0$$

would look as follows:

SOURCE		df	SS	MS	F
Substances $\begin{cases} (1, 2) \text{ vs. } (3, 4) \\ \text{remainder} \end{cases}$		1	179.776	179.776	18.444 ($P < 0.001$)
		2	70.099	Irrelevant	—
Error		36	350.900	9.747	
Total		39	600.775		

This partition follows from the fact that

$$SS(\mathcal{L}_3) = \frac{(4.24)^2}{[(\tfrac{1}{2})^2 + (\tfrac{1}{2})^2 + (-\tfrac{1}{2})^2 + (-\tfrac{1}{2})^2]/10} = 179.776$$

Finally, since $\mathcal{L}_1$ and $\mathcal{L}_3$ are *orthogonal*, we can represent the *independent* contributions of these two contrasts to SST in one ANOVA table, as follows:

SOURCE		df	SS	MS	F
Substances	(1, 3) vs. (2, 4)	1	42.025	42.025	4.312 (N.S.)
	(1, 2) vs. (3, 4)	1	179.776	179.776	18.444 ($P<0.001$)
	remainder	1	28.074	Irrelevant	—
Error		36	350.900	9.747	
Total		39	600.775		

Here

$$SS(\mathcal{L}_1) = \frac{\left(\dfrac{25.9+20.0}{2} - \dfrac{22.2+19.6}{2}\right)^2}{[(\tfrac{1}{2})^2 + (-\tfrac{1}{2})^2 + (\tfrac{1}{2})^2 + (-\tfrac{1}{2})^2]/10} = 42.025$$

It is important to point out that it would not be valid to present a partitioned ANOVA table simultaneously including partitions due to $\mathcal{L}_2$ and also either or both $\mathcal{L}_1$ and $\mathcal{L}_3$. This is because $\mathcal{L}_2$ is not orthogonal to these contrasts, and so its sum of squares does not represent a separate and independent contribution to SST; this can easily be seen by noting from the above tables that

$$SST \neq SS(\mathcal{L}_1) + SS(\mathcal{L}_2) + SS(\mathcal{L}_3)$$

As one final example to illustrate the use of orthogonal contrasts, it is often necessary to be able to assess whether or not the sample means exhibit evidence of a *trend* of some sort; this need often arises in the situation where the treatments (or populations) being studied represent, for example, different levels of the same factor (e.g., like different concentrations of the same material or different temperature or pressure settings). In such situations it is of interest to be able to quantify how the sample means vary with changes in the level of the factor; that is, one would like to know if the change in mean response takes place in a linear, quadratic, or other way as the level of the factor is increased or decreased.

A qualitative first step in assessing such a trend is to plot the observed treatment means as a function of the factor levels, the purpose being to obtain some general idea of any pattern that is present. Although standard regression techniques can then be used to quantify any apparent trends suggested by such plots, our goal here is not really to fit a regression model but is rather to reach a decision regarding the general trend in the means in a manner that is more statistical than is an opinion formed by simply examining a plot of the means.

[A standard regression approach would involve treating the independent variable (treatments) as an interval variable and using the actual value of the variable at each treatment setting. To test for linear trend using regression, therefore, the appropriate model would be $Y = \beta_0 + \beta_1 X + E$, where X denotes the (interval) treatment variable. Using this model, the test for linear trend is the usual test for zero slope. This test is, in general, not equivalent to the test for linear trend in *mean* response using orthogonal polynomials described in this section except when the pure error mean square is used in place of the residual mean square in the denominator of the F statistic for testing for zero slope; this is because the regression pure error mean square and the one-way ANOVA error mean square are identical. It also follows that the usual test for lack of fit of the straight-line regression model is equivalent to the test for nonlinear trend in mean response discussed in this section. (See Problem 10 at the end of the chapter for further considerations in this regard.)]

This may be accomplished by determining how much of the sum of squares due to treatments (SST) would be associated with each of the terms (linear, quadratic, cubic, etc.) in a polynomial regression. If the various levels of the treatment or factor being studied are *equally spaced*, this determination is best carried out using the method of *orthogonal polynomials* [for a discussion, see Armitage (1971)].

To illustrate the use of orthogonal polynomials, let us again focus on the potency data of Table 17.7. Further, let us suppose that the four substances actually represent four equally spaced concentrations of some toxic material, with substance 1 being the least concentrated solution and substance 4 the most concentrated solution. For example, substance 1 might represent a 10% solution of the toxic material, substance 2 a 20% solution, substance 3 a 30% solution, and substance 4 a 40% solution. In this case, a plot of the four sample means versus concentration would take the form shown in Figure 17.5. This plot suggests at least a linear, and possibly a quadratic, relationship between concentration and response, and this general impression can be quantified via the use of orthogonal polynomials. In particular, since there are $k = 4$ sample means, it is possible to fit up to a third-order $(k - 1 = 3)$ polynomial to these means; this cubic model (with four terms) would pass through all four points on the graph just given, thus explaining all the variation in the four sample means (or, equivalently, in

FIGURE 17.5   *Plot of the sample means versus concentration*

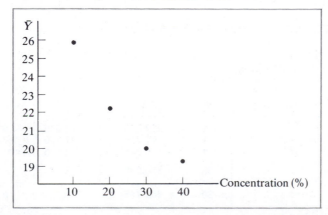

SST). Also, because of the equal spacing of the concentrations, it is possible via the use of orthogonal polynomials to define three orthogonal contrasts of the four sample means, one (say, $\mathcal{L}_l$) measuring the strength of the linear component of the third-degree polynomial, one (say, $\mathcal{L}_q$) the quadratic component contribution, and one (say, $\mathcal{L}_c$) the cubic component effect. The sums of squares associated with these three orthogonal contrasts each have one degree of freedom, are statistically independent, and satisfy the relationship

$$SST = SS(\mathcal{L}_l) + SS(\mathcal{L}_q) + SS(\mathcal{L}_c)$$

Without going into any further explanation here, it can be shown that the coefficients of these three particular orthogonal contrasts are as follows:

CONTRAST	COEFFICIENT OF:				CALCULATED VALUE OF CONTRAST
	$\bar{Y}_1 = 25.9$	$\bar{Y}_2 = 22.2$	$\bar{Y}_3 = 20.0$	$\bar{Y}_4 = 19.6$	
$\mathcal{L}_l$	−3	−1	+1	+3	−21.1
$\mathcal{L}_q$	+1	−1	−1	+1	+3.3
$\mathcal{L}_c$	−1	+3	−3	+1	+0.3

It should be clear that condition (17.25) holds for these three sets of coefficients, which is sufficient for mutual orthogonality here since the n_i's are all equal. Sets of orthogonal polynomial coefficients for other values of k are generally documented in most standard applied statistics texts.

Now, from (17.26), the sums of squares for these three particular contrasts are

$$SS(\mathcal{L}_l) = \frac{(-21.1)^2}{[(-3)^2 + (-1)^2 + (1)^2 + (3)^2]/10} = 222.605$$

$$SS(\mathcal{L}_q) = \frac{(+3.3)^2}{[(1)^2 + (-1)^2 + (-1)^2 + (1)^2]/10} = 27.225$$

$$SS(\mathcal{L}_c) = \frac{(+0.3)^2}{[(-1)^2 + (3)^2 + (-3)^2 + (1)^2]/10} = 0.045$$

Note that

$$SST = 249.875 = 222.605 + 27.225 + 0.045$$

Finally, to assess the significance of these sums of squares, we form the following partitioned ANOVA table:

SOURCE	df	SS	MS	F
Substances $\begin{cases} \mathcal{L}_l \\ \mathcal{L}_q \\ \mathcal{L}_c \end{cases}$	1 1 1	222.605 27.225 0.045	222.605 27.225 0.045	22.838 ($P < 0.001$) 2.793 ($0.10 < P < 0.25$) < 1 (N.S.)
Error	36	350.900	9.747	
Total	39	600.775		

In summary, then, it is clear from the F tests above that the relationship between potency (as measured by the amount of injected material needed to cause death) and concentration is a strong linear one, with no real evidence of higher-order effects.

PROBLEMS

1. Five treatments for fever blisters, including a placebo, were randomly assigned to 30 patients with such blisters. The following data give for each treatment the number of days from initial appearance of the blisters until healing is complete.

TREATMENT	NUMBER OF DAYS REQUIRED FOR COMPLETE HEALING
Placebo (1)	5, 8, 7, 7, 10, 8
2	4, 6, 6, 3, 5, 6
3	6, 4, 4, 5, 4, 3
4	7, 4, 6, 6, 3, 5
5	9, 3, 5, 7, 7, 6

(a) Compute the sample means and standard deviations for each treatment.
(b) Determine the appropriate ANOVA table for the data given.
(c) Are there significant differences among the effects of the five treatments with regard to the healing of fever blisters? In other words, test H_0: $\mu_1 = \mu_2 = \mu_3 = \mu_4 = \mu_5$ against H_A: at least two treatments have different population means.
(d) What are the estimates of the true effects $(\mu_i - \mu)$ of the treatments? Verify that the sum of these estimated effects is zero. (*Note:* μ_i is the population mean for the ith treatment and $\mu = \frac{1}{5}\sum_{i=1}^{5}\mu_i$ is the overall population mean.)
(e) Using dummy variables, give an appropriate regression model which describes this experiment. Give two possible ways to define these dummy variables (one way using 0's and 1's and the other way using 1's and -1's), and describe for each of these coding schemes how the regression coefficients are related to the population means μ_1, μ_2, μ_3, μ_4, μ_5, and μ.
(f) Carry out the Scheffé, Tukey, and LSD multiple-comparison procedures concerning pairwise differences between means for the data of this problem using the recommended procedure described in Section 17.7. Also, compare the widths of the confidence intervals obtained by the three procedures.

2. The following data are replicate measurements of the sulfur dioxide concentration in each of three cities:

City I: 2, 1, 3
City II: 4, 6, 8
City III: 2, 5, 2

(a) Determine the appropriate ANOVA table for simultaneously comparing the mean sulfur dioxide concentrations in the three cities.
(b) Test whether the three cities differ significantly in mean sulfur dioxide concentration levels.
(c) What is the estimated effect associated with each city?
(d) State precisely the appropriate ANOVA fixed-effects model for these data.
(e) State precisely, using dummy variables, the regression model that corresponds to the fixed-effects model in part (d). What is the relationship between the coefficients in the regression model and the effects in the ANOVA model?

(f) Using the t distribution, find a 90% confidence interval for the true difference between the effects of cities I and II (making sure to use the best estimate of σ^2 provided by the data).

3. Each of three chemical laboratories performed four replicate determinations of the concentration of suspended particulate matter in a certain area using the "Hi-Vol" method of analysis. The data are presented in tabular form as shown.

LAB I	LAB II	LAB III
4	2	5
4	2	2
6	5	5
10	3	8

(a) Determine the appropriate ANOVA table for these data.
(b) Test the null hypothesis of no differences among the laboratories.
(c) With large-scale interlaboratory studies, one is usually concerned with making inferences about a large population of laboratories of which only a random sample of labs (e.g., laboratories I, II, III) is available for investigation. In such a case, describe the appropriate *random-effects* model for the data.
(d) Two quantities of particular interest in a large-scale interlaboratory study are the *repeatability* (i.e., a measure of the variability among replicate measurements within a single laboratory) and the *reproducibility* (i.e., a measure of the variability between results from different laboratories). Using the random-effects model defined in part (c), define what you think are reasonable measures of repeatability and reproducibility, and obtain estimates of the quantities you have defined using the available data.

4. Ten randomly selected mental institutions were examined to determine the effects of three different antipsychotic drugs on patients with the same types of symptoms. Each institution used one and only one of the three drugs exclusively for a 1-year period. The proportion of treated patients in each institution who were discharged after 1 year of treatment is as follows for each drug used:

Drug 1: 0.10, 0.12, 0.08, 0.14 ($\bar{Y}_1 = 0.11$, $S_1 = 0.0192$)

Drug 2: 0.12, 0.14, 0.19 ($\bar{Y}_2 = 0.15$, $S_2 = 0.0361$)

Drug 3: 0.20, 0:25. 0.15 ($\bar{Y}_3 = 0.20$, $S_3 = 0.0500$)

(a) Determine the appropriate ANOVA table for this data set.
(b) Test to see if there are significant differences among drugs with regard to the average proportion of patients discharged.
(c) What other factors should possibly be considered when comparing the effects of the three drugs?
(d) What basic ANOVA assumptions might possibly be violated here?

5. Suppose that a random sample of five active members in each of four political parties in a certain western European country were given a questionnaire purported to measure (on a 100-point scale) the extent of "general authoritarian attitude toward interpersonal relationships." The means and standard deviations of the authoritarianism scores for each party are given in the table.

	PARTY 1	PARTY 2	PARTY 3	PARTY 4
$\bar{Y}_i$	85	80	95	50
S_i	6	7	4	10
n_i	5	5	5	5

(a) Determine the appropriate ANOVA table for this data set.

(b) Test to see if there are significant differences among parties with respect to mean authoritarianism scores.

(c) State, using dummy variables, an appropriate regression model for the above experimental situation.

(d) Apply Tukey's method of multiple comparisons to determine those pairs in which the means are significantly different from one another. (Use $\alpha = 0.05$.)

6. A psychosociological questionnaire was administered to a random sample of 200 persons on an island in the South Pacific going through increasing westernization over the past 30 years. From the questionnaire data, each of the 200 persons was classified into one of three groups according to the discrepancy between the amount of prestige in that person's traditional culture and the amount of prestige in the modern (westernized) culture. The three groups were called HI-POS, NO-DIF, and HI-NEG. Also, using the questionnaire data, a measure of "anomie" (i.e., social disorientation), denoted as Y, was determined on a 100-point scale, with the summarized results given in the table.

	n_i	$\bar{Y}_i$	S_i
HI-POS	50	65	9
NO-DIF	75	50	11
HI-NEG	75	55	10

(a) Determine the appropriate ANOVA table.

(b) Test whether the three different categories of prestige discrepancy have significantly different sample mean "anomie" scores.

(c) How would you test whether there is a significant difference between the NO-DIF category and the other two categories combined?

7. In an effort to determine whether reading skills of the average high school graduate have declined over the past 10 years, the average verbal college aptitude scores (VSAT) were compared for a random sample of five big-city high schools for the years 1965, 1970, and 1975. The results are as follows:

HS:	1	2	3	4	5
1965	550	560	535	545	555
1970	545	560	528	532	541
1975	536	552	526	527	530

(a) Determine the sample means for each year.

(b) Comment on the independence assumption needed for using one-way ANOVA with regard to the data given.

(c) Determine the one-way ANOVA table for these data.

(d) Test by means of one-way ANOVA whether there are any significant differences among the three VSAT average scores.

(e) Use Scheffé's method to locate any significant differences between pairs of means. (Use $\alpha = 0.05$.)

8. Three persons (denoted A, B, and C), claiming to have unusual psychic ability, underwent ESP tests at an eastern U.S. psychic research institute. On each of the five randomly selected days, each person was asked to specify for 26 pairs of cards whether both cards in a given pair were of the same color or not. The number of correct answers resulting are given as follows:

PAYS:	1	2	3	4	5
A	20	22	20	21	18
B	24	21	18	22	20
C	16	18	14	13	16

(a) Determine the mean score for each person and interpret the results.

(b) Test whether the three persons have significantly different ESP ability.

(c) Carry out Scheffé's multiple-comparison procedure to determine which pairs of persons, if any, are significantly different in ESP ability.

(d) On the basis of the results in parts (b) and (c), can one conclude that any of these persons has statistically significant ESP ability? Explain.

9. The average generation times for four different influenza virus strains were determined using six cultures for each strain. The data are summarized as given in the table.

	STRAIN A	STRAIN B	STRAIN C	STRAIN D
$\bar{Y}$	420.3	330.7	540.4	450.8
S	30.22	28.90	31.08	33.29

(a) Test whether the true mean generation time is different among the four strains.

(b) What is the appropriate ANOVA table for these data?

(c) Use Tukey's multiple-comparisons procedure to identify where any differences among the means lie. (Use $\alpha = 0.05$.)

10. Three replicate water samples were taken at each of four locations in a river to determine whether the quantity of dissolved oxygen, a measure of water pollution, varied from one location to another (the higher the level of pollution, the lower will be the dissolved oxygen reading). Location 1 was adjacent to the wastewater discharge point for a certain industrial plant, and locations 2, 3, and 4 were selected at points 10, 20, and 30 miles downstream from this discharge point. The data appear in the table. Y_{ij} denotes the value of the dissolved oxygen

LOCATION	DISSOLVED OXYGEN CONTENT (Y_{ij})	MEAN ($\bar{Y}_i$)
1	4, 5, 6	5
2	6, 6, 6	6
3	7, 8, 9	8
4	8, 9, 10	9

content for the jth replicate at location i ($j = 1, 2, 3$; $i = 1, 2, 3, 4$), and $\bar{Y}_i$ denotes the mean of the three replicates taken at location i.

(a) Do the data provide sufficient evidence to suggest a difference in mean dissolved oxygen content among the four locations? (Use $\alpha = 0.05$.) Make sure to write down the appropriate ANOVA table.

(b) If μ_i is the true mean level of dissolved oxygen at location i ($i = 1, 2, 3, 4$), test the null hypothesis

$$H_0: -3\mu_1 - \mu_2 + \mu_3 + 3\mu_4 = 0$$

versus

$$H_A: -3\mu_1 - \mu_2 + \mu_3 + 3\mu_4 \neq 0$$

at the 2% level; the quantity $(-3\mu_1 - \mu_2 + \mu_3 + 3\mu_4)$ is a contrast based on orthogonal polynomials (see Section 17.8), which can be shown to be a measure of the linear relationship between "locations" (as thought of in terms of four equally spaced distances of 0, 10, 20, and 30 miles downstream from the plant) and "dissolved oxygen content."

(c) Another way to quantify the strength of this linear relationship would be to fit by least squares the model

$$Y = \beta_0 + \beta_1 X + E$$

where

$$X = \begin{cases} 0 & \text{for location 1} \\ 10 & \text{for location 2} \\ 20 & \text{for location 3} \\ 30 & \text{for location 4} \end{cases}$$

Fitting such a regression model to the $n = 12$ data point yields the accompanying ANOVA table. Use this table to perform a test of $H_0: \beta_1 = 0$ at the 2% significance level.

SOURCE		df		SS	
Regression		1		29.40	
Residual	lack of fit	10	2	6.60	0.60
	pure error		8		6.00

(d) Noting that the regression model in part (c) is actually saying that $\mu_i = \beta_0 + \beta_1 X_i$, show that the hypothesis tested in part (b) is equivalent to the hypothesis tested in part (c).

(e) Can you supply a reason why the two test statistics calculated in parts (b) and (c) do *not* have the same numerical value? Can you suggest a reasonable modification of the test in part (c) that will yield the same F value as that obtained in part (b)?

(f) Using the results of part (b), a test for nonlinear trend in mean response can be obtained by subtracting the sum of squares for the contrast

$$L = -3\bar{Y}_1 - \bar{Y}_2 + \bar{Y}_3 + 3\bar{Y}_4$$

from the sum of squares for treatments and then dividing this difference by the appropriate degrees of freedom to yield an F statistic of the form

$$F(\text{nonlinear trend}) = \frac{[SST - SS(\mathcal{L})]/df}{MSE}$$

Carry out this test based on the results obtained in parts (a) and (b). (Use $\alpha = 0.05$.)

(g) Carry out the usual regression lack-of-fit test for adequacy of the straight-line model fit in part (c) using $\alpha = 0.05$. Does the value of the F statistic equal the value obtained in part (f)?

RANDOMIZED BLOCKS: SPECIAL CASE OF TWO-WAY ANOVA

18.1 PREVIEW

In Chapter 17 we considered the simplest kind of ANOVA problem, that involving a single factor (or independent variable). We now focus on the two-factor case, which is generally referred to as two-way ANOVA. This extension to include an additional factor is by no means trivial. In fact, we will devote three chapters (Chapters 18, 19, and 20) to consideration of different aspects of the two-factor case. In this first chapter we shall describe how a two-factor situation may be classified according to the type of "data pattern." We shall then restrict our attention to a specific type of pattern, which will lead us to consideration of the randomized blocks design, the main topic of this chapter.

18.1.1 TWO-WAY DATA PATTERNS

Several different types of data patterns for two-way ANOVA are illustrated in Figure 18.1. Each of these tables describes a two-factor situation with four levels of factor 1 (the "row" factor) and three levels of factor 2 (the "column" factor). The Y's in each table correspond to individual observations on a single dependent variable Y. The number of Y's in a given cell is denoted by n_{ij} for the ith level of factor 1 and the jth level of factor 2. The marginal total for the ith row is denoted by $n_{i\cdot}$ and for the jth column by $n_{\cdot j}$. The total number of observations is denoted as $n_{\cdot\cdot}$.

The simplest two-factor pattern, which is illustrated in Figure 18.1a, arises when there is a single observation in each cell (i.e., $n_{ij} = 1$ for all i, j). It is this pattern which incorporates the randomized blocks design to be discussed in this chapter, although there are other ways in which such "single observation per cell" data may arise.

A second type of pattern, illustrated in Figure 18.1b, occurs when there are equal numbers of observations in each cell. Here, $n_{ij} = 4$ for all i, j. Chapter 19 will focus on this "equal-replications" situation.

FIGURE 18.1　*Some two-way data patterns for a(4×3) table*

(a) *Single observation per cell ($n_{ij} = 1$)*

FACTOR 2

Y	Y	Y
Y	Y	Y
Y	Y	Y
Y	Y	Y

FACTOR 1

(b) *Equal replications per cell ($n_{ij} = 4$)*

FACTOR 2

YYYY	YYYY	YYYY
YYYY	YYYY	YYYY
YYYY	YYYY	YYYY
YYYY	YYYY	YYYY

FACTOR 1

(c) *Equal replications by column, proportionate replications by row ($n_{ij} = n_{.j}/4$)*

FACTOR 2

YYYY	YY	YYY	$n_{1.} = 9$
YYYY	YY	YYY	$n_{2.} = 9$
YYYY	YY	YYY	$n_{3.} = 9$
YYYY	YY	YYY	$n_{4.} = 9$

$n_{.1} = 16$　$n_{.2} = 8$　$n_{.3} = 12$

FACTOR 1

(d) *Proportionate row and column replications ($n_{ij} = n_{i.}n_{.j}/n_{..}$)*

FACTOR 2

YYYY	YY	YYY	$n_{1.} = 9$
YYYY YYYY	YY YY	YYY YYY	$n_{2.} = 18$
YYYY YYYY YYYY	YY YY YY	YYY YYY YYY	$n_{3.} = 27$
YYYY YYYY	YY YY	YYY YYY	$n_{4.} = 18$

$n_{.1} = 32$　$n_{.2} = 16$　$n_{.3} = 24$ | $n_{..} = 72$

FACTOR 1

(e) *Nonsystematic replications*

FACTOR 2

YY	YYY	YYY YYY	$n_{1.} = 11$
YYY	YYYY	YY	$n_{2.} = 9$
Y	YYY	YYYY	$n_{3.} = 8$
YYYYY	YY	Y	$n_{4.} = 8$

$n_{.1} = 11$　$n_{.2} = 12$　$n_{.3} = 13$ | $n_{..} = 36$

FACTOR 1

The patterns given in Figure 18.1c–e present different statistical analysis problems than do those exhibited in parts a and b.[1] We shall discuss these analysis considerations in Chapter 20. The common property of the latter three patterns is that all cells do not have the same number of observations. Unequal cell numbers often arise in observational studies for which the levels of certain factors are determined after, rather than before, the data have been collected.

For the pattern in Figure 18.1c, cells in the same column have the same number of observations, whereas cells in the same row are in the ratio 4 : 2 : 3. For this table each of the four cell frequencies in the jth column is equal to the same fraction of the

[1] In Chapters 18, 19, and 20 we shall assume that each cell in a table contains *at least* one observation. When there are cells with no observations, the analysis required is considerably more complicated. The reader is referred to texts by Ostle (1963), Peng (1967), and Armitage (1971) for further discussion in this regard.

corresponding total column frequency (i.e., $n_{ij} = n_{\cdot j}/4$ in this case). Note, for example, that $n_{\cdot 1}/4 = 16/4 = 4$, which is the number of observations in any cell in column 1.

For Figure 18.1d the cells in a given column are in the ratio $1:2:3:2$, whereas the cells in a given row are in the ratio $4:2:3$. This pattern results because n_{ij} is determined as

$$n_{ij} = \frac{n_{i\cdot}n_{\cdot j}}{n_{\cdot\cdot}}$$

which means that any cell frequency can be obtained by multiplying the corresponding row and column marginal frequencies together and then dividing by the total number of observations. Thus, for cell (1, 2) in Figure 18.1d, we have $n_1 \cdot n_{\cdot 2}/n_{\cdot\cdot} = 9(16)/72 = 2$, which equals n_{12}. Similarly, for cell (4, 3), $n_4 \cdot n_{\cdot 3}/n_{\cdot\cdot} = 18(24)/72 = 6$, which equals n_{43}.

For the pattern illustrated in Figure 18.1e, there is no mathematical rule for describing the pattern of cell frequencies, and so we say that such a pattern is "nonsystematic." As we will see in Chapter 20, patterns c and d differ from the irregular pattern e with regard to the ANOVA procedures required for analysis. For patterns c and d, the computational procedure may be carried out in the same way as for the case involving an "equal number of observations per cell." For the nonsystematic case, a different procedure is required.

18.1.2 SINGLE-OBSERVATION-PER-CELL CASE

A two-way table with a single observation in each cell can arise in a number of different experimental situations. For instance, consider the following three examples:

1. Six hypertensive individuals, matched pairwise by age and sex, are randomly assigned (within each pair) to either a treatment or a control group. For each individual, a measure of change in self-perception of health is determined after a 1-year period. The main question of interest is whether or not the true mean change in self-perception is different for the treatment group than for the control group.

2. Six growth-inducing treatment combinations are randomly assigned to six mice from the same litter. The treatment combinations are defined by the cross-classification of the levels of two factors: factor A (drug A1 or placebo A0) and factor B [drug B2 (high dose), drug B1 (low dose), and placebo B0]. The dependent variable of interest is weight gain 1 week after treatment is initiated. The questions to be considered include (a) whether or not the effect of drug A1 is different from that of placebo A0, (b) whether or not there are differences among the effects of drugs B1 and B2 and placebo B0, and (c) whether or not the drug A1 effect differs from that of placebo A0 in the same way at each level of factor B.

3. Scores of satisfaction with medical care are recorded for six hypertensive patients assigned to one of six categories depending on whether the nurse practitioner assigned to the patient was measured to have high or low autonomy (factor A) and high, medium, or low knowledge of hypertension (factor B). The main questions of interest are (a) whether mean satisfaction scores differ between patients with high autonomy nurses and those with low autonomy nurses, and (b) whether or not these differences in mean satisfaction scores differ in the same way at each level of knowledge grouping.

FIGURE 18.2   *Different experimental situations resulting in two-way tables with a single observation per cell*

(a)

	PAIR 1	PAIR 2	PAIR 3
TREATMENT	Y	Y	Y
CONTROL	Y	Y	Y

Y = change in self-perception of health after 1 year

(b)

	PLACEBO B0	DRUG B1	DRUG B2
PLACEBO A0	Y	Y	Y
DRUG A1	Y	Y	Y

Y = weight gain after 1 week

(c)

	LOW KNOWLEDGE	MEDIUM KNOWLEDGE	HIGH KNOWLEDGE
LOW AUTONOMY	Y	Y	Y
HIGH AUTONOMY	Y	Y	Y

Y = patient satisfaction with medical care

Each of these experiments may be represented by a (2×3) two-way table as given in Figure 18.2.

In the figure the third example (table c) is one involving two factors whose levels (or categories) were determined *after* the data were gathered. Such a study is often referred to as an *observational study* rather than as an *experiment*, since the latter term is usually reserved for studies involving factors whose levels are decided upon beforehand. Epidemiological, sociological, and psychological studies are more often than not of an observational rather than of an experimental nature. For such studies, the levels of the various factors of interest are determined after consideration of the frequency distributions of these factors based on the observed data.[2] For example c, the autonomy groupings were determined after the frequency distribution of autonomy scores on nurses was considered. Similarly, the knowledge categories were determined using the observed knowledge scores. Thus, one patient was in each of the six groups because of a posteriori (rather than a priori) considerations. In actual practice it may not often be possible to arrange things so nicely, especially when large samples are involved. That is why most such observational-type studies are of the "unequal-number-of-observations-per-cell" variety.

The second example in Figure 18.2 (table b) involves two factors whose levels were determined before the data were collected; and the resulting six treatment combinations could be viewed, in one sense, as representing the different levels of a

[2] In this chapter we shall focus on the case where the levels of each factor are considered *fixed*; nevertheless, even if one or both factors were considered as random, the tests of hypotheses of interest, as with one-way ANOVA, would be computed in exactly the same way as in the fixed-factor case. This is not so in the "more-than-one-observation-per-cell" case, as discussed in Chapter 19.

single factor (called "treatment combination") which have been randomly assigned to the six individuals. Although it may be necessary because of a limited amount of experimental material or because of prohibitive cost to apply or assign each treatment combination to only a single experimental unit, considerably more information would be obtained if several replications were made at each treatment combination.

This brings us to Figure 18.2a, which is of the general type on which we shall focus in this chapter. As with table (b), this example represents a *designed* rather than an *observational* study. However, it differs from tables (b) and (c) in the manner of allocation of individuals to cells (i.e., treatment combinations). The allocation in table (a) was done by randomization separately within each pair rather than, say, by randomization among all six individuals, ignoring any pairing. Another feature unique to table (a) is that the effect of only one of the two factors involved (in this case "treatment group") is of primary interest. The other factor, "pair group" (with three levels corresponding to the three pairs), is being used only to help make more precise the comparison between the effects of the treatment and control groups. That is, if pair matching (on age and sex) is used and significant differences are found between the change scores for treatment and control groups, such differences could not solely be attributed, for example, to one group being older or having a different sex composition than the other. The pairing, therefore, serves to eliminate or block out the noise affecting the comparison of treatment and control groups due to the confounding effects of age and sex. Such pairs are often referred to as *blocks* and the associated experimental design is called a *randomized-blocks design*.

The analysis required for data as in table (a) is described in most introductory statistics texts. Since two groups are involved (treatment and control) and since matching has been done, the generally recommended method of analysis involves the use of the "paired-difference *t* test," which is based on using the differences in (change) scores within pairs. That is, the key test statistic involved is of the form $T = \bar{d}\sqrt{n}/S_d$, where $\bar{d}$ is the difference between the treatment group mean and the control group mean, S_d is the standard deviation of the difference scores for all pairs, and n is the number of pairs [three in table (a)]. This statistic has the *t* distribution with $n-1$ degrees of freedom under H_0 (so that the critical region is of the form $|T| \geq t_{n-1, 1-\alpha/2}$ for a two-sided test of the null hypothesis of no difference in true average change score for treatment and control groups).

Actually, it can be shown that the paired-difference *t* test can be looked upon as a special case of the general *F* test used in a *randomized-blocks ANOVA*, involving more than two treatments per block. In fact, this randomized-blocks *F* test represents a generalization of the paired-difference *t* test in the same way that the one-way ANOVA *F* test is a generalization of the two-sample *t* test.

18.2 EQUIVALENT ANALYSES OF A MATCHED-PAIRS EXPERIMENT

Suppose that we make the matched-pairs example of the previous section more realistic by considering the data given in Table 18.1, which involves 15 pairs of individuals matched on age and sex. The main inferential question for these data concerns whether or not the treatment group has a mean change score which is significantly different from that of the control group. Stated in terms of population

TABLE 18.1 *Matched-pairs design concerning change scores in self-perception of health (Y) among hypertensives*

	PAIR NUMBER															TOTAL	MEAN
	1	2	3	4	5	6	7	8	9	10	11	12	13	14	15		
Treatment	10	12	8	8	13	11	15	16	4	13	2	15	5	6	8	146	9.73
Control	6	5	7	9	10	12	9	8	3	14	6	10	1	2	1	103	6.87
Total	16	17	15	17	23	23	24	24	7	27	8	25	6	8	9	249	8.30
Difference	4	7	1	−1	3	−1	6	8	1	−1	−4	5	4	4	7	43	2.86

means, the null hypothesis is given by

$$H_0: \mu_T = \mu_C$$

where μ_T and μ_C denote the treatment- and control-group population means, respectively. The alternative hypothesis would be given by

$$H_A: \mu_T \neq \mu_C$$

if it were not theorized in advance that one particular group would have a higher or lower population mean than the other. (If, however, it was thought a priori that the treatment group would have a significantly higher mean change score than the control group, the alternative would be one-sided: $H_A: \mu_T > \mu_C$.)

18.2.1 PAIRED-DIFFERENCE t TEST

One method for testing the null hypothesis H_0 was described in the previous section, the paired-difference t test. To make this test we first determine from Table 18.1 that $\bar{d} = \bar{Y}_T - \bar{Y}_C = 2.86$ and

$$S_d^2 = \frac{1}{14}\left[\sum_{i=1}^{15} d_i^2 - \frac{\left(\sum_{i=1}^{15} d_i\right)^2}{15}\right] = 12.695$$

where d_i is the observed difference between the treatment- and control-group scores for the ith pair. Then the test statistic is computed to be

$$T = \frac{\bar{d}\sqrt{n}}{S_d} = \frac{2.86\sqrt{15}}{\sqrt{12.695}} = 3.109$$

Since $t_{14,0.995} = 2.976$ and $t_{14,0.9995} = 4.140$, the P value for this two-sided test is given by

$$0.001 < P < 0.01$$

We would therefore reject H_0 and conclude that the mean change score for the treatment group is significantly different from that of the control group.

18.2.2 RANDOMIZED-BLOCKS F TEST

Another way to test the null hypothesis H_0: $\mu_T = \mu_C$ for the data of Table 18.1 is to use an F test, which is based on the ANOVA table given in Table 18.2.

This ANOVA table differs in general form from that used for one-way ANOVA in that the total sum of squares has now been partitioned into three components instead of just two (between and within). The "treatments" component here is similar to the "between" (i.e., treatment) source in the one-way ANOVA case. The "error" component here is analogous to the "within" component in the one-way ANOVA case, because it is this source that is used to provide an estimate of the population variance σ^2. Finally, we have a "pairs" (or "blocks") component in Table 18.2, for which there is no corresponding component in the one-way ANOVA case.

Thus, whereas the total sum of squares in one-way ANOVA may be split up into *two* components, as indicated by the equation

$$SS(total) = SS(between) + SS(within)$$

the total sum of squares in a randomized blocks ANOVA can be partitioned into *three* meaningful components, as reflected in the expression

$$SS(total) = SS(treatments) + SS(blocks) + SS(error) \qquad (18.1)$$

The last two components on the right-hand side of (18.1) can be looked upon as representing a partition of the experimental error (or within) sum of squares associated with one-way ANOVA (i.e., with the blocking ignored). By separating out the blocking effect, a more precise estimate of experimental error is hopefully obtained, which is not contaminated by any "noise" due to the effects of the blocking variables (in our case, age and sex).

The computation of the sums of squares in expression (18.1) is a fairly straightforward exercise. To obtain SS(treatments), for example, we divide the sum of the squares of the treatment totals by the number of pairs (i.e., blocks) and then subtract the usual correction factor (which is the squared total of all observations divided by the total number of observations). In other words, using the row totals in Table 18.1, we obtain

$$SS(treatments) = \frac{1}{15}[(146)^2 + (103)^2] - \frac{(249)^2}{30}$$

$$= 2{,}128.33 - 2{,}066.70$$

$$= 61.63$$

TABLE 18.2   *ANOVA table for matched-pairs data of Table 18.1*

SOURCE	SS	df	MS	F
Treatments	61.63	1	61.63	9.68 ($0.005 < P < 0.01$)
Pairs (blocks)	391.80	14	27.97	4.39 ($0.001 < P < 0.005$)
Error	89.17	14	6.37	
Total	542.30	29		

The SS(blocks) is obtained similarly by dividing the sum of the squares of the block totals by the number of treatments and then subtracting off the correction factor:

$$SS(blocks) = \frac{1}{2}[(16)^2 + (17)^2 + (15)^2 + \cdots + (8)^2 + (9)^2] - \frac{(249)^2}{30}$$

$$= 2{,}458.50 - 2{,}066.70$$

$$= 391.80$$

The total sum of squares is obtained by subtracting the correction factor from the total of the squares of all the observations:

$$SS(total) = \sum_{\text{all observations}} (\text{observation})^2 - (\text{correction factor})$$

$$= [(10)^2 + (6)^2 + (12)^2 + (5)^2 + \cdots + (8)^2 + (1)^2] - \frac{(249)^2}{30}$$

$$= 2{,}609.00 - 2{,}066.70$$

$$= 542.30$$

We then obtain SS(error) by difference:

$$SS(error) = SS(total) - SS(treatments) - SS(blocks)$$

$$= 542.30 - 61.63 - 391.80$$

$$= 89.17$$

With these sums of squares now computed, the next step is to determine the degrees of freedom associated with each sum of squares. The degrees of freedom for "treatments" are always equal to one less than the number of treatments (in our example, $2 - 1 = 1$). The degrees of freedom for "blocks" are equal to one less than the number of blocks (in our example, $15 - 1 = 14$). The "total sum of squares" degrees of freedom are equal to one less than the number of observations (in our example, $30 - 1 = 29$). And, finally, the degrees of freedom for "error" are obtained by subtraction (i.e., $29 - 14 - 1 = 14$). Alternatively, the "error df" in a randomized-blocks analysis can be obtained as the product of the treatment df and the block df:

$$error\ df = (\text{treatment df})(\text{block df}) = 1(14) = 14$$

The mean squares, as usual, are obtained by dividing the sums of squares by their corresponding degrees of freedom. And, finally, the F statistic for the test of the null hypothesis of no differences among the treatments is given by the formula

$$F = \frac{MS(treatments)}{MS(error)} \tag{18.2}$$

In particular, to test H_0: $\mu_T = \mu_C$ for the data of Table 18.1, we calculate

$$F = \frac{61.63/1}{89.17/14} = \frac{61.63}{6.37} = 9.68$$

which is the observed value of an F with 1 and 14 degrees of freedom under H_0. Thus, H_0 is rejected at significance level α if the observed value of F is greater than $F_{1,14,1-\alpha}$. Since $F_{1,14,0.99} = 8.86$ and $F_{1,14,0.995} = 11.06$, the P value for this test satisfies the inequality $0.005 < P < 0.01$. Since this is quite small, it is reasonable to reject H_0 and to conclude that the treatment group has a significantly different mean change score from that of the control group.

Thus, the conclusions reached via the paired difference t test and the randomized blocks F test are exactly the same: reject H_0. This is not due to mere coincidence but to the fact that the two tests are completely equivalent. In fact, it can be shown mathematically that the square of a paired-difference T statistic with ν degrees of freedom is exactly equal to a randomized-blocks F statistic with 1 and ν degrees of freedom (i.e., $F_{1,\nu} = T_{\nu}^2$), and that $F \geq F_{1,\nu,1-\alpha}$ whenever $|T| \geq t_{\nu,1-\alpha/2}$, and vice versa.

In our example,

$$T_{14}^2 = (3.109)^2 = 9.67 = F_{1,14}$$

and

$$t_{14,0.995}^2 = (2.977)^2 = 8.86 = F_{1,14,0.99}$$

An F test of the null hypothesis "H_0: no significant differences among the blocks" may also be performed. This test is not of primary interest, since the very use of a randomized-blocks design is based on the a priori assumption that there is significant block-to-block variation. Nevertheless, an a posteriori F test may be used to check on the reasonableness of this assumption. The test statistic to be used in this case is given by

$$F = \frac{MS(blocks)}{MS(error)}$$

which, for the example of Table 18.1, has the F distribution under H_0 with 14 df in the numerator and 14 df in the denominator. From Table 18.2 this F statistic is computed to be 4.39, with a P value satisfying $0.001 < P < 0.005$. The conclusion for this test, as expected, is to reject the null hypothesis of no block differences.

18.3 PRINCIPLE OF BLOCKING

For the matched-pairs example that we have been considering, we have indicated that the primary reason for "pairing up" the data was to avoid having the confounding factors age and sex blur the comparison between the treatment and control groups. Another way of saying this is that the use of the matched-pairs design represented an attempt to account for the fact that the experimental units (i.e., the subjects) were not "homogeneous" with regard to factors (other than "treatment group") that were likely to affect the response variable. The key point here is not simply that subjects of different age and sex are different, but more precisely that age and sex are likely to affect the response variable. In another experimental situation it might be that age and sex would not be important covariables and so would not have to be controlled or adjusted for.

FIGURE 18.3   *Steps used in forming randomized blocks*

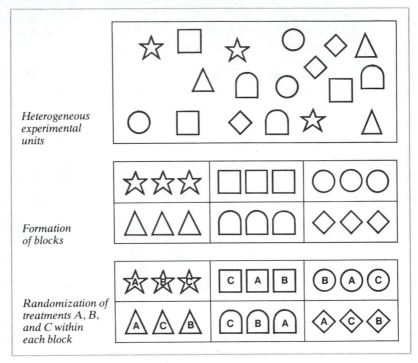

One motivation, then, for using a randomized-blocks design comes from the fact that the experimental units under study are heterogeneous relative to certain concomitant variables that affect the response variable but are not of primary interest. In such a case, to use a randomized blocks design requires taking the following two steps: (1) *the experimental units (e.g., people, animals) which are "homogeneous" are collected together to form a "block," and* (2) *the various treatments are assigned at random to the experimental units within each block*.

These steps are illustrated in Figure 18.3, where six blocks are formed, each consisting of three homogeneous experimental units. Three treatments (labeled A, B, and C, say) are then assigned at random to the three units within each block.

Figure 18.4, on the other hand, provides an example of incorrect blocking, which could result in one type of experimental unit getting predominantly one kind of treatment and another type possibly not getting that treatment at all. In Figure 18.4, for

FIGURE 18.4   *Example of incorrect blocking*

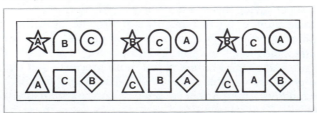

example, the experimental unit type ☆ was assigned treatment B twice, whereas experimental units types ○ and △ were not assigned treatment B at all. If the blocking had been done correctly, every distinct type of experimental unit would have been assigned each treatment exactly once.

With regard to our matched-pairs design of Table 18.1, if we had not blocked on age, the treatment conceivably could have been assigned mostly to older subjects, and the control group might have consisted mostly of younger subjects. Then, any differences that might be found in such a case could very well be due entirely to age differences in the two groups and not to the effect of the treatment itself.

18.4 ANALYSIS OF A RANDOMIZED-BLOCKS EXPERIMENT

In this section we shall describe the appropriate analysis for a randomized-blocks experiment. We shall focus our attention on the case where both factors (i.e., treatments and blocks) are considered as *fixed*, although the tests of hypotheses of interest are computed in exactly the same way even if one or both factors are considered as *random*; that is, practically speaking, it does not matter how the factors are defined.[3]

18.4.1 DATA CONFIGURATION

Table 18.3 gives the general data layout for a randomized-blocks design involving k treatments and b blocks. In this table Y_{ij} denotes the value of the observation on the dependent variable Y corresponding to the ith treatment in the jth block (e.g., Y_{23} denotes the value of the observation associated with treatment 2 in block 3). The (row) total for treatment i is denoted by T_i; that is, $T_i = \sum_{j=1}^{b} Y_{ij}$. The (column) total for block j is denoted by B_j; that is, $B_j = \sum_{i=1}^{k} Y_{ij}$. The grand total of all bk observations is $G =$

TABLE 18.3   *Data layout for a randomized-block design*

TREATMENT NUMBER	BLOCK NUMBER					TOTAL	MEAN
	1	2	3	$\cdots$	b		
1	Y_{11}	Y_{12}	Y_{13}	$\cdots$	Y_{1b}	T_1	$\bar{Y}_{1\cdot}$
2	Y_{21}	Y_{22}	Y_{23}	$\cdots$	Y_{2b}	T_2	$\bar{Y}_{2\cdot}$
3	Y_{31}	Y_{32}	Y_{33}	$\cdots$	Y_{3b}	T_3	$\bar{Y}_{3\cdot}$
$\vdots$	$\vdots$	$\vdots$	$\vdots$		$\vdots$	$\vdots$	$\vdots$
k	Y_{k1}	Y_{k2}	Y_{k3}	$\cdots$	Y_{kb}	T_k	$\bar{Y}_{k\cdot}$
Total	B_1	B_2	B_3	$\cdots$	B_b	G	—
Mean	$\bar{Y}_{\cdot 1}$	$\bar{Y}_{\cdot 2}$	$\bar{Y}_{\cdot 3}$	$\cdots$	$\bar{Y}_{\cdot b}$	—	$\bar{Y}_{\cdot\cdot}$

[3] In Chapter 19, which considers the situation of equal (≥ 2) replications per cell, the F tests required differ depending on the factor definitions.

$\sum_{i=1}^{k} \sum_{j=1}^{b} Y_{ij}$. Finally, the treatment (row) means are denoted by $\bar{Y}_{i\cdot}$ for the ith treatment, the block (column) means are denoted by $\bar{Y}_{\cdot j}$ for the jth block, and the grand mean is denoted by $\bar{Y}_{\cdot\cdot}$.

A special case of this general format was given in Table 18.1, where $k = 2$ and $b = 15$. Another example (with $k = 4$ and $b = 8$) is given in Table 18.4. This example (although based on artificial data) illustrates the type of information being considered in several ongoing large-scale U.S. intervention studies dealing with the risk factors associated with heart disease. Table 18.4 presents the data for a small experiment designed to assess the effects of four different cholesterol-reducing diets on persons identified to have hypercholesterolemia. In such a study it would seem logical to try to take into account (or adjust for) the effects of age, sex, and body size on the dependent variable Y, which is the "reduction in cholesterol level after one year." One way to do this is through the use of a randomized-blocks design, where the blocks are chosen to represent combinations of the various age–sex–body size categories of interest. Thus, in Table 18.4, block 1 consists of males over 50 years of age with a quetelet index [$= 100$(weight)/(height)^2] above 3.5; block 3 consists of males under 50 with a quetelet index above 3.5; and so on. For each block of subjects, the four diets are randomly assigned to the sample of four persons in the block. Each subject is then followed for one year, after which the change in cholesterol level (Y) is recorded.

The primary research question of interest in this study concerns whether or not there are significant differences among the average reductions in cholesterol level achieved by the four diets. From inspection of Table 18.4 it can be seen that diet 1 appears to be the best, since it is associated with the largest average reduction (12.57 units). This statement, nevertheless, needs to be examined statistically by determining whether or not the observed differences among the mean reductions for the four diets can be attributed solely to chance. The randomized-blocks F test provides us with a method for making such a statistical evaluation.

Further inspection of Table 18.4 indicates that there are considerable differences among the block means. This is to be expected, since the very reason for blocking was based on the assumption that different blocks (or, equivalently, different categories of the covariates age, sex, and body size) would have different effects on the response. We can, nevertheless, perform a statistical test to satisfy ourselves that such block differences are statistically significant. In fact, if this test proves to be nonsignificant, it

TABLE 18.4   *Randomized-blocks experiment for comparing the effects of four cholesterol-reducing diets on persons with hypercholesterolemia ($Y =$ cholesterol level reduction after 1 year)*

TREAT-MENT (DIET) NUMBER	BLOCK NUMBER 1 (MALE, AGE>50, QUET>3.5)	2 (MALE, AGE>50, QUET<3.5)	3 (MALE, AGE<50, QUET>3.5)	4 (MALE, AGE<50, QUET<3.5)	5 (FEMALE, AGE>50, QUET>3.5)	6 (FEMALE, AGE>50, QUET<3.5)	7 (FEMALE, AGE<50, QUET>3.5)	8 (FEMALE, AGE<50, QUET<3.5)	TOTAL	MEAN
1	11.2	6.2	16.5	8.4	14.1	9.5	21.5	13.2	100.6	12.57
2	9.3	4.1	14.2	6.9	14.2	8.9	15.2	10.1	82.9	10.36
3	10.4	5.1	14.0	6.2	11.1	8.4	17.3	11.2	83.7	10.46
4	9.0	4.9	13.7	6.1	11.8	8.4	15.9	9.7	79.5	9.94
Total	39.9	20.3	58.4	27.6	51.2	35.2	69.9	44.2	346.7	—
Mean	9.97	5.07	14.60	6.90	12.80	8.80	17.47	11.05	—	10.83

means that we probably should not have used these blocks in the first place, since their use cost us degrees of freedom for estimating σ^2 and consequently gave us a less sensitive comparison of treatment effects.

18.4.2 HYPOTHESES TESTED IN A RANDOMIZED-BLOCKS ANALYSIS

The primary null hypothesis of interest in a randomized-blocks analysis is the equality of treatment means[4]:

$$H_0: \mu_1. = \mu_2. = \cdots = \mu_k. \tag{18.3}$$

where $\mu_i.$ denotes the population mean response associated with the ith treatment. The alternative hypothesis may be stated

H_A: not all the $\mu_i.$'s are equal

If performing this test leads to rejection of H_0, it becomes of interest to determine where the important differences are among the treatment effects. One qualitative way of making such a determination is simply to look at the observed treatment means and to make visual comparisons. Thus, from Table 18.4, rejection of the null hypothesis of no differences among the diets leads to the visual conclusion that diet 1 is the best in reducing cholesterol level, that diets 2 and 3 are next in line and have about equal success, and that diet 4 is the worst of the four. To verify such suspicions statistically, appropriate tests can be made using *multiple-comparisons techniques*, as described in Chapter 17.

As mentioned earlier, another null hypothesis sometimes of interest is that there are no significant differences among the block means. Since the use of a ran-domized-blocks design is based on the a priori conviction that the block means will be different, such a hypothesis is generally only tested to check that the blocking was justified.

18.4.3 *F* TEST FOR THE EQUALITY OF TREATMENT MEANS

The test of $H_0: \mu_1. = \mu_2. = \cdots = \mu_k.$ is performed using the following F statistic (defined in terms of the notation of Table 18.3):

$$F = \frac{MST}{MSE} \tag{18.4a}$$

where

$$MST = \frac{1}{k-1}\left(\frac{1}{b}\sum_{i=1}^{k} T_i^2 - \frac{G^2}{bk}\right) \tag{18.4b}$$

[4] Again, we point out that we are considering the null hypothesis for the fixed-effects case. As for one-way ANOVA, this hypothesis would be stated differently if the treatment effects were considered as random, but the test of hypothesis would be computed exactly the same way.

and[5]

$$MSE = \frac{1}{(k-1)(b-1)}\left(\sum_{i=1}^{k} \sum_{j=1}^{b} Y_{ij}^2 - \frac{1}{k} \sum_{j=1}^{b} B_j^2 - \frac{1}{b} \sum_{i=1}^{k} T_i^2 + \frac{G^2}{bk} \right) \qquad (18.4c)$$

The assumptions required for this test are as follows:

1. The observations are statistically independent of one another.
2. Each observation is selected from a population that is normally distributed.
3. Each observation is selected from a population with variance σ^2 (i.e., *variance homogeneity* is assumed).
4. There is no "block–treatment" interaction effect; that is, the true extent to which treatments differ is the same regardless of the block considered. (We will discuss this in more detail in Section 18.5.)

If H_0 is true, the F statistic in (18.4) has the F distribution with $(k-1)$ df in the numerator and $(k-1)(b-1)$ df in the denominator. Thus, for a given α, we would reject H_0 and conclude that not all treatments have the same effect on the response provided that

$$F \geq F_{k-1,(k-1)(b-1),1-\alpha}$$

For the data of Table 18.4, we find that

$$MST = \frac{1}{4-1}\left\{ \frac{1}{8}[(100.6)^2 + (82.9)^2 + (83.7)^2 + (79.5)^2] - \frac{(346.7)^2}{32} \right\}$$

$$= 11.187$$

$$MSE = \frac{1}{(4-1)(8-1)}\left\{ [(11.2)^2 + (9.3)^2 + (10.4)^2 + \cdots + (11.2)^2 + (9.7)^2] \right.$$

$$- \frac{1}{4}[(9.97)^2 + (5.07)^2 + \cdots + (11.05)^2]$$

$$\left. - \frac{1}{8}[(100.6)^2 + (82.9)^2 + (83.7)^2 + (79.5)^2] + \frac{(346.7)^2}{32} \right\}$$

$$= 0.939$$

so that

$$F = \frac{11.187}{0.939} = 11.914$$

[5] Although (18.4) provides an explicit expression for MSE, in practice one would ordinarily calculate MSE as

$$MSE = \frac{1}{(k-1)(b-1)}(TSS - SST - SSB)$$

where TSS, SST, and SSB are, respectively, the total, treatment, and block sums of squares.

Since $F_{3,21,0.999} = 7.94$, the P value for this test satisfies $P < 0.001$. Thus, the null hypothesis H_0 should be rejected and the conclusion be made that there are significant differences among the four diets.

18.4.4 F TEST FOR THE EQUALITY OF BLOCK MEANS

As previously mentioned, this test is rarely used except as an a posteriori check that blocking was effective. To perform this test, the following F statistic is calculated:

$$F = \frac{\text{MSB}}{\text{MSE}} \tag{18.5a}$$

where

$$\text{MSB} = \frac{1}{b-1}\left(\frac{1}{k}\sum_{j=1}^{b} B_j^2 - \frac{G^2}{bk}\right) \tag{18.5b}$$

and where MSE is calculated as before (see footnote 5).

Under the null hypothesis "H_0: no differences among the true block means," the F statistic (18.5) has the F distribution with $(b-1)$ and $(k-1)(b-1)$ df. So H_0 is rejected at significance level α when F exceeds $F_{b-1,(k-1)(b-1),1-\alpha}$.

For the data in Table 18.4, this test yields the following F:

$$F = \frac{66.123}{0.939} = 70.419$$

The P value for this test satisfies $P \ll 0.001$, so H_0 is rejected, as expected.

18.5 ANOVA TABLE FOR A RANDOMIZED-BLOCKS EXPERIMENT

The procedures necessary to perform the tests of equality of treatment and block means can be summarized in ANOVA Table 18.5. For the data given in Table 18.4, the corresponding ANOVA table is given by Table 18.6. From these ANOVA tables it can be seen that the total (corrected) sum of squares TSS is split up into the three components SST(treatments), SSB(blocks), and SSE(error) using the *fundamental equation*

$$\text{TSS(total)} = \text{SST(treatments)} + \text{SSB(blocks)} + \text{SSE(error)} \tag{18.6}$$

Equivalent formulas for the total sum of squares are:

$$\text{TSS(total)} = \sum_{i=1}^{k}\sum_{j=1}^{b}(Y_{ij} - \bar{Y}_{..})^2 = \sum_{i=1}^{k}\sum_{j=1}^{b} Y_{ij}^2 - \frac{G^2}{bk} \tag{18.7}$$

TABLE 18.5 *ANOVA table for a randomized-blocks experiment (with k treatments and b blocks)*

SOURCE	df	SS	MS	F
Treatments	$k-1$	$\text{SST} = \frac{1}{b} \sum_{i=1}^{k} T_i^2 - \frac{G^2}{bk}$	$\text{MST} = \frac{\text{SST}}{k-1}$	MST/MSE
Blocks	$b-1$	$\text{SSB} = \frac{1}{k} \sum_{j=1}^{b} B_j^2 - \frac{G^2}{bk}$	$\text{MSB} = \frac{\text{SSB}}{b-1}$	MSB/MSE
Error	$(k-1)(b-1)$	$\text{SSE} = \text{TSS} - \text{SST} - \text{SSB}$	$\text{MSE} = \frac{\text{SSE}}{(k-1)(b-1)}$	
Total	$kb-1$	$\text{TSS} = \sum_{i=1}^{k} \sum_{j=1}^{b} Y_{ij}^2 - \frac{G^2}{bk}$		

TABLE 18.6 *ANOVA table for data of Table 18.4*

SOURCE	df	SS	MS	F
Treatments	3	33.561	11.187	11.914 ($P<0.001$)
Blocks	7	462.860	66.123	70.419 ($P \ll 0.001$)
Error	21	19.711	0.939	
Total	31	516.132		

The far-right-hand-side formula in (18.7) is preferable for computational purposes. As is the case in any ANOVA situation, TSS measures the total unexplained variation in the data, corrected for the grand mean. The degrees of freedom associated with TSS are given by $kb-1$. For the data of Table 18.4, TSS is computed to be

$$\text{TSS} = [(11.2)^2 + (9.3)^2 + (10.4)^2 + (9.0)^2 + (6.2)^2 + \cdots + (11.2)^2 + (9.7)^2]$$

$$-\frac{(346.7)^2}{32} = 516.132$$

The treatment sum of squares SST is defined as

$$\text{SST} = b \sum_{i=1}^{k} (\bar{Y}_{i.} - \bar{Y}_{..})^2 = \frac{1}{b} \sum_{i=1}^{k} T_i^2 - \frac{G^2}{bk} \tag{18.8}$$

The formula on the far-right-hand side of (18.8) is the preferred computational formula. The expression in the middle of (18.8) illustrates why SST reflects the variation among the treatment means, since its basic components are of the form $(\bar{Y}_{i.} - \bar{Y}_{..})$, which is the difference between the ith treatment mean and the grand mean. For the data of Table

18.4, SST is computed as

$$SST = \frac{1}{8}[(100.6)^2 + (82.9)^2 + (83.7)^2 + (79.5)^2] - \frac{(346.7)^2}{32}$$

$$= 33.561$$

The treatment sum of squares has $k - 1$ degrees of freedom.

The block sum of squares SSB is defined as

$$SSB = k \sum_{j=1}^{b} (\bar{Y}_{\cdot j} - \bar{Y}_{\cdot\cdot})^2 = \frac{1}{k} \sum_{j=1}^{b} B_j^2 - \frac{G^2}{bk} \qquad (18.9)$$

with degrees of freedom equal to $b - 1$. Again, the far-right-hand side of (18.9) is the computational formula, whereas the middle expression provides an explanation of why SSB measures variation among blocks. For the data of Table 18.4, we get

$$SSB = \frac{1}{4}[(39.9)^2 + (20.3)^2 + (58.4)^2 + (27.6)^2 + (51.2)^2 + (35.2)^2 + (69.9)^2$$

$$+ (44.2)^2] \quad - \frac{(346.7)^2}{32} = 462.860$$

Finally, the residual sum of squares SSE is given by

$$SSE = \sum_{i=1}^{k} \sum_{j=1}^{b} (Y_{ij} - \bar{Y}_{i\cdot} - \bar{Y}_{\cdot j} + \bar{Y}_{\cdot\cdot})^2 = TSS - SST - SSB \qquad (18.10)$$

with degrees of freedom equal to $(k-1)(b-1)$.

Applying the computational formula on the far-right-hand side of (18.10) to our example yields

$$SSE = 516.132 - 33.561 - 462.860$$

$$= 19.711$$

The complexity of the middle expression in (18.10) indicates that SSE is not easily recognizable as an estimate of σ^2. In fact, its basic component $(Y_{ij} - \bar{Y}_{i\cdot} - \bar{Y}_{\cdot j} + \bar{Y}_{\cdot\cdot})$ can be written in the form

$$(Y_{ij} - \bar{Y}_{\cdot j}) - (\bar{Y}_{i\cdot} - \bar{Y}_{\cdot\cdot})$$

which is an estimate of the *difference between the effect of the* ith *treatment relative to the average effect of all treatments in the* jth *block* (i.e., $Y_{ij} - \bar{Y}_{\cdot j}$) *and the overall effect of the* ith *treatment relative to the overall mean* (i.e., $\bar{Y}_{i\cdot} - \bar{Y}_{\cdot\cdot}$). Actually, SSE here measures block–treatment interaction, in the sense that SSE will be large when the treatment effects vary from block to block. Although we will discuss the concept of "interaction" more thoroughly in Chapter 19, it is imperative to note here that *the use of a randomized-blocks design requires the assumption that no such block–treatment interaction exists*, and that, as a result, the residual variation reflected in SSE can be attributed solely to experimental error.[6] It is mandatory that this assumption not be violated because, otherwise, there would be no way to obtain an unbiased estimate of σ^2 that could be used in the denominator of the F statistic. After all, for each block in a randomized-blocks design, no treatment is applied more than once, making it impossible to obtain a "pure error" estimate of σ^2 associated with a particular treatment in a given block. When we consider the two-way ANOVA case with more than one observation per cell, we will see that a "pure error" estimate of σ^2 can be developed by utilizing the available information on within-cell variability.

Each mean-square term in Table 18.5 is, as usual, obtained by dividing the corresponding sum-of-squares term by its degrees of freedom. Also, the F statistics are formed as the ratios of mean-square terms, with MSE in the denominator in each case (regardless of whether each factor is considered as fixed or random).

18.6 REGRESSION MODELS FOR A RANDOMIZED-BLOCKS EXPERIMENT

The randomized-blocks experiment, like the one-way ANOVA situation, can be described by either a *regression* model or a classical ANOVA *effects* model. The regression formulation, which we consider in this section, is technically equivalent to the fixed-effects ANOVA approach in terms of estimation of the unknown parameters in each model. Nevertheless, for testing purposes, mean-square terms in the ANOVA table obtained from the fit of a proper regression model can be used to compute appropriate F statistics regardless of whether the actual ANOVA model is fixed, random, or mixed.

An appropriate regression model for the randomized-blocks experiment should contain $k-1$ dummy variables for the k treatments and $b-1$ dummy variables for the b blocks. One such model formulation is given as follows:

$$Y = \mu + \sum_{i=1}^{k-1} \alpha_i X_i + \sum_{j=1}^{b-1} \beta_j Z_j + E$$

where

$$X_i = \begin{cases} -1 & \text{if treatment } k \\ 1 & \text{if treatment } i \\ 0 & \text{otherwise} \end{cases} \quad \text{and} \quad Z_j = \begin{cases} -1 & \text{if block } b \\ 1 & \text{if block } j \\ 0 & \text{otherwise} \end{cases} \tag{18.11}$$

$(i = 1, 2, \ldots, k-1; j = 1, 2, \ldots, b-1)$.

[As in the one-way ANOVA case, other coding schemes for the independent variables are possible; for example, we may let

$$X_i = \begin{cases} 1 & \text{if treatment } i \\ 0 & \text{otherwise} \end{cases} \qquad Z_j = \begin{cases} 1 & \text{if block } j \\ 0 & \text{otherwise} \end{cases}$$

---

[6] A method for testing the validity of the assumption of no block–treatment interaction, due to Tukey (1949), is described in Problem 3(f) at the end of the chapter.

For any such coding scheme, the F tests obtained from the fit of the regression model are exactly equivalent to the randomized-blocks F tests described previously. The regression coefficients, however, represent different functions of the cell population means when the coding schemes are changed.]

The regression coefficients in this model may, as in the one-way ANOVA situation considered in Chapter 17, be expressed in terms of underlying cell (i.e., block–treatment combination) means. To see this we need to consider the matrix of population cell means associated with a general randomized-blocks layout, as presented in Table 18.7. In this table, the (cell) mean for the ith treatment in the jth block is denoted by μ_{ij}, the mean for the ith treatment (averaged over the b blocks) is denoted by $\mu_{i.}$, the mean for the jth block (averaged over the k treatments) is denoted by $\mu_{.j}$, and the overall mean is denoted by $\mu_{..}$ That is, $\mu_{i.}$, $\mu_{.j}$, and $\mu_{..}$ satisfy

$$\mu_{i.} = \frac{1}{b} \sum_{j=1}^{b} \mu_{ij}, \quad \mu_{.j} = \frac{1}{k} \sum_{i=1}^{k} \mu_{ij}, \quad \mu_{..} = \frac{1}{bk} \sum_{i=1}^{k} \sum_{j=1}^{b} \mu_{ij}$$

For model (18.11), the coefficients α_i and β_j are then expressible as follows:

$$\mu = \mu_{..}$$
$$\alpha_i = \mu_{i.} - \mu_{..}, \quad i = 1, 2, \ldots, k-1$$
$$\beta_j = \mu_{.j} - \mu_{..}, \quad j = 1, 2, \ldots, b-1 \qquad (18.12)$$
$$-\sum_{i=1}^{k-1} \alpha_i = \mu_{k.} - \mu_{..}$$
$$-\sum_{j=1}^{b-1} \beta_j = \mu_{.b} - \mu_{..}$$

Thus, the coefficient μ is the overall mean, the coefficient α_i represents the difference between the ith treatment mean and the overall mean, and the coefficient β_j represents the difference between the jth block mean and the overall mean. Furthermore, the negative sum of the α_i (i.e., $-\sum_{i=1}^{k-1} \alpha_i$) gives the difference between the kth treatment mean and the overall mean, and the negative sum of the β_j (i.e., $-\sum_{j=1}^{b-1} \beta_j$) gives the difference between the bth block mean and the overall mean.

TABLE 18.7   *Matrix of cell means for a randomized-blocks layout*

TREATMENT (i)	BLOCK (j)					TREATMENT MEANS
	1	2	3	$\cdots$	b	
1	μ_{11}	μ_{12}	μ_{13}	$\cdots$	μ_{1b}	$\mu_{1.}$
2	μ_{21}	μ_{22}	μ_{23}	$\cdots$	μ_{2b}	$\mu_{2.}$
3	μ_{31}	μ_{32}	μ_{33}	$\cdots$	μ_{3b}	$\mu_{3.}$
$\vdots$	$\vdots$	$\vdots$	$\vdots$		$\vdots$	$\vdots$
k	μ_{k1}	μ_{k2}	μ_{k3}	$\cdots$	μ_{kb}	$\mu_{k.}$
Block means	$\mu_{.1}$	$\mu_{.2}$	$\mu_{.3}$	$\cdots$	$\mu_{.b}$	$\mu_{..}$

As with one-way ANOVA, the F tests resulting from the fit of regression model (18.11), or from any other properly coded regression model, for testing the hypotheses of equality of treatment means ($H_0: \mu_1. = \mu_2. = \cdots = \mu_k.$) and of equality of block means ($H_0: \mu._1 = \mu._2 = \cdots = \mu._b$) are exactly the same as obtained for the ANOVA procedures presented earlier. That is, the multiple-partial F test of $H_0: \alpha_1 = \alpha_2 = \cdots = \alpha_{k-1} = 0$ under model (18.11) provides exactly the same F value as that given by

$$F = \frac{\text{MST}}{\text{MSE}}$$

of (18.4). Similarly, the multiple-partial F test of $H_0: \beta_1 = \beta_2 = \cdots = \beta_{b-1} = 0$ in model (18.11) yields exactly the same F value as given by

$$F = \frac{\text{MSB}}{\text{MSE}}$$

of (18.5).

Thus, it really makes no difference which, if any, regression model (i.e., coding scheme) is used, if we are only interested in performing the above "global" F tests. Furthermore, if we want to make certain specific comparisons of means, we can always calculate such comparisons directly without having to use a regression model.

18.7 FIXED-EFFECTS ANOVA MODEL FOR A RANDOMIZED-BLOCKS EXPERIMENT

If both the block and treatment effects are considered to be fixed, a "classical" ANOVA model may be written in terms of such "effects" as was done in the one-way ANOVA case. The effects for a randomized-blocks experiment are defined in terms of differences between a given treatment mean and the overall mean (i.e., $\mu_i. - \mu..$) and differences between a given block mean and the overall mean (i.e., $\mu._j - \mu..$). The fixed-effects model may be written in the form

$$Y_{ij} = \mu + \alpha_i + \beta_j + E_{ij}, \qquad i = 1, 2, \ldots, k; \, j = 1, 2, \ldots, b \tag{18.13}$$

where

Y_{ij} = observed response associated with the ith treatment in the jth block

$\mu = \mu..$ = grand (overall) mean

$\alpha_i = (\mu_i. - \mu..)$ = effect of treatment i

$\beta_j = (\mu._j - \mu..)$ = effect of block j

$E_{ij} = (Y_{ij} - \mu - \alpha_i - \beta_j)$ = error component associated with the ith treatment in the jth block

The key characteristic of model (18.13) is that the effect of any given treatment is the same regardless of the block and that, similarly, the effect of any given block is the same regardless of the treatment. Another way of saying this is that, to determine the mean response for a given cell, it is only necessary to know the treatment effect and block effect associated with that cell, without having to consider the particular contribution of the cell itself. Still a third way of saying this is that there is no block–treatment interaction. A model acknowledging such an interaction would be of the form

$$Y_{ij} = \mu + \alpha_i + \beta_j + \gamma_{ij} + E_{ij}$$

which contains the (interaction) term γ_{ij} specific to cell (i, j).

A few additional properties of the fixed-effects model (18.13) are worth noting. First, it can be shown that

$$\sum_{i=1}^{k} \alpha_i = 0 \quad \text{and} \quad \sum_{j=1}^{b} \beta_j = 0$$

since $\alpha_i = \mu_{i \cdot} - \mu_{\cdot \cdot}$ and $\beta_j = \mu_{\cdot j} - \mu_{\cdot \cdot \cdot}$.

Another property of interest is that the various treatment and block effects can be simply estimated from the data. These estimated effects are given by

$$\hat{\alpha}_i = \bar{Y}_{i \cdot} - \bar{Y}_{\cdot \cdot} \quad \text{and} \quad \hat{\beta}_j = \bar{Y}_{\cdot j} - \bar{Y}_{\cdot \cdot}$$

A final property of the fixed-effects model (18.13) is that it corresponds in structure to the regression model given by (18.11). That is, the coefficients α_1 through α_{k-1} of model (18.11) correspond to the first $k-1$ treatment effects in model (18.13). Also, the coefficients β_1 through β_{b-1} correspond to the first $b-1$ block effects in model (18.13). Finally, the negative sums $-\sum_{i=1}^{k-1} \alpha_i$ and $-\sum_{j=1}^{b-1} \beta_j$ represent the effects of the kth treatment and the bth block, respectively.

PROBLEMS
1. An experiment was conducted by a private research corporation to investigate the toxic effects of three chemicals (I, II, and III) used in the tire-manufacturing industry. In this experiment 1-inch squares of skin on rats were treated with the chemicals and then scored from 0 to 10, depending on the degree of irritation. Three adjacent 1-inch squares were marked on the back of each of eight rats, and each of the three chemicals was applied to each rat. The data are as shown in the table.

CHEMICAL	RAT NUMBER								TOTAL
	1	2	3	4	5	6	7	8	
I	6	9	6	5	7	5	6	6	50
II	5	9	9	8	8	7	7	7	60
III	3	4	3	6	8	5	5	6	40
Total	14	22	18	19	23	17	18	19	150

(a) Recognizing this as a randomized-blocks design, what are the blocks and what are the treatments?
(b) Determine the appropriate ANOVA table for the data set given.
(c) Do the data provide sufficient evidence to indicate a significant difference in the toxic effects of the three chemicals?

(d)   Using a confidence interval of the form

$$(\bar{Y}_{1.} - \bar{Y}_{2.}) \pm t_{\nu,1-\alpha/2} \sqrt{MSE(1/n_1 + 1/n_2)}$$

where ν is the degrees of freedom, find a 98% confidence interval for the true difference in the toxic effects of chemicals I and II.

(e)   Using the ANOVA table obtained in part (b), provide a reasonable measure of the proportion of total variation explained by the particular statistical model used in analyzing this data set.

(f)   State the fixed-effects ANOVA model and the corresponding regression model for this analysis.

(g)   State the assumptions on which the validity of the analysis depends.

2.   In a study of the psychosocial changes in individuals participating in a community-based blood pressure intervention program, individuals who were clinically identified as hypertensives were randomly assigned to one of two treatment groups (assume that there are n individuals in each group). Group 1 was given the *usual care* provided for hypertensives by existing community facilities, whereas group 2 was given the *special care* provided by the intervention study team. Among the variables measured on each individual were: SP1, an index of the individual's self-perception of health immediately after being identified as hypertensive but before assignment to one of the two groups; SP2, an index of the individual's self-perception of health 1 year after assignment to one of the two groups; AGE; and SEX. Restricting attention to these variables only, one main research question of interest concerns whether or not the change in self-perception of health after 1 year will be greater for individuals in group 2 than for those in group 1. To examine this question, several different analytical approaches are possible, depending on the choice of dependent variable and on how the variables SP1, AGE, and SEX are treated in the analysis. Some of these approaches are:

(1)   Matching pairwise on AGE and SEX (which we assume is possible) and then performing a paired-difference t test to determine whether or not the mean of group 1 change scores is significantly different from the mean of group 2 change scores. (*Note*: This is equivalent to doing a randomized-blocks analysis, where the blocks are the pairs of individuals.)

(2)   Matching pairwise on AGE and SEX and then performing a regression analysis with the change score $Y = SP2 - SP1$ as the dependent variable and with SP1 as one of the independent variables.

(3)   Matching pairwise on AGE and SEX and then performing a regression analysis with SP2 as the dependent variable and with SP1 as one of the independent variables.

(4)   Controlling for AGE and SEX (without any prior matching) via analysis of covariance, with the change score $Y = SP2 - SP1$ as the dependent variable.

(5)   Controlling for AGE and SEX via analysis of covariance, with the change score $Y = SP2 - SP1$ as the dependent variable and with SP1 as one of the independent variables.

(6)   Controlling for AGE and SEX via analysis of covariance, with SP2 as the dependent variable and with SP1 as one of the independent variables:
   (a)   What is the appropriate regression model associated with each of the above six approaches? (Make sure to define your variables carefully.)

    (b) For each of the regression models above, state the appropriate null hypothesis (in terms of regression coefficients) for testing for group differences with respect to self-perception scores. Indicate for each regression model how to set up the appropriate ANOVA table to carry out the desired test.

    (c) Assuming that you have decided to match pairwise on AGE and SEX, which of the above regression models would you prefer to use and why? [*Note*: Actually, you have only two models to choose from, since both models (2) and (3) will produce exactly the same test statistic for comparing the two groups.]

3. An experiment was conducted at the University of North Carolina to see whether the BOD test for water pollution is biased by the presence of copper. In this BOD test the amount of dissolved oxygen in a sample of water is measured at the beginning and at the end of a 5-day period; the difference in dissolved oxygen content is ascribed to the action of bacteria on the impurities in the sample and is called the *biochemical oxygen demand* (BOD). The question is whether dissolved copper retards the bacterial action and results in an artificially low response for the test.

    The following data are partial results from this experiment. The three samples are from different sources. They are split up into five subsamples, and the concentration of copper ion in each subsample is given. The BOD measurements are given for each (subsample, copper-ion concentration) combination.

SAMPLE	COPPER-ION CONCENTRATION (ppm)					MEAN
	0	0.1	0.3	0.5	0.75	
1	210	195	150	148	140	168.60
2	194	183	135	125	130	158.40
3	138	98	89	90	85	100.00
Mean	180.67	158.67	124.67	121.00	118.33	140.67

    (a) Using dummy variables and treating the copper-ion concentration as a categorical (or nominal) variable, give an appropriate regression model for this experiment. Is this a randomized-blocks experiment?

    (b) If copper-ion concentration is treated as an interval (continuous) variable, one appropriate regression model would be given by

$$Y = \beta_0 + \beta_1 Z_1 + \beta_2 Z_2 + \beta_3 X + \beta_4 Z_1 X + \beta_5 Z_2 X + E$$

where

$$Z_1 = \begin{cases} 1 & \text{for sample 1} \\ 0 & \text{for sample 2,} \\ -1 & \text{for sample 3} \end{cases} \quad Z_2 = \begin{cases} 0 & \text{for sample 1} \\ 1 & \text{for sample 2} \\ -1 & \text{for sample 3} \end{cases}$$

and X = copper-ion concentration. Comment on the advantages and disadvantages of using the models in parts (a) and (b). Which model would you prefer to use and why?

(c) Compare (without doing any statistical tests) the average BOD responses at the various copper-ion concentration levels.

(d) Use the table, which is based on a randomized-blocks analysis, to test (at $\alpha = 0.05$) the null hypothesis that copper concentration has no effect on the BOD test.

SOURCE	df	SS	MS	F
Samples	2	12,980.9330	6,490.4667	56.886
Concentrations	4	9,196.6667	2,299.1667	20.130
Error	8	913.7332	114.2166	

(e) Based on the ANOVA table and on the observed block means, does blocking appear to be justified?

(f) The randomized-blocks analysis assumes that the relative differences in BOD responses at different copper-ion concentration levels are the same regardless of the sample used; in other words, there is no "concentration-level-by-sample" interaction. One method (see Tukey, 1949) for testing whether such an interaction effect actually exists is called *Tukey's test for additivity*, which addresses the null hypothesis

H_0: no interaction exists (i.e., the model is "additive" in the block and treatment effects)

versus the alternative

H_A: the model is not "additive," but there exists a transformation $f(Y)$ which "removes" the nonadditivity in the model for Y

Tukey's test statistic is given by

$$F = \frac{SSN}{(SSE - SSN)/[(k-1)(b-1)-1]}$$

[which is distributed as $F_{1,(k-1)(b-1)-1}$ under H_0], where

$$SSN = \frac{\left[\sum\limits_{i=1}^{k}\sum\limits_{j=1}^{b} Y_{ij}(\bar{Y}_{i\cdot} - \bar{Y}_{\cdot\cdot})(\bar{Y}_{\cdot j} - \bar{Y}_{\cdot\cdot})\right]^2}{\sum\limits_{i=1}^{k}(\bar{Y}_{i\cdot} - \bar{Y}_{\cdot\cdot})^2 \sum\limits_{j=1}^{b}(\bar{Y}_{\cdot j} - \bar{Y}_{\cdot\cdot})^2}$$

using the notation in this chapter. [Since the numerator df for this test statistic is 1, the square root of F will have the t distribution with $(k-1)(b-1)-1$ df.] Given that the computed t statistic is $T = 2.090$ in Tukey's test for additivity, is there significant evidence of nonadditivity? (Use $\alpha = 0.05$.)

(g) Use the accompanying computer output based on fitting the multiple regression model given in part (b) to test whether there is evidence of a significant effect due to copper concentration (i.e., test $H_0: \beta_3 = \beta_4 = \beta_5 = 0$). Multiple $R^2 = 0.888$.

VARIABLE	REGRESSION COEFFICIENT	BETA COEFFICIENT
(constant)	167.200	
Z_1	32.513	0.67659
Z_2	16.972	0.35319
X	−80.389	−0.55585
Z_1X	−13.877	−0.12337
Z_2X	−12.844	−0.11419

SOURCE	df	MS
Z_1	1	11,765.00
$Z_2\|Z_1$	1	1,216.00
$X\|Z_1, Z_2$	1	7,134.60
$Z_1X\|Z_1, Z_2, X$	1	303.30
$Z_2X\|Z_1, Z_2, X, Z_1X$	1	91.07
Error	9	286.83

4. For the data in Problem 7 of Chapter 17, carry out a randomized-blocks analysis, treating the high schools as blocks, to test whether there are significant differences among the mean VSAT scores for the years 1965, 1970, and 1975. Do the results obtained from this randomized-blocks analysis differ from those previously obtained from the one-way ANOVA?

5. For the data in Problem 8 of Chapter 17, carry out a randomized-blocks analysis, treating the days as blocks, to test whether the three persons have significantly different ESP ability. Comment concerning whether or not blocking on days seems appropriate.

6. The promotional policies of four companies in a certain industry were compared to determine whether there were any differences among the companies with respect to the discrepancy between black and white promotion rates. Data on the variable *rate discrepancy*, defined as

$$d = \hat{p}_W - \hat{p}_B$$

where

$$\hat{p}_W = \frac{\text{(number of whites promoted) (100)}}{\text{number of whites eligible for promotion}}$$

and

$$\hat{p}_B = \frac{\text{(number of blacks promoted)(100)}}{\text{number of blacks eligible for promotion}}$$

were obtained for the four companies in each of three different 2-year periods, and these data are presented here in tabular form.

PERIOD	COMPANY			
	1	2	3	4
1	3	5	5	4
2	4	4	3	5
3	8	12	10	9

(a) Using dummy variables, write down an appropriate regression model for this data set. Is this a randomized-blocks design?

(b) Use the accompanying table to test whether there are any significant differences among the average rate discrepancies for the four companies. Tukey's T for testing additivity = 1.893 with 5 df.

SOURCE	df	SS	MS	F
Periods	2	84.5000	42.2500	33.800
Companies	3	6.0000	2.0000	1.600
Error	6	7.5000	1.2500	

(c) Does Tukey's test indicate for these data that a "removable" interaction effect is present?

(d) If one finds no significant differences among the rate discrepancies for the four companies, would this support the contention that none of the companies is discriminatory in promotional policy?

(e) Comment on the suitability of this analysis in view of the fact that the response variable is a difference in proportions.

7. Suppose that, in a study to compare body sizes of three genotypes of fourth-instar silkworm,[7] the mean length (in millimeters) for separately reared cocoons of heterozygous (HET), homozygous (HOM), and wild (WLD) silkworms were determined at five laboratory sites; the data are given in the table.

VARIABLE	SITE				
	1	2	3	4	5
HOM	29.87	28.16	32.08	30.84	29.44
HET	32.51	30.82	34.17	33.46	32.99
WLD	35.76	33.14	36.29	34.95	35.89

(a) Assuming that this is a randomized-blocks type of experiment, what are the blocks and what are the treatments?

(b) Comment on whether or not you think a randomized-blocks analysis is appropriate for this experiment.

(c) Carry out an appropriate analysis of the data for this experiment and state your conclusions. Make sure to present the ANOVA table and to state the null hypothesis for each test performed.

[7] Adapted from a study by Sokal and Karlen (1964).

CHAPTER 19

TWO-WAY ANOVA WITH EQUAL CELL NUMBERS

19.1 PREVIEW

In this chapter we shall consider the analysis of the simplest two-way data layout involving more than one observation per cell, that is, the layout for which the number of observations in any given cell is at least two and is exactly the same as the number in any other cell (as previously illustrated in Figure 18.1b). We will see in this chapter that having equal cell numbers makes for a straightforward analysis involving only slightly more involved calculations than those for a randomized-blocks experiment. On the other hand, when there are unequal cell numbers, the analysis is much more complicated (see Chapter 20).

Two-way layouts with equal cell numbers are rarely seen in *observational studies* but are often generated by design in *experimental studies* where the investigator has a priori control regarding the choice of levels of the factors and the allocation of subjects to the various factor combinations. Such two-way layouts with equal cell numbers can be obtained by:

1. *Blocking* so that several (but equal numbers of) observations on each treatment occur in each block.
2. *Stratifying* according to the levels of the two factors of interest and then sampling within each stratum.
3. *Forming treatment combinations* (i.e., cells) and then allocating these combinations to individuals.

The choice of design would depend, of course, on the study characteristics. If, for example, we wanted to eliminate the effects of a confounding factor, we could use blocking to do this. If, on the other hand, we were interested in measuring the respiratory function of industrial workers in different plants (factor 1) subject to different environmental exposure levels (factor 2), a stratified sampling procedure would seem to be appropriate. Or, if we were interested in the effects of combinations of different doses of two different drugs, it would be appropriate to randomly assign the different drug combinations to different groups of subjects.

Regardless of the specific experimental design used, the importance of having more than one observation at each combination of factor levels should be apparent. For example, considering the respiratory-function example mentioned above, if only one person from a given plant subject to a certain environmental exposure level was examined, there would be no direct way to determine how other persons in the same circumstances would differ in response from that individual. Similarly, for the drug example, if no more than one individual received a specific treatment combination of drugs, there would be no way to assess the variation in response among persons receiving that same treatment combination. Thus, a major reason for having more than one observation at each combination of factor levels (i.e., in each *cell*) is to be able to compute a pure estimate of experimental error (σ^2), "pure" in the sense of being a "within-cell" measure.

The use of a randomized-blocks design (with a single observation per cell) precludes the possibility of obtaining such a "within-cell" estimate of σ^2. However, if the blocking does *only* what it is assumed to do (i.e., eliminate the effects of confounding factors), it is still possible to obtain an estimate of σ^2 (although not pure) from what would ordinarily measure block–treatment interaction (which is assumed not to exist).

The detection of an interaction effect between two factors, although not of interest for the randomized-blocks design (in which the blocks are not awarded the status of being considered as the levels of a factor or independent variable), is an important reason for having repeated observations in each cell in two-way layouts. This notion of interaction will be further elaborated on later in the chapter.

EXAMPLE   In Table 19.1 the CMI data of Table 17.3 from Daly's study (1973) has been categorized according to the levels of two factors:

Factor 1: percent black in the surrounding neighborhood (PSN); level 1 = low ($\leq 50\%$), level 2 = high ($> 50\%$)

Factor 2: number of households in Turnkey neighborhood (NHT); level 1 = low (≤ 100), level 2 = high (> 100)

Each of the four cells in Table 19.1 represents some combination of a level of factor 1 and a level of factor 2. There are 25 observations in each cell, which constitute random samples of 25 women heads of household selected from the four Turnkey neighborhoods as defined by the stratification of the two factors NHT and PSN.

[It may be argued that the categorization scheme in Table 19.1 is inappropriate if the neighborhood with 98 households (Cherryview) should really be considered large in size. That is, the cut point for dichotomizing the variable NHT should possibly be lower than 100. Such a decision is often based on subjective considerations and is by no means an easy one to make. Realizing this problem, we nevertheless have proceeded to assume that this categorization scheme is reasonable.]

We have already seen for these data that the *F* test for one-way ANOVA, which treats the four cells of Table 19.1 as the levels of a single-factor "neighborhood," was nonsignificant. Thus, it was concluded that the mean CMI scores for the four neighborhoods, when compared simultaneously, are not significantly different from one another.

TABLE 19.1 *CMI scores for a sample of women from four Turnkey neighborhoods categorized by percent black in surrounding neighborhood (PSN) and by number of households in the Turnkey neighborhood (NHT)*

NUMBER OF HOUSEHOLDS IN TURNKEY NEIGHBORHOOD (NHT)	PERCENT BLACK IN SURROUNDING NEIGHBORHOOD (PSN)		TOTAL
	LOW (≤50%)	HIGH (>50%)	
Low (≤100)	(Cherryview) 49, 12, 28, 24, 16, 28, 21, 48, 30, 18, 10, 10, 15, 7, 6, 11, 13, 17, 43, 18, 6, 10, 9, 12, 12 ($n_{11} = 25$, $\bar{Y}_{11.} = 18.92$)	(Easton) 13, 10, 20, 20, 22, 14, 10, 8, 21, 35, 17, 23, 17, 23, 83, 21, 17, 41, 20, 25, 49, 41, 27, 37, 57 ($n_{12} = 25$, $\bar{Y}_{12.} = 26.84$)	$n_{1.} = 50$ $\bar{Y}_{1..} = 22.88$
High (>100)	(Northhills) 20, 31, 19, 9, 7, 16, 11, 17, 9, 14, 10, 5, 15, 19, 29, 23, 70, 25, 6, 62, 2, 14, 26, 7, 55 ($n_{21} = 25$, $\bar{Y}_{21.} = 20.84$)	(Morningside) 5, 1, 44, 11, 4, 3, 14, 2, 13, 68, 34, 40, 36, 40, 22, 25, 14, 23, 26, 11, 20, 4, 16, 25, 17 ($n_{22} = 25$, $\bar{Y}_{22.} = 20.72$)	$n_{2.} = 50$ $\bar{Y}_{2..} = 20.78$
Total	$n_{.1} = 50$, $\bar{Y}_{.1.} = 19.88$	$n_{.2} = 50$, $\bar{Y}_{.2.} = 23.78$	$n_{..} = 100$ $\bar{Y}_{...} = 21.83$

This result would justifiably tend to influence the researcher not to expect much from further analysis, especially if no additional variables were to be taken into account. On the other hand, should the one-way ANOVA *F* test have led to the conclusion that the four neighborhoods had significantly different mean CMI scores, it would be of considerable interest to examine the nature of these differences. For example, do neighborhoods with a high percentage of blacks in the surrounding environs have significantly smaller mean CMI scores than those with a low percentage of blacks in surrounding environs? Or do neighborhoods with a large number of households have significantly smaller mean CMI scores than neighborhoods with a small number of households? Or, does the amount and direction of the difference in CMI scores between neighborhoods of different size depend significantly on the racial makeup of the surrounding environs (i.e., is there an "interaction" effect)?

We shall proceed to demonstrate how these questions can be answered using two-way ANOVA. In fact, in spite of the nonsignificance of the one-way ANOVA *F* test, it would be of interest to perform a two-way ANOVA anyhow in order to quantify the separate effects of the factors PSN and NHT, and, even more important, to examine the possibility of an interaction between these two factors.

19.2 USE OF A TABLE OF CELL MEANS

An important first step in examining a two-way layout should always be the construction of a table of cell means. For our CMI data, such a table is presented in Table 19.2.

TABLE 19.2.   *Cell means for CMI data*

NHT	PSN LOW	PSN HIGH	ROW MEANS
	LOW	HIGH	
Low	$\hat{\mu}_{11} = \bar{Y}_{11.} = 18.92$	$\hat{\mu}_{12} = \bar{Y}_{12.} = 26.84$	$\hat{\mu}_{1.} = \bar{Y}_{1..} = 22.88$
High	$\hat{\mu}_{21} = \bar{Y}_{21.} = 20.84$	$\hat{\mu}_{22} = \bar{Y}_{22.} = 20.72$	$\hat{\mu}_{2.} = \bar{Y}_{2..} = 20.78$
Column Means	$\hat{\mu}_{.1} = \bar{Y}_{.1.} = 19.88$	$\hat{\mu}_{.2} = \bar{Y}_{.2.} = 23.78$	$\hat{\mu}_{..} = \bar{Y}_{...} = 21.83$

From the table it can be observed that:

1.   The mean CMI score for low NHT is larger than for high NHT:

$$\hat{\mu}_{1.} - \hat{\mu}_{2.} = \bar{Y}_{1..} - \bar{Y}_{2..} = 22.88 - 20.78 = 2.08$$

(This comparison measures what is called the *main effect of NHT.*)

2.   The mean CMI score for low PSN is smaller than for high PSN:

$$\hat{\mu}_{.1} - \hat{\mu}_{.2} = \bar{Y}_{.1.} - \bar{Y}_{.2.} = 19.88 - 23.78 = -3.90$$

(This comparison measures what is called the *main effect of PSN.*)

3.   There is little difference between high PSN and low PSN when NHT is high:

$$\hat{\mu}_{22} - \hat{\mu}_{21} = \bar{Y}_{22.} - \bar{Y}_{21.} = 20.72 - 20.84 = -0.12$$

whereas there is considerable difference between high PSN and low PSN when NHT is low:

$$\hat{\mu}_{12} - \hat{\mu}_{11} = \bar{Y}_{12.} - \bar{Y}_{11.} = 26.84 - 18.92 = 7.92$$

(These two comparisons measure what is called the *interaction between NHT and PSN.*)

Observation 1 suggests that persons from small Turnkey neighborhoods might not be as healthy as persons from large Turnkey neighborhoods (remembering that the lower the CMI score the healthier), which was as Daly theorized.

Observation 2 suggests that persons from Turnkey neighborhoods with a high percentage of blacks in the surrounding neighborhood might not be as healthy as persons from Turnkey neighborhoods with a low percentage of blacks in the surrounding neighborhood. This observation is counter to Daly's theory.

Observation 3 suggests that (a) there is little difference between neighborhoods with high and low black percentages in the surroundings when the Turnkey neighborhood size is large, whereas (b) there is considerable difference between neighborhoods with high and low black percentages in the surroundings when the size of the Turnkey neighborhood is small.

Another way of describing the "interaction" effect pointed out in observation 3 is that (a) when PSN is low, persons from neighborhoods with low NHT seem to be healthier than persons from neighborhoods with high NHT, but that (b) when PSN is high, persons from neighborhoods with high NHT seem to be healthier than persons from neighborhoods with low NHT.

Of all the remarks above, clearly the most important is that of observation 3, which suggests the possibility of some kind of *interaction* between NHT and PSN. That is, any difference in the health of persons from different PSN categories seems to depend on what NHT category is being considered. Or, equivalently, any difference between persons from different NHT categories appears to depend on the PSN category.

Nevertheless, we must remember that the differences found above were obtained from a sample and that, consequently, such differences could have occurred solely by chance. In other words, it is necessary to determine whether or not the above differences are statistically significant, and this can be done using two-way ANOVA.

19.2.1 FIXED, RANDOM, OR MIXED MODEL

To determine the appropriate significance tests for a two-way ANOVA, it is first necessary to specify whether each of the two factors is fixed or random. Although such a specification in the one-way ANOVA situation altered *only* the statement of the null and alternative hypotheses and not the form of the F test used, how the factors are classified in the two-way case affects the F test as well.

In fact, there are three different cases to be considered, depending on the classification of the two factors:

1. Fixed-effects case: both factors are fixed.
2. Random-effects case: both factors are random.
3. Mixed-effects case: one factor is fixed and the other is random.

With regard to our example (i.e., Table 19.1), the classification of the factors NHT and PSN is not straightforward but depends on one's point of view. The *fixed-effects case* would be applicable if the researcher is interested only in the particular Turnkey neighborhoods selected for study, or (in terms of the two factors NHT and PSN) does not wish to make inferences to neighborhoods of different sizes or to different black percentages for surrounding neighborhoods. The *random-effects case* would apply if the Turnkey neighborhood sizes chosen are considered representative of a larger population of sizes of interest and the black percentages are representative of a larger population of black percentages of interest. The *mixed-effects case* would be applicable if one of the factors is considered to be fixed and the other to be random. Of these three cases it is our opinion that the *random-effects case* best represents the true situation. Nevertheless, the appropriate analysis for each case will be discussed.

19.2.2 TWO-WAY ANOVA TABLE FOR THE DATA OF TABLE 19.1

Table 19.3 gives the two-way ANOVA table layout for the CMI data of Table 19.1. There are four sources of variation in this table, corresponding to the two main effects (for NHT and PSN, respectively), the interaction effect, and the error variation.

TABLE 19.3 *ANOVA table for CMI data of Table 19.1*

SOURCE	df	SS	MS	F(fixed)	F(random)	$F\left(\dfrac{\text{NHT fixed}}{\text{PSN random}}\right)$	$F\left(\dfrac{\text{NHT random}}{\text{PSN fixed}}\right)$
NHT	1	110.25	110.25	$0.43_{(1,96)}$	$0.27_{(1,1)}$	$0.27_{(1,1)}$	$0.43_{(1,96)}$
PSN	1	380.25	380.25	$1.50_{(1,96)}$	$0.94_{(1,1)}$	$1.50_{(1,96)}$	$0.94_{(1,1)}$
NHT $\times$ PSN (interaction)	1	404.01	404.01	$1.60_{(1,96)}$	$1.60_{(1,96)}$	$1.60_{(1,96)}$	$1.60_{(1,96)}$
Error	96	24,302.00	253.14				
Total	99	25,196.51					

Corresponding to these four sources, there are three null hypotheses that may be tested:

1. H_0: no main effect of NHT.
2. H_0: no main effect of PSN.
3. H_0: no interaction effect between NHT and PSN.

Each of these hypotheses can be stated more precisely in terms of population cell means and/or variances, depending on whether the fixed-, random-, or mixed-effects case applies. For example, when the fixed-effects case is considered, the null hypotheses may be given in terms of cell means (see Table 19.2) as follows:

1. H_0: $\mu_{1.} = \mu_{2.}$ (no main effect of NHT).
2. H_0: $\mu_{.1} = \mu_{.2}$ (no main effect of PSN).
3. H_0: $\mu_{22} - \mu_{21} - \mu_{12} + \mu_{11} = 0$ (no interaction effect between NHT and PSN).

More will be said about the null hypotheses being tested in the fixed-, random-, and mixed-effects cases in Sections 19.3 and 19.7. Also, we shall describe later (Section 19.3) how the SS and df terms are determined for the general two-way ANOVA case. For now, we focus entirely on the *F* statistics given in Table 19.3, which differ according to the factor classification schemes. The two numbers in parentheses next to any particular *F* statistic indicate the appropriate degrees of freedom to be used for that *F* test. None of the tests turns out to be significant, as might be expected from the previous results for one-way ANOVA.

Fixed-effects tests

　　Each *F* test for the fixed-effects case *always* involves dividing the MS for the particular source being considered by the *MS*(*error*). The degrees of freedom correspond to the particular mean squares that are used. Thus,

$$F(\text{NHT}) = \frac{MS(\text{NHT})}{MS(\text{error})} = \frac{110.25}{253.14} = 0.43_{(1,96)}$$

$$F(\text{PSN}) = \frac{MS(\text{PSN})}{MS(\text{error})} = \frac{380.25}{253.14} = 1.50_{(1,96)}$$

$$F(\text{interaction}) = \frac{MS(\text{interaction})}{MS(\text{error})} = \frac{404.01}{253.14} = 1.60_{(1,96)}$$

Random-effects tests

In the random-effects case, the F test for each main effect involves dividing the MS for the particular main effect being considered by the $MS(interaction)$. Again, the degrees of freedom are based on the particular mean squares being used. Thus,

$$F(\text{NHT}) = \frac{\text{MS(NHT)}}{\text{MS(interaction)}} = \frac{110.25}{404.01} = 0.27_{(1,1)}$$

$$F(\text{PSN}) = \frac{\text{MS(PSN)}}{\text{MS(interaction)}} = \frac{380.25}{404.01} = 0.94_{(1,1)}$$

The F test for interaction is the same for the random-effects case as for the fixed-effects case.

Mixed-effects tests (*NHT fixed, PSN random*)

In the mixed-effects case, the F test for the main effect of the fixed factor involves dividing the MS for that factor by the $MS(interaction)$. The F test for the main effect of the random factor involves dividing the MS for that factor by the $MS(error)$. Thus, for the case when NHT is fixed and PSN is random, we compute

$$F(\text{NHT}) = \frac{\text{MS(NHT)}}{\text{MS(interaction)}} = \frac{110.25}{404.01} = 0.27_{(1,1)}$$

$$F(\text{PSN}) = \frac{\text{MS(PSN)}}{\text{MS(error)}} = \frac{380.25}{253.14} = 1.50_{(1,96)}$$

The F test for interaction is the same as for the fixed-effects and random-effects cases.

Mixed-effects tests (*NHT random, PSN fixed*)

In this situation the fixed and random factors have simply been reversed from the previous case. It thus follows that

$$F(\text{NHT}) = \frac{\text{MS(NHT)}}{\text{MS(error)}} = \frac{110.25}{253.14} = 0.43_{(1,96)}$$

$$F(\text{PSN}) = \frac{\text{MS(PSN)}}{\text{MS(interaction)}} = \frac{380.25}{404.01} = 0.94_{(1,1)}$$

Again, the F test for interaction is the same as for the other cases.

19.3 GENERAL METHODOLOGY

In this section we shall describe the data configuration, computational formulas, and ANOVA table for the general *balanced* (i.e., equal cell numbers) two-way situation, for which there are r levels of one factor (which we call the *row factor*), c levels of the other factor (which we call the *column factor*), and n observations in each of the rc cells.

19.3.1 GENERAL DATA LAYOUT FOR TWO-WAY ANOVA

As with one-way ANOVA, the computations necessary for a two-way ANOVA are easily performed using an ordinary desk calculator when the data are suitably arranged. Table 19.4 gives a useful way of presenting the data for the general two-way situation when there are equal cell numbers. Table 19.5 gives the corresponding table of (sample) cell means.

In Table 19.4 we have used three subscripts to differentiate among the individual observations. The first two subscripts index the row and column (i.e., the cell), and the third subscript denotes the observation number within the given cell. For example, Y_{122} denotes the second observation in cell (1, 2) corresponding to row 1 and column 2.

TABLE 19.4   *General data layout for two-way ANOVA with equal cell numbers*

ROW FACTOR	COLUMN FACTOR 1	COLUMN FACTOR 2	...	COLUMN FACTOR c	ROW TOTALS
1	$(Y_{111}, Y_{112}, \ldots, Y_{11n})$ T_{11}	$(Y_{121}, Y_{122}, \ldots, Y_{12n})$ T_{12}	...	$(Y_{1c1}, Y_{1c2}, \ldots, Y_{1cn})$ T_{1c}	R_1
2	$(Y_{211}, Y_{212}, \ldots, Y_{21n})$ T_{21}	$(Y_{221}, Y_{222}, \ldots, Y_{22n})$ T_{22}	...	$(Y_{2c1}, Y_{2c2}, \ldots, Y_{2cn})$ T_{2c}	R_2
⋮	⋮	⋮	⋮	⋮	⋮
r	$(Y_{r11}, Y_{r12}, \ldots, Y_{r1n})$ T_{r1}	$(Y_{r21}, Y_{r22}, \ldots, Y_{r2n})$ T_{r2}	...	$(Y_{rc1}, Y_{rc2}, \ldots, Y_{rcn})$ T_{rc}	R_r
Column Totals	C_1	C_2	...	C_c	G

TABLE 19.5   *Sample cell means for two-way ANOVA*

ROW FACTOR	COLUMN FACTOR 1	COLUMN FACTOR 2	...	COLUMN FACTOR c	ROW MEANS
1	$\bar{Y}_{11\cdot}$	$\bar{Y}_{12\cdot}$	...	$\bar{Y}_{1c\cdot}$	$\bar{Y}_{1\cdot\cdot}$
2	$\bar{Y}_{21\cdot}$	$\bar{Y}_{22\cdot}$	...	$\bar{Y}_{2c\cdot}$	$\bar{Y}_{2\cdot\cdot}$
⋮	⋮	⋮		⋮	⋮
r	$\bar{Y}_{r1\cdot}$	$\bar{Y}_{r2\cdot}$	...	$\bar{Y}_{rc\cdot}$	$\bar{Y}_{r\cdot\cdot}$
Column Means	$\bar{Y}_{\cdot1\cdot}$	$\bar{Y}_{\cdot2\cdot}$	...	$\bar{Y}_{\cdot c\cdot}$	$\bar{Y}_{\cdot\cdot\cdot}$

In general, Y_{ijk} denotes the kth observation in the (i, j)th cell of the table. Also, the cell total for the (i, j)th cell is denoted by T_{ij}, the ith row total is R_i, the jth column total is C_j, and the grand total is G. In other words,

$$R_i = \sum_{j=1}^{c} \sum_{k=1}^{n} Y_{ijk}, \qquad C_j = \sum_{i=1}^{r} \sum_{k=1}^{n} Y_{ijk}, \qquad G = \sum_{i=1}^{r} \sum_{j=1}^{c} \sum_{k=1}^{n} Y_{ijk}$$

In Table 19.5 we have denoted the mean of the n observations in cell (i, j) by $\bar{Y}_{ij\cdot}$. This sample mean estimates the population cell mean μ_{ij}. Also, the ith row mean is $\bar{Y}_{i\cdot\cdot}$, the jth column mean is $\bar{Y}_{\cdot j\cdot}$, and the grand (overall) mean is $\bar{Y}_{\cdot\cdot\cdot}$. Thus, we have

$$\bar{Y}_{ij\cdot} = \frac{1}{n} \sum_{k=1}^{n} Y_{ijk}, \qquad \bar{Y}_{i\cdot\cdot} = \frac{R_i}{cn}, \qquad \bar{Y}_{\cdot j\cdot} = \frac{C_j}{rn}, \qquad \bar{Y}_{\cdot\cdot\cdot} = \frac{G}{rcn}$$

EXAMPLE In our earlier example (Table 19.1), $r = c = 2$ and $n = 25$. Tables 19.6 and 19.7 give the data layout and table of cell means for an example for which $r = 3$, $c = 3$, and $n = 12$. This example (although artificial) considers one kind of data set being examined in occupational health studies aimed at evaluating the health status of industrial workers. The dependent variable here is forced expiratory volume (FEV), which is a measure of respiratory function. Very low FEV indicates possible respiratory dysfunction, whereas high FEV indicates good respiratory function. In this example observations are taken on $n = 12$ individuals in each of three plants in a given industry whose occupational history may be characterized by exposure primarily to one of three toxicological substances. Thus, we have two factors, each with three levels. The categories of the row factor

TABLE 19.6 *Forced expiratory volumes (FEV) classified by plant and toxicological exposure*

PLANT NUMBER	TOXIC SUBSTANCE			ROW TOTALS
	A	B	C	
1	4.64, 5.92, 5.25 6.17, 4.20, 5.90 5.07, 4.13, 4.07 5.30, 4.37, 3.76 ($T_{11} = 58.78$)	3.21, 3.17, 3.88 3.50, 2.47, 4.12 3.51, 3.85, 4.22 3.07, 3.62, 2.95 ($T_{12} = 41.57$)	3.75, 2.50, 2.65 2.84, 3.09, 2.90 2.62, 2.75, 3.10 1.99, 2.42, 2.37 ($T_{13} = 32.98$)	$R_1 = 133.33$
2	5.12, 6.10, 4.85 4.72, 5.36, 5.41 5.31, 4.78, 5.08 4.97, 5.85, 5.26 ($T_{21} = 62.81$)	3.92, 3.75, 4.01 4.64, 3.63, 3.46 4.01, 3.39, 3.78 3.51, 3.19, 4.04 ($T_{22} = 45.33$)	2.95, 3.21, 3.15 3.25, 2.30, 2.76 3.01, 2.31, 2.50 2.02, 2.64, 2.27 ($T_{23} = 32.37$)	$R_2 = 140.51$
3	4.64, 4.32, 4.13 5.17, 3.77, 3.85 4.12, 5.07, 3.25 3.49, 3.65, 4.10 ($T_{31} = 49.56$)	4.95, 5.22, 5.16 5.35, 4.35, 4.89 5.61, 4.98, 5.77 5.23, 4.86, 5.15 ($T_{32} = 61.52$)	2.95, 2.80, 3.63 3.85, 2.19, 3.32 2.68, 3.35, 3.12 4.11, 2.90, 2.75 ($T_{33} = 37.65$)	$R_3 = 148.73$
Column Totals	$C_1 = 171.15$	$C_2 = 148.42$	$C_3 = 103.00$	$G = 422.57$

TABLE 19.7 *Cell means for FEV data of Table 19.6*

PLANT NUMBER	TOXIC SUBSTANCE			ROW MEANS
	A	B	C	
1	4.90	3.46	2.75	3.70
2	5.23	3.78	2.70	3.90
3	4.13	5.13	3.13	4.13
Column Means	4.75	4.12	2.86	3.91

(PLANT) are labeled 1, 2, and 3 in Table 19.6. The categories of the column factor (TOXSUB) are labeled A, B, and C. Among the questions of interest here are:

1. Does the mean FEV level differ among plants (i.e., is there a main effect due to PLANT)?

2. Does the mean FEV level differ among exposure categories (i.e., is there a main effect due to TOXSUB)?

3. Do the differences in mean FEV levels among plants depend on the exposure category, and vice versa (i.e., is there an interaction effect between PLANT and TOXSUB)?

A preliminary evaluation regarding these questions can be made by examining the cell means in Table 19.7. Of the three plants, plant 1 has the lowest mean FEV (3.70), followed by plant 2 (3.90), and then plant 3 (4.13). This suggests that the workers in plant 1 might have poorer respiratory health than those in plant 2, and so on. Nevertheless, if the 3.70 value for plant 1 is considered clinically normal, then, despite the differences observed, all plants would be given a "clean bill of health." Furthermore, these differences might have occurred solely by chance (i.e., might not be statistically significant).

With regard to toxicological exposure, it can be seen from Table 19.7 that exposure to substance C is associated with the poorest respiratory health (2.86), whereas exposures to substances B (4.12) and A (4.75) are associated with considerably better respiratory health. Again, it is necessary to decide whether the 2.86 value should be considered meaningfully low in a practical sense, and to determine whether the differences among substances A, B, and C are statistically significant.

Finally, it can be observed from Table 19.7 that the differences among plants depend somewhat on the toxicological exposure being considered. For example, when considering toxic substance A, plant 3 has the lowest mean FEV (4.13). However, for toxic substance B, plant 1 has the lowest mean (3.46); and, for toxic substance C, plant 2 has the lowest (2.70). Furthermore, the magnitude of the differences among plants also varies with the toxic substance. For toxic substance B, the difference between the highest and lowest plant means is $5.13 - 3.46 = 1.67$, whereas for toxic substances A and C the maximum differences are smaller ($5.23 - 4.13 = 1.10$ and $3.13 - 2.70 = 0.43$, respectively). Such fluctuations in these differences suggest the possibility of a significant *interaction effect*, although this must be verified statistically.

19.3.2 ANOVA TABLE FOR TWO-WAY ANOVA

Table 19.8 gives the general form of the two-way ANOVA table when there are r levels of the row factor and c levels of the column factor. Table 19.9 gives the corresponding ANOVA table associated with the FEV data of Table 19.6. From these tables it can be seen that the total (corrected) sum of squares (TSS) has been split up into the four components SSR (rows), SSC (columns), SSRC (R×C interaction), and SSE (error), based on the following fundamental equation:

TSS(total) = SSR(rows) + SSC(columns) + SSRC(R×C interaction) + SSE(error)

or

$$\sum_{i=1}^{r} \sum_{j=1}^{c} \sum_{k=1}^{n} (Y_{ijk} - \bar{Y}...)^2$$

$$= \sum_{i=1}^{r} \sum_{j=1}^{c} \sum_{k=1}^{n} (\bar{Y}_{i..} - \bar{Y}...)^2 + \sum_{i=1}^{r} \sum_{j=1}^{c} \sum_{k=1}^{n} (\bar{Y}_{.j.} - \bar{Y}...)^2$$

$$+ \sum_{i=1}^{r} \sum_{j=1}^{c} \sum_{k=1}^{n} (\bar{Y}_{ij.} - \bar{Y}_{i..} - \bar{Y}_{.j.} + \bar{Y}...)^2 + \sum_{i=1}^{r} \sum_{j=1}^{c} \sum_{k=1}^{n} (Y_{ijk} - \bar{Y}_{ij.})^2 \qquad (19.1)$$

or, equivalently,

$$\sum_{i=1}^{r} \sum_{j=1}^{c} \sum_{k=1}^{n} (Y_{ijk} - \bar{Y}...)^2$$

$$= cn \sum_{i=1}^{r} (\bar{Y}_{i..} - \bar{Y}...)^2 + rn \sum_{j=1}^{n} (\bar{Y}_{.j.} - \bar{Y}...)^2$$

$$+ n \sum_{i=1}^{r} \sum_{j=1}^{c} (\bar{Y}_{ij.} - \bar{Y}_{i..} - \bar{Y}_{.j.} + \bar{Y}...)^2 + \sum_{i=1}^{r} \sum_{j=1}^{c} \sum_{k=1}^{n} (Y_{ijk} - \bar{Y}_{ij.})^2$$

Computational formulas for the sums of squares for each of these sources are given in the SS column of Table 19.8. These formulas involve the use of the grand total (G), the row totals (R_i), the column totals (C_j), the cell totals (T_{ij}), and the individual observations (Y_{ijk}). An efficient stepwise procedure for making these calculations is as follows:

1. Calculate

$$TSS = \sum_{i=1}^{r} \sum_{j=1}^{c} \sum_{k=1}^{n} Y_{ijk}^2 - G^2/rcn$$

$$SSR = \frac{1}{cn} \sum_{i=1}^{r} R_i^2 - G^2/rcn$$

$$SSC = \frac{1}{rn} \sum_{j=1}^{c} C_j^2 - G^2/rcn \qquad (19.2)$$

2. Calculate the sum of squares for cells as

$$SS(cells) = \frac{1}{n} \sum_{i=1}^{r} \sum_{j=1}^{c} T_{ij}^2 - \frac{G^2}{rcn}$$

[This sum of squares, given by the general formula

$$SS(cells) = \sum_{i=1}^{r} \sum_{j=1}^{r} \sum_{k=1}^{n} (\bar{Y}_{ij.} - \bar{Y}...)^2$$

is a measure of the variation in the cell means about the overall mean.]

TABLE 19.8 *General (balanced) two-way ANOVA table*

SOURCE	df	SS	MS	F(fixed)	F(random)	$F\left(\dfrac{\text{row fixed}}{\text{col. random}}\right)$	$F\left(\dfrac{\text{row random}}{\text{col. fixed}}\right)$
Row (main effect)	$r-1$	$\text{SSR} = \dfrac{1}{cn}\sum\limits_{i=1}^{r} R_i^2 - \dfrac{G^2}{rcn}$	$\text{MSR} = \text{SSR}/(r-1)$	MSR/MSE	MSR/MSRC	MSR/MSRC	MSR/MSE
Column (main effect)	$c-1$	$\text{SSC} = \dfrac{1}{rn}\sum\limits_{j=1}^{c} C_j^2 - \dfrac{G^2}{rcn}$	$\text{MSC} = \text{SSC}/(c-1)$	MSC/MSE	MSC/MSRC	MSC/MSE	MSC/MSRC
Row × column (interaction)	$(r-1)(c-1)$	$\text{SSRC} = \dfrac{1}{n}\sum\limits_{i=1}^{r}\sum\limits_{j=1}^{c} T_{ij}^2$ $- \text{SSR} - \text{SSC} - \dfrac{G^2}{rcn}$	$\text{MSRC} =$ $\text{SSRC}/(r-1)(c-1)$	MSRC/MSE	MSRC/MSE	MSRC/MSE	MSRC/MSE
Error	$rc(n-1)$	$\text{SSE} = \text{TSS} - \text{SSR} - \text{SSC} - \text{SSRC}$	$\text{MSE} = \text{SSE}/rc(n-1)$				
Total	$rcn-1$	$\text{TSS} = \sum\limits_{i=1}^{r}\sum\limits_{j=1}^{c}\sum\limits_{k=1}^{n} Y_{ijk}^2 - \dfrac{G^2}{rcn}$					

TABLE 19.9 *Two-way ANOVA for FEV data of Table 19.6*

SOURCE	df	SS	MS	F(fixed)	F(random)	$F\left(\dfrac{\text{PLANT fixed}}{\text{TOXSUB random}}\right)$	$F\left(\dfrac{\text{PLANT random}}{\text{TOXSUB fixed}}\right)$
PLANT	2	3.299	1.649	$\dfrac{1.649}{0.2684}=6.14^{**}$	$\dfrac{1.649}{6.128}=0.27$	$\dfrac{1.649}{6.128}=0.27$	$\dfrac{1.649}{0.2684}=6.14^{**}$
TOXSUB	2	66.889	33.445	$\dfrac{33.445}{0.2684}=124.60^{**}$	$\dfrac{33.445}{6.128}=5.46$	$\dfrac{33.445}{0.2684}=124.60^{**}$	$\dfrac{33.445}{6.128}=5.46$
PLANT × TOXSUB	4	24.510	6.128	$\dfrac{6.128}{0.2684}=22.83^{**}$	$\dfrac{6.128}{0.2684}=22.83^{**}$	$\dfrac{6.128}{0.2684}=22.83^{**}$	$\dfrac{6.128}{0.2684}=22.83^{**}$
Error	99	26.576	0.2684				
Total	107	121.274					

3. Calculate SSRC by subtraction as

$$\text{SSRC} = \text{SS(cells)} - \text{SSR} - \text{SSC} \tag{19.3}$$

4. Calculate SSE by subtraction as

$$\text{SSE} = \text{TSS} - \text{SSR} - \text{SSC} - \text{SSRC} = \text{TSS} - \text{SS(cells)} \tag{19.4}$$

For the FEV data of Table 19.6, the procedure is illustrated as follows:

1.
$$\text{TSS} = \sum_{i=1}^{r} \sum_{j=1}^{c} \sum_{k=1}^{n} Y_{ijk}^2 - \frac{G^2}{rcn}$$

$$= \underbrace{[(4.64)^2 + (5.92)^2 + \cdots + (2.90)^2 + (2.75)^2]}_{\text{a sum of 108 terms}} - \frac{(422.57)^2}{108}$$

$$= 1{,}774.657 - 1{,}653.383$$

$$= 121.274$$

$$\text{SSR(PLANT)} = \frac{1}{cn} \sum_{i=1}^{r} R_i^2 - \frac{G^2}{rcn}$$

$$= \tfrac{1}{36}[(133.33)^2 + (140.51)^2 + (148.73)^2] - 1{,}653.383$$

$$= 3.299$$

$$\text{SSC(TOXSUB)} = \frac{1}{rn} \sum_{j=1}^{c} C_j^2 - \frac{G^2}{rcn}$$

$$= \tfrac{1}{36}[(171.15)^2 + (148.42)^2 + (103.00)^2] - 1{,}653.383$$

$$= 66.889$$

2.
$$\text{SS(cells)} = \frac{1}{n} \sum_{i=1}^{r} \sum_{j=1}^{c} T_{ij}^2 - \frac{G^2}{rcn}$$

$$= \tfrac{1}{12}[(58.78)^2 + (41.57)^2 + \cdots + (61.52)^2 + (37.65)^2] - 1{,}653.383$$

$$= 94.698$$

3.
$$\text{SSRC(PLANT} \times \text{TOXSUB)} = \text{SS(cells)} - \text{SSR} - \text{SSC}$$

$$= 94.698 - 3.299 - 66.889$$

$$= 24.510$$

4.
$$\text{SSE} = \text{TSS} - \text{SS(cells)}$$

$$= 121.274 - 94.698$$

$$= 26.576$$

The degrees of freedom associated with these sums of squares are as follows:

$$\left\{ \begin{array}{l} \text{SSR has } (r-1) \text{ df} \\ \text{SSC has } (c-1) \text{ df} \\ \text{SSRC has } (r-1)(c-1) \text{ df} \\ \text{SSE has } rc(n-1) \text{ df} \\ \text{TSS has } (rcn-1) \text{ df} \end{array} \right. \tag{19.5}$$

Each mean-square term is obtained (as usual) by dividing the corresponding sum of squares by its associated degrees of freedom. The appropriate F statistics to use, as discussed earlier, will depend on the classification of the row and column factors as being fixed or random. These are described in the next section.

19.4 F TESTS FOR TWO-WAY ANOVA

The *null hypotheses* of interest for two-way ANOVA, as well as the basic *statistical assumptions* required for validly testing them, can be stated quite generally so as to encompass the four possible types of situations resulting from each factor possibly being classified as fixed or random.[1] These are given as follows:

The null hypotheses

1. $H_0(R)$: *There is no row-factor (main) effect* (i.e., there are no differences among the effects of the levels of the row factor).
2. $H_0(C)$: *There is no column-factor (main) effect* (i.e., there are no differences among the effects of the levels of the column factor).
3. $H_0(RC)$: *There is no interaction effect between rows and columns* (i.e., the row-level effects within any one column are the same as within any other column, and the column-level effects within any one row are the same as within any other row).

The assumptions

1. All observations are *statistically independent* of one another.
2. Each observation comes from a *normally distributed population*.
3. Each observation has the same population variance (i.e., there is the usual assumption of *variance homogeneity*).

As previously stated, the choice of appropriate F statistics depends on how the row and column factors have been classified as being fixed or random. We will see later, when specifying the different statistical models for two-way ANOVA, that the mean-square term associated with a given source of variation will estimate different quantities, depending on the classification of the row and column factors as being fixed or random, and that this accounts for the different denominators used in the various F tests of two-way ANOVA.

Nevertheless, regardless of the factor classification scheme, the F statistic used to test $H_0(RC)$ of no row × column interaction is always of the form

$$F(RC) = \frac{MSRC}{MSE}$$

with $(r-1)(c-1)$ and $rc(n-1)$ degrees of freedom.[2]

---

[1] However, the null hypotheses are quite different when more *precisely* stated in terms of population cell means and/or variances.

[2] $F(RC)$ denotes the F test of $H_0(RC)$. Similarly, $F(R)$ and $F(C)$ will denote the F tests of $H_0(R)$ and $H_0(C)$, respectively.

The tests for main effects, however, differ as follows with respect to the factor classification scheme:

1. *Rows and columns fixed*. Divide the mean squares for rows and columns by the mean square for error:

$$F(R) = \frac{MSR}{MSE}$$

with $(r-1)$ and $rc(n-1)$ df's;

$$F(C) = \frac{MSC}{MSE}$$

with $(c-1)$ and $rc(n-1)$ df's.

2. *Rows and columns random*. Divide the mean squares for rows and columns by the mean square for interaction:

$$F(R) = \frac{MSR}{MSRC}$$

with $(r-1)$ and $(r-1)(c-1)$ df's;

$$F(C) = \frac{MSC}{MSRC}$$

with $(c-1)$ and $(r-1)(c-1)$ df's.

3. *Rows fixed and columns random*. Divide the mean square for rows by the mean square for interaction:

$$F(R) = \frac{MSR}{MSRC}$$

with $(r-1)$ and $(r-1)(c-1)$ df's. Divide the mean square for columns by the mean square for error:

$$F(C) = \frac{MSC}{MSE}$$

with $(c-1)$ and $rc(n-1)$ df's.

4. *Rows random and columns fixed*. Divide the mean square for rows by the mean square for error:

$$F(R) = \frac{MSR}{MSE}$$

with $(r-1)$ and $rc(n-1)$ df's. Divide the mean square for columns by the mean square for interaction:

$$F(C) = \frac{MSC}{MSRC}$$

with $(c-1)$ and $(r-1)(c-1)$ df's.

With regard to the FEV data, the classification of the factors would, as in the previous examples, depend on the point of view of the researcher. If, for example, the plants and toxicological substances were the only ones of interest, both factors would be considered as fixed. However, if the plants were considered to be a sample from a large population of plants of interest and if the toxicological substances were representative of a population of toxic agents of interest, both factors would be considered to be random. Also, the classification would be mixed if one of these factors was considered fixed and the other random.

We will not attempt here to defend any particular choice of classification scheme for the factors in the FEV example, especially since the example is artificial to begin with. However, it is important to notice from Table 19.9 that the decisions regarding certain null hypotheses will be different depending on the way in which the factors are classified. This can be seen from Table 19.9 as follows:

1. *Both factors fixed*. Both main effects are significant, since $F(\text{PLANT}) = 6.14^{**}$ (with 2 and 99 df's) and $F(\text{TOXSUB}) = 124.60^{**}$ (with 2 and 99 df's).

2. *Both factors random*. Neither main effect is significant, since $F(\text{PLANT}) = 0.27$ (with 2 and 4 df's) and $F(\text{TOXSUB}) = 5.46$ (with 2 and 4 df's).

3. *PLANT fixed, TOXSUB random*. The PLANT main effect is not significant and the TOXSUB main effect is significant, since $F(\text{PLANT}) = 0.27$ (with 2 and 4 df's) and $F(\text{TOXSUB}) = 124.60^{**}$ (with 2 and 99 df's).

4. *PLANT random, TOXSUB fixed*. The PLANT main effect is significant and the TOXSUB main effect is not significant, since $F(\text{PLANT}) = 6.14^{**}$ (with 2 and 99 df's) and $F(\text{TOXSUB}) = 5.46$ (with 2 and 4 df's).

Nevertheless, despite these differences among the main-effect test results, the most important finding from this analysis is that the interaction effect is significant, the F statistic being

$$F(\text{PLANT} \times \text{TOXSUB}) = 22.83^{**} \text{ (with 4 and 99 df's)}$$

In Section 19.6 we shall discuss the interpretation of such interaction effects. For now it suffices to say that the presence of this (PLANT × TOXSUB) interaction means that it does not make much sense to talk about the separate or independent effects (i.e., main effects) of PLANT and TOXSUB on FEV, since there is strong evidence that these factors do not affect FEV independently of one another.

19.5 REGRESSION MODEL FOR FIXED-EFFECTS TWO-WAY ANOVA

In this section we shall describe a particular regression model[3] and a related "classical" fixed-effects ANOVA model for two-way ANOVA when both factors are considered to

---

[3] Several other alternative regression models are definable, of course, depending on the coding choice for the dummy variables. The regression model given here is the one most commonly used because of its natural connection with the "classical" fixed-effects two-way ANOVA model.

TABLE 19.10   *Table of population cell means for two-way layout*

ROWS	COLUMNS 1	2	$\cdots$	c	ROW MEANS
1	μ_{11}	μ_{12}	$\cdots$	μ_{1c}	$\mu_{1\cdot}$
2	μ_{21}	μ_{22}	$\cdots$	μ_{2c}	$\mu_{2\cdot}$
$\vdots$	$\vdots$	$\vdots$		$\vdots$	$\vdots$
r	μ_{r1}	μ_{r2}	$\cdots$	μ_{rc}	$\mu_{r\cdot}$
Column Means	$\mu_{\cdot1}$	$\mu_{\cdot2}$	$\cdots$	$\mu_{\cdot c}$	$\mu_{\cdot\cdot}$

be fixed.[4] As in the one-way ANOVA and randomized-blocks ANOVA cases, a regression model for two-way ANOVA can be interpreted in terms of the cell, marginal, and overall means associated with the two-way layout (see Table 19.10). In the table

$$\mu_{i\cdot} = \frac{1}{c} \sum_{j=1}^{c} \mu_{ij}, \qquad i = 1, 2, \ldots, r$$

$$\mu_{\cdot j} = \frac{1}{r} \sum_{i=1}^{r} \mu_{ij}, \qquad j = 1, 2, \ldots, c$$

$$\mu_{\cdot\cdot} = \frac{1}{rc} \sum_{i=1}^{r} \sum_{j=1}^{c} \mu_{ij}$$

19.5.1 REGRESSION MODEL

When there are r rows and c columns, a regression model can be formulated involving $(r-1)$ dummy variables for the row factor, $(c-1)$ dummy variables for the column factor, and $(r-1)(c-1)$ interaction dummy variables constructed by forming products of each of the row dummy variables with each of the column dummy variables. Such a model can be expressed as follows:

$$Y = \mu + \sum_{i=1}^{r-1} \alpha_i X_i + \sum_{j=1}^{c-1} \beta_j Z_j + \sum_{i=1}^{r-1} \sum_{j=1}^{c-1} \gamma_{ij} X_i Z_j + E \qquad (19.6)$$

where

$$X_i = \begin{cases} -1 & \text{for level } r \text{ of the row factor} \\ 1 & \text{for level } i \text{ of the row factor} \\ 0 & \text{otherwise} \end{cases}$$

[4] Random-effects and mixed-effects models are discussed in detail in Section 19.7.

and

$$Z_j = \begin{cases} -1 & \text{for level } c \text{ of the column factor} \\ 1 & \text{for level } j \text{ of the column factor} \\ 0 & \text{otherwise} \end{cases}$$

($i = 1, 2, \ldots, r-1; j = 1, 2, \ldots, c-1$).

The formulas relating the coefficients α_i, β_j, and γ_{ij} to the various means of Table 19.10 are given as follows:

$$\begin{cases} u = \mu_{..} \\[4pt] \alpha_i = \mu_{i.} - \mu_{..}, & i = 1, \ldots, r-1 \\[4pt] \beta_j = \mu_{.j} - \mu_{..}, & j = 1, \ldots, c-1 \\[4pt] \gamma_{ij} = \mu_{ij} - \mu_{i.} - \mu_{.j} + \mu_{..}, & i = 1, \ldots, r-1; j = 1, \ldots, c-1 \\[4pt] -\sum_{i=1}^{r-1} \alpha_i = \mu_{r.} - \mu_{..} \\[4pt] -\sum_{j=1}^{c-1} \beta_j = \mu_{.c} - \mu_{..} \\[4pt] -\sum_{i=1}^{r-1} \gamma_{ij} = \mu_{rj} - \mu_{r.} - \mu_{.j} + \mu_{..}, & j = 1, \ldots, c-1 \\[4pt] -\sum_{j=1}^{c-1} \gamma_{ij} = \mu_{ic} - \mu_{i.} - \mu_{.c} + \mu_{..}, & i = 1, \ldots, r-1 \end{cases} \tag{19.7}$$

As with the other ANOVA-regression analogies made in earlier chapters, the same F tests as given in Table 19.8 when both factors are fixed can be obtained using the appropriate multiple-partial F tests concerning subsets of the coefficients in the regression model (19.6). Specifically, the multiple-partial F test of H_0: $\alpha_1 = \cdots = \alpha_{r-1} = 0$ for model (19.6) yields exactly the same F statistic as that used in standard (balanced) two-way fixed-effects ANOVA for testing the significance of the row-factor main effect (i.e., $F = \text{MSR/MSE}$). Similarly, the multiple-partial F test of H_0: $\beta_1 = \cdots = \beta_{c-1} = 0$ in model (19.6) yields exactly the same F statistic as that used in standard (balanced) two-way fixed-effects ANOVA for testing the significance of the column-factor main effect (i.e., $F = \text{MSC/MSE}$). Finally, the multiple-partial F test of H_0: $\gamma_{ij} = 0$ ($i = 1, \ldots, r-1$ and $j = 1, \ldots, c-1$) is identical to the (balanced) two-way fixed-effects ANOVA F test for interaction (i.e., $F = \text{MSRC/MSE}$).

19.5.2 CLASSICAL TWO-WAY FIXED-EFFECTS ANOVA MODEL

When both factors are considered fixed, there are three types of effects to be considered:

1. *Row-factor main effects*, which are the differences between the various row means and the overall mean (i.e., $\mu_{i.} - \mu_{..}$, $i = 1, 2, \ldots, r$).

2. *Column-factor main effects*, which are the differences between the various column means and the overall mean (i.e., $\mu_{.j} - \mu_{..}$, $j = 1, 2, \ldots, c$).

3. *Interaction effects*, which are differences between differences, of the form $(\mu_{ij} - \mu_{i.}) - (\mu_{.j} - \mu_{..})$ or $(\mu_{ij} - \mu_{.j}) - (\mu_{i.} - \mu_{..})$, $i = 1, 2, \ldots, r$ and $j = 1, 2, \ldots, c$.

The "classical" two-way fixed-effects ANOVA model involving such effects is of the following form:

$$Y_{ijk} = \mu + \alpha_i + \beta_j + \gamma_{ij} + E_{ijk} \tag{19.8}$$

where

$$\mu = \mu_{..} = \text{overall mean}$$

$$\alpha_i = \mu_{i.} - \mu_{..} = \text{effect of row } i$$

$$\beta_j = \mu_{.j} - \mu_{..} = \text{effect of column } j$$

$$\gamma_{ij} = \mu_{ij} - \mu_{i.} - \mu_{.j} + \mu_{..} = \text{interaction effect associated with cell } (i, j)$$

$$E_{ijk} = Y_{ijk} - \mu - \alpha_i - \beta_j - \gamma_{ij} = \text{error (or residual) associated with the } k\text{th}$$
$$\text{observation in cell } (i, j)$$

$(i = 1, 2, \ldots, r; j = 1, 2, \ldots, c; k = 1, 2, \ldots, n)$.

The following relationships are clearly satisfied by the effects in the model above:

$$\sum_{i=1}^{r} \alpha_i = 0, \qquad \sum_{j=1}^{c} \beta_j = 0, \qquad \sum_{i=1}^{r} \gamma_{ij} = 0, \qquad \sum_{j=1}^{c} \gamma_{ij} = 0 \tag{19.9}$$

It is also clear when comparing the regression coefficients in model (19.6) with the ANOVA effects in model (19.8) that the models are completely equivalent.

Finally, it should be pointed out that each of the effects in model (19.8) can be simply estimated using sample means, as follows:

$$\begin{cases} \hat{\mu} = \bar{Y}_{...} \\ \hat{\alpha}_i = \bar{Y}_{i..} - \bar{Y}_{...}, & i = 1, 2, \ldots, r \\ \hat{\beta}_j = \bar{Y}_{.j.} - \bar{Y}_{...}, & j = 1, 2, \ldots, c \\ \hat{\gamma}_{ij} = \bar{Y}_{ij.} - \bar{Y}_{i..} - \bar{Y}_{.j.} + \bar{Y}_{...}, & i = 1, 2, \ldots, r; j = 1, 2, \ldots, c \end{cases} \tag{19.10}$$

19.6 INTERACTIONS IN TWO-WAY ANOVA

In this section we shall describe several ways to look at the concept of *interaction* in the context of two-way ANOVA. We focus, for convenience, on the fixed-effects case. Nevertheless, even though the parameters involved and the test statistics used for making inferences in the fixed-effects case will be different from those used in the random- and mixed-effects cases, the interpretations regarding interactions will be generally the same regardless of whether the factors are fixed or random.

19.6.1 CONCEPT OF INTERACTION

Generally speaking, an interaction exists between two factors if the relationship among the effects associated with the levels of one factor differs according to the levels of the second factor. Another way of saying this is that an interaction represents an effect due to the joint influence of two factors, over and above the effects of each factor considered separately.

More specifically, there are three equivalent ways of describing or representing an interaction in statistical terms if we consider the two-way table of cell means in the fixed-effects case and if we consider the various ways of writing the statistical model in this situation.

METHOD 1   *Interaction as a difference in differences of means*

In the context of two-way ANOVA, an interaction exists between the row and column factors if any of the following equivalent statements is true:

1. For some pair of columns, the difference between the means in these columns for a given row is *not equal* to the difference between these means for some other row. For example, for rows 1 and 2 and columns 1 and 2, $\mu_{11} - \mu_{12} \neq \mu_{21} - \mu_{22}$; or

2. For some pair of rows, the difference between the means in these rows for a given column is *not equal* to the difference between these means for some other column. For example, for rows 1 and 2 and columns 1 and 2, $\mu_{11} - \mu_{21} \neq \mu_{12} - \mu_{22}$; or

3. For some cell in the table, the difference between that cell mean and its associated marginal row mean is *not equal* to the difference between its associated marginal column mean and the overall mean. For example, for the (i, j)th cell, $\mu_{ij} - \mu_{i\cdot} \neq \mu_{\cdot j} - \mu_{\cdot\cdot}$, or $\mu_{ij} - \mu_{i\cdot} - \mu_{\cdot j} + \mu_{\cdot\cdot} \neq 0$; or

4. For some cell in the table, the difference between that cell mean and its associated marginal column mean is *not equal* to the difference between its associated marginal row mean and the overall mean. For example, $\mu_{ij} - \mu_{\cdot j} \neq \mu_{i\cdot} - \mu_{\cdot\cdot}$, or $\mu_{ij} - \mu_{i\cdot} - \mu_{\cdot j} + \mu_{\cdot\cdot} \neq 0$.

Another way of saying all this is that when there is *no interaction*, the relationship among the column effects (β_j's) is the same regardless of the row being considered, and vice versa. Also, since from statements 3 and 4

$$\mu_{ij} - \mu_{i\cdot} - \mu_{\cdot j} + \mu_{\cdot\cdot} = 0$$

when there is no interaction, it should be apparent that

$$MSRC = \frac{n}{(r-1)(c-1)} \sum_{i=1}^{r} \sum_{j=1}^{c} (Y_{ij} - \bar{Y}_{i\cdot\cdot} - \bar{Y}_{\cdot j\cdot} + \bar{Y}_{\cdots})^2$$

which estimates

$$\frac{n}{(r-1)(c-1)} \sum_{i=1}^{r} \sum_{j=1}^{c} (\mu_{ij} - \mu_{i\cdot} - \mu_{\cdot j} + \mu_{\cdot\cdot})^2$$

will be small when there is no interaction and will be large when there is interaction.

METHOD 2   *Interaction as an effect in the fixed-effects model*

An interaction exists if the *appropriate* fixed-effects ANOVA model is of the form

$$Y_{ijk} = \mu + \alpha_i + \beta_j + \gamma_{ij} + E_{ijk}$$

where $\gamma_{ij} \neq 0$ for at least one (i, j) pair.

[Representations 1 and 2 are completely equivalent. For example, if there is no interaction, then $\mu_{ij} = \mu + \alpha_i + \beta_j$, so that $\mu_{1j} - \mu_{2j} = (\mu + \alpha_1 + \beta_j) - (\mu + \alpha_2 + \beta_j) = \alpha_1 - \alpha_2$, which is independent of j. Thus, $\mu_{11} - \mu_{21} = \mu_{12} - \mu_{22} = \cdots = \mu_{1c} - \mu_{2c}$.]

METHOD 3 *Interaction as a term in a regression model*

An interaction exists if the *appropriate* regression model (using dummy variables) contains a term that involves the product (or, in general, any function) of variables from different factors, for example, if the appropriate model is of the form

$$Y = \mu + \sum_{i=1}^{r-1} \alpha_i X_i + \sum_{j=1}^{c-1} \beta_j Z_j + \sum_{i=1}^{r-1}\sum_{j=1}^{c-1} \gamma_{ij} X_i Z_j + E$$

where at least one of the γ_{ij} is not zero.

[When $r = c = 2$, the model simplifies to

$$Y = \mu + \alpha_1 X_1 + \beta_1 Z_1 + \gamma_{11} X_1 Z_1 + E$$

where

$$X_1 = \begin{cases} -1 & \text{if level 2 of the row factor} \\ 1 & \text{if level 1 of the row factor} \end{cases}$$

$$Z_1 = \begin{cases} -1 & \text{if level 2 of the column factor} \\ 1 & \text{if level 1 of the column factor} \end{cases}$$

Then, $\mu = \mu_{..}$, $\alpha_1 = \mu_{1.} - \mu_{..}$, $\beta_1 = \mu_{.1} - \mu_{..}$, and $\gamma_{11} = \mu_{11} - \mu_{1.} - \mu_{.1} + \mu_{..}$.

Representation 3 is equivalent to the other two representations provided that both independent variables (i.e., factors) are considered *nominal* and so are represented by dummy variables. If, however, both independent variables are *continuous*, a regression model with any kind of product term would exhibit an interaction effect of a somewhat different type not necessarily characterized by a nonzero difference of mean differences.]

19.6.2 SOME HYPOTHETICAL EXAMPLES

We shall now consider some (hypothetical) two-way tables of *population* cell means illustrating different patterns with regard to interaction. These tables will pertain to the example in Section 19.2, for which the factors were NHT and PSN and the dependent variable was CMI score. Subsequently, we shall examine the table of sample cell means actually obtained (Table 19.2), keeping in mind that the statistical test for interaction may negate whatever tentative trends are suggested by the sample means. We shall also examine the example of Section 19.3 in this light.

Row and column main effects but no interaction effect

Table 19.11 presents three alternative layouts, each representing the general situation in which there is *both* a row main effect and a column main effect but no

TABLE 19.11   *Main effects but no interaction*

(a)

NHT	PSN		
	LOW	HIGH	
Low	26	23	24.5
High	20	17	18.5
	23	20	21.5

(b)

NHT	PSN		
	LOW	HIGH	
Low	18	26	22
High	20	28	24
	19	27	23

(c)

NHT	PSN		
	LOW	HIGH	
Low	18	26	22
High	16	24	20
	17	25	21

interaction effect. Keep in mind that each of these tables gives *population* (and not sample) mean values, so there is no sampling variation to consider.

The main effects are reflected in the differences between the marginal row means and between the marginal column means in each subtable. The lack of an interaction effect can be established via comparison of the differences among the cell means, as discussed in Section 19.6.1. From Table 19.11a, for example, we have $\mu_{11} - \mu_{12} = \mu_{21} - \mu_{22}$, since $26 - 23 = 3 = 20 - 17$. Also, for this table, $\mu_{11} - \mu_{1.} - \mu_{.1} + \mu_{..} = 26 - 24.5 - 23 + 21.5 = 0$, and similar terms associated with the other three cells in the table are also zero. Furthermore, for this table, the model (19.8) can be shown to have the specific structure:

(a) $\mu_{ij} = 21.5 + \alpha_i + \beta_j$

where

$$\alpha_i = \begin{cases} 3 & \text{if } i = 1 \\ -3 & \text{if } i = 2 \end{cases} \quad \text{and} \quad \beta_j = \begin{cases} 1.5 & \text{if } j = 1 \\ -1.5 & \text{if } j = 2 \end{cases}$$

Note that this model does not involve any γ_{ij} term (i.e., there is no interaction term in the model). Thus, we have

$$\mu_{11} = 21.5 + 3 + 1.5 = 26$$

$$\mu_{12} = 21.5 + 3 - 1.5 = 23$$

$$\mu_{21} = 21.5 - 3 + 1.5 = 20$$

$$\mu_{22} = 21.5 - 3 - 1.5 = 17$$

The models for tables b and c are also "no-interaction" models; they have the particular forms

(b) $\mu_{ij} = 23 + \alpha_i + \beta_j$

where

$$\alpha_i = \begin{cases} -1 & \text{if } i = 1 \\ 1 & \text{if } i = 2 \end{cases} \quad \text{and} \quad \beta_j = \begin{cases} -4 & \text{if } j = 1 \\ 4 & \text{if } j = 2 \end{cases}$$

TABLE 19.12 *One main effect and no interaction*

(a)

NHT	PSN LOW	PSN HIGH	
Low	18	26	22
High	18	26	22
	18	26	22

(b)

NHT	PSN LOW	PSN HIGH	
Low	19	19	19
High	24	24	24
	21.5	21.5	21.5

and (c) $\mu_{ij} = 21 + \alpha_i + \beta_j$

where

$$\alpha_i = \begin{cases} 1 & \text{if } i = 1 \\ -1 & \text{if } i = 2 \end{cases} \quad \text{and} \quad \beta_j = \begin{cases} -4 & \text{if } j = 1 \\ 4 & \text{if } j = 2 \end{cases}$$

Exactly one main effect and no interaction effect

This situation is depicted in Table 19.12. Table 19.12a contains a main effect due to PSN but no main effect due to NHT. Table 19.12b contains a main effect due to NHT but no main effect due to PSN. One can also verify that there is no PSN × NHT interaction.

Same-direction interaction

Three examples of "same-direction" interaction are given in Table 19.13. Focusing on Table 19.13a we can see that $\mu_{11} - \mu_{12} = 26 - 23 = 3$, whereas $\mu_{21} - \mu_{22} = 20 - 13 = 7$. Also, $\mu_{11} - \mu_{1\cdot} = 26 - 24.5 = 1.5$ and $\mu_{\cdot 1} - \mu_{\cdot\cdot} = 23 - 20.5 = 2.5$, so that $\mu_{11} - \mu_{1\cdot} - \mu_{\cdot 1} + \mu_{\cdot\cdot} = -1.0$. The other interactions of the general form $(\mu_{ij} - \mu_{i\cdot} - \mu_{\cdot j} + \mu_{\cdot\cdot})$ are similarly determined to be either +1.0 (when $i = 1$, $j = 2$ or when $i = 2$, $j = 1$) or −1.0 (when $i = 2$, $j = 2$).

These hypothetical results would indicate that in *both* low and high NHT neighborhoods, persons in friendly surroundings (i.e., high PSN) are healthier (i.e., have a lower CMI score) than persons in unfriendly surroundings (i.e., low PSN), but that the extent of this difference would be greater when there is a large number of

TABLE 19.13 *Same-direction interaction*

(a)

NHT	PSN LOW	PSN HIGH	
Low	26	23	24.5
High	20	13	16.5
	23	18	20.5

(b)

NHT	PSN LOW	PSN HIGH	
Low	18	26	22
High	20	36	28
	19	31	25

(c)

NHT	PSN LOW	PSN HIGH	
Low	18	26	22
High	12	24	18
	15	25	20

households (i.e., high NHT) than when there is a small number of households (i.e., low NHT). In other words, then, at each level of NHT, the difference between the PSN level effects is in the "same direction" (i.e., high PSN is associated with a lower mean CMI score than is low PSN), but the magnitude of the difference depends on the NHT level. This is what is meant by "same-direction interaction."

The model for Table 19.13a may be given as follows:

$$\mu_{ij} = 20.5 + \alpha_i + \beta_j + \gamma_{ij}$$

where

$$\alpha_i = \begin{cases} 4.0 & \text{if } i = 1 \\ -4.0 & \text{if } i = 2 \end{cases}, \quad \beta_j = \begin{cases} 2.5 & \text{if } j = 1 \\ -2.5 & \text{if } j = 2 \end{cases},$$

$$\gamma_{ij} = \begin{cases} -1.0 & \text{if } i = 1, j = 1 \\ 1.0 & \text{if } i = 1, j = 2 \\ 1.0 & \text{if } i = 2, j = 1 \\ -1.0 & \text{if } i = 2, j = 2 \end{cases}$$

Thus,

$$\begin{cases} \mu_{11} = 20.5 + 4.0 + 2.5 - 1.0 = 26 \\ \mu_{12} = 20.5 + 4.0 - 2.5 + 1.0 = 23 \\ \mu_{21} = 20.5 - 4.0 + 2.5 + 1.0 = 20 \\ \mu_{22} = 20.5 - 4.0 - 2.5 - 1.0 = 13 \end{cases}$$

The reader may verify that the same type of model holds for tables b and c.

Reverse interaction:

Two examples of *reverse interaction* are given in Table 19.14, where by the term "reverse" we mean that the direction of the difference between two cell means for one row (column) is opposite or reversed from the direction of the difference between the corresponding cell means for some other row (column).

Focusing on Table 19.14a, we can see that $\mu_{11} - \mu_{12} = 18 - 26 = -8$, whereas $\mu_{21} - \mu_{22} = 22 - 20 = +2$. Also, $\mu_{21} - \mu_{2.} = 22 - 21 = +1$ and $\mu_{.1} - \mu_{..} = 20 - 21.5 = -1.5$, so that $\mu_{21} - \mu_{2.} - \mu_{.1} + \mu_{..} = 2.5$.

TABLE 19.14   *Reverse interaction*

(a)

	PSN		
NHT	LOW	HIGH	
Low	18	26	22
High	22	20	21
	20	23	21.5

(b)

	PSN		
NHT	LOW	HIGH	
Low	26	22	24
High	18	24	21
	22	23	22.5

These hypothetical results indicate (for this table) that for neighborhoods with a small number of households (low NHT), persons in unfriendly surroundings (low PSN) are healthier than are persons in friendly surroundings, but that for neighborhoods with a large number of households, this situation is reversed. In other words, the difference between the effects of the high and low PSN levels is positive for low NHT but is negative for high NHT (i.e., there is a reversal in sign), and this is what we mean by "reverse interaction."

The model in this case is given as follows:

$$\mu_{ij} = 21.5 + \alpha_i + \beta_j + \gamma_{ij}$$

where

$$\alpha_i = \begin{cases} 0.5 & \text{if } i = 1 \\ -0.5 & \text{if } i = 2 \end{cases}, \qquad \beta_j = \begin{cases} -1.5 & \text{if } j = 1 \\ 1.5 & \text{if } j = 2 \end{cases}$$

$$\gamma_{ij} = \begin{cases} -2.5 & \text{if } i = 1, j = 1 \\ 2.5 & \text{if } i = 1, j = 2 \\ 2.5 & \text{if } i = 2, j = 1 \\ -2.5 & \text{if } i = 2, j = 2 \end{cases}$$

Thus,

$$\begin{cases} \mu_{11} = 21.5 + 0.5 - 1.5 - 2.5 = 18 \\ \mu_{12} = 21.5 + 0.5 + 1.5 + 2.5 = 26 \\ \mu_{21} = 21.5 - 0.5 - 1.5 + 2.5 = 22 \\ \mu_{22} = 21.5 - 0.5 + 1.5 - 2.5 = 20 \end{cases}$$

It can be shown that a reverse interaction is reflected in Table 19.14b as well.

19.6.3 INTERACTION EFFECTS FOR CMI DATA OF TABLE 19.2

The table of sample cell means actually obtained for the CMI example of Section 19.2 is given in Table 19.15. From this table the following comparisons of differences of means can be made:

$$\bar{Y}_{11\cdot} - \bar{Y}_{12\cdot} = 18.92 - 26.84 = -7.92, \qquad \text{whereas } \bar{Y}_{21\cdot} - \bar{Y}_{22\cdot} = 20.84 - 20.72 = 0.12$$

$$\bar{Y}_{11\cdot} - \bar{Y}_{21\cdot} = 18.92 - 20.84 = -1.92, \qquad \text{whereas } \bar{Y}_{12\cdot} - \bar{Y}_{22\cdot} = 26.84 - 20.72 = 6.12$$

$$\bar{Y}_{11\cdot} - \bar{Y}_{1\cdot\cdot} - \bar{Y}_{\cdot 1\cdot} + \bar{Y}_{\cdots} = 18.92 - 22.88 - 19.88 + 21.83 = -2.01$$

$$\bar{Y}_{12\cdot} - \bar{Y}_{1\cdot\cdot} - \bar{Y}_{\cdot 2\cdot} + \bar{Y}_{\cdots} = 26.84 - 22.88 - 23.78 + 21.83 = 2.01$$

$$\bar{Y}_{21\cdot} - \bar{Y}_{2\cdot\cdot} - \bar{Y}_{\cdot 1\cdot} + \bar{Y}_{\cdots} = 20.84 - 20.78 - 19.88 + 21.83 = 2.01$$

$$\bar{Y}_{22\cdot} - \bar{Y}_{2\cdot\cdot} - \bar{Y}_{\cdot 2\cdot} + \bar{Y}_{\cdots} = 20.72 - 20.78 - 23.78 + 21.83 = -2.01$$

TABLE 19.15   *Cell means for CMI data of Table 19.1*

NHT	PSN LOW	PSN HIGH	
Low	18.92	26.84	22.88
High	20.84	20.72	20.78
	19.88	23.78	21.83

These comparisons suggest the possibility of a reverse type of interaction effect for these data. More specifically, it is indicated that (a) for small Turnkey neighborhoods, persons from friendly surroundings (high PSN) appear to have worse health (higher mean CMI scores) than persons from unfriendly surroundings, but that (b) there is little difference in mean CMI scores for large Turnkey neighborhoods. As mentioned earlier in Section 19.2, this pattern is counter to that expected by Daly. However, these observed differences are subject to sampling variation (i.e., they are sample values and not population values), and, as we know, the test for interaction for these data is not significant.

19.6.4   INTERACTION EFFECTS FOR FEV DATA OF TABLE 19.6

The table of sample cell means for the FEV example (Table 19.6) is as shown in Table 19.16. This table of means is slightly more difficult to interpret than the one for the CMI data, because now there are three rows and columns instead of two. Nevertheless, we can see immediately by observation that the relative magnitudes of the means vary from column to column. For example, for TOXSUB A, the order of PLANTS by increasing mean FEV is 3, 1, and 2. For TOXSUB B, on the other hand, the order is 1, 2, 3. And, for TOXSUB C, the order is 2, 1, 3. These differences in ordering are indicative of the presence of interaction, the significance of which has been established earlier. The following comparisons of cell means should help in interpreting the nature of this significant interaction effect:

$$\bar{Y}_{11\cdot} - \bar{Y}_{12\cdot} = 4.90 - 3.46 = +1.44, \quad \text{whereas}$$

$$\bar{Y}_{31\cdot} - \bar{Y}_{32\cdot} = 4.13 - 5.13 = -1.00,$$

$$\bar{Y}_{21\cdot} - \bar{Y}_{31\cdot} = 5.23 - 4.13 = +1.10 \quad \text{whereas}$$

$$\bar{Y}_{22\cdot} - \bar{Y}_{32\cdot} = 3.78 - 5.13 = -1.35$$

$$\bar{Y}_{21\cdot} - \bar{Y}_{\cdot1\cdot} = 5.23 - 4.75 = +0.48, \quad \text{whereas}$$

$$\bar{Y}_{2\cdot\cdot} - \bar{Y}_{\cdots} = 3.90 - 3.91 = -0.01$$

The set of interaction effects of the form $\hat{\gamma}_{ij} = \bar{Y}_{ij\cdot} - \bar{Y}_{i\cdot\cdot} - \bar{Y}_{\cdot j\cdot} + \bar{Y}_{\cdots}$ is given in Table 19.17. These patterns demonstrate that some plants are associated with better respiratory health than others for one kind of toxic exposure but are worse for other kinds of exposures. It would not be possible, therefore, to conclude that one plant was better "overall" than another. Rather, the differences in respiratory health among plants depend on which toxic substance is being considered.

TABLE 19.16   *Cell means for FEV data of Table 19.6*

PLANT	TOXSUB			
	A	B	C	
1	4.90	3.46	2.75	3.70
2	5.23	3.78	2.70	3.90
3	4.13	5.13	3.13	4.13
Column Means	4.75	4.12	2.86	3.91

TABLE 19.17   *Interaction effects ($\hat{\gamma}_{ij}$'s) for the FEV data*

PLANT	TOXSUB			ROW TOTAL
	A	B	C	
1	0.35	−0.45	0.10	0.00
2	0.49	−0.34	−0.15	0.00
3	−0.84	0.78	0.06	0.00
Column Total	0.00	0.00	0.00	0.00

19.7   RANDOM- AND MIXED-EFFECTS TWO-WAY ANOVA MODELS

In this section we present the classical two-way ANOVA statistical models appropriate when *both factors are random* and when *one factor is fixed and the other is random*. We will specify the appropriate null hypotheses of interest and the expected mean squares (EMS) associated with each model, the EMS for a particular source being the true average (population) value of the MS term in the ANOVA table.

19.7.1   RANDOM-EFFECTS MODEL

When both factors are random, the two-way ANOVA model is given as follows:

$$Y_{ijk} = \mu + A_i + B_j + C_{ij} + E_{ijk} \tag{19.11}$$

where A_i, B_j, C_{ij}, and E_{ijk} are mutually independent random variables satisfying

$$\begin{cases} A_i \frown N(0, \sigma_R^2) \\ B_j \frown N(0, \sigma_C^2) \\ C_{ij} \frown N(0, \sigma_{RC}^2) \\ E_{ijk} \frown N(0, \sigma^2) \end{cases}$$

$(i = 1, 2, \ldots, r; j = 1, 2, \ldots, c; k = 1, 2, \ldots, n)$.

19.7.2 MIXED-EFFECTS MODEL WITH FIXED ROW FACTOR AND RANDOM COLUMN FACTOR

One particular model[5] is

$$Y_{ijk} = \mu + \alpha_i + B_j + C_{ij} + E_{ijk} \tag{19.12}$$

where each α_i is a constant such that $\sum_{i=1}^{r} \alpha_i = 0$, and where B_j, C_{ij}, and E_{ijk} are random variables satisfying $\sum_{i=1}^{r} C_{ij} = 0$ for each j and

$$\begin{cases} B_j \frown N(0, \sigma_C^2) \\ C_{ij} \frown N(0, (r-1)\sigma_{RC}^2/r) \\ E_{ijk} \frown N(0, \sigma^2) \end{cases}$$

$(i = 1, 2, \ldots, r; j = 1, 2, \ldots, c; k = 1, 2, \ldots, n)$. Also, $\text{cov}(C_{ij}, C_{i'i}) = -\sigma_{RC}^2/r$ for $i \neq i'$, and all other covariances are zero.

19.7.3 MIXED-EFFECTS MODEL WITH RANDOM ROW FACTOR AND FIXED COLUMN FACTOR

One particular model (see footnote 5) is

$$Y_{ijk} = \mu + A_i + \beta_j + C_{ij} + E_{ijk} \tag{19.13}$$

where each β_j is a constant such that $\sum_{j=1}^{c} \beta_j = 0$, and where A_i, C_{ij}, and E_{ijk} are random variables satisfying $\sum_{j=1}^{c} C_{ij} = 0$ for each i and

$$\begin{cases} A_i \frown N(0, \sigma_R^2) \\ C_{ij} \frown N(0, (c-1)\sigma_{RC}^2/c) \\ E_{ijk} \frown N(0, \sigma^2) \end{cases}$$

$(i = 1, 2, \ldots, r; j = 1, 2, \ldots, c; k = 1, 2, \ldots, n)$. Also, $\text{cov}(C_{ij}, C_{ij'}) = -\sigma_{RC}^2/c$ for $j \neq j'$, and all other covariances are zero.

19.7.4 NULL HYPOTHESES AND EXPECTED MEAN SQUARES FOR TWO-WAY ANOVA MODELS

Table 19.18 gives (for fixed-, random-, and mixed-effects models) the specific null hypotheses being tested regarding row main effects, column main effects, and interaction effects. Table 19.19 gives the expected mean square (EMS) for each factor in each of the models. These two tables emphasize why different F statistics are required for testing the various hypotheses of interest. In this regard, the primary consideration involves the choice of the appropriate *denominator* mean squares to use in the various F statistics. The numerator mean square *always* corresponds to the factor being considered; for example, if the factor is *rows*, the numerator mean square is MSR, regardless of the type of model. Similarly, if the factor is *columns* or *interaction*, the numerator mean square is MSC or MSRC, respectively. *The denominator mean square,*

[5] There are several alternative ways to define a two-way mixed model; for an excellent discussion regarding these various models, see Hocking (1973).

TABLE 19.18   *Null hypotheses for two-way ANOVA*

SOURCE	FIXED-EFFECTS	RANDOM-EFFECTS	MIXED-EFFECTS (rows fixed, columns random)	MIXED-EFFECTS (rows random, columns fixed)
		MODEL TYPE		
Rows	$\alpha_1 = \alpha_2 = \cdots = \alpha_r = 0$	$\sigma_R^2 = 0$	$\alpha_1 = \cdots = \alpha_r = 0$	$\sigma_R^2 = 0$
Columns	$\beta_1 = \beta_2 = \cdots = \beta_c = 0$	$\sigma_C^2 = 0$	$\sigma_C^2 = 0$	$\beta_1 = \cdots = \beta_c = 0$
Interaction	$\gamma_{ij} = 0$ for all i, j	$\sigma_{RC}^2 = 0$	$\sigma_{RC}^2 = 0$	$\sigma_{RC}^2 = 0$

TABLE 19.19   *Expected mean squares for two-way ANOVA (r rows, c columns, n replications per cell)*

SOURCE	FIXED-EFFECTS	RANDOM-EFFECTS	MIXED-EFFECTS (rows fixed, columns random)	MIXED-EFFECTS (rows random, columns fixed)
		MODEL TYPE		
Rows	$\sigma^2 + cn \sum_{i=1}^{r} \dfrac{\alpha_i^2}{r-1}$	$\sigma^2 + n\sigma_{RC}^2 + cn\sigma_R^2$	$\sigma^2 + n\sigma_{RC}^2 + cn \sum_{i=1}^{r} \dfrac{\alpha_i^2}{r-1}$	$\sigma^2 + cn\sigma_R^2$
Columns	$\sigma^2 + rn \sum_{j=1}^{c} \dfrac{\beta_j^2}{c-1}$	$\sigma^2 + n\sigma_{RC}^2 + rn\sigma_C^2$	$\sigma^2 + rn\sigma_C^2$	$\sigma^2 + n\sigma_{RC}^2 + rn \sum_{j=1}^{c} \dfrac{\beta_j^2}{c-1}$
Interaction	$\sigma^2 + n \sum_{i=1}^{r} \sum_{j=1}^{c} \dfrac{\gamma_{ij}^2}{(r-1)(c-1)}$	$\sigma^2 + n\sigma_{RC}^2$	$\sigma^2 + n\sigma_{RC}^2$	$\sigma^2 + n\sigma_{RC}^2$
Error	σ^2	σ^2	σ^2	σ^2

however, is chosen to correspond to that EMS to which the numerator EMS reduces under the null hypothesis of interest. For example, when testing for significant row effects in a *random-effects model*, the numerator EMS for this test, $(\sigma^2 + n\sigma_{RC}^2 + cn\sigma_R^2)$ from Table 19.19, reduces to $(\sigma^2 + n\sigma_{RC}^2)$ under H_0: $\sigma_R^2 = 0$. This requires that the denominator mean square be MSRC, since the EMS of MSRC under the random-effects model is exactly $(\sigma^2 + n\sigma_{RC}^2)$.

In this way, the ratio of expected mean squares

$$\frac{\text{EMS(R)}}{\text{EMS(RC)}} = \frac{\sigma^2 + n\sigma_{RC}^2 + cn\sigma_R^2}{\sigma^2 + n\sigma_{RC}^2}$$

reduces to $(\sigma^2 + n\sigma_{RC}^2)/(\sigma^2 + n\sigma_{RC}^2) = 1$ under H_0: $\sigma_R^2 = 0$, so the F statistic MSR/MSRC is the ratio of two estimates of the same variance under H_0.

As another example, let us consider the F test for significant row effects based on the *mixed-effects model with the row factor fixed and column factor random*. The test statistic in this case, $F = $ MSR/MSRC, concerns the following ratio of EMS's (see Table 19.19):

$$\frac{\text{EMS(R)}}{\text{EMS(RC)}} = \frac{\sigma^2 + n\sigma_{RC}^2 + cn \sum_{i=1}^{r} \alpha_i^2/(r-1)}{\sigma^2 + n\sigma_{RC}^2}$$

Under H_0: $\alpha_1 = \alpha_2 = \cdots = \alpha_r = 0$, this ratio simplifies to $(\sigma^2 + n\sigma_{RC}^2)/(\sigma^2 + n\sigma_{RC}^2) = 1$. Thus, the F statistic is the ratio of two estimates of the same variance under H_0.

As a final example, we consider the F test for significant row effects based on the *mixed-effects model with the row factor random and column factor fixed*. The test statistic is $F = $ MSR/MSE, which concerns

$$\frac{\text{EMS(R)}}{\text{EMS(E)}} = \frac{\sigma^2 + cn\sigma_R^2}{\sigma^2}$$

Under H_0: $\sigma_R^2 = 0$, this ratio simplifies to $\sigma^2/\sigma^2 = 1$, as desired.

PROBLEMS

1. The following data come from an animal experiment designed to investigate whether levorphanol produces a reduction in stress as reflected in the cortical sterone level. There were five animals in each of the four treatment groups, and the data are given in the table.

CONTROL	LEVORPHANOL ONLY	EPINEPHRINE ONLY	LEVORPHANOL AND EPINEPHRINE
1.90	0.82	5.33	3.08
1.80	3.36	4.84	1.42
1.54	1.64	5.26	4.54
4.10	1.74	4.92	1.25
1.89	1.21	6.07	2.57

(a) These data may be analyzed by means of two-way ANOVA. What are the two factors?

(b) Classify each factor as being either fixed or random.

(c) Rearrange the data into a two-way table appropriate for analysis by means of two-way ANOVA.

(d) Form the table of sample means and comment.

(e) Determine the appropriate ANOVA table for this data set.

(f) Analyze the data to determine whether there are significant main effects due to levorphanol and epinephrine and whether there is a significant interaction effect between epinephrine and levorphanol.

2. The table gives the performance competency scores for a random sample of family nurse practitioners (FNP's) with different specialties from hospitals in three cities.

(a) Classify each factor as being either fixed or random and justify your classifications.

(b) Form the table of sample means (you may use the computer printout given) and then comment.

(c) Using the computer results given, compute the appropriate F statistics for each of the four possible factor classification schemes (i.e., both factors fixed, both random, and one fixed, one random).

SPECIALTY	CITY 1	CITY 2	CITY 3
Pediatrics	91.7, 74.9 88.2, 79.5	86.3, 88.1 92.0, 69.5	82.3, 78.7 89.8, 84.5
Obstetrics and Gynecology	80.1, 76.2 70.3, 89.5	71.3, 73.4 76.9, 87.2	90.1, 65.6 74.6, 79.1
Diabetes and Hypertension	71.5, 49.8 55.1, 75.4	80.2, 76.1 44.2, 50.5	48.7, 54.4 60.1, 70.8

(d) Analyze the data based on each of the possible factor classification schemes. How do the results compare?

(e) Find, using Scheffé's method as described in Chapter 17, a 95% confidence interval for the true difference in mean scores between pediatric FNP's and ob-gyn FNP's.

(f) Assuming each factor to be fixed, state a regression model appropriate for the two-way ANOVA table and provide estimates of the regression coefficients associated with the factor main effects using the sample means obtained in part (b).

Computer results for Problem 2

SPECIALTY	CITY	N	MEAN	STANDARD DEVIATION
PEDIATRI	1	4	8.3575E+01	7.73019E+00
PEDIATRI	2	4	8.3975E+01	9.93894E+00
PEDIATRI	3	4	8.3825E+01	4.64569E+00
OBSGYN	1	4	7.9025E+01	8.06200E+00
OBSGYN	2	4	7.7200E+01	7.05550E+00
OBSGYN	3	4	7.7350E+01	1.01858E+01
DIABHYP	1	4	6.2950E+01	1.24184E+01
DIABHYP	2	4	6.2750E+01	1.80452E+01
DIABHYP	3	4	5.8500E+01	9.42868E+00
Overall mean (means of means):			7.43500E+01	

Analysis of variance

SOURCE	df	SS	MS
SPEC	2	3.2299E+03	1.6149E+03
CITY	2	2.4542E+01	1.2271E+01
SPEC×CITY	4	3.4537E+01	8.6342E+00
Error	27	2.9022E+03	1.0749E+02

3. The table gives the average patient waiting time in minutes for a sample of 16 physicians classified by type of practice and type of physician.

(a) Classify each factor as being either fixed or random and justify your classification scheme.

(b) Using the computer results given, compute the F statistics corresponding to each of the four possible factor classification schemes.

(c) Discuss the analysis of the data when both factors are considered to be fixed.

PHYSICIAN TYPE	TYPE OF PRACTICE	
	GROUP	SOLO
General practitioner	15, 20, 25, 20	20, 25, 30, 25
Specialist	30, 25, 30, 35	25, 20, 30, 30

(d) What is the estimate of the (fixed) effect due to GP's, the (fixed) effect due to group practice, and the interaction effect $(\mu_{11} - \mu_{12} - \mu_{21} + \mu_{22})$, where μ_{ij} denotes the cell mean in the ith row and jth column of the table of cell means.

(e) Interpret the interaction effect that is observed.

(f) What is an appropriate regression model for this two-way ANOVA?

(g) How might one modify the model in part (f) to reflect the conclusions made in part (c)?

Computer results for Problem 3

PHYSTP	TYPRAC	N	MEAN	STANDARD DEVIATION
GP	GROUP	4	2.0000E + 01	4.08248E + 00
GP	SOLO	4	2.5000E + 01	4.08248E + 00
SPEC	GROUP	4	3.0000E + 01	4.08248E + 00
SPEC	SOLO	4	2.6250E + 01	4.78714E + 00
Overall mean (mean of means):			2.53125E + 01	

Analysis of variance

SOURCE	df	SS	MS
PHYSTP	1	1.2656E + 02	1.2656E + 02
TYPRAC	1	1.5625E + 00	1.5625E + 00
PHYSTP × TYPRAC	1	7.6563E + 01	7.6563E + 01
Error	12	2.1875E + 02	1.8229E + 01

4. A study was undertaken to measure and compare the sexist attitudes of students at various types of colleges. Random samples of 10 undergraduate seniors of each sex were selected from each of three types of colleges. A questionnaire was then administered to each student, from which a "degree of sexism"

COLLEGE TYPE	MALE	FEMALE
Coed with 75% or more males	50, 35, 37, 32, 46, 38, 36, 40, 38, 41	38, 27, 34, 30, 22, 32, 26, 24, 31, 33
Coed with less than 75% males	30, 29, 31, 27, 22, 20, 31, 22, 25, 30	28, 31, 28, 26, 20, 24, 31, 24, 31, 26
Not coed	45, 40, 32, 31, 26, 28, 39, 27, 37, 35	40, 35, 32, 29, 24, 26, 36, 25, 35, 35

score was determined (where the higher the score, the more sexist the attitude). Here, "degree of sexism" reflected the extent to which a student considered males and females to have different "life roles." The resulting data are given in the table.

(a) Form the table of cell means and interpret the results obtained (see the computer printout).
(b) Using the computer results given, compute the F statistics corresponding to a model with both factors considered fixed.
(c) Discuss the analysis of the data for this fixed-effects model case.

Computer results for Problem 4

TYPCOL	SEX	N	MEAN	STANDARD DEVIATION
75PLCOED	MALE	10	3.9300E+01	5.31350E+00
75PLCOED	FEMALE	10	2.9700E+01	4.92274E+00
LS75COED	MALE	10	2.6700E+01	4.16467E+00
LS75COED	FEMALE	10	2.6900E+01	3.63471E+00
NOCOED	MALE	10	3.4000E+01	6.27163E+00
NOCOED	FEMALE	10	3.1700E+01	5.41705E+00
Overall mean (mean of means):			3.13833E+01	

Analysis of variance

SOURCE	df	SS	MS
TYPCOL	2	6.5743E+02	3.2872E+02
SEX	1	2.2815E+02	2.2815E+02
TYPCOL×SEX	2	2.5930E+02	1.2965E+02
Error	54	1.3653E+03	2.5283E+01

5. Random samples of 100 persons awaiting trial on felony charges were selected from rural, urban, and suburban court locations in each of two states, one (state 1) in the northeastern United States and the other (state 2) in the South. The table summarizes the data on the time (in months) between arrest and the beginning of trial for these random samples.

STATE	COURT LOCATIONS		
	RURAL	SUBURBAN	URBAN
1	$n = 100$ $\bar{Y} = 3.4, S = 1.3$	$n = 100$ $\bar{Y} = 5.8, S = 1.2$	$n = 100$ $\bar{Y} = 6.8, S = 1.5$
2	$n = 100$ $\bar{Y} = 2.4, S = 1.5$	$n = 100$ $\bar{Y} = 3.5, S = 1.7$	$n = 100$ $\bar{Y} = 4.7, S = 1.7$

(a) Do the sample means in the table suggest that the average waiting times for state 1 vary by court location differently than they do for state 2; in other words, is there an interaction effect?

(b) Analyze these data using the ANOVA table and assuming that both factors are fixed.

SOURCE	df	SS	MS
State	1	486.00	486.00
Court location	2	826.33	413.17
Interaction	2	49.00	24.50
Error	594	1,327.591	2.235

(c) Define an appropriate regression model for this two-way ANOVA.

(d) How might one revise the model in part (c) and the associated ANOVA table in order to investigate whether there is a linear trend in waiting time with the degree of urbanization (defined by treating the categories rural, suburban, and urban on an original scale)? What difficulty does one encounter when considering such a model?

6. An experiment was conducted at a large state university to determine whether two different instructional methods for teaching a beginning statistics course would result in different levels of cognitive achievement. One instructional method involved the use of a self-instructional format, including a sequence of slide-tape presentations; the other method utilized the standard lecture format. The 100 students who registered for the course were randomly assigned to one of four sections, corresponding to the combinations of one of the two methods with one of two instructors. The results obtained from identical final exams given to each section are summarized in the table.

INSTRUCTOR	METHOD	
	LECTURE	SELF-INSTRUCTION
A	$n = 25$ $\bar{Y} = 71.2$ $S = 13.8$	$n = 25$ $\bar{Y} = 80.2$ $S = 12.1$
B	$n = 25$ $\bar{Y} = 73.8$ $S = 11.7$	$n = 25$ $\bar{Y} = 77.5$ $S = 14.1$

(a) What do the results suggest concerning the comparative effects of the two instructional methods?

(b) Classify each factor as being either fixed or random, and explain your classification scheme.

(c) Using the ANOVA table, perform the appropriate F tests for each of the four types of factor classification schemes possible. Compare the conclusions reached under each factor classification scheme.

(d) What are some factors that should be controlled for in this experiment?

(e) Given a continuous variable C to be controlled for, write down an appropriate regression model for this data set which takes C into account. What general method of analysis is characterized by such a model?

Analysis of variance

SOURCE	df	SS	MS
INSTRUC	1	6.2500E−02	6.2500E−02
METHOD	1	1.0081E+03	1.0081E+03
INSTRUC×METHOD	1	1.7556E+02	1.7556E+02
Error	96	1.6141E+04	1.6814E+02

7. The table presents data on the uric acid level found in the bloodstreams of mongoloids and in the bloodstreams of normal control subjects or nonmongoloid mentally retarded subjects. All subjects were between the ages of 21 and 25.

SUBJECTS	MALES	FEMALES
Mongoloid	5.84, 6.30, 6.95, 5.92, 7.94	4.90, 6.95, 6.73, 5.32, 4.81
Others	5.50, 6.08, 5.12, 7.58, 6.78	4.94, 7.20, 5.22, 4.60, 3.88

Analyze these data using the ANOVA table to determine whether there is evidence of a higher uric acid level in the mongoloid group, making sure to characterize any sex relationships that exist.

SOURCE	df	SS	MS
Groups	1	1.1329	1.1329
Sex	1	4.4746	4.4746
Interaction	1	0.4802	0.4802
Error	16	17.3150	1.0822

8. An experiment was conducted to investigate the survival of diplococcus pneumonia bacteria in chick embryos under relative humidities (RH) of 0, 25, 50, and 100% and under temperatures (TEMP) of 10, 20, 30 and 40°C using 10 chicks for each RH–TEMP combination.[6] The partially completed ANOVA table is as given.

SOURCE	df	MS
RH		2.010
TEMP		7.816
Interaction		1.642
Error		0.775
Total		

(a) Should the two factors, RH and TEMP, be considered as fixed or random? Explain.

[6] Adapted from a study by Price (1954).

(b) Carry out the analysis of variance for both the fixed-effects case and the random-effects case. Are your conclusions different in the two cases?

(c) Write down both the fixed-effects and the random-effects models that could describe this experiment.

(d) Provide a regression model using dummy variables that can be used to obtain the results in the ANOVA table.

(e) What regression model would be appropriate for describing the relationship of RH and TEMP to survival time (Y) if the independent variables are to be treated intervally rather than nominally?

9. The diameters (Y) of three species of pine trees were compared at each of four locations using samples of five trees per species at each location. The data are given in the table.

SPECIES	LOCATION			
	1	2	3	4
A	23	25	21	14
	15	20	17	17
	26	21	16	19
	13	16	24	20
	21	18	27	24
B	28	30	19	17
	22	26	24	21
	25	26	19	18
	19	20	25	26
	26	28	29	23
C	18	15	23	18
	10	21	25	12
	12	22	19	23
	22	14	13	22
	13	12	22	19

(a) Comment on whether each of the two factors should be considered as fixed or random.

(b) Use the partially-completed ANOVA table to carry out your analysis, first considering both factors fixed and then considering a mixed model with locations as random. Compare your conclusions.

SOURCE	df	SS
Species		314.10
Locations		55.80
Interaction		103.10
Error		945.60

TWO-WAY ANOVA WITH UNEQUAL CELL NUMBERS

20.1 PREVIEW

When we first began our discussion of two-way ANOVA (in Chapter 18), we pointed out (by means of Figure 18.1) several ways to classify a two-factor problem according to the observed data pattern. We have already described methods for handling the single-observation-per-cell case (Chapter 18) and the equal-cell-number case (Chapter 19). In this chapter we shall present procedures for analyzing two-factor patterns having unequal cell numbers. This latter case exhibits special statistical-analysis problems

TABLE 20.1   *Data layout for the unequal-cell-number case (two-way ANOVA)*

ROW FACTOR	COLUMN FACTOR				ROW MARGINALS
	COLUMN 1	COLUMN 2	...	COLUMN c	
Row 1	$Y_{111}, Y_{112}, \ldots, Y_{11n_{11}}$ (sample size $= n_{11}$) (cell mean $= \bar{Y}_{11\cdot}$)	$Y_{121}, Y_{122}, \ldots, Y_{12n_{12}}$ (sample size $= n_{12}$) (cell mean $= \bar{Y}_{12\cdot}$)	...	$Y_{1c1}, Y_{1c2}, \ldots, Y_{1cn_{1c}}$ (sample size $= n_{1c}$) (cell mean $= \bar{Y}_{1c\cdot}$)	$n_{1\cdot}$ $\bar{Y}_{1\cdot\cdot}$
Row 2	$Y_{211}, Y_{212}, \ldots, Y_{21n_{21}}$ (sample size $= n_{21}$) (cell mean $= \bar{Y}_{21\cdot}$)	$Y_{221}, Y_{222}, \ldots, Y_{22n_{22}}$ (sample size $= n_{22}$) (cell mean $= \bar{Y}_{22\cdot}$)	...	$Y_{2c1}, Y_{2c2}, \ldots, Y_{2cn_{2c}}$ (sample size $= n_{2c}$) (cell mean $= \bar{Y}_{2c\cdot}$)	$n_{2\cdot}$ $\bar{Y}_{2\cdot\cdot}$
$\vdots$	$\vdots$	$\vdots$	...	$\vdots$	$\vdots$
Row r	$Y_{r11}, Y_{r12}, \ldots, Y_{r1n_{r1}}$ (sample size $= n_{r1}$) (cell mean $= \bar{Y}_{r1\cdot}$)	$Y_{r21}, Y_{r22}, \ldots, Y_{r2n_{r2}}$ (sample size $= n_{r2}$) (cell mean $= \bar{Y}_{r2\cdot}$)	...	$Y_{rc1}, Y_{rc2}, \ldots, Y_{rcn_{rc}}$ (sample size $= n_{rc}$) (cell mean $= \bar{Y}_{rc\cdot}$)	$n_{r\cdot}$ $\bar{Y}_{r\cdot\cdot}$
Column Marginals	$n_{\cdot 1}$ $\bar{Y}_{\cdot 1\cdot}$	$n_{\cdot 2}$ $\bar{Y}_{\cdot 2\cdot}$	...	$n_{\cdot c}$ $\bar{Y}_{\cdot c\cdot}$	$n_{\cdot\cdot}$ $\bar{Y}_{\cdot\cdot\cdot}$

not characteristic of patterns with equal cell frequencies. In dealing with these problems we will find it necessary to distinguish between nonsystematic unequal cell number data patterns, which require special analysis, and proportional cell-frequency patterns (see Figures 18.1c and d), which can be analyzed as already outlined in Chapter 19.

The general data configuration for the unequal-cell-number case in two-way ANOVA is presented in Table 20.1. A numerical example is given in Table 20.2, which we shall focus on throughout most of this chapter. In the table

$Y_{ijk} = k$th observation in the cell associated with the ith row and jth column

$n_{ij} = $ number of observations in the cell associated with the ith row and jth column

Also,

$$\bar{Y}_{ij\cdot} = \frac{1}{n_{ij}} \sum_{k=1}^{n_{ij}} Y_{ijk}$$

$$\bar{Y}_{i\cdot\cdot} = \frac{1}{n_{i\cdot}} \sum_{j=1}^{c} \sum_{k=1}^{n_{ij}} Y_{ijk}, \qquad \text{where } n_{i\cdot} = \sum_{j=1}^{c} n_{ij}$$

$$\bar{Y}_{\cdot j\cdot} = \frac{1}{n_{\cdot j}} \sum_{i=1}^{r} \sum_{k=1}^{n_{ij}} Y_{ijk}, \qquad \text{where } n_{\cdot j} = \sum_{i=1}^{r} n_{ij}$$

$$\bar{Y}_{\cdots} = \frac{1}{n_{\cdot\cdot}} \sum_{i=1}^{r} \sum_{j=1}^{c} \sum_{k=1}^{n_{ij}} Y_{ijk}, \qquad \text{where } n_{\cdot\cdot} = \sum_{i=1}^{r} \sum_{j=1}^{c} n_{ij}$$

TABLE 20.2   *Example: satisfaction with medical care (Y) by patient worry and affective communication between patient and physician*

AFFECTIVE COMMUNICATION	WORRY		ROW MARGINALS
	NEGATIVE	POSITIVE	
High	2, 5, 8, 6, 2, 4, 3, 10 $(n_{11}=8)$ $(\bar{Y}_{11\cdot}=5)$	7, 5, 8, 6, 3, 5, 6, 4, 5, 6, 8, 9 $(n_{12}=12)$ $(\bar{Y}_{12\cdot}=6)$	$n_{1\cdot}=20$ $\bar{Y}_{1\cdot\cdot}=5.6$
Medium	4, 6, 3, 3 $(n_{21}=4)$ $(\bar{Y}_{21\cdot}=4)$	7, 7, 8, 6, 4, 9, 8, 7 $(n_{22}=8)$ $(\bar{Y}_{22\cdot}=7)$	$n_{2\cdot}=12$ $\bar{Y}_{2\cdot\cdot}=6$
Low	8, 7, 5, 9, 9, 10, 8 6, 8, 10 $(n_{31}=10)$ $(\bar{Y}_{31\cdot}=8)$	5, 8, 6, 6, 9, 7, 7, 8 $(n_{32}=8)$ $(\bar{Y}_{32\cdot}=7)$	$n_{3\cdot}=18$ $\bar{Y}_{3\cdot\cdot}=7.56$
Column Marginals	$n_{\cdot 1}=22$ $\bar{Y}_{\cdot 1}=6.18$	$n_{\cdot 2}=28$ $\bar{Y}_{\cdot 2}=6.57$	$n_{\cdot\cdot}=50$ $\bar{Y}_{\cdots}=6.40$

The unequal-cell-number case arises quite frequently in *observational studies*. In such studies, typically

1. All the variables of interest are not categorized in advance of data collection;

and/or

2. New variables are often considered after data collection;

and/or

3. When all the variables are separately categorized, it is often not practical or even possible to control in advance how the various categories of variables will combine to form combinations of categories of interest.

The unequal-cell-number case can also arise in *experimental studies* when (a posteriori) consideration is given to variables other than those of primary interest, even if the design based on the primary variables calls for equal cell numbers. Furthermore, unequal cell numbers will generally result whenever there are missing data points, which, for example, can occur due to study dropouts, incomplete records, and other assorted reasons.

The example presented in Table 20.2 is derived from Thompson's study (1972) concerning the relationship of patient perception of pregnancy and physician–patient communication to patient satisfaction with medical care. Two main variables of interest were the patient's WORRY and a measure of affective communication (AFFCOM). These variables were developed from scales based on questionnaires administered to patients and their physicians. Based on the distribution of scores obtained, the WORRY variable was grouped into the categories "positive" and "negative," and the AFFCOM variable was grouped into the categories "high," "medium," and "low." Table 20.2 presents artificial data of this type, showing satisfaction-with-medical-care scores (TOTSAT) classified according to these six combinations of levels of the factors WORRY and AFFCOM.

As can be seen from the table, the categorization scheme used for the two independent variables leads to a two-way table with unequal cell numbers. For WORRY, there are 22 negatives and 28 positives; for AFFCOM, there are 20 high, 12 medium, and 18 low scores. When the separate categories for the two variables are considered together, the resulting six categories have different cell sample sizes, ranging from 4 (medium AFFCOM, negative WORRY) to 12 (high AFFCOM, positive WORRY).

20.2 PROBLEM WITH UNEQUAL CELL NUMBERS: NONORTHOGONALITY

The key statistical concept associated with the special analytical problems in the unequal-cell-number case pertains to the *nonorthogonality* of the sums of squares usually used to describe the sources of variation in a two-way ANOVA table. To explain what we mean by "orthogonality," we first note that the general formulas for these sums of squares (given in Section 19.3 for the equal-cell-number case) in terms of unequal cell numbers are as follows:

$$SSR = \sum_{i=1}^{r} \sum_{j=1}^{c} \sum_{k=1}^{n_{ij}} (\bar{Y}_{i..} - \bar{Y}_{...})^2, \qquad SSC = \sum_{i=1}^{r} \sum_{j=1}^{c} \sum_{k=1}^{n_{ij}} (\bar{Y}_{.j.} - \bar{Y}_{...})^2,$$

$$SSRC = \sum_{i=1}^{r} \sum_{j=1}^{c} \sum_{k=1}^{n_{ij}} (\bar{Y}_{ij.} - \bar{Y}_{i..} - \bar{Y}_{.j.} + \bar{Y}_{...})^2, \qquad (20.1)$$

$$SSE = \sum_{i=1}^{r} \sum_{j=1}^{c} \sum_{k=1}^{n_{ij}} (Y_{ijk} - \bar{Y}_{ij.})^2, \qquad TSS = \sum_{i=1}^{r} \sum_{j=1}^{c} \sum_{k=1}^{n_{ij}} (Y_{ijk} - \bar{Y}_{...})^2$$

These formulas for SSR, SSC, and SSRC are often referred to as the *unconditional* sums of squares for rows, columns, and interaction, respectively; by "unconditional" we mean that each of these sums of squares may be separately defined from basic principles to describe the variability associated with the estimated effects $(\bar{Y}_{i..} - \bar{Y}_{...})$, $(\bar{Y}_{.j.} - \bar{Y}_{...})$, and $(\bar{Y}_{ij.} - \bar{Y}_{i..} - \bar{Y}_{.j.} + \bar{Y}_{...})$ for rows, columns, and interaction, respectively. (There is an equivalent way to illustrate the meaning of the term "unconditional" using regression-analysis methodology, as we shall soon see.)

If the collection of sums of squares in (20.1) are *orthogonal*, the following fundamental equation holds:

$$SSR + SSC + SSRC + SSE = TSS$$

That is, the terms on the left-hand side must partition the total sum of squares into nonoverlapping sources of variation.

We have already seen that this fundamental equation holds true for the equal-cell-number case (Chapter 19). Unfortunately, with unequal cell numbers, the unconditional sums of squares will no longer represent completely separate (i.e., orthogonal, sources of variation, so that we have

$$\boxed{\text{unequal cell numbers}} \Rightarrow \boxed{SSR + SSC + SSRC + SSE \neq TSS}$$

To see why this is so, it helps to consider the general regression formulation for two-way ANOVA, which incorporates the unequal-cell-number case as well as the equal-cell-number case; the general regression equation is

$$Y = \mu + \sum_{i=1}^{r-1} \alpha_i X_i + \sum_{j=1}^{c-1} \beta_j Z_j + \sum_{i=1}^{r-1} \sum_{j=1}^{c-1} \gamma_{ij} X_i Z_j + E \qquad (20.2)$$

where μ, α_i, β_j, and γ_{ij} are regression coefficients, and X_i and Z_j are appropriately defined dummy variables. Recall that the general form of the *fundamental regression equation* for this model may be written

$$\text{total sum of squares} = SS(\text{regression}) + SS(\text{error})$$

where

$$\text{total sum of squares} = \sum (Y_\ell - \bar{Y})^2 = TSS$$

$$SS(\text{regression}) = \sum (\hat{Y}_\ell - \bar{Y})^2 = SS \text{ reg}(X_1, \ldots, X_{r-1}; Z_1, \ldots, Z_{c-1}; X_1 Z_1, \ldots, X_{r-1} Z_{c-1})$$

$$SS(\text{error}) = \sum (Y_\ell - \hat{Y}_\ell)^2 = SSE,$$

and where the summation is over all $n_{..}$ observations.

Now, using the *extra-sum-of-squares principle* (see Chapter 10), we can partition SS(regression) in various ways to emphasize the contribution due to adding sets of variables to a regression model already containing other sets of variables. In particular, we can further partition the fundamental regression equation above as follows with regard to model (20.2):

$$TSS = SS\ reg\ \overbrace{(X_1, \ldots, X_{r-1})}^{R} + SS\ reg\ \overbrace{(Z_1, \ldots, Z_{c-1}}^{C}\ |\ \overbrace{X_1, \ldots, X_{r-1})}^{R}$$

$$+ SS\ reg\ \overbrace{(X_1Z_1, \ldots, X_{r-1}Z_{c-1}}^{RC}\ |\ \overbrace{X_1, \ldots, X_{r-1}, Z_1, \ldots, Z_{c-1})}^{R, C}$$
$$+ SSE \tag{20.3}$$

On the other hand, if we wish to enter the column effects into the model first, the equation would be as follows:

$$TSS = SS\ reg\ \overbrace{(Z_1, \ldots, Z_{c-1})}^{C} + SS\ reg\ \overbrace{(X_1, \ldots, X_{r-1}}^{R}\ |\ \overbrace{Z_1, \ldots, Z_{c-1})}^{C}$$

$$+ SS\ reg\ \overbrace{(X_1Z_1, \ldots, X_{r-1}Z_{c-1}}^{RC}\ |\ \overbrace{X_1, \ldots, X_{r-1}, Z_1, \ldots, Z_{c-1})}^{R, C}$$
$$+ SSE \tag{20.4}$$

As suggested by (20.3) and (20.4), it can be shown that

$$SS\ reg\ (X_1, \ldots, X_{r-1}) \equiv SSR$$

$$SS\ reg\ (Z_1, \ldots, Z_{c-1}) \equiv SSC \tag{20.5}$$

$$SS\ reg\ (X_1Z_1, \ldots, X_{r-1}Z_{c-1}) \equiv SSRC$$

where SSR, SSC, and SSRC are the unconditional sums of squares given by (20.1). For example, we can express (20.3) and (20.4) as

$$SSR + SS(C|R) + SS(RC|R, C) + SSE = TSS$$

and

$$SSC + SS(R|C) + SS(RC|R, C) + SSE = TSS$$

respectively. Note that both of these equations involve "conditional" sums of squares. However, when all the cell sample sizes are equal, it is also true that

$$\boxed{\begin{array}{l} \text{equal} \\ \text{cell} \\ \text{numbers} \end{array}} \Rightarrow \begin{cases} SSR = SS(R|C) \\ SSC = SS(C|R) \\ SSRC = SS(RC|R, C) \end{cases}$$

In other words, when all the cell sample sizes are equal, the extra sums of squares are not actually affected by variables already in the model, and the following holds:

$$\boxed{\begin{array}{l}\text{equal}\\\text{cell}\\\text{numbers}\end{array}} \Rightarrow SSR + SSC + SSRC + SSE = TSS \tag{20.6}$$

However, in the unequal-cell-number case, we have the following:

$$\boxed{\begin{array}{l}\text{unequal}\\\text{cell}\\\text{numbers}\end{array}} \Rightarrow \begin{cases} SSR \neq SS(R|C) \\ SSC \neq SS(C|R) \\ SSRC \neq SS(RC|R, C) \end{cases}$$

In other words, in the unequal-cell-number case, (20.6) does not hold and it is necessary to consider such expressions as (20.3) and (20.4), which reflect the importance of the order in which the effects are entered into the model. As we shall see in Section 20.4, the unequal-cell-number case is best handled by using regression analysis to carry out the two-way ANOVA calculations.

There is, however, one exception to this, which occurs when there are proportional cell frequencies satisfying

$$n_{ij} = \frac{n_i.n_{\cdot j}}{n_{\cdot\cdot}} \tag{20.7}$$

When (20.7) holds, it turns out that the following statement can be made:

$$\boxed{n_{ij} = \frac{n_i.n_{\cdot j}}{n_{\cdot\cdot}}} \Rightarrow \begin{cases} SSR = SS(R|C) \\ SSC = SS(C|R) \\ SSRC \neq SS(RC|R, C) \end{cases}$$

Thus, although (20.6) still does not hold in this case, (20.3) and (20.4) simplify to the single equation

$$SSR + SSC + SS(RC|R, C) + SSE = TSS \tag{20.8}$$

Thus, (20.8) contains only one term, $SS(RC|R, C)$, which is different from the terms in (20.6). This sum of squares, however, can be easily obtained by subtraction, as is clear from (20.8). Thus, *when the proportional cell frequency allocation of (20.7) is used, the standard equal-cell-number ANOVA calculations can be performed*, without the need to resort to regression analysis or alternative methods (e.g., the method of unweighted means, to be described in the next section).

To summarize, we have the flow diagram for two-way ANOVA shown in Figure 20.1.

FIGURE 20.1   *Flow diagram for two-way ANOVA*

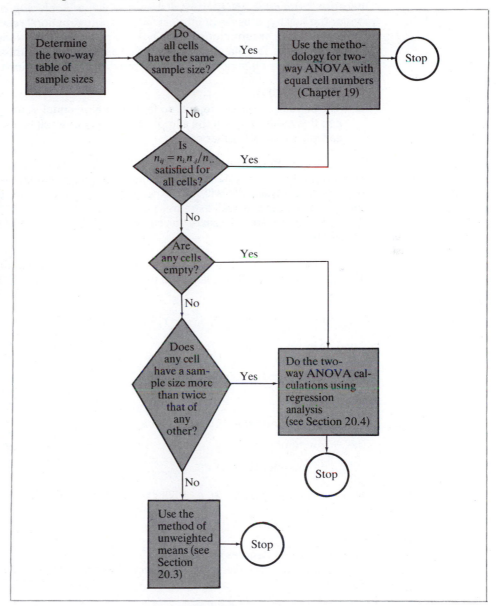

20.3 METHOD OF UNWEIGHTED MEANS

One method for analyzing a two-way data array with unequal cell numbers involves constructing a new two-way table involving *only* the cell means for the original data set and then analyzing this new table as a two-way data layout with one observation (the cell mean) in each cell.[1] This method is only an *approximate* procedure, however, because it provides ANOVA test statistics which are only *approximately F* statistics under the null hypotheses of interest.[2] The "approximation" results from the fact that one of the basic assumptions of ANOVA, homoscedasticity, does not hold when this method is used. The effect of violating this assumption is almost negligible (i.e., the approximation is quite good), however, provided two conditions are satisfied:

1. No cell is empty.
2. The cell sample sizes are not too far from being equal. A rough rule of thumb is to use this method only if the sample size in any one cell is not more than twice the sample size in any other cell.

A look at the sample sizes in Table 20.2 indicates that condition 2 is not satisfied, since $n_{21} = 4$ and $n_{12} = 12$. Thus, in this example, it would be best to perform a regression analysis. We will, nevertheless, analyze these data using the method of unweighted means as well as regression analysis.

Four steps are necessary for carrying out the method of unweighted means. These are:

Step 1. Compute the usual (pooled) within-cell estimate of experimental error. This estimate is calculated using the formula

$$\text{MSE} = \frac{1}{n_{..} - rc} \sum_{i=1}^{r} \sum_{j=1}^{c} \sum_{k=1}^{n_{ij}} (Y_{ijk} - \bar{Y}_{ij.})^2$$

$$= \frac{1}{n_{..} - rc} \sum_{i=1}^{r} \sum_{j=1}^{c} \left[\sum_{k=1}^{n_{ij}} Y_{ijk}^2 - \left(\sum_{k=1}^{n_{ij}} Y_{ijk} \right)^2 \bigg/ n_{ij} \right]$$

where

$n_{..} = $ total number of observations

$r = $ number of rows

$c = $ number of columns

In our example (Table 20.2), $n_{..} = 50$, $r = 3$, $c = 2$, and MSE is calculated to be

$$\text{MSE} = 3.4091$$

with $n_{..} - rc = 44$ degrees of freedom.

Step 2. Compute the estimated average variance of the cell means, $\hat{\sigma}_{av}^2$, using the formula

$$\hat{\sigma}_{av}^2 = \frac{1}{rc} (\text{MSE}) \sum_{i=1}^{r} \sum_{j=1}^{c} \frac{1}{n_{ij}} \qquad (20.9)$$

[1] The discussion of the method of unweighted means, as well as the regression approach of the next section, treats both factors as fixed. Nevertheless, consideration of mixed and random models is also possible and is discussed briefly in Section 20.5.

[2] An alternative analysis of means is that known as the "weighted squares of means" [see Searle (1971)]; it is computationally complex, but provides *exact F* tests.

[$\hat{\sigma}^2_{av}$ can be equivalently written as

$$\hat{\sigma}^2_{av} = \frac{MSE}{\text{harmonic mean of the } n_{ij}}$$

where this harmonic mean is given by

$$\frac{1}{1/rc \sum\limits_{i=1}^{r} \sum\limits_{j=1}^{c} 1/n_{ij}}]$$

To provide some motivation for (20.9), note that any cell mean $\bar{Y}_{ij.}$ has a variance equal to σ^2/n_{ij}, and these variances will not all be equal unless all the n_{ij} are equal. Then the average variance of these rc means would be defined as

$$\sigma^2_{av} = \frac{1}{rc} \sum_{i=1}^{r} \sum_{j=1}^{c} \frac{\sigma^2}{n_{ij}} = \frac{1}{rc} \sigma^2 \sum_{i=1}^{r} \sum_{j=1}^{c} \frac{1}{n_{ij}}$$

and the estimate (20.9) is obtained by substituting MSE for σ^2. If conditions 1 and 2 stated earlier are satisfied, the assumption that each mean has the same variance (estimated by $\hat{\sigma}^2_{av}$), although not actually true, holds to a reasonable degree of approximation.

In our example (Table 20.2), the estimated average variance is computed as follows:

$$\hat{\sigma}^2_{av} = \frac{1}{3(2)} (MSE) \left(\frac{1}{8} + \frac{1}{12} + \frac{1}{4} + \frac{1}{8} + \frac{1}{10} + \frac{1}{8} \right)$$

$$= \frac{97}{720} (MSE)$$

$$= \frac{MSE}{7.42} = \frac{3.4091}{7.42}$$

$$= 0.4594$$

Step 3. Form the table of cell means. In our example this table is given by

AFFCOM	WORRY		TOTAL
	NEGATIVE	POSITIVE	
High	$\bar{Y}_{11.} = 5$	$\bar{Y}_{12.} = 6$	$T_{1.} = 11$
Medium	$\bar{Y}_{21.} = 4$	$\bar{Y}_{22.} = 7$	$T_{2.} = 11$
Low	$\bar{Y}_{31.} = 8$	$\bar{Y}_{32.} = 7$	$T_{3.} = 15$
Total	$T_{.1} = 17$	$T_{.2} = 20$	$T_{..} = 37$

Notice that, in this table, the row (column) marginal totals (the $T_{i.}$'s and $T_{.j}$'s) are obtained simply by adding up the cell means in each row (column); the grand total $T_{..} = 37$ is the sum of the six cell means. These marginals are clearly not the same as those based on the original table (Table 20.2). This is because we are treating these cell means as if they were the only observations available, whereas in Table 20.2 there are multiple observations in each cell.

Step 4. Analyze the table of cell means using two-way, one-observation-per-cell ANOVA, with $\hat{\sigma}_{av}^2$ as the denominator in all F tests. The appropriate analysis-of-variance table associated with this step is given in general by Table 20.3 and for our example by Table 20.4.

The sums of squares given in Table 20.4 are computed from the table of cell means as follows:

$$SS(AFFCOM) = \frac{1}{2}[(11)^2 + (11)^2 + (15)^2] - \frac{(37)^2}{6} = 5.33$$

$$SS(WORRY) = \frac{1}{3}[(17)^2 + (20)^2] - \frac{(37)^2}{6} = 1.50$$

$$SS(AFFCOM \times WORRY) = [(5)^2 + (4)^2 + (8)^2 + (6)^2 + (7)^2 + (7)^2] - \frac{(37)^2}{6} - 5.33 - 1.50$$

$$= 4.00$$

The conclusions based on Table 20.4 are that AFFCOM has a highly significant main effect (since an F of 5.81 with df's of 2 and 44 has $P < 0.01$), and that the AFFCOM $\times$ WORRY interaction is mildly significant (since an F of 4.35 with df's of 2 and 44 has $P < 0.025$). Thus, if the researcher prefers to be conservative in the choice of significance level, the primary conclusion from this ANOVA would be that patient

TABLE 20.3   *ANOVA table based on the method of unweighted means*

SOURCE	df	SS	MS	F RATIO
Rows	$r-1$	$SSR = \frac{1}{c} \sum_{i=1}^{r} T_{i.}^2 - \frac{T_{..}^2}{rc}$	$MSR = \frac{SSR}{r-1}$	$\frac{MSR}{\hat{\sigma}_{av}^2}$
Columns	$c-1$	$SSC = \frac{1}{r} \sum_{j=1}^{c} T_{.j}^2 - \frac{T_{..}^2}{rc}$	$MSC = \frac{SSC}{c-1}$	$\frac{MSC}{\hat{\sigma}_{av}^2}$
Interaction	$(r-1)(c-1)$	$SSRC = \sum_{i=1}^{r} \sum_{j=1}^{c} \bar{Y}_{ij.}^2$ $- \frac{T_{..}^2}{rc} - SSR - SSC$	$MSRC = \frac{SSRC}{(r-1)(c-1)}$	$\frac{MSRC}{\hat{\sigma}_{av}^2}$
Error	$n_{..} - rc$	$\cdots$	$\hat{\sigma}_{av}^2$	

($T_{i.}$ = sum of cell means in the ith row, $T_{.j}$ = sum of cell means in the jth column, $T_{..}$ = sum of all cell means)

TABLE 20.4   *Application of method of unweighted means to satisfaction with medical care data of Table 20.2*

SOURCE	df	SS	MS	F RATIO
AFFCOM	2	5.33	2.6650	5.81**
WORRY	1	1.50	1.5000	3.27
AFFCOM × WORRY	2	4.00	2.0000	4.35*
Error	44	· · ·	0.4594	

satisfaction with medical care differs only according to the level of affective patient–physician communication. If, on the other hand, the researcher works at the 0.05 significance level, the conclusion would be that patient satisfaction depends on the combined effect of AFFCOM and WORRY; and from Table 20.2, this interaction is such that when the AFFCOM level is high or medium, patients with a positive score on WORRY are more satisfied with care than patients with a negative score; whereas when the AFFCOM level is low, patients with negative WORRY are more satisfied than patients with positive WORRY.

20.3.1 ALTERNATIVE ANOVA TABLE FOR THE METHOD OF UNWEIGHTED MEANS

Instead of using the form of ANOVA table given in Table 20.3, some researchers prefer to use an alternative form, for which *all* the SS and MS terms are multiplied by the harmonic mean

$$\frac{1}{\frac{1}{rc} \sum_{i=1}^{r} \sum_{j=1}^{c} \frac{1}{n_{ii}}}$$

The primary reason for doing this is that the error term for this alternative ANOVA table turns out to be MSE, as can be seen from (20.9).

Thus, the alternative table (with MSE as the error term) is more like the usual two-way ANOVA table for equal cell-sample sizes. Nevertheless, it is important to point out that using this alternative table does not affect the *F* statistics. This is because *both* MS terms in any *F* statistic of Table 20.3 are multiplied by the same harmonic mean; e.g., the *F* statistic for AFFCOM in the alternative ANOVA Table 20.5

TABLE 20.5   *Alternative ANOVA table for the method of unweighted means*

SOURCE	df	SS	MS	F
AFFCOM	2	39.588	19.794	5.81**
WORRY	1	11.134	11.134	3.27
AFFCOM × WORRY	2	29.691	14.845	4.35*
Error	44	150.000	3.4091	

is given by

$$F_{(AFFCOM)} = \frac{19.794}{3.4091} = \frac{(2.6650)(7.42)}{(0.4594)(7.42)} = \frac{2.6650}{0.4594} = 5.81$$

which was previously obtained in Table 20.4.

In any case, it is this alternative form of ANOVA table that is most often given by computer programs performing the method of unweighted means.

20.4 REGRESSION APPROACH WHEN THE CELL SAMPLE SIZES ARE UNEQUAL

As described in Section 20.2, the general regression model that is applicable to both the equal and unequal cell number cases is given by

$$Y = \mu + \sum_{i=1}^{r-1} \alpha_i X_i + \sum_{j=1}^{c-1} \beta_j Z_j + \sum_{i=1}^{r-1}\sum_{j=1}^{c-1} \gamma_{ij} X_i Z_j + E$$

where the $\{X_i\}$ and $\{Z_j\}$ are sets of dummy variables representing the r levels of the row factor and the c levels of the column factor, respectively. In general, *any* two-way ANOVA problem can be analyzed by a regression approach utilizing such a model. However, in the unequal-cell-number situation, there still remains the decision as to in what order the effects (row, column, or interaction) should be entered into the regression model. There are actually *four* alternative orderings worth mentioning; the computational components of these four orderings are described in general in Table 20.6 and are illustrated in Table 20.7 for the special case $r = 3$ and $c = 2$.

As discussed earlier, if all cells contain the same number of observations, the order in which the various effects are entered does not affect the different F tests; in other words, methods 1 through 4 given in Table 20.6 yield *exactly* the same results. If the special condition $n_{ij} = n_{i.}n_{.j}/n_{..}$ is satisfied, methods 1 through 3 yield exactly the same results.

In general, with respect to Tables 20.6 and 20.7, the test for interaction is the same regardless of which method is used. However, the tests chosen for the main effects depend on whether the investigator prefers to enter the effects sequentially (as in methods 1 and 2) or prefers to test for the significance of a main effect controlling either for the remaining main effect or for all remaining effects (including interaction).[3]

Of the four methods, method 3 is the most widely recommended [e.g., see Applebaum and Cramer (1973)]. Furthermore, these authors have pointed out regarding method 3 that the logical flow of decisions made in the analysis should not directly correspond to the actual order in which the computations are usually performed (as presented in Tables 20.6 and 20.7). In particular, they recommend a three-step procedure regarding method 3 that begins by testing for significant interaction using $F(X_1 Z_1, \ldots, X_{r-1} Z_{c-1} | X_1, \ldots, X_{r-1}, Z_1, \ldots, Z_{c-1})$, proceeds only if this test is nonsignificant to consider both $F(X_1, \ldots, X_{r-1} | Z_1, \ldots, Z_{c-1})$ and $F(Z_1, \ldots, Z_{c-1} | X_1, \ldots, X_{r-1})$, and then proceeds further only if the latter two tests are nonsignificant to test for the main effect of each factor ignoring the other using $F(X_1, \ldots, X_{r-1})$ and $F(Z_1, \ldots, Z_{c-1})$.

---

[3] We do not recommend method 4 since the main effect test results depend on the choice of coding scheme used for the dummy variables; see Speed and Hocking (1976) for further discussion in this regard.

TABLE 20.6   Alternative methods for regression approach to two-way unequal-cell-numbers ANOVA

METHOD	EFFECT	H_0	REGRESSION MODEL	TEST PROCEDURE
1	Row	$\alpha_1 = \alpha_2 = \cdots = \alpha_{r-1} = 0$	$Y = \mu + \sum \alpha_i X_i + E$	Usual one-way ANOVA F test [or, equivalently, $F(X_1, \ldots, X_{r-1})$ for overall regression]
	Column	$\beta_1 = \beta_2 = \cdots = \beta_{c-1} = 0$	$Y = \mu + \sum \alpha_i X_i + \sum \beta_j Z_j + E$	Multiple–partial $F(Z_1, \ldots, Z_{c-1} \mid X_1, \ldots, X_{r-1})$
	Interaction	$\gamma_{ij} = 0$ for all i, j	$Y = \mu + \sum \alpha_i X_i + \sum \beta_j Z_j + \sum\sum \gamma_{ij} X_i Z_j + E$	Multiple–partial $F(X_i Z_1, \ldots, X_{r-1} Z_{c-1} \mid X_1, \ldots, X_{r-1}, Z_1, \ldots, Z_{c-1})$
2	Column	$\beta_1 = \beta_2 = \cdots = \beta_{c-1} = 0$	$Y = \mu + \sum \beta_j Z_j + E$	Usual one-way ANOVA F test [or, equivalently, $F(Z_1, \ldots, Z_{c-1})$ for overall regression]
	Row	$\alpha_1 = \alpha_2 = \cdots = \alpha_{r-1} = 0$	$Y = \mu + \sum \alpha_i X_i + \sum \beta_j Z_j + E$	Multiple–partial $F(X_1, \ldots, X_{r-1} \mid Z_1, \ldots, Z_{c-1})$
	Interaction	$\gamma_{ij} = 0$ for all i, j	$Y = \mu + \sum \alpha_i X_i + \sum \beta_j Z_j + \sum\sum \gamma_{ij} X_i Z_j + E$	Multiple–partial $F(X_i Z_1, \ldots, X_{r-1} Z_{c-1} \mid X_1, \ldots, X_{r-1}, Z_1, \ldots, Z_{c-1})$
3	Row	$\alpha_1 = \alpha_2 = \cdots = \alpha_{r-1} = 0$	$Y = \mu + \sum \alpha_i X_i + \sum \beta_j Z_j + E$	Multiple–partial $F(X_1, \ldots, X_{r-1} \mid Z_1, \ldots, Z_{c-1})$
	Column	$\beta_1 = \beta_2 = \cdots = \beta_{c-1} = 0$	$Y = \mu + \sum \alpha_i X_i + \sum \beta_j Z_j + E$	Multiple–partial $F(Z_1, \ldots, Z_{c-1} \mid X_1, \ldots, X_{r-1})$
	Interaction	$\gamma_{ij} = 0$ for all i, j	$Y = \mu + \sum \alpha_i X_i + \sum \beta_j Z_j + \sum\sum \gamma_{ij} X_i Z_j + E$	Multiple–partial $F(X_i Z_1, \ldots, X_{r-1} Z_{c-1} \mid X_1, \ldots, X_{r-1}, Z_1, \ldots, Z_{c-1})$
4	Row	$\alpha_1 = \alpha_2 = \cdots = \alpha_{r-1} = 0$	$Y = \mu + \sum \alpha_i X_i + \sum \beta_j Z_j + \sum\sum \gamma_{ij} X_i Z_j + E$	Multiple–partial $F(X_1, \ldots, X_{r-1} \mid Z_1, \ldots, Z_{c-1}, X_1 Z_1, \ldots, X_{r-1} Z_{c-1})$
	Column	$\beta_1 = \beta_2 = \cdots = \beta_{c-1} = 0$	$Y = \mu + \sum \alpha_i X_i + \sum \beta_j Z_j + \sum\sum \gamma_{ij} X_i Z_j + E$	Multiple–partial $F(Z_1, \ldots, Z_{c-1} \mid X_1, \ldots, X_{r-1}, X_1 Z_1, \ldots, X_{r-1} Z_{c-1})$
	Interaction	$\gamma_{ij} = 0$ for all i, j	$Y = \mu + \sum \alpha_i X_i + \sum \beta_j Z_j + \sum\sum \gamma_{ij} X_i Z_j + E$	Multiple–partial $F(X_i Z_1, \ldots, X_{r-1} Z_{c-1} \mid X_1, \ldots, X_{r-1}, Z_1, \ldots, Z_{c-1})$

TABLE 20.7 Alternative methods for regression approach to satisfaction-with-medical-care data of Table 20.2 (r = 3 and c = 2)

METHOD	EFFECT	H_0	REGRESSION MODEL*	TEST PROCEDURE
1	Row	$\alpha_1 = \alpha_2 = 0$	$Y = \mu + \alpha_1 X_1 + \alpha_2 X_2 + E$	Overall $F(X_1, X_2)$
	Column	$\beta_1 = 0$	$Y = \mu + \alpha_1 X_1 + \alpha_2 X_2 + \beta_1 Z_1 + E$	Partial $F(Z_1|X_1, X_2)$
	Interaction	$\gamma_{11} = \gamma_{21} = 0$	$Y = \mu + \alpha_1 X_1 + \alpha_2 X_2 + \beta_1 Z_1 + \gamma_{11} X_1 Z_1 + \gamma_{21} X_2 Z_1 + E$	Multiple–partial $F(X_1 Z_1, X_2 Z_2 | X_1, X_2, Z_1)$
2	Column	$\beta_1 = 0$	$Y = \mu + \beta_1 Z_1 + E$	Overall $F(Z_1)$
	Row	$\alpha_1 = \alpha_2 = 0$	$Y = \mu + \alpha_1 X_1 + \alpha_2 X_2 + \beta_1 Z_1 + E$	Multiple–partial $F(X_1, X_2|Z_1)$
	Interaction	$\gamma_{11} = \gamma_{21} = 0$	$Y = \mu + \alpha_1 X_1 + \alpha_2 X_2 + \beta_1 Z_1 + \gamma_{11} X_1 Z_1 + \gamma_{21} X_2 Z_1 + E$	Multiple–partial $F(X_1 Z_1, X_2 Z_1 | X_1, X_2, Z_1)$
3	Row	$\alpha_1 = \alpha_2 = 0$	$Y = \mu_1 + \alpha_1 X_1 + \alpha_2 X_2 + \beta_1 Z_1 + E$	Multiple–partial $F(X_1, X_2|Z_1)$
	Column	$\beta_1 = 0$	$Y = \mu + \alpha_1 X_1 + \alpha_2 X_2 + \beta_1 Z_1 + E$	Partial $F(Z_1|X_1, X_2)$
	Interaction	$\gamma_{11} = \gamma_{21} = 0$	$Y = \mu + \alpha_1 X_1 + \alpha_2 X_2 + \beta_1 Z_1 + \gamma_{11} X_1 Z_1 + \gamma_{21} X_2 Z_1 + E$	Multiple–partial $F(X_1 Z_1, X_2 Z_1 | X_1, X_2, Z_1)$
4	Row	$\alpha_1 = \alpha_2 = 0$	$Y = \mu + \alpha_1 X_1 + \alpha_2 X_2 + \beta_1 Z_1 + \gamma_{11} X_1 Z_1 + \gamma_{21} X_2 Z_1 + E$	Multiple–partial $F(X_1, X_2|Z_1, X_1 Z_1, X_2 Z_1)$
	Column	$\beta_1 = 0$	$Y = \mu + \alpha_1 X_1 + \alpha_2 X_2 + \beta_1 Z_1 + \gamma_{11} X_1 Z_1 + \gamma_{21} X_2 Z_1 + E$	Partial $F(Z_1|X_1, X_2, X_1 Z_1, X_2 Z_1)$
	Interaction	$\gamma_{11} = \gamma_{21} = 0$	$Y = \mu + \alpha_1 X_1 + \alpha_2 X_2 + \beta_1 Z_1 + \gamma_{11} X_1 Z_1 + \gamma_{21} X_2 Z_1 + E$	Multiple–partial $F(X_1 Z_1, X_2 Z_1 | X_1, X_2, Z_1)$

*Y = TOTSAT, $X_1 = \begin{cases} -1 & \text{if high AFFCOM} \\ 1 & \text{if low AFFCOM} \\ 0 & \text{if medium AFFCOM} \end{cases}$, $X_2 = \begin{cases} -1 & \text{if high AFFCOM} \\ 1 & \text{if medium AFFCOM} \\ 0 & \text{if low AFFCOM} \end{cases}$, $Z_1 = \begin{cases} -1 & \text{if negative WORRY} \\ 1 & \text{if positive WORRY} \end{cases}$

20.4.1 NUMERICAL ILLUSTRATION OF METHODS 1, 2, AND 3

For the satisfaction-with-medical-care data of Table 20.2, the analysis results using a regression approach are presented below for methods 1, 2, and 3.

METHOD 1   *ANOVA table*

SOURCE	df	SS	MS	F RATIO	P VALUE
X_1, X_2	2	38.756	19.378	$4.97_{(2,47)}$	$0.01 < P < 0.025$
$Z_1 \mid X_1, X_2$	1	5.861	5.861	$1.52_{(1,46)}$	$0.10 < P < 0.25$
$X_1 Z_1, X_2 Z_1 \mid X_1, X_2, Z_1$	2	27.384	13.692	$4.02_{(2,44)}$	$0.01 < P < 0.025$
Residual	44	149.996	3.409	—	—

AFFCOM effect

$$F(X_1, X_2) = \frac{38.756/2}{(149.996 + 27.384 + 5.861)/47} = \frac{19.378}{3.899} = 4.97$$

with 2 and 47 df, which yields $0.01 < P < 0.025$ *(borderline significance).*

WORRY effect

$$F(Z_1 \mid X_1, X_2) = \frac{5.861}{(149.996 + 27.384)/46} = \frac{5.861}{3.856} = 1.520$$

with 1 and 46 df, which yields $0.10 < P < 0.25$ *(not significant).*

Interaction effect

$$F(X_1 Z_1, X_2 Z_1 \mid X_1, X_2, Z_1) = \frac{27.384/2}{149.996/44} = \frac{13.692}{3.409} = 4.02$$

with 2 and 44 df, which yields $0.01 < P < 0.025$ *(borderline significance).*

METHOD 2   *ANOVA table*

SOURCE	df	SS	MS	F RATIO	P VALUE
Z_1	1	1.870	1.870	$0.41_{(1,48)}$	$P > 0.25$
$X_1, X_2 \mid Z_1$	2	42.746	21.373	$5.54_{(2,46)}$	$0.005 < P < 0.01$
$X_1 Z_1, X_2 Z_1 \mid X_1, X_2, Z_1$	2	27.384	13.692	$4.02_{(2,44)}$	$0.01 < P < 0.025$
Residual	44	149.996	3.409		

WORRY effect

$$F(Z_1) = \frac{1.870}{(149.996 + 27.384 + 42.746)/48} = \frac{1.870}{4.586} = 0.41$$

with 1 and 48 df, which yields $P > 0.25$ *(not significant)*.

AFFCOM effect

$$F(X_1, X_2|Z_1) = \frac{42.746/2}{(149.996 + 27.384)/46} = \frac{21.373}{3.856} = 5.54$$

with 2 and 46 df, which yields $0.005 < P < 0.01$ *(significant)*.

Interaction effect

$$F(X_1 Z_1, X_2 Z_1 | X_1, X_2, Z_1) = 4.02$$

with 2 and 44 df, which yields $0.01 < P < 0.025$ *(borderline significance)*.

METHOD 3   *Computationally*, this method utilizes the ANOVA tables of methods 1 and 2.

AFFCOM effect

$$F(X_1, X_2|Z_1) = 5.54, \text{ with 2 and 46 df, which yields } 0.005 < P < 0.01 \text{ (significant)}.$$

WORRY effect

$$F(Z_1|X_1, X_2) = 1.520, \text{ with 1 and 46 df, which yields } 0.10 < P < 0.25 \text{ (not significant)}.$$

Interaction effect

$$F(X_1 Z_1, X_2 Z_1 | X_1, X_2, Z_1) = 4.02 \text{ with 2 and 44 df, which yields } 0.01 < P < 0.025 \text{ (borderline significance)}.$$

The recommended way to look at these results for method 3 would be first to consider the test for interaction, which is significant at the 0.05 but not at the 0.01 level. If one does not feel that this is indicative of a strong (i.e., highly significant) interaction effect, one could then consider the two main-effect tests, which suggest a highly significant AFFCOM main effect and no WORRY main effect. These conclusions are essentially the same as those obtained in Section 20.3 based on the method of unweighted means, although it was previously pointed out that the latter method was not altogether appropriate in this instance because of the large variation in the cell sample sizes.

20.5   RANDOM- AND MIXED-EFFECTS MODELS

The method of unweighted means (Section 20.3) can easily be adapted to deal with the random- and mixed-effects model situations. All that is necessary is to use the same

ANOVA format as given in Table 19.8 but to use the MS terms of Table 20.3. For example, to test for a significant row effect in the random-effects model, we simply use

$$F = \frac{MSR}{MSRC}$$

where MSR and MSRC are obtained from the unweighted-means approach. Other more detailed approaches for dealing with random- and mixed-effects models for two-way data arrays with unequal cell numbers are discussed, for example, in Searle (1971).

20.6 HIGHER-WAY ANOVA

It should not be surprising to learn that ANOVA, as a special case of regression analysis, can be generalized to allow for the consideration of any number of factors (i.e., independent variables).

We contend, nevertheless, that too much emphasis on complex ANOVA models and associated testing procedures is likely to be unwarranted, especially for researchers in the health, medical, social, and behavioral sciences. This is because:

1. As more independent variables are considered, it becomes more unlikely to want to treat them all as nominal variables.

2. Even if all independent variables are treated as nominal, either sufficient numbers of observations are not available in all cells (e.g., there are some empty cells) or the equal-cell-number situation cannot be arranged.

TABLE 20.8 *General three-way ANOVA table for equal-cell sample sizes (see Ostle, 1963)**

SOURCE	df	MS	F (all fixed)	F (A, B fixed, C random)	F (A, B random, C fixed)	F (all random)
A (a levels)	$a-1$	MSA	MSA/MSE	MSA/MSAC	MSA/MSAB	No exact test
B (b levels)	$b-1$	MSB	MSB/MSE	MSB/MSBC	MSB/MSAB	No exact test
C (c levels)	$c-1$	MSC	MSC/MSE	MSC/MSE	No exact test	No exact test
AB	$(a-1)(b-1)$	MSAB	MSAB/MSE	MSAB/MSABC	MSAB/MSE	MSAB/MSABC
AC	$(a-1)(c-1)$	MSAC	MSAC/MSE	MSAC/MSE	MSAC/MSABC	MSAC/MSABC
BC	$(b-1)(c-1)$	MSBC	MSBC/MSE	MSBC/MSE	MSBC/MSABC	MSBC/MSABC
ABC	$(a-1)(b-1)(c-1)$	MSABC	MSABC/MSE	MSABC/MSE	MSABC/MSE	MSABC/MSE
Error	$n-abc$	MSE				
Total	$n-1$					

* Those cases for which there is no exact F test result from the lack of equivalent numerator and denominator expected mean squares under the null hypothesis. Four other factor categorizations have been omitted from this table: (1) A fixed, B and C random; (2) B fixed, A and C random; (3) A and C fixed, B random; and (4) B and C fixed, A random. The F statistics for each of these cases, nevertheless, can be derived from one of the cases in the table by an appropriate permutation of the letters A, B, and C. For example, when considering A fixed, B and C random, switch A with C in the source and third F columns.

Methods are available, however, for designing and analyzing experimental studies in which only a *fraction* of the total number of possible cells need be used, but which still permit the researcher to estimate the effects of primary interest. The reader is referred to texts by Ostle (1963), Snedecor and Cochran (1967), and Peng (1967) for applied treatments of such methods.

In general, however, regression analysis should be the predominant analysis method of choice for higher-way ANOVA situations, especially since so much research in the health, medical, social, and behavioral sciences is observational in nature. We thus will not further extend our discussion to three-way or higher ANOVA situations, but rather suggest the use of regression analysis for such situations. Nevertheless, since the three-way ANOVA case is more prevalent than other higher-way cases, we present as Table 20.8 for reference purposes the general three-way ANOVA table for the equal-cell-number situation. We have also provided an exercise at the end of the chapter dealing with the three-way case.

PROBLEMS 1. Consider hypothetical data based on a study concerning the effects of rapid cultural change on blood pressure levels for native citizens of an island in Micronesia. Blood pressures were taken on a random sample of 30 males over age 40 from a certain province who commuted to work in the nearby westernized capital city. These persons were also given a sociological question - naire from which their social rankings in both their traditional and modern (i.e., westernized) cultures were determined. The results are summarized in the table.

MODERN RANK (factor A)	TRADITIONAL RANK (factor B)			
	HI	MED	LO	
HI	130 140 135	150 145	175 160 170	165 155
MED	145 140 150	150 160 155	165 155 165	170 160
LO	180 160 145	155 140 135	125 130 110	

(a) Discuss the table of sample means for this data set.

(b) Analyze this data set by the method of unweighted means, making sure to give the appropriate ANOVA table.

(c) Give an appropriate regression model for this data set, treating the two factors as nominal variables.

(d) Using the attached regression ANOVA tables (where X pertains to factor A and Z to factor B), carry out two different main-effect tests for each factor, and also test for interaction. Compare the results of the two main-effect tests made for each factor. Also, compare the regression results with the results obtained using the method of unweighted means.

(e) How might one modify the regression model given in part (c) to be able to quantify any trends in blood pressure levels in terms of increasing social rankings for the two factors? (Note that this requires assigning numerical values to the categories of each factor.) What difficulty does one encounter in defining such a model?

Regression results for Problem 1

SOURCE	df	MS
X_1	1	469.17985
$X_2\|X_1$	1	508.52217
$Z_1\|X_1, X_2$	1	187.97673
$Z_2\|X_1, X_2, Z_1$	1	7.54570
$X_1Z_1\|X_1, X_2, Z_1, Z_2$	1	3,925.29395
$X_1Z_2\|X_1, X_2, Z_1, Z_2, X_1Z_1$	1	9.70621
$X_2Z_1\|X_1, X_2, Z_1, Z_2, X_1Z_1, X_1Z_2$	1	633.17613
$X_2Z_2\|X_1, X_2, Z_1, Z_2, X_1Z_1, X_1Z_2, X_2Z_1$	1	2.67593
Residual	21	75.83333

SOURCE	df	MS
Z_1	1	278.59213
$Z_2\|Z_1$	1	22.71129
$X_1\|Z_1, Z_2$	1	391.77041
$X_2\|Z_1, Z_2, X_1$	1	480.15062
$X_1Z_1\|Z_1, Z_2, X_1, X_2$	1	3,925.29395
$X_1Z_2\|Z_1, Z_2, X_1, X_2, X_1Z_1$	1	9.70621
$X_2Z_1\|Z_1, Z_2, X_1, X_2, X_1Z_1, X_1Z_2$	1	633.17613
$X_2Z_2\|Z_1, Z_2, X_1, X_2, X_1Z_1, X_1Z_2, X_2Z_1$	1	2.67593
Residual	21	75.83333

2. A study was conducted to assess the combined effects of patient attitude and patient–physician communication on patient satisfaction with medical care during pregnancy. A random sample of 110 pregnant women under private physician care was followed from the time of first visit with the physician until delivery. Using specially devised questionnaires, the following variables were measured for each patient: Y = satisfaction score, X_1 = attitude score, X_2 = communication score. Each score was developed as an interval variable, but there was some question as to whether the analysis should treat the attitude and/or communication scores as nominal variables.

    (a) What would be an appropriate regression model for describing the joint effect of X_1 and X_2 on Y if it is desired to allow for the possibility of an interaction between communication and attitude and if all variables are treated as interval variables?

    (b) What would be an appropriate regression model (using dummy variables) if one still wished to allow for an interaction effect, but desired only to compare *high* values versus *low* values (i.e., to make group comparisons) for both the communication and attitude variables? What kind of analysis of variance model would this regression model correspond to?

    (c) When would one prefer the model in part (a) to that in (b), and vice versa?

    (d) If both independent variables are treated nominally as in part (b), would one expect the associated 2×2 table to have equal numbers in each of the four cells?

3. The data in the table represent the numbers of illness absences during a certain time period for 53 male factory workers from two departments who have been classified by age. It is of interest for these data to inquire whether the observed difference in the mean number of absences for the two departments is significant, controlling for differences in the age distributions in the two departments.

DEPARTMENT	NUMBER OF ABSENCES	FREQUENCY BY AGE GROUP (years)		
		30–39	40–49	50–59
A	0	1	2	3
	1	1	3	2
	2	2	3	6
	3	1	2	3
		5	10	14
B	0	4	2	3
	1	3	7	0
	2	1	2	0
	3	0	1	1
		8	12	4

(a) Determine the appropriate (2×3) table of mean numbers of illness absences for the six departments by age-group categories.

(b) Compute the harmonic mean of the sample sizes for the six cells and the average variance of the six cell means.

(c) Analyze the data using the method of unweighted means (the needed computer results are given here). What conclusions do you draw?

(d) How appropriate is this approach as compared to a regression approach for analyzing this data set?

(e) What particular ANOVA assumption might be violated for this data set?

(f) How would you analyze this data set using regression analysis (i.e., what model would you use, what hypothesis of primary interest would you test, and how would you set up your ANOVA table to perform the test)?

Analysis of variance

SOURCE	df	SS	MS
DEPT	1	5.8393E+00	5.8393E+00
AGE GROUP	2	3.5940E−01	1.7970E−01
DEPT×AGE GROUP	2	8.8164E−01	4.4082E−01
Error	47	4.9206E+01	1.0469E+00

4. The secretary of health in a certain country requested his statistician to determine whether or not there was any evidence that ethnic group *A* had different-sized families (on the average) than ethnic group *B*. The statistician selected independent random samples of 25- and 30-year-old women from each of the two groups and then determined how many children each woman had. Means and standard errors were calculated and the results are given in the accompanying table.

The statistician wrote a memorandum to the secretary stating that: "The mean number of children per woman is significantly higher in group *A* than in group *B*. The 95% confidence interval for the true difference in mean numbers of children is 0.6 ± 0.2."

(a) Do you agree with the statistician's conclusion? Explain your answer.

(b) Assuming that you desired to analyze these data yourself, describe briefly two alternative procedures that you might use. Is any sophisticated statistical analysis really needed to answer the secretary's question? Explain.

	AGE (years)		SUMMARY STATISTICS
	25	30	
Number of women			
Group A	100	200	300
Group B	200	100	300
Mean number of children			
Group A	1.8	3.6	3.0
Group B	1.8	3.6	2.4
Standard error of the mean			
Group A	0.070	0.072	0.072
Group B	0.050	0.100	0.071

5. A crime victimization study was undertaken in a medium-sized southern city to determine the effects of being a crime victim on confidence in law enforcement authority and in the legal system itself. A questionnaire was administered to a stratified random sample of 40 city residents, from which was obtained data on the number of times victimized, a measure of social class status, and a measure of the respondent's confidence in law enforcement and in the legal system. The data are as given in the table.

NUMBER OF TIMES VICTIMIZED	SOCIAL CLASS STATUS		
	LO	MED	HI
0	4, 14, 15, 19, 17, 17, 16	7, 10, 12, 15, 16	8, 19, 10, 17
1	2, 7, 18	6, 19, 12, 12	7, 6, 5, 3, 16
2+	7, 8, 2, 11, 12	1, 2, 4	4, 2, 8, 9

(a) Determine the table of sample means and comment on any patterns noted.
(b) Analyze this data set using the ANOVA computer results given.
(c) How would you analyze this data set using the regression computer output?
(d) What ANOVA assumption(s) might not hold for these data?

Analysis of variance

SOURCE	df	SS	MS
VICTIM	2	4.0000E+02	2.0000E+02
SCLS	2	2.2739E+01	1.1370E+01
VICTIM×SCLS	4	1.0993E+02	2.7483E+01
Error	31	7.0408E+02	2.2712E+01

Regression results

SOURCE*	df	MS
Z_1	1	44.03235
$Z_2|Z_1$	1	1.03496
$X_1|Z_1, Z_2$	1	395.75734
$X_2|Z_1, Z_2, X_1$	1	0.06778
$X_1Z_1|Z_1, Z_2, X_1, X_2$	1	1.68985
$X_1Z_2|Z_1, Z_2, X_1, X_2, X_1Z_1$	1	3.31635
$X_2Z_1|Z_1, Z_2, X_1, X_2, X_1Z_1, X_1Z_2$	1	0.40190
$X_2Z_2|Z_1, Z_2, X_1, X_2, X_1Z_1, X_1Z_2, X_2Z_1$	1	94.59353
Residual	31	22.71229

SOURCE	df	MS
X_1	1	407.86993
$X_2|X_1$	1	0.52174
$Z_1|X_1, X_2$	1	27.98766
$Z_2|X_1, X_2, Z_1$	1	4.51309
$X_1Z_1|X_1, X_2, Z_1, Z_2$	1	1.68985
$X_1Z_2|X_1, X_2, Z_1, Z_2, X_1Z_1$	1	3.31635
$X_2Z_1|X_1, X_2, Z_1, Z_2, X_1Z_1, X_1Z_2$	1	0.40190
$X_2Z_2|X_1, X_2, Z_1, Z_2, X_1Z_1, X_1Z_2, X_2Z_1$	1	94.59353
Residual	31	22.71229

* X pertains to "number of times victimized"; Z pertains to "social class status."

6. The effect of a new antidepressant drug on reduction in severity of syndrome was studied in manic–depressive patients at two state mental hospitals. In each hospital all such patients were randomly assigned to either a treatment (new drug) or control (old drug) group. The results of this experiment are summarized in the table, where a high mean score indicates more of a lowering in depression level than does a low mean score.

HOSPITAL	TREATMENT GROUP	CONTROL GROUP
A	$n = 25$ $\bar{Y} = 8.5$ $S = 1.3$	$n = 31$ $\bar{Y} = 4.6$ $S = 1.8$
B	$n = 25$ $\bar{Y} = 2.3$ $S = 0.9$	$n = 31$ $\bar{Y} = -1.7$ $S = 1.1$

(a) Short of performing any statistical tests, interpret the means in the table.

(b) Analyze the data using the method of unweighted means. Is the new drug effective? (Make sure to give the appropriate ANOVA table.)

(c) What regression model is appropriate for analyzing the data? For this model, describe how to test whether there is a significant effect due to the new drug.

7. A study was conducted by a television network to evaluate the viewing characteristics of adult females in a certain state. Each individual in a stratified random sample of 480 women was sent a questionnaire; the strata were formed on the basis of the following three factors: season (winter, summer), location (eastern, central, western), residence (rural, urban). Results on the total reported average TV viewing time (hours per day) are summarized in the accompanying table of sample means and standard deviations.

RESIDENCE/LOCATION	SUMMER			WINTER			MARGINALS	
	n	$\bar{Y}$	S	n	$\bar{Y}$	S	n	$\bar{Y}$
Rural								
East	40	2.75	1.340	40	4.80	0.851	80	3.78
Central	40	2.75	1.380	40	4.85	0.935	80	3.80
West	40	2.65	1.180	40	4.78	0.843	80	3.71
Marginals	120	2.72		120	4.81		240	3.76
Urban								
East	40	3.38	0.958	40	3.65	0.947	80	3.52
Central	40	3.15	1.130	40	4.50	0.743	80	3.83
West	40	3.65	0.779	40	4.05	0.781	80	3.85
Marginals	120	3.39		120	4.07		240	3.73
Marginals	240	3.06		240	4.44		480	3.75

(a) Suppose that the questionnaire contained items concerning certain additional factors such as occupation (categorized as housewife, blue-collar worker, white-collar worker, professional), age (categorized as 20–34, 35–50, over 50), and number of children (categorized as 0, 1–2, 3+). Comment on the likelihood of having equal cell numbers when carrying out an analysis of variance considering these additional variables.

(b) Examine the table of sample means with regard to main effects and interactions. (You may want to form two-factor summary tables for looking at two-factor interactions.)

(c) Using the analysis-of-variance computer results given, carry out appropriate F tests and discuss your results.

(d) State a regression model appropriate for obtaining information equivalent to the analysis-of-variance results presented.

Analysis of variance

SOURCE	df	SS	MS
RESID (rural, urban)	1	1.3333E−01	1.3333E−01
REGION (east, central, west)	2	2.5527E+00	1.2763E+00
SEASON (summer, winter)	1	2.2963E+02	2.2963E+02
RESID × REGION	2	3.3247E+00	1.6623E+00
RESID × SEASON	1	6.0492E+01	6.0492E+01
REGION × SEASON	2	7.2247E+00	3.6123E+00
RESID × REGION × SEASON	2	6.7460E+00	3.3730E+00
Error	468	4.7821E+02	1.0218E+00

8. Suppose that the data given in the table were obtained by an investigator studying the influence of estrogen injections on change in pulse rate of adolescent chimpanzees.

Male $\begin{cases}\text{Control:} & 5.1, -2.3, 4.2, 3.8, 3.2, -1.5, 6.1, -2.5, 1.9, -3.0, -2.8, 1.7 \\ \text{Estrogen:} & 15.0, 6.2, 4.1, 2.3, 7.6, 14.8, 12.3, 13.1, 3.4, 8.5, 11.2, 6.9\end{cases}$

Female $\begin{cases}\text{Control:} & -2.3, -5.8, -1.5, 3.8, 5.5, 1.6, -2.4, 1.9 \\ \text{Estrogen:} & 7.3, 2.4, 6.5, 8.1, 10.3, 2.2, 12.7, 6.3\end{cases}$

(a) What are the factors in this experiment and should they be designated as fixed or random?
(b) Demonstrate that the proportional cell frequency situation for two-way ANOVA holds for this problem; that is, is $n_{ij} = n_i.n._j/n..$ for each of the four cells?
(c) Using the following table of sample means:

	CONTROL	ESTROGEN
Male	1.158333	8.783333
Female	0.100000	6.975000

and the general formulas

$$\text{SS rows} = \sum_{i=1}^{r}\sum_{j=1}^{c}\sum_{k=1}^{n_{ij}} (\bar{Y}_{i..} - \bar{Y}_{...})^2,$$

$$\text{SS columns} = \sum_{i=1}^{r}\sum_{j=1}^{c}\sum_{k=1}^{n_{ij}} (\bar{Y}_{.j.} - \bar{Y}_{...})^2,$$

$$\text{SS cells} = \sum_{i=1}^{r}\sum_{j=1}^{c}\sum_{k=1}^{n_{ij}} (\bar{Y}_{ij.} - \bar{Y}_{...})^2$$

analyze the data for this problem using the usual methodology for equal-cell-number two-way ANOVA and the fact that SS error = 530.24078 (i.e., compute SS rows, SS columns, and SS cells directly, and then obtain SS interaction by subtraction).
(d) Analyze the data for this problem using the method of unweighted means and compare your results with those obtained in part (c).
(e) What regression model is appropriate for analyzing this data set?
(f) Using the regression analysis results given, check whether SS rows = SS reg(SEX) = SS reg(SEX|TREATMENT) and SS columns = SS reg(TREATMENT) = SS reg(TREATMENT|SEX), where SS rows and SS columns are as obtained in part (c). What has been demonstrated here?

SOURCE*	df	SS
SEX	1	19.72667
TREATMENT|SEX	1	536.55619
SEX×TREATMENT|SEX, TREATMENT	1	1.35000
Residual	36	530.24078

SOURCE	df	SS
TREATMENT	1	536.55619
SEX|TREATMENT	1	19.72667
SEX×TREATMENT|SEX, TREATMENT	1	1.35000
Residual	36	530.24078

$$* \text{SEX} = \begin{cases} -1 & \text{if male} \\ 1 & \text{if female} \end{cases}, \quad \text{TREATMENT} = \begin{cases} -1 & \text{if control} \\ 1 & \text{if estrogen} \end{cases}$$

CHAPTER 21 FACTOR ANALYSIS

21.1 PREVIEW

Factor analysis is a multivariable method that has as its aim the explanation of relationships among several difficult-to-interpret, correlated variables in terms of a few conceptually meaningful, relatively independent factors. Pictorially, this purpose of factor analysis is represented by Figure 21.1, in which the mass of several overlapping circles of various shades is reconstituted into two relatively nonoverlapping circles with different shading patterns.

FIGURE 21.1 *General purpose of factor analysis*

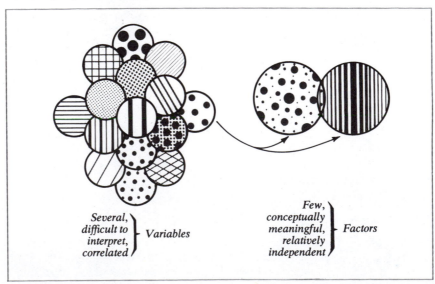

Several, difficult to interpret, correlated } Variables

Few, conceptually meaningful, relatively independent } Factors

In Chapter 1 we introduced the reader to factor analysis by example using the Ponape study, where an application of this method led to the construction of two measures of cultural incongruity. We also gave a brief general description of factor analysis in Chapter 2, in which we pointed out that the method involves the construction of new "factors" that may be used as independent or dependent variables in later analyses. Thus, factor analysis is frequently used as a data-analysis tool in conjunction with other methods such as regression or discriminant analysis.

In this chapter we shall describe in some detail the essential terminology and mechanics of factor analysis. We shall motivate our discussion of this subject with an example based on a study by James and Kleinbaum (1976) concerning the relationship between socioecologic stress and hypertension-related mortality in North Carolina.

EXAMPLE   Prior research [e.g., Stamler (1967) and Harburg et al. (1973)] has shown that black Americans tend to have higher average blood pressure levels than white Americans and that the death rate due to hypertension and related disorders is considerably higher among blacks than among whites. The reasons for these differences are not fully understood, although most researchers attribute them to a combination of genetic and socioenvironmental factors, including differential access to good medical care. A recent study by James and Kleinbaum (1976) considered the socioenvironmental aspect by focusing on the issue of whether or not blacks residing in high-stress areas would tend to have higher hypertension-related mortality rates than either blacks in low-stress areas or whites in high- or low-stress areas. Using 86 counties in North Carolina as their sampling units, James and Kleinbaum computed a 3-year hypertension-related death rate for the period 1959–1961 for each county and then related these county rates to an index of socioecologic stress for each county which was derived by factor analysis. The primary question of interest was whether or not high-stress counties had significantly higher mortality rates than low-stress counties.

The socioecologic stress index was constructed using 15 variables (whose 1960 values were obtained separately by race for each county) which appeared to reflect the economic and social well-being of the counties during the 1959–1961 period. These 15 variables and their code names are given in Table 21.1. Also identified in this table are two factors that were to result from the factor analysis and which were used in combination to define the index of socioecologic stress. These two factors are called the *socioeconomic status* (SES) factor and the *socioinstability* (SIS) factor. As seen in Table 21.1, certain variables (e.g., per capita income, median years' education) were expected to be more related to SES than SIS; such SES-type variables are labeled ⊙ in the table. On the other hand, other variables (e.g., juvenile deliquency, correction school measures) were expected to be SIS-type variables, labeled ⊠. Also, some variables (e.g., unemployment) were expected to be related to both the SES and SIS factors. It was expected, therefore, that each of the factors emerging from the factor analysis would correlate highly with those variables thought a priori to be related to that factor and would be essentially uncorrelated with those variables hypothesized to be unrelated to that factor. Later in this chapter we shall examine whether or not this actually occurred.

[Although in this study the two factors of primary interest had been identified conceptually a priori to the factor analysis, such an a priori identification is not a prerequisite for applying this method. In fact, it is sometimes the case that the researcher does not a priori have the resulting factors firmly in mind and wishes to utilize factor analysis to help characterize the meaningful factors describing the data.]

TABLE 21.1   *Variables used in the James–Kleinbaum study of 1960 North Carolina counties*

⊙	1	Per capita income	PCI
⊙	2	Median years of education	MED
⊙⊠	3	Percent unemployment	UEM
⊙	4	Percent families earning over $8,000/year	P8K+
⊙	5	Percent male white-collar jobs	WCM
⊙	6	Percent male blue-collar jobs	BC
⊙⊠	7	Percent families earning under $3,000/year	P3K−
⊠	8	Percent females separated or divorced	PSDF
⊠	9	Juvenile-delinquency-index males	JDM
⊠	10	Juvenile-delinquency-index females	JDF
⊠	11	Percent of males in correction school	CSM
⊠	12	Percent of females in correction school	CSF
⊠	13	Percent of males in prisons	PM
⊠	14	Homicide rate	HR
⊠	15	Percent of children under 18 not with parent	PCBH

FIGURE 21.2   *Purpose of factor analysis in the James–Kleinbaum study*

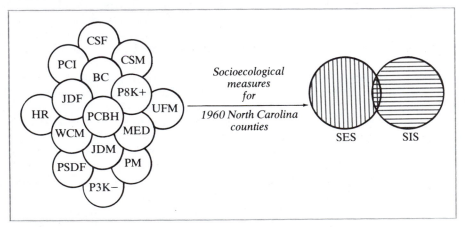

Figure 21.2 illustrates the purpose of the use of factor analysis in the James–Kleinbaum study. The circles on the left denote the 15 intercorrelated variables considered in the study, which are used to construct the two relatively independent factors (SES and SIS) on the right.

21.2   BASIC TERMINOLOGY OF FACTOR ANALYSIS

Although a long list of factor-analytic terms could be presented to encompass both the general and the very specific methodological points concerning factor analysis, we think that one can more clearly understand this method by focusing on only four basic terms: (1) factor loadings, (2) factor cosines, (3) factor weights (or factor score

TABLE 21.2   *Basic terminology of factor analysis (all data for nonwhites in North Carolina counties, 1960)*

factor loadings
= correlations between factors and variables

VARIABLES	FACTOR	
	SES	SIS
PCI	0.88	0.22
MED	0.88	-0.07
UEM	0.03	-0.03
P8K+	0.65	0.03
WCM	0.56	0.23
BC	0.81	0.12
P3K-	-0.88	-0.03
PSDF	0.18	0.75
JDM	0.34	0.73
JDF	0.44	0.62
CSM	0.38	0.60
CSF	0.50	0.35
PM	-0.06	0.71
HR	-0.41	0.26
PCBH	-0.12	0.75

factor weights (score coefficients)
= weights assigned to variables to determine factor scores

VARIABLE	FACTOR	
	SES	SIS
PCI	0.19	-0.00
MED	0.22	-0.10
UEM	0.01	-0.01
P8K+	0.15	-0.05
WCM	0.11	0.03
BC	0.18	-0.03
P3K-	-0.21	0.07
PSDF	-0.02	0.24
JDM	0.01	0.22
JDF	0.05	0.17
CSM	0.04	0.17
CSF	0.09	0.07
PM	-0.08	0.25
HR	-0.12	0.13
PCBH	-0.10	0.27

factor cosines
= correlations between factors

FACTOR	FACTOR	
	SES	SIS
SES	1.0	0.17
SIS	0.17	1.0

standardized factor scores
= standardized values of new variables (the factors) defined as weighted sums of the original standardized variables.

COUNTY	FACTOR	
	SES	SIS
1	0.87	-0.57
2	1.04	-0.52
3	-0.72	-0.08
....	....	....
85	-1.04	0.94
86	0.25	-0.29

coefficients), and (4) factor scores. Examples of these terms using data from the James–Kleinbaum study are given in Table 21.2.

21.2.1 FACTOR LOADINGS

Factor loadings describe the correlations between the factors emerging from a factor analysis and the original variables used in the construction of the factors. As shown in Table 21.2, the loadings associated with a given factor-analytic solution can be represented by a (matrix) display, where the numbers in each column are the correlations of a specific factor with the original variables. The primary use of such a matrix is to pinpoint those variables that are highly correlated (i.e., "load high") with a given factor, so that the factor can be conceptually interpreted. For example, PCI loads high (0.88) with SES, as originally hypothesized, whereas PSDF (0.75) loads high with SIS. We shall provide a more thorough discussion of these factor-analysis results later. For now, the important thing to remember is that *a factor loading is a correlation of a variable with a factor.*

21.2.2 FACTOR COSINES

A factor cosine is also a correlation, but, in contrast to a factor loading, relates one factor to another factor. Such correlations are important because they quantify the degree to which different factors are related. Ideally, as indicated earlier, one hopes that the factors will be relatively uncorrelated; that is, it is hoped that the factor cosines will be relatively close to zero. If this is so, each factor can be thought of as representing a distinctly different underlying component of information contained in the original set of variables. On the other hand, if two factors are very *highly* correlated, each would be describing essentially the same component of information and so only one of them would have to be considered. This desire for independence, however, does not necessarily call for complete independence among factors. If two factors are considered on empirical or theoretical grounds to be necessarily related to some extent, there should be allowance for some amount of correlation between the two factors. For example, empirical evidence from other studies about socioeconomic status (SES) and socioinstability (SIS) suggested that these factors would be negatively correlated, with low SES scores corresponding to high SIS scores and high SES scores corresponding to low SIS scores. As Table 21.2 shows, the (positive) direction of the factor cosine actually obtained (0.17) was in the opposite direction to that predicted, although the amount of correlation was nonsignificant.

21.2.3 FACTOR WEIGHTS

A factor weight is not a correlation, but rather is a number assigned to a variable (usually standardized into Z-score form) for use in determining the scores for a given factor. If one compares the matrix of factor *weights* in Table 21.2 with that for factor *loadings*, one should notice that the corresponding entries are not equal. Nevertheless, for a given column (i.e., factor), high factor loadings tend to correspond to high weights, and this generally will be the case for any factor-analysis solution. Thus, factor weights and factor loadings give similar information, except that they are measured on different scales and are used for different purposes, the weights to compute factor scores and the loadings to describe correlations.

21.2.4 FACTOR SCORES

A factor score is a specific value of a factor calculated for a particular sampling unit and is formed as a weighted sum of the values of the (usually standardized) variables for that sampling unit. Since the sampling units for the James–Kleinbaum study are 86 counties in North Carolina, there will be 86 factor scores for each factor, each score corresponding to a specific county. In Table 21.2, the scores for any one factor (e.g., SES) are scaled to Z-score form, so that a county (e.g., county 2) with a positive score (e.g., 1.04 on SES) is above average with regard to that factor, whereas a county with a negative score (e.g., -1.03 on SES for county 85) is below average with regard to that factor.

21.3 MATHEMATICAL DEFINITION OF A FACTOR

Let us examine the mathematical structure of a factor more closely. A factor F is a (weighted) linear combination of the original variables:

$$F = \sum_{j=1}^{p} w_j X_j = w_1 X_1 + w_2 X_2 + \cdots + w_p X_p$$

where the w's are the factor weights (to be estimated from the data) and the X's are the original variables (generally expressed in standardized form). When there are k factors under consideration (say, $F_1, F_2, \ldots, F_k$), the ith of these factors is denoted as

$$F_i = \sum_{j=1}^{p} w_{ij} X_j = w_{i1} X_1 + w_{i2} X_2 + \cdots + w_{ip} X_p, \qquad i = 1, 2, \ldots, k$$

As an illustration, from Table 21.2 (where $p = 15$), the (estimated) SES factor is $[0.19(PCI) + 0.22(MED) \times 0.01(UEM) \times 0.15(P8K\times) + 0.11(WCM) \times 0.18(BC) - 0.21(P3K-) - 0.02(PSDF) \times 0.01(JDM) + 0.05(JDF) + 0.04(CSM) + 0.09(CSF) - 0.08(PM) - 0.12(HR) - 0.10(PCBH)]$; and the (estimated) SIS factor is $[-0.00(PCI) - 0.10(MED) - 0.01(UEM) \times 0.05(P8K+) \times 0.03(WCM) - 0.03(BC) + 0.07(P3K-) + 0.24(PSDF) + 0.22(JDM) + 0.17(JDF) + 0.17(CSM) + 0.07(CSF) + 0.25(PM) + 0.13(HR) + 0.27(PCBH)]$. Note that for the SES factor the weights are higher (in absolute value) for most of the SES-type variables (the first seven variables) than for the SIS-type variables; the reverse is true for the SIS factor.

Now, let us distinguish between a "factor" and a "factor score." The term *factor* refers to the general expression for F given earlier, which takes a specific value for each sampling unit selected. The term *factor score* refers to a specific value of the factor obtained by plugging into the general expression for F the values of the original variables for the particular sampling unit selected. When the X's are standardized (e.g., so that they are dimensionless), it makes sense to say that the higher the factor weight for a given variable, the more that variable contributes to the overall factor score and the higher the corresponding factor loading.

It should be mentioned that the factor weights calculated for whites (which have not been presented in Table 21.2) turned out, as expected, to be quite different from the corresponding weights calculated for nonwhites, since different values of the original variables were obtained for each group. The problem of whether to use a separate set of weights for each racial group rather than a single set for both groups was a methodological issue faced by James and Kleinbaum, and the choice of using separate sets of weights reflected their goal of obtaining the most conceptually meaningful factors for each group. A single set of weights for both groups was actually determined, but the factors resulting were found to have extremely poor "construct validity" (i.e., the factor loadings obtained were not conceptually meaningful on either empirical or theoretical grounds).

21.4 ANALYSIS STRATEGY FOR THE JAMES–KLEINBAUM STUDY

Another important methodological issue in the James–Kleinbaum study concerned how to use the factor scores to test the primary research question of interest: Do high-stress counties have higher mortality rates than low-stress counties? The steps involved in the analysis strategy finally adopted were as follows:

1. The factor scores for the counties were first rank-ordered separately for each race. This yielded four orderings of 86 scores, one ordering for each combination of factor (SES and SIS) and race (white and nonwhite).

2. For each ordering, the county factor scores were divided at the median score into two groups, one group representing counties with "high" scores on the factor and the other representing counties with "low" scores on the factor. This yielded four groupings:

SES–W	SIS–W	SES–NW	SIS–NW
HI	HI	HI	HI
LO	LO	LO	LO

3. For each race, the two HI–LO groups were then cast into a two-way layout (see Figure 21.3) consisting of the following four cells: HI–SES, HI–SIS; HI–SES, LO–SIS; LO–SES, HI–SIS; and LO–SES, LO–SIS. Of these four cells, there were two of primary interest: "HI–SES, LO–SIS" (which was designated the *low-stress group*) and "LO–SES, HI–SIS" (which was designated the *high-stress group*). The rationale for these designations was that a county considered to be high in socioeconomic status (SES) and to have low socioinstability (SIS) was a "healthy" county in terms of social stress and so was hypothesized to have low mortality. On the other hand, a county low in socioeconomic status and high in socioinstability was considered an "unhealthy" county and so was hypothesized to have high mortality. The other two cells (HI–SES, HI–SIS) and (LO–SES, LO–SIS), because of their "intermediate" nature with regard to level of stress, were not considered of principal interest in the analysis.

FIGURE 21.3   *Diagrammatic representation of the analysis strategy for the James–Kleinbaum study*

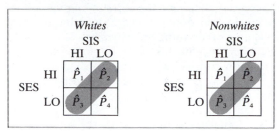

4.   For each of the cells in these two-way tables (one for whites and the other for nonwhites), crude hypertension-related 3-year mortality rates were computed for certain age–sex specific groups (we shall report on only those rates computed for males between 45 and 54 years of age).

    The primary comparison of interest was between $\hat{P}_2$ and $\hat{P}_3$, the "off-diagonal" rates in each of the two-way tables, as illustrated in Figure 21.3.

21.5 METHODOLOGIC STEPS IN FACTOR ANALYSIS

Having briefly described some of the basic factor-analysis terminology and some of the important methodologic issues involved in our example, we are now ready to discuss in detail the factor-analytic method using this example. The method generally proceeds in four steps.[1]

    The *first step* involves setting up the data for input. To this point we have described the data as the set of values of the original variables $X_1, X_2, \ldots, X_p$ for each of the units of study (in the James–Kleinbaum example, $p = 15$ and the units of study are 86 counties in North Carolina). Actually, although considering the data in this form is sufficient, the factor-analytic method requires only the correlation matrix among the variables.[2] Many factor-analysis computer programs, therefore, give the user the option to input the correlation matrix if it has already been computed elsewhere (rather than the original data). This option is particularly useful if the

[1] There are, however, factor-analytic approaches (e.g., the orthogonal power vector method) which could not be described as four-step procedures. We will not discuss such approaches here, but the reader is referred to the text by Overall and Klett (1972).
[2] Recall from Chapter 11 that the correlation matrix **R** has the form

$$\mathbf{R} = \begin{array}{c} \\ X_1 \\ X_2 \\ \vdots \\ X_p \end{array} \begin{array}{cccc} X_1 & X_2 & \cdots & X_p \\ \begin{bmatrix} 1 & r_{12} & \cdots & r_{1p} \\ r_{12} & 1 & \cdots & r_{2p} \\ \vdots & \vdots & \ddots & \vdots \\ r_{1p} & r_{2p} & \cdots & 1 \end{bmatrix} \end{array}$$

researcher is concerned about costs, because direct use of the correlation matrix involves less storage space and less computer time.

There is one option concerning the correlation matrix **R** that the researcher must consider at this first step. This option concerns whether or not to replace the 1's on the diagonal of **R** with some other numbers, which are called *communalities*. The need to consider this option is why the first step is often associated with the expression "prepare the correlation matrix," where the word "prepare" relates to the choice of these communalities. In the next section we shall discuss the implications surrounding this option in more detail.

The *second step* involves using the correlation matrix **R** to determine a set of initial factors. This is usually accomplished by the method of principal components, which we shall describe in Section 21.7. The important point to mention here is that the determination of these initial factors in this second step serves to accomplish the first two of the *three goals of factor analysis*: (1) parsimony; (2) approximate independence; and (3) conceptual meaningfulness.

Parsimony is achieved by representing the information contained in several original variables in terms of a much smaller number of factors. *Approximate independence* is achieved by constructing the factors in such a way so that they are essentially statistically independent.

[Actually, the initial factors, if constructed as principal components, are completely independent rather than only approximately independent. The term "approximate" becomes meaningful after the third step, depending on the procedure used at that step.]

Unfortunately, although the initial factors provide a useful first impression, they usually are difficult to interpret directly without some further manipulation. Consequently, a *third step*, which involves *rotation*, is usually required to achieve *conceptual meaningfulness*. The concept of rotation will be discussed in detail in Section 21.8.

After rotation has been performed, the researcher will have arrived at the set of factor weights to be used to determine the factor scores. This is what is done in the *fourth step*. The researcher will obtain a set of factor scores on the units of study for each derived factor, which may then be used as (newly constructed) variables in further analyses. We will examine the factor scores for the James–Kleinbaum study in Section 21.9 and will describe in Section 21.10 how these scores were used to test the primary research hypothesis of interest for that study.

21.6 STEP 1: PREPARATION OF THE CORRELATION MATRIX

As described in the previous section, factor analysis works on the correlation matrix **R** of the original variables to obtain the desired factors. This should be an intuitively appealing approach, since the relationships among the variables, which we seek to succinctly describe, are represented by the information in the correlation matrix.

[A simple *matrix* equation describes how the **R** matrix is worked on; in particular, this equation is of the form

$$\mathbf{RW} = \mathbf{L}$$

where **W** stands for a matrix of factor weights and **L** stands for a matrix of factor loadings. Although this matrix equation can be interpreted in a purely mathematical way, we hope that the reader can view the expression here in a conceptual way, without being concerned with the matrix mathematics (see Appendix C). What this equation says quite succinctly is that the factor-analytic method determines a weight matrix **W** that can be applied to the correlation matrix **R** to get a desirable factor-loading matrix **L** (by desirable, we mean an **L** matrix that reflects parsimony, approximate independence, and conceptual meaningfulness). This particular **L** can be thought of as resulting from a two-stage procedure. The first stage (or, equivalently, the second step in our four-step procedure) would involve the determination (e.g., by means of principal-components analysis) of an initial weight matrix $\mathbf{W_I}$ and an initial factor loading matrix $\mathbf{L_I}$, such that $\mathbf{RW_I = L_I}$. The second stage (or third step) would involve the use of a rotation matrix **T** to generate the desired factor loading matrix $\mathbf{L(=L_I T)}$ and the corresponding weight matrix $\mathbf{W\ (=W_I T)}$.]

In fact, it is only through the off-diagonal elements that the relationships among the variables are reflected. The diagonal elements (which are 1's in the **R** matrix) are, in fact, superfluous to the purpose of factor analysis. This suggests an alternative approach: Replace these 1's with some other numbers (called *communalities*) and then factor-analyze the resulting "adjusted" correlation matrix, the purpose being to find a more parsimonious and more conceptually meaningful factor-analytic solution than could be obtained using the "unadjusted" **R**.

Of course, there are some questions associated with the use of such an approach. What values for the communalities should be used? Does such an alteration of **R** affect the basic structure and/or meaning of the original variables and, if so, is the goal of the analysis subverted?

Researchers involved in the study of factor analysis have provided some theoretical answers to both of these questions. We shall briefly discuss these answers below; we wish to point out here, however, that many applied researchers using factor analysis have not been entirely satisfied with these theoretical explanations and consequently have, for the most part, worked with the original correlation matrix. James and Kleinbaum, in their study, considered three different correlation matrices, the original and two adjusted correlation matrices. They eventually settled on that factor solution based on the original **R** both because the use of **R** provided the only conceptually meaningful solution and because they could not really justify theoretically the use of an adjusted correlation matrix.

21.6.1 THEORETICAL DISCUSSION OF COMMUNALITIES

Communalities can be incorporated into the theoretical framework of factor analysis by consideration of the underlying structure of the original variables $X_1, X_2, \ldots, X_p$. In Section 21.3 we considered structure in terms of how *a factor* was constructed in terms of the original variables; that is, we considered the expression

$$F_i = \sum_{j=1}^{p} w_{ij} X_j = w_{i1} X_1 + w_{i2} X_2 + \cdots + w_{ip} X_p$$

where F_i and w_{ij} denote the ith factor and the factor weight associated with the jth variable in the ith factor, respectively.

Here we wish to consider structure in terms of how a variable is described by the factors. That is, rather than having F_i on the left side of the equation, we now wish to put a variable (say, X_j) on the left side. In particular, the *general factor-analysis model* which describes X_j (in standardized form) can be written

$$X_j = (\lambda_{1j}F_1 + \lambda_{2j}F_2 + \cdots + \lambda_{kj}F_k) + U_j = C_j + U_j$$

where $F_1, F_2, \ldots, F_k$ are the k factors (which are present in or common to the expressions for all the X's); $\lambda_{1j}, \lambda_{2j}, \ldots, \lambda_{kj}$ are the factor loadings; and U_j is a random component unique to X_j and which is (statistically) independent of the F's. In words, any variable can be represented as the sum of a linear combination of the F's (namely, C_j) and a random quantity (namely, U_j) unique to that variable.

Furthermore, the variance of X_j (which is equal to 1 since X_j is in standardized form) can be written as the sum of two variances:

$$\text{Var}(X_j) = 1 = \sigma^2_{C_j} + \sigma^2_{U_j}$$

where $\sigma^2_{C_j}$ is the variance of the linear combination C_j of the (common) factors and where $\sigma^2_{U_j}$ is the variance of the (unique) component U_j.

It is $\sigma^2_{C_j}$ which we call the *communality* of the variable X_j since this quantity measures the information (in terms of variance) that the variable has in *common* (through the common factors) with all the other variables. Thus, replacing the 1's on the diagonal of the correlation matrix **R** with some other numbers can be shown to be theoretically equivalent to using estimates of the $\sigma^2_{C_j}$'s. Thus, using communalities is equivalent to restricting attention to the common parts (i.e., the C_j's) of the variables (i.e., ignoring the unique components). Keeping 1's on the diagonal, on the other hand, can be viewed as focusing attention on the original variables in their entirety and/or assuming that the variables have no unique components at all.

Thus, the choice of whether or not to use communalities depends on whether the researcher wishes to consider only the common parts of the variables or whether the researcher wishes to work directly with the original variables. Because such considerations are difficult to quantify directly, many researchers tend to prefer using the original correlation matrix **R**, as did James and Kleinbaum in their study.

21.6.2 TWO METHODS FOR ESTIMATING COMMUNALITIES

The two methods most often used to estimate communalities[3] are as follows:

METHOD 1   For each variable, use the squared multiple correlation coefficient (R^2) relating that variable to all other variables in the set.

METHOD 2   For each variable, use the largest (in absolute value) off-diagonal element in **R** associated with that variable (i.e., use the largest correlation involving that variable).

The reader is referred to other texts on factor analysis [e.g., Overall and Klett (1972)] for a discussion of the rationale behind these methods.

[3] Both methods will yield communalities which are less than 1 in value. In general, the true communality for a given variable will be a number between R^2 and 1, where R^2 is defined as in method 1.

TABLE 21.3   Correlation matrices for the James–Kleinbaum study

Correlation matrix—whites

	1(PCI)	2(MED)	3(UEM)	4(PK8+)	5(WCM)	6(BC)	7(PK3−)	8(PSDF)	9(JDM)	10(JDF)	11(CSM)	12(CSF)	13(PM)	14(HR)	15(PCBH)
(PCI)1	1.000														
(MED)2	0.646	1.000													
(UEM)3	−0.168	−0.227	1.000												
(PK8+)4	0.931	0.673	−0.201	1.000											
(WCM)5	0.734	0.865	−0.040	0.765	1.000										
(BC)6	0.286	−0.281	0.278	0.096	−0.170	1.000									
(PK3−)7	−0.905	−0.492	0.129	−0.799	−0.567	−0.481	1.000								
(PSDF)8	0.649	0.225	0.207	0.592	0.484	0.447	−0.650	1.000							
(JDM)9	0.629	0.396	0.025	0.574	0.534	0.303	−0.596	0.566	1.000						
(JDF)10	0.442	0.172	0.084	0.396	0.274	0.439	−0.477	0.483	0.625	1.000					
(CSM)11	0.237	0.034	0.175	0.199	0.245	0.348	−0.270	0.480	0.369	0.397	1.000				
(CSF)12	0.250	0.076	0.241	0.219	0.269	0.395	−0.253	0.451	0.401	0.567	0.694	1.000			
(PM)13	0.158	0.038	0.138	0.189	0.225	0.005	−0.206	0.310	0.308	0.206	0.406	0.192	1.000		
(HR)14	−0.137	−0.176	0.209	−0.135	−0.138	0.137	0.100	0.127	−0.009	0.052	0.071	0.050	−0.005	1.000	
(PCBH)15	−0.158	−0.175	0.265	−0.103	0.003	0.111	0.116	0.402	−0.007	0.071	0.363	0.351	0.235	0.306	1.000

Correlation matrix—nonwhites

	1	2	3	4	5	6	7	8	9	10	11	12	13	14	15
1	1.000														
2	0.720	1.000													
3	0.010	0.053	1.000												
4	0.633	0.446	0.118	1.000											
5	0.519	0.587	0.311	0.433	1.000										
6	0.802	0.702	0.043	0.436	0.377	1.000									
7	−0.930	−0.693	0.088	−0.568	−0.420	−0.732	1.000								
8	0.425	0.289	0.111	0.197	0.473	0.295	−0.269	1.000							
9	0.542	0.337	−0.021	0.251	0.437	0.464	−0.387	0.579	1.000						
10	0.590	0.376	−0.044	0.402	0.441	0.496	−0.463	0.519	0.881	1.000					
11	0.537	0.415	−0.166	0.247	0.301	0.458	−0.404	0.575	0.581	0.499	1.000				
12	0.533	0.437	−0.225	0.257	0.235	0.496	−0.456	0.370	0.427	0.492	0.570	1.000			
13	0.276	−0.011	−0.189	0.117	0.132	0.083	−0.213	0.432	0.463	0.397	0.445	0.226	1.000		
14	−0.144	−0.292	0.022	−0.070	−0.126	−0.170	0.142	0.008	−0.073	−0.118	−0.142	−0.220	0.200	1.000	
15	0.178	−0.024	0.224	0.104	0.206	0.209	0.017	0.559	0.433	0.366	0.359	0.245	0.283	0.182	1.000

21.6.3 EXAMINATION OF THE CORRELATION MATRICES IN THE JAMES–KLEINBAUM STUDY

Table 21.3 gives the two correlation matrices (one for whites and the other for nonwhites) that were used in the James–Kleinbaum study. There are several markings on these matrices (letters, circles, boxes) which need some interpretation. These markings are intended to indicate how a correlation matrix can be examined to give the researcher a preliminary idea of what to expect from a factor analysis. For the James–Kleinbaum study, where the goal of the factor analysis was to construct an SES index and an SIS index from several correlated variables, the ideal correlation matrix for such a purpose would be of the form shown in Figure 21.4.

That is, it was hoped that (1) the SES-type variables would be highly intercorrelated, (2) the SIS-type variables would be highly intercorrelated, and (3) the correlations among the SES-type and SIS-type variables would be close to zero. If these conditions were satisfied, it could be reasoned that the entire set of original variables were separated into two essentially independent factors, one describing SES and the other describing SIS.

The differences between this ideal pattern and the one actually obtained are highlighted by the markings in Table 21.3. The triangular-shaped boxes marked SES enclose the set of intercorrelations of SES-type variables, which were expected to be high. Similarly, the triangular-shaped boxes marked SIS identify the set of expected high SIS intercorrelations. A look inside these boxes indicates that most of these correlations are reasonably high (for both whites and nonwhites), although some correlations are quite close to zero.

The correlations or sets of correlations which are circled are those correlations between SES-type and SIS-type variables which did not conform to the ideal; that is, they are the correlations which were hoped to be negligible but which actually turned out to be considerably different from zero. The fact that there are several circles within both matrices indicates that the final factor solutions for both whites and nonwhites are not likely to be entirely satisfactory. This will be borne out later when we examine the factor loadings obtained from the analysis.

[A factor that does not measure what it was expected to measure is often described as "lacking construct validity." In the James–Kleinbaum example, this could occur if the computed factors are describing phenomena

FIGURE 21.4   *Ideal correlation matrix structure*

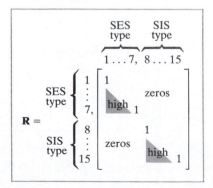

different from those usually measured by indices like SES and SIS or if the variables themselves are not strongly associated with SES and/or SIS measures.]

It should be pointed out that rarely, if ever, does factor analysis produce an ideal result. On the other hand, the fact that most of the SES-variable intercorrelations are fairly high, most of the SIS-variable intercorrelations are also reasonably high, and several of the SES-variable, SIS-variable correlations are somewhat close to zero suggests that the factor solutions are not likely to be bad enough to preclude their use in testing the study hypothesis of primary interest.

21.7 STEP 2: DETERMINING INITIAL FACTORS BY PRINCIPAL-COMPONENTS ANALYSIS

Initial factors are usually determined by the method of *principal-components analysis*.[4] The term "principal components" is the name given to the set of factors resulting from the application of this method.

21.7.1 TOTAL VARIATION IN $X_1, X_2, \ldots, X_p$

In order to describe the method of principal components, we first need to introduce the concept of the *total variation* in the data with regard to the variables $X_1, X_2, \ldots, X_p$. The total variation is mathematically defined to be the sum of the sample variances of the k variables[5]:

$$\text{total variation} = S_1^2 + S_2^2 + \cdots + S_p^2$$

where S_j^2 is the sample variance of X_j, $j = 1, 2, \ldots, p$.

Conceptually, the total variation is a measure of the amount of "uncertainty" associated with the observations on all p variables. By uncertainty, we refer to how much the observations on the units of study differ from one another. For example, if all the observations on a given variable are exactly the same, there would be no uncertainty about what value to expect for that variable, whereas if the observations are quite different from one another, there would be considerable uncertainty about what value to expect.

21.7.2 DEFINITION OF PRINCIPAL COMPONENTS

The purpose of principal-components analysis is to determine factors (i.e., principal components) in such a way so as to explain as much of the total variation in the data as possible with as few of these factors as possible.

---

[4] The term *principal-axis method* is sometimes used instead of "principal-components analysis," particularly when communalities are used.
[5] If, as is usually the case, the variables are in standardized form (so that $S_j^2 = 1$ for every j), the total variation is simply equal to p, the number of variables.

The *first principal component*, PC(1), is that weighted linear combination of the variables which accounts for the largest amount of the total variation in the data. That is, PC(1) is that linear combination of the X's, say

$$PC(1) = w_{(1)1}X_1 + w_{(1)2}X_2 + \cdots + w_{(1)p}X_p$$

where the weights $w_{(1)1}, w_{(1)2}, \ldots, w_{(1)p}$ have been chosen so as to maximize the quantity

$$\frac{\text{variance of PC(1)}}{\text{total variation}}$$

In other words, no other linear combination of the X's will have as large a variance as PC(1).

[When the X's are in standardized form (so that the analysis is based on the correlation matrix), the proportion of the total variation in the data accounted for by PC(1) is

$$\frac{\text{variance of PC(1)}}{p}$$

Also, the weights are chosen subject to the restriction $\sum_{j=1}^{p} w_{(1)j}^2 = 1$ in order that the variance of PC(1) will not exceed the total variation.]

The *second principal component*, PC(2), is that weighted linear combination of the variables which is uncorrelated with PC(1) and which accounts for the maximum amount of the remaining total variation not already accounted for by PC(1). In other words,

$$PC(2) = w_{(2)1}X_1 + w_{(2)2}X_2 + \cdots + w_{(2)p}X_p$$

is that linear combination of the X's which has the largest variance of all linear combinations which are uncorrelated with PC(1).

In general, the ith *principal component* $PC(i)$ is that linear combination

$$PC(i) = w_{(i)1}X_1 + w_{(i)2}X_2 + \cdots + w_{(i)p}X_p$$

which has the largest variance of all linear combinations which are uncorrelated with all of the previously determined $i - 1$ principal components. Actually, it is possible to determine as many principal components as there are original variables. However, in most practical applications, most of the total variation in the data is usually accounted for by the first few components. Furthermore, these components are chosen to be mutually uncorrelated. Thus, the analytic goals of parsimony and independence are quite often achieved via this method.

21.7.3 INTERPRETATION OF PRINCIPAL COMPONENTS

As was pointed out in Section 21.5, principal components are often difficult to interpret directly, and, as a result, further manipulation (e.g., via rotation) is usually required. In any case, it is often found that the first principal component PC(1) represents an overall measure of the information contained in all the variables.[6] Such a general index usually has *large factor loadings* (in absolute value) on almost all the variables. Consequently, it is usually difficult to pin a specific label on such a factor with regard to its being more related to one particular interpretable subset of the variables than to another. Nevertheless, if the primary aim of factor analysis is data reduction by index construction, with no major emphasis on interpretability, the final factor solution is likely to consist of principal components themselves.

On the other hand, if it is desired to find meaningful underlying factors that describe the variation in a set of variables, then another step, involving the technique of *rotation*, is generally required to achieve this goal.

21.7.4 PRINCIPAL-COMPONENTS SOLUTIONS FOR THE JAMES–KLEINBAUM STUDY

The principal-components solutions for both whites and nonwhites in the James–Kleinbaum study are given in Table 21.4. As can be seen from the table, the first two principal components for whites explain 57% of the total variation, while for nonwhites 55% of the total variation is explained by the first two principal components. The

TABLE 21.4   *Principal-components solutions in the James–Kleinbaum study*

	PRINCIPAL COMPONENTS			
	WHITES		NONWHITES	
VARIABLES	SES	SIS	SES	SIS
PCI	0.90	−0.31	0.90	−0.26
MED	0.60	−0.61	0.72	−0.50
UEM	0.00	0.58	0.01	−0.04
P8K+	0.85	−0.37	0.58	−0.31
WCM	0.76	−0.38	0.63	−0.10
BC	0.37	0.55	0.78	−0.31
P3K−	−0.86	0.16	−0.78	0.42
PSDF	0.79	0.32	0.64	0.49
JDM	0.79	0.05	0.76	0.39
JDF	0.65	0.30	0.79	0.26
CSM	0.51	0.55	0.72	0.27
CSF	0.54	0.55	0.66	0.02
PM	0.35	0.25	0.41	0.58
HR	−0.06	0.42	−0.18	0.41
PCBH	0.09	0.63	0.38	0.63
Variance of principal component	5.70	2.85	6.18	2.13
Proportion of total variation explained by principal component	(0.38)	(0.19)	(0.41)	(0.14)

[6] The second principal component often can be interpreted as a contrast between a particular subset of variables and the remaining subset.

proportions 0.38, 0.19, 0.41, and 0.14 are easily obtained from the variances above them (5.70, 2.85, 6.18, and 2.13, respectively) by dividing each of the latter numbers by $p = 15$. This, of course, follows because the variables have been standardized prior to analysis.

Notice also in Table 21.4 that the first component for both whites and nonwhites involves mostly high loadings. We will see in the next section how rotation techniques can be used to provide more interpretable factors by reducing the number of high loadings associated with each factor.

21.8 STEP 3: ROTATION OF INITIAL FACTORS

Rotation is a method of altering the initial factors in order to achieve more interpretability. For example, in rotating the first two principal components obtained in the James–Kleinbaum study, it was hoped that each of the resulting rotated factors would be more interpretable in terms of some meaningful subset of the original variables (e.g., the SES-type variables would ideally be highly correlated with one rotated factor but not with the other rotated factor, and vice versa for the SIS-type variables).

The primary objective of obtaining conceptually meaningful factors by rotation may be translated into more quantitative terms via the concept of *simple structure*. A factor structure is considered to be simple if each of the original variables relates highly to only one factor and each factor can be identified as representing what is common to a relatively small number of variables. Thus, simple structure is said to be achieved when, for each factor, the factor loadings for most variables are near zero and the remaining factor loadings are relatively large. If so, the factor can be conceived as describing the variation shared in common by the subset of variables highly related to it and not describing the variation in the other variables.

It is important, nevertheless, to realize that obtaining simple structure by rotation does not guarantee that the variables that "hang together" on a given factor will describe a conceptually meaningful factor. It is always possible that a relatively good simple structure will still yield factors which are difficult to interpret. Nevertheless, without such simple structure, the interpretation of factors is virtually impossible.

The two best ways to describe how rotation attempts to achieve simple structure are (1) *geometrically*, by rotation of the coordinate axes; and (2) *numerically*, by improving the structure of the factor loadings.

21.8.1 GEOMETRICAL ILLUSTRATION OF THE PURPOSE OF ROTATION

Figure 21.5 illustrates geometrically how rotation may help to achieve simple structure.[7] The figure portrays an essentially ideal result due to rotation, and, consequently, is not based on the James–Kleinbaum data. In the figure there are as many dots as there are variables, and each dot corresponds to a particular variable. The coordinates

---

[7] The rotation of *two* factors is all that is illustrated here. When more than two factors are rotated, rotation is performed pairwise until satisfactory simple structure is achieved for all factors under consideration.

FIGURE 21.5   *Illustration of the purpose of rotation [The goal here is to rotate the axes so that (1) each dot is close to only one of the two rotated axes (simple structure) and (2) dots close to the same rotated axis define a meaningful factor]*

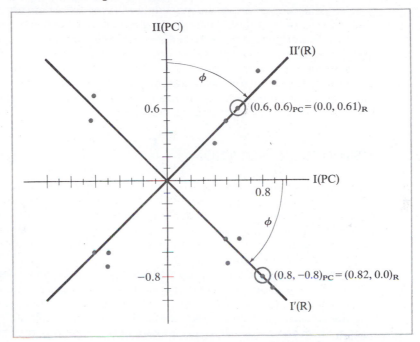

associated with each dot (or variable) are the two factor loadings on that variable for the two factors chosen to be rotated. The values of these coordinates, of course, are based on the two axes used to define the scales of measurement, which represent the two factors being considered for rotation. If these axes are rotated, we shall then have defined two new rotated factors.

Since the initial factors determined are the principal components, the prerotation axes represent two principal components. These are labeled I(PC) and II(PC) in Figure 21.5 for the first and second components, respectively, being considered. In the figure, for example, the coordinates of the two dots that are circled are (0.6, 0.6) and (0.8, −0.8), based on axes I(PC) and II(PC). The first 0.6 is the factor loading on a particular variable for the first component, and the second 0.6 is the factor loading on that same variable for the second component. Similarly, the 0.8 and −0.8 are the factor loadings on another variable for the first and second components, respectively.

The second set of coordinates given for each of these two circled dots are relative to a new pair of axes, labeled as I′(R) and II′(R). These two axes were determined by rotating the original axes clockwise (through an angle ϕ).

Notice what has been achieved by this rotation. Each one of these circled dots is now close to only one of the two new (rotated) axes. In fact, the coordinates of the two circled dots have now been changed as follows:

$$(0.6, 0.6)_{PC} \rightarrow (0.0, 0.61)_R$$

and

$$(0.8, \ -0.8)_{PC} \rightarrow (0.82, \ 0.0)_R$$

Similarly, the coordinates of the other dots will tend to be high for one coordinate and close to zero for the other. Thus, simple structure has been achieved!

The importance of this accomplishment in conceptual terms comes from the fact that the dots (i.e., the variables) can now be seen to be clustered into two subgroups, one subgroup lying close to one rotated axis and the other subgroup lying close to the other rotated axis. Since these new axes represent new (rotated) factors, we are now in the position of being able to interpret each new factor in terms of the particular subgroup of variables lying close to that factor.

21.8.2 METHODS OF ROTATION

In performing a rotation, there are two ways in which the axes may be rotated. First, the axes may be kept in the same orientation to one another during rotation, so that they are still perpendicular after rotation (i.e., there is a 90° angle between the two new axes); this is called *orthogonal rotation*. Second, each axis may be rotated independently, so that they are *not* necessarily perpendicular to one another after rotation; this is called *oblique rotation*.

The difference between these two types of rotation can be described in terms of the angles of rotation. In general, there are two such angles: the angle between the original first axis (I) and its corresponding rotated axis (I'), and the angle between the original second axis (II) and its rotated axis (II'). Under orthogonal rotation (because the axes are kept perpendicular to one another during rotation), only one angle has to be specified—this is the angle ϕ in Figure 21.6. Under oblique rotation, two angles need to be specified—these are the angles ϕ_1 (between axes I and I') and ϕ_2 (between II and II') in Figure 21.6.

An important statistical difference between orthogonal and oblique rotation is that the factors resulting from the orthogonal rotation of principal components will remain statistically uncorrelated (i.e., the factor cosines will all be zero), whereas factors resulting from an oblique rotation will usually be correlated to some extent (i.e.,

FIGURE 21.6   *Orthogonal and oblique rotation*

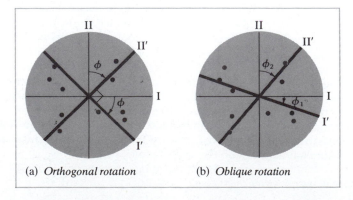

(a) *Orthogonal rotation*   (b) *Oblique rotation*

some or all of the factor cosines will be nonzero). To generate statistically uncorrelated factors is a desirable goal, primarily because of the advantages associated with representing a complex set of interrelationships among several correlated variables in terms of a few uncorrelated indices. Another desirable property of orthogonal rotation is that the amount of the total variation accounted for by the factors under consideration is unaffected by the rotation. For example, under orthogonal rotation of the first two principal components (for whites) obtained in the James–Kleinbaum study, the two rotated factors account for 57% of the total variation, which is precisely the percentage accounted for by the first two principal components.

Unfortunately, however, the use of just orthogonal rotation may result in not finding the best set of rotated factors. Often, the researcher can reason on an empirical basis what characteristics the factors should be measuring, and the use of orthogonal rotation often is not sufficient to determine factors with the desired attributes. Since the primary goals of rotation are "simple structure" and "meaningful factors," these goals would more likely be achieved if oblique as well as orthogonal rotations were considered. In fact, by permitting the factor axes to become *oblique*, it is frequently possible to arrive at much more interpretable factors.

Although perhaps the most simplistic approach to rotation would be to examine the data graphically and then decide upon the proper rotation visually, this involves considerable subjectivity. Because of this, rotation is generally performed using computerized algorithms which are based on well-defined quantitative criteria. Nevertheless, we do recommend that the researcher consider looking at the data graphically, since a geometric picture often provides additional insight.

21.8.3 COMPUTER ALGORITHMS FOR ORTHOGONAL ROTATION

Three algorithms for orthogonal rotation which are available as options in most factor-analysis computer programs are the varimax, quartimax, and equimax methods. Varimax is the most often used of these methods. The essential differences among these three methods are as follows: *varimax* attempts to achieve simple structure with respect to the *columns* of the factor-loading matrix; *quartimax* attempts to achieve simple structure with respect to the *rows* of the factor-loading matrix; and *equimax* attempts to achieve simple structure with respect to both the *rows and columns* of the factor-loading matrix. For more detailed descriptions of these methods, we refer the reader to other texts on factor analysis [e.g., see Overall and Klett (1972) and Rummel (1970)].

21.8.4 COMPUTER ALGORITHMS FOR OBLIQUE ROTATION

A large number of computer algorithms have been developed to perform oblique rotation. Those most conveniently available are the *oblimin*, *quartimin*, *biquartimin*, and *covarimin* algorithms. All of these represent algorithms designed to satisfy various types of simple structure criteria. Unfortunately, no one algorithm always produces a superior solution, so that several different algorithms may need to be tried on the same data set. [See Rummel (1970) and Harman (1960) for detailed descriptions of these algorithms.]

The user of computer algorithms for factor analysis should also be aware of some additional complexities of oblique rotation not present for orthogonal rotation. In

FIGURE 21.7   *Pattern loadings*

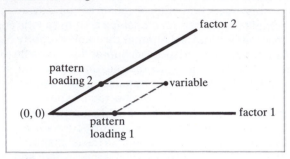

particular, when considering the results of an oblique rotation performed using some computer algorithm, it is necessary to know that there are two alternative representations for the factor loadings [i.e., the coordinates of the points (or variables) with respect to the rotated axes] depending on how each point is projected onto the rotated axes. One alternative, yielding what are called *pattern loadings* (Figure 21.7) is based on projecting each point onto each rotated axis by lines *parallel to these two axes* (so that the factor loadings are then defined as the two projected coordinates. The other alternative, yielding what are called *structure loadings* (Figure 21.8) is based on projecting each point onto each rotated axis by lines *perpendicular to these two axes*. As the diagrams suggest, corresponding pattern and structure loadings will generally be different under oblique rotation. Consequently, some computer programs (e.g., SPSS) print out two factor-loading matrices (called the *pattern matrix* and the *structure matrix*) for the same data set. Other programs (e.g., BMD) let the user specify a priori which of these matrices is to be printed out.

An important difference between pattern and structure loadings is that pattern loadings are not really correlation coefficients between variables and factors, whereas structure loadings do represent such correlations. Nevertheless, the pattern matrix is often much more useful than the structure matrix with regard to the interpretation of the rotated factors.

An additional characteristic of oblique rotation procedures associated with the use of factor-analysis programs involves an option to consider an adjusted (usually called *reference*) solution obtained from the original (or *primary*) oblique solution.

FIGURE 21.8   *Structure loadings*

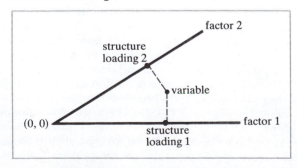

FIGURE 21.9   *Relationship between reference axes and primary axes*

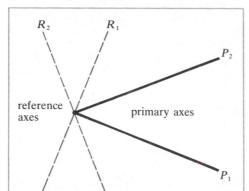

This reference solution is obtained by determining two new adjusted axes which are perpendicular to the primary axes (see Figure 21.9). As with the primary axes, there are pattern and structure matrices associated with the reference axes. In contrast with the primary loadings, however, the reference pattern loadings are correlation coefficients (as were the primary structure loadings), whereas the reference structure loadings (like the primary pattern loadings) are not correlations but are nevertheless more often useful for making interpretations about the nature of the rotated factors. Some programs (e.g., BMD), in fact, only print out either the primary pattern or the reference structure solution and ignore the primary structure and the reference pattern solutions, because the former two solutions are generally more useful for interpretation purposes.

21.8.5   ROTATION IN THE JAMES–KLEINBAUM STUDY

Figures 21.10 and 21.11 illustrate what was achieved by oblique rotation of the first two principal components in the James–Kleinbaum study. In each of these graphs, the SES-type variables are designated by dots (•), whereas the SIS-type variables are designated by dotted X's (×). Those variables that are not strongly associated with either (rotated) factor are circled; these variables will be identified when we examine the factor-loading matrices.

Although Figures 21.10 and 21.11 illustrate *oblique* rotation, examination of these graphs indicates that the rotated axes are not far from being perpendicular. This suggests that orthogonal rotation would likely yield similar results, which indeed was the case.

For both whites and nonwhites, notice that the results of rotation are far from ideal, since the dots (•) do not closely hug one rotated axis and the dotted X's (×) do not closely hug the other rotated axis in either figure. Nevertheless, it can still be seen that the SES-type variables are at least "approximately" clustered along one axis and the SIS-type variables are somewhat similarly clustered along the other axis.

FIGURE 21.10   *Oblique rotation; whites, North Carolina counties, 1960*
(×, SIS-type variable; •, SES-type variable)

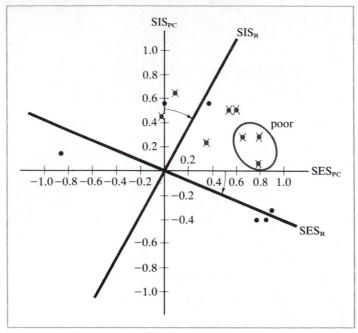

FIGURE 21.11   *Oblique rotation: nonwhites, North Carolina counties, 1960*
(×, SIS-type variable; •, SES-type variable)

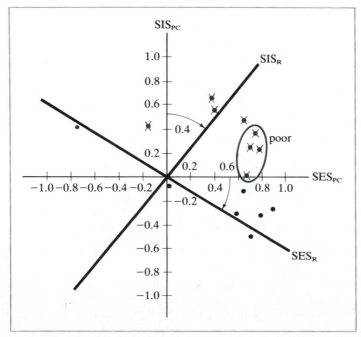

21.8.6 EXAMINATION OF FACTOR-LOADING MATRICES

The results of rotation are usually quantified and evaluated through consideration of the factor-loading matrices. The goal of simple structure is represented in terms of a factor-loading matrix which contains high loadings on only one (or very few) factor(s) for each variable and/or high loadings on only a few variables for each factor, with close-to-zero loadings otherwise.

Table 21.5 illustrates the idealized simple structure hoped for in the James–Kleinbaum factor analysis. Note that the SES factor was expected to load high on the SES-type variables and to load zero on the SIS-type variables; the SIS factor, on the other hand, was expected to load high on the SIS-type variables and to load zero on the SES-type variables. Some variables, such as BC and P3K−, did not have clearly predictable factor loadings and were so designated with question marks.

Comparing the factor loadings of initial factors versus rotated factors

Tables 21.6 and 21.7 illustrate how rotation helps to improve the factor-loading matrix with regard to the goal of simple structure. In these tables, the factor loadings for the principal-components solution and for the oblique (reference)-factor solution are presented for the data on whites and nonwhites from the James–Kleinbaum study. For each racial group, the principal components are compared with the rotated factors with regard to simple structure; loadings that would be considered poor in comparison with the idealized loadings have been circled. *The main feature to notice here is that the number of circled factor loadings has been considerably reduced by rotation.* Thus, although the rotated factors do not have perfect simple structure, they are nevertheless much more interpretable as SES and SIS factors after rotation than before.

Comparing various rotational methods

Tables 21.8 and 21.9 describe the results of three different methods of rotation: reference oblique, orthogonal varimax, and orthogonal quartimax. James and Kleinbaum felt it appropriate to consider different methods of rotation before deciding on the best simple structure solution.

TABLE 21.5 *Idealized factor loadings for the James–Kleinbaum study*

VARIABLE	SES	SIS
PCI	High(+)	0
MED	High(+)	0
UEM	High(−)	High(+)
P8K+	High(+)	0
WCM	High(+)	0
BC	High(+)	?
P3K−	High(−)	?
PSDF	0	High(+)
JDM	0	High(+)
JDF	0	High(+)
CSM	0	High(+)
CSF	0	High(+)
PM	0	High(+)
HR	0	High(+)
PCBH	0	High(+)

TABLE 21.6 *Principal components versus oblique-rotated factor loadings for whites in North Carolina counties, 1960*

VARIABLE	PRINCIPAL COMPONENTS		OBLIQUE FACTORS	
	SES	SIS	SES	SIS
PCI	0.90	−0.31	0.95	0.04
MED	0.60	(−0.61)	0.82	−0.34
UEM	(0.00)	0.58	−0.27	0.54
P8K+	0.85	−0.37	0.93	−0.03
WCM	0.76	(−0.38)	0.86	−0.06
BC	0.37	0.55	(0.06)	0.65
P3K−	−0.86	0.16	−0.84	−0.17
PSDF	(0.79)	0.32	(0.55)	0.59
JDM	(0.79)	(0.05)	(0.67)	0.34
JDF	(0.65)	0.30	(0.44)	0.53
CSM	(0.51)	0.55	0.19	0.71
CSF	(0.54)	0.55	0.21	0.71
PM	(0.35)	0.25	0.19	0.37
HR	−0.06	0.42	−0.25	0.37
PCBH	0.09	0.63	−0.22	0.62

Examination of Tables 21.8 and 21.9 shows that, as expected, *there is little difference among the results obtained for each of these rotations.* For whites, the same number of poor loadings is circled for each rotational method; for nonwhites, the oblique solution has one less circled loading than either of the other two solutions. Also, speaking qualitatively, it appears that the factor solution for whites is somewhat better in terms of simple structure than the factor solution for nonwhites.

TABLE 21.7 *Principal components versus oblique-rotated factor loadings for nonwhites in North Carolina counties, 1960*

VARIABLE	PRINCIPAL COMPONENTS		OBLIQUE FACTORS	
	SES	SIS	SES	SIS
PCI	0.90	−0.26	0.88	0.22
MED	0.72	(−0.50)	0.88	−0.07
UEM	(0.01)	(−0.04)	(0.03)	(−0.03)
P8K+	0.58	−0.31	0.65	0.03
WCM	0.63	−0.10	0.56	0.23
BC	0.78	−0.31	0.81	0.12
P3K−	−0.78	0.42	−0.88	−0.03
PSDF	(0.64)	0.49	0.18	0.75
JDM	(0.76)	0.39	(0.34)	0.73
JDF	(0.79)	0.26	(0.44)	0.62
CSM	(0.72)	0.27	(0.38)	0.60
CSF	(0.66)	(0.02)	(0.50)	0.35
PM	(0.41)	0.58	−0.06	0.71
HR	−0.18	0.41	−0.41	0.26
PCBH	(0.38)	0.63	−0.12	0.75

TABLE 21.8 *Factor loadings for three different rotational methods; whites, North Carolina counties, 1960*

VARIABLE	REFERENCE OBLIQUE		ORTHOGONAL VARIMAX		ORTHOGONAL QUARTIMAX	
	SES	SIS	SES	SIS	SES	SIS
PCI	0.95	0.04	0.95	0.10	0.95	0.10
MED	0.82	−0.34	0.80	−0.30	0.80	−0.29
UEM	−0.27	0.54	−0.24	0.52	−0.24	0.52
P8K+	0.93	−0.03	0.93	0.02	0.93	0.03
WCM	0.86	−0.06	0.85	−0.01	0.85	−0.02
BC	(0.06)	0.65	(0.10)	0.66	(0.10)	0.66
P3K−	−0.84	−0.17	−0.85	−0.22	−0.85	−0.22
PSDF	(0.55)	0.59	(0.59)	0.62	(0.58)	0.62
JDM	(0.67)	0.34	(0.69)	0.38	(0.69)	0.38
JDF	(0.44)	0.53	(0.47)	0.55	(0.46)	0.55
CSM	0.19	0.71	0.23	0.72	0.22	0.72
CSF	0.21	0.71	0.25	0.73	0.25	0.73
PM	0.19	0.37	0.21	0.38	0.20	0.38
HR	−0.25	0.37	−0.23	0.36	−0.23	0.36
PCBH	−0.22	0.62	−0.19	0.60	−0.19	0.60

Because an oblique rotation was considered preferable on empirical grounds and also simply because little difference in factor loadings was found when other methods of rotation were used, James and Kleinbaum decided to consider as the final factor loadings those obtained by means of the (reference) oblique rotation method; the values of these factor loadings for whites and nonwhites are given in the first two columns of Tables 21.8 and 21.9.

TABLE 21.9 *Factor loadings for three different rotational methods; nonwhites, North Carolina counties, 1960*

VARIABLE	REFERENCE OBLIQUE		ORTHOGONAL VARIMAX		ORTHOGONAL QUARTIMAX	
	SES	SIS	SES	SIS	SES	SIS
PCI	0.88	0.22	0.88	(0.34)	0.92	0.19
MED	0.88	−0.07	0.87	0.05	0.87	−0.09
UEM	(0.03)	(−0.03)	(0.03)	(−0.03)	(0.03)	(−0.03)
P8K+	0.65	0.03	0.65	0.12	0.66	0.01
WCM	0.56	0.23	0.56	0.31	0.61	0.21
BC	0.81	0.12	0.80	0.23	0.83	0.10
P3K−	−0.88	−0.03	−0.87	−0.15	−0.88	−0.01
PSDF	0.18	0.75	0.21	0.78	(0.33)	0.73
JDM	(0.34)	0.73	(0.36)	0.78	(0.48)	0.71
JDF	(0.44)	0.62	(0.46)	0.69	(0.57)	0.60
CSM	(0.38)	0.60	(0.40)	0.68	(0.50)	0.58
CSF	(0.50)	0.35	(0.51)	0.42	(0.51)	0.33
PM	−0.06	0.71	−0.04	0.70	0.08	0.70
HR	−0.41	0.26	−0.40	0.21	−0.36	0.27
PCBH	−0.12	0.75	−0.09	0.73	0.03	0.74

21.9 STEP 4: DETERMINATION OF THE FACTOR SCORES

Table 21.10 gives the factor weights associated with the factor loadings obtained via (reference) oblique rotation in the James–Kleinbaum study. As previously described, a factor score is a numerical value of a factor F obtained by substituting specific values for the (standardized) X's into the expression

$$F = w_1 X_1 + w_2 X_2 + \cdots + w_p X_p$$

In computing such scores in the James–Kleinbaum study, variable values and factor weights for whites naturally were used to compute the SES and SIS factor scores for whites, and the variable values and factor weights for nonwhites were used to compute the SES and SIS scores for nonwhites.

The computation of these factor scores allows for the analysis strategy described in Section 21.4 to be carried out in order to test the major study question: Do high-stress counties have higher mortality rates than low-stress counties?

Recall that the steps involved in the analysis of these factor scores are as follows:

1. Rank-order the factor scores (by factor) separately for each race.
2. Divide each ordered set of scores into a high-score group and a low-score group.
3. For each race separately, form a two-way table based on the HI and LO groups defined for each factor:

		SIS	
		HI	LO
SES	HI	$\hat{P}_1$	$\hat{P}_2$
	LO	$\hat{P}_3$	$\hat{P}_4$

TABLE 21.10   *Factor weights from (reference) oblique rotation in the James–Kleinbaum study*

VARIABLE	WHITES		NONWHITES	
	SES	SIS	SES	SIS
PCI	0.19	−0.02	0.19	−0.00
MED	0.18	−0.14	0.22	−0.10
UEM	−0.07	0.18	0.01	−0.01
P8K+	0.19	−0.04	0.15	−0.05
WCM	0.17	−0.05	0.11	0.03
BC	−0.01	0.20	0.18	−0.03
P3K−	−0.16	−0.02	−0.21	0.07
PSDF	0.09	0.16	−0.02	0.24
JDM	0.12	0.08	0.01	0.22
JDF	0.07	0.15	0.05	0.17
CSM	0.01	0.21	0.04	0.17
CSF	0.02	0.21	0.09	0.07
PM	0.02	0.11	−0.08	0.25
HR	−0.06	0.13	−0.12	0.13
PCBH	−0.07	0.20	−0.10	0.27

4. Compute crude hypertension-related mortality rates for each of the cells in the table and then compare $\hat{P}_2$ (low-stress rate) with $\hat{P}_3$ (high-stress rate).

Tables 21.11 and 21.12 summarize the numerical results of this procedure up through the third step for males, 45–54 years of age. These tables give the standardized factor scores for each of the four "stress" groups defined in the third step. Mean factor scores are also given for each group.

21.10 STUDY RESULTS

James and Kleinbaum conjectured that the crude hypertension-related death rate for *nonwhite* males in high-stress counties would be significantly higher than that in low-stress counties. Level of socioecologic stress was not expected, however, to mediate these rates for *white* males.

TABLE 21.11 *Standardized factor scores in the four stress groups: white males, 45–54 years of age, North Carolina counties, 1960*

	N = 25		N = 18		N = 18		N = 25	
	HI SES	HI SIS	HI SES	LO SIS	LO SES	HI SIS	LO SES	LO SIS
	1.51	0.03	0.25	−0.61	−0.13	0.71	−1.00	−0.37
	1.09	1.37	0.63	−0.79	−0.83	0.53	−0.64	−0.38
	0.79	1.54	0.45	−0.25	−1.08	0.40	−0.41	−1.70
	0.15	2.75	2.58	−0.25	−0.35	0.99	−0.59	−0.35
	0.23	0.91	0.59	−1.84	−0.14	−0.04	−0.57	−1.52
	0.84	0.73	0.59	−0.15	−0.84	0.06	−0.24	−0.76
	1.07	0.11	3.09	−0.42	−0.35	−0.03	−0.90	−0.10
	0.32	0.76	0.25	−1.54	−0.24	0.12	−1.05	−1.12
	2.73	0.35	1.75	−1.31	−0.84	0.13	−0.75	−0.43
	0.50	1.39	0.76	−0.79	−1.10	1.92	−0.63	−1.65
	2.29	0.72	0.39	−1.29	−0.36	0.07	−1.13	−1.84
	0.33	0.55	−0.11	−0.30	−0.54	1.41	−1.08	−1.53
	0.40	0.48	0.01	−0.73	−0.45	0.91	−1.31	−0.52
	0.58	0.76	2.77	−0.27	−0.50	0.56	−1.64	−0.87
	0.34	0.61	−0.07	−1.32	−0.16	2.24	−0.38	−0.74
	0.45	0.30	−0.08	−0.68	−1.44	2.70	−0.57	−0.71
	0.30	0.23	0.38	−0.38	−1.40	0.30	−0.92	−0.37
	0.47	0.22	0.14	−0.48	−1.19	0.97	−1.18	−0.23
	2.07	1.37					−0.65	−1.66
	0.23	0.00					−0.33	−0.84
	0.08	1.20					−1.11	−1.10
	−0.06	1.54					−0.25	−0.32
	0.03	0.69					−1.27	−0.59
	0.74	0.58					−0.61	−0.24
	0.12	0.59					−0.86	−0.39
Total	17.66	19.78	13.95	−13.40	−11.94	13.73	−20.07	−20.33
Mean	0.71	0.79	0.78	−0.74	−0.66	0.76	−0.80	−0.81

TABLE 21.12 *Factor scores in the four stress groups: nonwhite males, 45–54 years of age, North Carolina counties, 1960*

	N = 23		N = 20		N = 20		N = 23	
	HI SES	HI SIS	HI SES	LO SIS	LO SES	HI SIS	LO SES	LO SIS
	1.40	1.63	0.87	−0.57	−0.72	−0.08	−1.02	−0.74
	0.51	1.52	1.04	−0.52	−1.30	0.33	−0.48	−0.61
	1.31	0.12	1.55	−1.27	−0.60	0.51	−0.24	−1.02
	1.40	0.51	1.11	−0.18	−1.05	0.25	−0.20	−1.28
	−0.09	0.42	0.49	−1.34	−1.03	0.02	−0.80	−1.29
	1.24	0.22	0.92	−1.06	−0.91	0.39	−0.30	−0.25
	1.21	0.46	1.70	−1.28	−1.62	0.04	−0.64	−0.58
	1.53	2.19	−0.04	−0.30	−0.99	0.28	−0.29	−1.44
	1.93	2.68	1.28	−1.24	−1.46	0.11	−0.61	−0.55
	0.81	0.38	1.23	−1.30	−1.10	0.82	−1.67	−0.40
	2.47	1.06	0.96	−0.20	−0.37	0.72	−1.35	−0.26
	0.16	2.61	0.57	−0.78	−1.05	0.56	−0.21	−0.69
	0.15	0.08	0.40	−0.52	−0.45	−0.13	−1.67	−0.28
	−0.10	1.46	1.03	−0.45	−1.33	0.37	−0.33	−0.83
	1.61	2.38	0.82	−0.74	−1.29	0.35	−1.21	−0.60
	−0.06	0.99	0.05	−0.87	−0.93	−0.13	−0.57	−0.82
	0.94	3.29	0.23	−0.44	−0.79	0.36	−0.62	−0.84
	−0.02	0.22	1.50	−1.75	−0.32	0.28	−0.87	−1.07
	1.05	−0.14	0.17	−0.83	−0.86	−0.03	−0.55	−0.25
	0.53	0.90	0.25	−0.29	−1.04	0.94	−0.79	−0.67
	1.29	0.05					−1.04	−0.36
	0.75	1.72					−1.24	−0.66
	−0.02	1.34					−0.21	−0.65
Total	20.00	26.09	16.13	−15.93	−19.21	5.96	−16.91	−16.14
Mean	0.87	1.13	0.81	−0.80	−0.96	0.30	−0.74	−0.70

TABLE 21.13 *Crude hypertension-related mortality rates in the four stress groups: white males, 45–54 years of age, North Carolina counties, 1960*

SES	SIS		MARGINALS
	HIGH	LOW	
High	No. counties = 25 No. deaths = 75 PAR = 79,852 Mortality rate = 926.7×10^{-6}	No. counties = 18 No. deaths = 30 PAR = 42,946 Mortality rate = 698.5×10^{-6}	846.91×10^{-6}
Low	No. counties = 18 No. deaths = 24 PAR = 23,940 Mortality rate = $1,002.5 \times 10^{-6}$	No. counties = 25 No. deaths = 32 PAR = 24,298 Mortality rate = $1,316.9 \times 10^{-6}$	$1,160.91 \times 10^{-6}$
Marginals	944.19×10^{-6}	922.01×10^{-6}	

TABLE 21.14 *Crude hypertension-related mortality rates in the four stress groups: nonwhite males, 45–54 years of age, North Carolina counties, 1960*

SES	SIS		MARGINALS
	HIGH	LOW	

High	No. counties = 23
	No. deaths = 156
	PAR = 22,190
	    Mortality rate = $7,030.1 \times 10^{-6}$
		No. counties = 20
		No. deaths = 18
		PAR = 4,540
		    Mortality rate = $\boxed{3,964.7 \times 10^{-6}}$
			$6,509.53 \times 10^{-6}$

Low	No. counties = 20
	No. deaths = 113
	PAR = 14,391
	    Mortality rate = $\boxed{7,852.1 \times 10^{-6}}$
		No. counties = 23
		No. deaths = 50
		PAR = 9,336
		    Mortality rate = $5,355.6 \times 10^{-6}$
			$6,869.81 \times 10^{-6}$

Marginals	$7,253.54 \times 10^{-6}$	$4,900.54 \times 10^{-6}$	

Tables 21.13 and 21.14 present the data necessary to test the appropriate hypotheses of interest. The rates of interest are

$$\text{whites} \begin{cases} \hat{P}_2 = 1,002.5 \text{ deaths per million} \\ \hat{P}_3 = 698.5 \text{ deaths per million} \end{cases}, \quad \text{nonwhites} \begin{cases} \hat{P}_2 = 7,852.1 \text{ deaths per million} \\ \hat{P}_3 = 3,964.7 \text{ deaths per million} \end{cases}$$

Based on these rates, the conjecture concerning nonwhite males was confirmed using a simple one-tailed Z test ($P < 0.005$). Furthermore, the high-stress rate was nearly twice as large as that for the low-stress group! For white males the difference between the two observed rates was not significant ($P > 0.10$), as expected.

Thus, the results obtained by James and Kleinbaum provide some evidence in support of the theory that socioecologic stress is a mediating factor in the determination of cerebrovascular disease for populations of the type considered in their study. Nevertheless, similar conclusions would have to be made independently by other researchers in order to solidify the findings of James and Kleinbaum.

Also, other methods of analysis of these data might also be considered, an example being the use of regression analysis with county death rate as the dependent variable and with the stress scores and other concomitant variables as independent variables. In any case the use of factor analysis has been shown here to be an important tool in data analysis.

PROBLEMS   1. Determine whether each of the following statements about the method of factor analysis is generally true or false:

(a) Factor analysis may be used as a method for data reduction.

(b) Factor analysis may be used to help determine the underlying dimensions that are measured by an instrument like a questionnaire.

(c) Factor analysis differs from regression analysis in that the former considers nominal dependent variables, whereas the latter considers continuous dependent variables.

(d) Factor analysis is inappropriate if the original (input) variables are measured on different scales (e.g., responses to item i in a questionnaire can range between 1 and 3, whereas responses to item j can range between 1 and 10).

(e) Mathematically speaking, a factor is a variable defined as the sum of all the original variables input into the factor analysis.

(f) Communalities are the 1's on the diagonal of the correlation matrix.

(g) The use of communalities assumes that each variable can be described entirely in terms of factors common to all variables.

(h) Inspection of the correlation matrix would suggest the possibility of a good factor solution if the set of original variables can be partitioned into mutually exclusive groups for which there are low correlations between variables within the same group and high correlations between variables from different groups.

(i) Factor loadings are correlations between factors.

(j) It is generally true that the higher (in absolute value) a factor weight is, the higher (in absolute value) is the factor loading.

(k) If a factor loading is highly negative, the variable involved is an important component of the associated factor.

(l) If the factor loadings for a given variable on all factors obtained are near zero, the removal of this variable followed by a second factor analysis on the remaining variables will always yield a better factor solution.

(m) The method of principal components usually results in factors that are conceptually meaningful.

(n) If the main goal of factor analysis is data reduction, the method of principal components can often be used without being followed by any rotation of the principal-component solutions.

(o) If the proportion of total variation explained by the first principal component is small (e.g., below 0.20), the use of the first principal component as an overall general factor is always preferable to the use of the unweighted average of the (standardized) original variables.

(p) There are as many principal components as there are original variables.

(q) The second principal component is uncorrelated with the first principal component and always explains a lesser amount of the total variation.

(r) Initial factors are rotated to achieve parsimony and independence.

(s) Oblique rotation of two factors should be preferred to orthogonal rotation if the two factors are considered to be relatively independent on either theoretical or empirical grounds.

(t) A simple-structure-factor solution, even if obtained, does not guarantee that the resulting factors are conceptually meaningful.

(u) In an oblique rotation of two factors, the starting (i.e., principal-component) axes must be rotated at two unequal angles of rotation.

(v) Geometrically speaking, the goal of any rotation is to rotate the original axes so that each point on the graph lies very close to only one of the two rotated axes.

(w) A factor solution has good construct validity if, for each factor, most variables have nearly zero loadings, yet a few variables have very high loadings.

(x) A factor obtained from a factor analysis has good construct validity if the size and direction of the factor loadings correspond favorably to the size and direction of such loadings as perceived theoretically or empirically by the investigator.

(y) If a factor solution has poor construct validity, the only possible explanation is that the factors obtained measure different dimensions than those perceived by the investigator.

2. A questionnaire containing 19 statements was developed by Arkin (1976) to measure sex-role orientation in women. Each item was scaled so that a response high on the scale reflected a modern or self-actualized sex-role orientation (i.e., role behavior not predetermined by sex) and a response low on the scale reflected a more traditional attitude (i.e., favoring strong differentiation between roles according to sex). To determine what, if any, were the underlying dimensions of sex-role orientation as measured by the instrument, factor analysis was applied to data collected using this instrument on 34 women reported to have an abnormal pap smear test for cervical cancer. (The basic aim of the study was to determine whether sex-role orientation was related to patient delay in pursuing further treatment for cancer.)

    One of the factor-analysis computer runs made on these data involved a reduced set of 13 items (after eliminating items with low loadings on all factors) with 1's on the diagonal of the correlation matrix. The loadings for the unrotated (i.e., principal components) solution and for the varimax rotated solution involving the first two factors are as follows:

LOADINGS FOR PRINCIPAL COMPONENTS			LOADINGS FOR VARIMAX ROTATED FACTORS		
ITEM NO.	FACTOR 1	FACTOR 2	ITEM NO.	FACTOR 1	FACTOR 2
1	0.770	−0.064	1	0.511	0.580
2	0.689	0.484	2	0.832	0.129
3	0.750	0.031	3	0.562	0.497
4	0.634	0.262	4	0.639	0.251
5	0.686	−0.440	5	0.190	0.793
6	0.661	0.398	6	0.753	0.172
7	0.682	−0.424	7	0.198	0.778
8	0.591	−0.433	8	0.126	0.721
9	0.581	−0.553	9	0.035	0.801
10	0.581	0.410	10	0.703	0.107
11	0.553	0.397	11	0.674	0.097
12	0.334	0.266	12	0.426	0.040
13	0.395	−0.294	13	0.081	0.486

(a) Based on the information provided, what would you say is the most serious drawback to the utility of the factor solution obtained in this study?

(b) How would you contrast the principal-components solution with the varimax rotated solution in regard to simple structure?

(c) Which items, if any, would you consider for possible elimination from the factor-analytic model?

(d) Using the rotated solution, determine the cluster of variables that best describes the first factor and the cluster that best describes the second factor. Can you conclude from the information provided whether or not any of the factors are conceptually meaningful?

(e) Plot each of the original 13 items as points on graph paper by using as the coordinates for each item the pair of factor loadings for the first two principal components associated with that item.

(f) Using the graph constructed in part (e), rotate your axis orthogonally until you have obtained what you think is the best fit to the plotted points and then draw your rotated axes on the same graph. Using these rotated axes as your new coordinate frame of reference, determine (roughly) the coordinates of at least three of the points on the graph. Compare these rotated loadings with their corresponding varimax loadings.

(g)   Using the graph in part (e), rotate your axes obliquely until you are satisfied with the fit to the points and then draw your oblique axes on the graph. Using these oblique axes as your new coordinate frame of reference, determine the coordinates of the same points you selected in part (f) and compare these rotated loadings with the loadings obtained in (f).

(h)   The number of factors to be used in describing the underlying dimensions in a data set is usually determined by considering the variances (also called *eigenvalues*) and the proportions of total variation explained (P_v) by each of the principal components. A table containing such information is presented for the principal-component solution given earlier (note that 13 variables always yield 13 principal-component factors).

PRINCIPAL COMPONENT NUMBERS	1	2	3	4	5	6
Eigenvalues	5.00	1.82	1.17	1.12	0.83	0.62
P_v	0.385	0.140	0.090	0.086	0.064	0.048

PRINCIPAL COMPONENT NUMBERS	7	8	9	10	11	12	13
Eigenvalues	0.59	0.55	0.46	0.34	0.23	0.16	0.10
P_v	0.046	0.042	0.036	0.026	0.018	0.013	0.007

Two rough rules for selecting the number of factors are:

(1)   Use as many factors as there are variances larger than 1, provided that the first principal-component factor explains at least 20% of the total variation in the data.

(2)   Use as many factors as there are conceptually meaningful rotated factors, provided that the first principal-component factor explains at least 20% of the total variation.

Based on these rules, what would you recommend to the investigator who finds only the first two factors to be conceptually meaningful in the example above?

(i)   For orthogonal rotation, the proportion of the total variation explained by each factor may always be computed from the factor loadings on that factor by summing the squares of these loadings and dividing by the number of original (standardized) variables considered in the analysis (i.e., $P_v = \sum_{i=1}^{p} l_i^2 / p$, where l_i denotes the loading on the ith variable, $i = 1, 2, \ldots, p$). Using this rule, determine the proportion of total variation explained by each of the rotated factors in the example given, and show that, even though these values are not the same as those for the unrotated factors (i.e., the principal components), their sum is the same as the sum for the unrotated factors (0.525).

3. Kleinbaum and Kleinbaum (1976a) developed an instrument containing 28 items to measure three basic dimensions of attitudes toward statistics of students enrolled in an introductory statistics course. The three attitudes were defined as (1) the *confidence* that the student has in his/her ability to learn, understand, or use statistics; (2) the *interest* that the student has in learning or using statistics; and

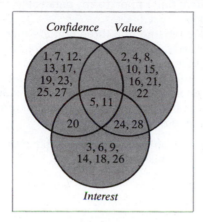

Oblique solution—Kleinbaum questionnaire on attitudes toward statistics

ITEM NO.	TIME 1			ITEM NO.	TIME 2		
	FACTOR 1	FACTOR 2	FACTOR 3		FACTOR 1	FACTOR 2	FACTOR 3
01	0.165	−0.839	−0.077	01	0.038	−0.897	−0.139
02	−0.583	0.123	0.209	02	−0.717	0.180	−0.068
03	−0.112	0.018	0.721	03	−0.291	−0.035	0.528
04	0.052	0.225	0.429	04	−0.258	−0.196	−0.574
05	−0.082	−0.051	0.471	05	−0.580	0.135	0.186
06	−0.532	0.060	0.060	06	−0.698	−0.315	−0.048
07	0.183	−0.821	0.225	07	0.091	−0.850	0.252
08	−0.683	0.199	0.052	08	−0.870	−0.053	−0.094
09	−0.205	−0.186	0.116	09	−0.585	−0.089	−0.056
10	−0.811	0.050	−0.136	10	−0.686	−0.074	0.085
11	−0.580	−0.069	0.238	11	−0.595	−0.007	0.180
12	0.232	−0.903	−0.039	12	0.082	−0.910	0.087
13	0.201	−0.874	0.015	13	0.066	−0.931	−0.058
14	−0.076	−0.510	0.519	14	−0.173	−0.393	0.476
15	−0.093	0.057	0.400	15	−0.320	−0.119	0.163
16	−0.779	0.129	−0.035	16	−0.825	0.101	−0.231
17	0.208	−0.100	0.818	17	0.056	−0.076	0.608
18	−0.084	0.042	0.508	18	−0.168	0.030	0.604
19	−0.275	−0.331	−0.075	19	−0.247	−0.217	0.242
20	−0.368	−0.070	0.202	20	−0.546	0.134	0.403
21	−0.228	−0.046	0.457	21	−0.611	0.006	−0.059
22	−0.872	0.065	0.041	22	−0.887	0.009	−0.048
23	−0.516	−0.529	−0.161	23	−0.430	−0.554	−0.024
24	−0.785	−0.133	−0.002	24	−0.784	0.212	0.211
25	−0.330	−0.555	−0.073	25	0.043	−0.469	0.481
26	−0.475	0.038	0.113	26	−0.423	0.361	0.396
27	−0.129	−0.301	0.433	27	0.001	−0.166	0.564
28	−0.442	−0.278	0.347	28	−0.364	−0.134	0.599

(3) the *value* that the student places on the importance or use of statistics. The instrument was constructed so that each of the 28 items was supposed to tap some aspect of at least one of the three attitude dimensions. Initial attitude dimension scores were then derived as the sums of the scores on those items associated with the same attitude dimension. In terms of item numbers, the three attitude dimension clusters of items (identified a priori to any data analysis) are given as shown in the diagram on page 409.

In order to evaluate whether the three attitude measures thus constructed actually reflected the underlying dimensions measured by the instrument, and to determine whether any items should be eliminated from the instrument, a factor analysis was undertaken using data collected on 47 students at different time points during the introductory course. Oblique factor loadings for three factors based on data obtained at the beginning of the course (time 1) and 2 weeks later (time 2) are presented in the table on page 409.

(a) Why does an oblique solution appear to be more appropriate than an orthogonal solution for this data set?

DIMENSION	ITEM NO.	r
C	1	0.607
C	5	0.419
C	7	0.693
C	11	0.399
C	12	0.662
C	13	0.678
C	17	0.494
C	19	0.437
C	20	0.397
C	23	0.635
C	25	0.572
C	27	0.598
I	3	0.608
I	5	0.564
I	6	0.470
I	9	0.485
I	11	0.733
I	14	0.590
I	18	0.433
I	20	0.531
I	24	0.684
I	26	0.535
I	28	0.714
V	2	0.725
V	4	0.297
V	5	0.545
V	8	0.682
V	10	0.615
V	11	0.695
V	15	0.389
V	16	0.686
V	21	0.582
V	22	0.800
V	24	0.731
V	28	0.635

(b) For each time point separately, label, *if possible*, each of the factors as *C* (for confidence), *I* (for interest), or *V* (for value) by consideration of the factor loadings in relation to the clusters of items suspected a priori to be associated with each of the three attitude dimensions. Discuss your choices of labels.

(c) Which items, if any, would you consider removing from the instrument entirely? (Explain.) How might you substantiate by further analysis that certain items should be eliminated?

(d) Based on the factor loadings given, how might you redefine or relabel the three clusters of items in view of the three factors obtained from the factor analysis?

(e) An alternative approach for assessing the construct validity of a newly developed instrument is called *item analysis*. This approach considers the correlations of a set of individual items with that particular factor defined as the sum of those items; items that correlate highly with a factor involving those items are considered to be appropriate components of that factor, whereas those items with small (or nonsignificant) correlations with the factor are considered to be poor items that should be removed.

For the data collected at time 1, the item-analysis results listed in the table on page 410 were obtained.

What conclusions can you draw from the results above about the construct validity of the three factors in general and about the utility of specific items in particular? (In answering this question, consider an *r* below 0.500 to be small.)

(f) What overall recommendations or conclusions do you have concerning the attitude instrument discussed in this problem?

4. For residents on an island in the South Seas experiencing rapid social change [e.g., the Ponape study (Patrick et al., 1974)], an index of cultural incongruity (called PIML) was developed by means of a factor analysis of data obtained from a sociological questionnaire administered to a random sample of such residents. This index was constructed by using the method of principal components to determine two factors, one measuring a person's preparedness for modern life (PML) and the other measuring his involvement in modern life (IML)—and then defining PIML as the difference between these two factors, PIML = PML − IML. The factors PML and IML are the first principal components based on data from two disjoint subsets of questionnaire items suspected (a priori) to be related to these factors. The factor loadings associated with each factor (based on working with the correlation matrix for a sample of 496 males) are given in the table.

FACTOR PML: "Preparedness for Modern Life"	
ITEM	FACTOR LOADING (1st PC)
1. Where grew up	0.397
2. Educational level	0.761
3. Years in town	0.656
4. Travel outside homeland	0.599 ·
5. Understand/read English	0.792
(proportion of total variation = 0.430)	

FACTOR IML: "Involvement in Modern Life"	
ITEM	FACTOR LOADING (1st PC)
6. Source of financial support	0.579
7. How often listen to news	0.327
8. Zone living in currently	0.604
9. Segment of economy	0.829
10. Occupational rank	0.739
(proportion of total variation = 0.408)	

(a) For each of the factors, would you recommend the removal of any items? (Explain the rationale behind your answer.)

(b) Based on the information given, comment on whether it would be more appropriate to use a simple average of the (standardized) scores for each item to represent the PML and IML indices or to use the principal components obtained.

(c) When the principal components based on the correlation matrix are used (without further rotation) as the factors, the factor weights for a given factor may be determined directly from the factor loadings by dividing each loading by the square root of the variance associated with that factor. (*Note:* If P_1 denotes the proportion of total variation explained by the first principal component, p denotes the total number of input variables, and V_1 denotes the variance associated with the first principal component, then $P_1 = V_1/p$ when the correlation matrix is used.)

   For each set of factor loadings presented above, determine the corresponding factor weights; that is, fill in the table.

FACTOR PML		FACTOR IML	
ITEM NO.	FACTOR WEIGHT	ITEM NO.	FACTOR WEIGHT
1		6	
2		7	
3		8	
4		9	
5		10	

(d) Based on these weights, compute and interpret the PIML incongruity scores based on the three sets of standardized item responses given.

PERSON	ITEM NO.									
	1	2	3	4	5	6	7	8	9	10
1	3.10	2.70	2.50	2.20	3.80	1.60	1.20	0.76	−0.45	−0.36
2	1.60	0.81	0.90	1.10	0.25	1.50	0.93	0.36	0.85	0.95
3	1.60	−0.85	−0.75	0.58	−1.72	3.70	2.60	2.40	2.30	3.10

5. The Brief Psychiatric Rating Scale (BPRS) is an instrument used by clinical psychologists for measurement and classification of persons with manifest psychopathology [see Overall and Klett (1972)]. In the table is presented a fictitious (rotated) factor-loading matrix for three factors based on a subset of the items contained in this scale.

ITEM	FACTOR 1	FACTOR 2	FACTOR 3
Somatic concern	0.15	0.40	−0.07
Anxiety	−0.06	0.70	−0.15
Conceptual disorganization	−0.12	−0.05	0.60
Guilt feelings	0.09	0.60	0.03
Tension	0.30	0.35	0.06
Depressive mood	0.01	0.75	0.18
Hostility	0.70	0.15	−0.08
Suspiciousness	0.65	0.10	0.35
Hallucinatory behavior	0.09	0.03	0.65
Uncooperativeness	0.50	−0.08	0.06
Unusual thought content	0.17	−0.04	0.80

Based on these loadings, identify three clusters of variables that describe the three factors and interpret each factor in conceptual terms (i.e., what type of psychopathology does each factor measure?). Also, what names would you use to describe these factors?

TWO-GROUP
DISCRIMINANT ANALYSIS

22.1 PREVIEW

In this chapter we shall be concerned with the problem of distinguishing (discriminating) between two populations on the basis of observations of a multivariable nature. We will have defined the two populations before the data are gathered and then will have available a sample of individuals from each population. The statistical problem consists in developing a rule or "discriminant function," based on the measurements obtained on each of these individuals, which will help us to assign some new individual to the correct population when it is not known from which of the two populations the individual comes.

The data obtained on each individual will invariably consist of observed values of a set of mutually correlated random variables, and the presence of these intercorrelations will necessitate consideration of the variables together rather than one at a time. The general approach to be adopted, as with regression analysis, is to construct in some *optimal* way a linear combination of these variables which would then be used for classification purposes. This allows the basic problem to be transformed from a complex multivariable one to an easier-to-handle univariable one, and the assignment of an individual to one of the two populations is then based simply on the value of the linear combination for that particular individual. The statistical manipulations associated with constructing a good (i.e., optimal) linear combination and with developing related methodology differ from those of regression analysis and are classed under the general heading "discriminant analysis." This particular statistical procedure was first introduced by R. A. Fisher (1936) as a statistical technique useful in taxonomic problems, and much statistical research in the area has gone on since that time [see Lachenbruch (1975) for a comprehensive review].

We previously (Section 2.3) distinguished discriminant analysis from regression analysis by pointing out that discriminant analysis involves a *nominal* dependent variable, whereas the classical regression analysis considers a *continuous* dependent variable. This distinction is somewhat oversimplified, since these two methods are actually based on different conceptual frameworks and require different statistical assumptions for inference-making purposes (e.g., classically, regression analysis requires normality of the *dependent* variable, whereas discriminant analysis requires multivariate normality of the *independent* variables). Nevertheless, the goals of these two methods are quite similar: both attempt to describe using a linear model the relationship between a dependent and several independent variables, one for the primary purpose of discrimination, the other for the primary purpose of prediction. Furthermore, as we shall see later in this chapter, the numerical results of a discriminant analysis can be computationally obtained using an appropriately defined regression formulation.

The development to be presented in this chapter will be confined to the two-population (or two-group) situation, in which case just one linear combination—or *discriminant function*—is needed. The extension to discrimination problems involving more than two groups (in which case more than one discriminant function is needed) can be found, for example, in Anderson (1958), Morrison (1967), and in Overall and Klett (1972), the latter offering a more applied discussion than the former two.

Before proceeding with the detailed development of the methodology of discriminant analysis, let us emphasize one important point. In order to perform a discriminant analysis, the groups to be delineated must be specified in advance of the collection and analysis of the data and without regard to the variables being studied. This philosophy is in contrast to that of classification-of-variables procedures (e.g., cluster analysis), which begin without the designation of the groups and attempt to form groups (e.g., clusters) as distinct as possible based on the data at hand [see Overall and Klett (1972)]. This chapter thus discusses discrimination procedures (groups specified a priori) and not classification of variables procedures (groups determined a posteriori).

22.2 REAL-LIFE EXAMPLES

To formalize somewhat the discriminant-analysis problem of interest to us, let us suppose that there are two populations, designated population 1 and population 2, and that we have sets of n_1 and n_2 individuals, respectively, which have been selected from each of these two populations. Let us suppose further that, for each individual, we have measured or observed values on p correlated random variables $X_1, X_2, \ldots, X_p$. The basic strategy in discriminant analysis is to form a linear combination of these variables, say

$$L = \beta_1 X_1 + \beta_2 X_2 + \cdots + \beta_p X_p$$

and then to assign a new individual to either group 1 or group 2 on the basis of the value of L obtained for that new individual.

Some real-life research investigations whose statistical analysis problems fit quite naturally into the above framework are as follows:

1. In a study of primary health care, it is of interest to be able to discriminate between those people with symptoms of illness who seek medical care (population 1, say) and those with symptoms who don't (population 2, say). The types of variables (the X's) to be used in describing differences between the two populations are measures concerning the duration of symptoms, the perceived seriousness of the symptoms, the amount of worry or anxiety concerning the symptoms, the person's feelings concerning the doctor's ability to relieve the complaint, the number of bed-loss days, and various socioeconomic and demographic variables. Discriminant-analysis techniques are useful in deciding which combination of these variables is most helpful in predicting accurately whether a person with symptoms will seek medical aid or not [see Hulka et al. (1972)].

2.   An epidemiologic study concerning the incidence of coronary heart disease (CHD) is designed to follow for a number of years a group of people initially free of CHD. At the start of the follow-up period, variables (the X's) concerning cholesterol level, blood pressure, occupation, height, weight, and certain socioeconomic and demographic variables are recorded for each individual. At the end of the follow-up period, it is determined whether a subject has developed CHD (population 1) or has not developed CHD (population 2). Techniques of discriminant analysis permit the identification of that combination of variables providing an accurate quantification of the risk of developing CHD [see Kleinbaum et al. (1971)].

3.   In a study of occupational health in a certain industry, the research goal is to determine whether the work history (e.g., the pattern of jobs held) for those employees (past or present) who now have leukemia (population 1) is different in any significant way from that of employees who do not have leukemia (population 2), after controlling for the effects of such possible confounding variables as race, sex, age, education, and smoking history, among others. Discriminant-analysis techniques are useful in assessing the importance of an employee's work history *relative* to that of several concomitant factors as a possible leukemia-associated agent [see McMichael et al. (1975)].

The above real-life examples emphasize most dramatically how many seemingly diverse research questions of a statistical nature can be viewed in the context of a problem in discriminant analysis. Clearly, the common thread running through all three of these examples is the need for a statistical rule for deciding which combination of a number of important variables provides the best discriminator between two defined populations. The calculation of that best linear combination is the subject of the next section.

22.3 CALCULATION OF THE DISCRIMINANT FUNCTION

Let us first establish some notation. Recall that we are given the existence of two populations and that we have observed values on p random variables $X_1, X_2, \ldots, X_p$ for each of the n_1 individuals selected from population 1 and for each of the n_2 from population 2. In particular, for the ith population ($i = 1, 2$), suppose that we let x_{ijk} denote the observed value of variable X_j ($j = 1, 2, \ldots, p$) for the kth sampled individual ($k = 1, 2, \ldots, n_i$). Thus, the set of variable values

$$\{x_{i1k}, x_{i2k}, \ldots, x_{ipk}\}$$

represents the group of measurements obtained for the kth individual selected from population i.

Now, recall from Section 22.2 that the idea was to develop a linear combination L of the variables, say

$$L = \beta_1 X_1 + \beta_2 X_2 + \cdots + \beta_p X_p$$

with values for $\beta_1, \beta_2, \ldots, \beta_p$ chosen so as to provide maximum discrimination between the two populations. What do we mean by "maximum discrimination"? Well, if the function L is going to discriminate between the two groups, we would hope that the variation in the values of L *between* the two groups would be much greater than the variation in the values of L *within* the two groups. This is not unlike the reasoning employed in analysis-of-variance procedures for detecting differences among population means.

More specifically, for any individual in the group of $(n_1 + n_2)$ individuals available, we can calculate (if we know the β's) the value of L for that individual. In particular, for the kth individual selected from population i, the associated value of L would be

$$L_{ik} = \beta_1 x_{i1k} + \beta_2 x_{i2k} + \cdots + \beta_p x_{ipk}$$

so, for each individual we can convert or transform from a set of p variable values to a single univariate score. Once this transformation is made, the problem can be viewed as one of distinguishing between the two populations on the basis of values $\{L_{11}, L_{12}, \ldots, L_{1n_1}\}$ from population 1 and values $\{L_{21}, L_{22}, \ldots, L_{2n_2}\}$ from population 2. In an analysis-of-variance framework, then, the total amount of variation in these scores is measured by

$$\sum_{i=1}^{2} \sum_{k=1}^{n_i} (L_{ik} - \bar{L})^2,$$

where

$$\bar{L} = \frac{1}{n_1 + n_2} \sum_{i=1}^{2} \sum_{k=1}^{n_i} L_{ik} = \frac{1}{n_1 + n_2} (n_1 \bar{L}_1 + n_2 \bar{L}_2)$$

And, as we know from our previous experience with analysis-of-variance procedures, the total sum of squares can be broken down into two interpretable components, a between-groups sum of squares

$$B = \sum_{i=1}^{2} n_i (\bar{L}_i - \bar{L})^2$$

and a within-group sum of squares

$$W = \sum_{i=1}^{2} \sum_{k=1}^{n_i} (L_{ik} - \bar{L}_i)^2$$

It is not too difficult to show that B can equivalently be written in the form

$$B = \frac{n_1 n_2}{n_1 + n_2} (\bar{L}_1 - \bar{L}_2)^2$$

so that B is large if the average value of L in group 1 is quite different from the average value of L in group 2. The statistic W is calculated by first obtaining the sum of squares of the L's about their mean within each group separately and then adding (or *pooling*) these within-group sums of squares together.

The ratio B/W, then, can be thought of as a measure of the discriminatory power of L, in the sense that the larger the value of B relative to W the more L is reflecting between-population variation (which is what we are interested in) as opposed to within-population variation.

Now, the function B/W depends on L and thus on the parameters $\beta_1, \beta_2, \ldots, \beta_p$. It is reasonable from our discussion up to this point to ask whether it is possible to choose particular values of $\beta_1, \beta_2, \ldots, \beta_p$, say $b_1, b_2, \ldots, b_p$ (which will depend, of course, on the observed x's) in order to *maximize* the function B/W. To maximize B/W with respect to $\beta_1, \beta_2, \ldots, \beta_p$ is certainly an intuitively appealing goal in choosing an "optimal" discriminator, but the question arises as to whether such a maximization procedure is mathematically tractable. The answer is, in fact, yes, and the techniques of calculus permit a fairly straightforward solution to the problem. Without going into all the mathematical details, we will simply present the solution to the maximization problem.

Let $\bar{x}_{ij} = \sum_{k=1}^{n_i} x_{ijk}/n_i$ be the observed mean value of variable j for the sample of n_i individuals from population i, and let

$$d_j = \bar{x}_{1j} - \bar{x}_{2j} \tag{22.1}$$

be the observed difference between the mean values of variable j in the two groups. Further, let

$$s_{jj'} = \frac{1}{n_1 + n_2 - 2} \sum_{i=1}^{2} \sum_{k=1}^{n_i} (x_{ijk} - \bar{x}_{ij})(x_{ij'k} - \bar{x}_{ij'}) \tag{22.2}$$

for $j, j' = 1, 2, \ldots, p$, so that (when $j = j'$) s_{jj} gives the usual *pooled sample variance* for the jth variable, and $s_{jj'}$ gives the *pooled sample covariance* between the variables j and j'. Then, form the matrix of such pooled estimates of population variances and covariances of the x's,

$$\mathbf{S} = \begin{bmatrix} s_{11} & s_{12} & \cdots & s_{1p} \\ s_{21} & s_{22} & \cdots & s_{2p} \\ \cdot & \cdot & & \cdot \\ \cdot & \cdot & & \cdot \\ \cdot & \cdot & & \cdot \\ s_{p1} & s_{p2} & \cdots & s_{pp} \end{bmatrix}$$

At this point it is informative to express B/W explicitly as a function of the β's, the d's, and the elements of the matrix $\mathbf{S}$. With some algebraic manipulation, one can show that B/W can equivalently be written in the form

$$B/W = \frac{\dfrac{n_1 n_2}{n_1 + n_2} \sum_{i=1}^{p} \sum_{j'=1}^{p} \beta_i \beta_{i'} d_j d_{j'}}{(n_1 + n_2 - 2) \sum_{j=1}^{p} \sum_{j'=1}^{p} \beta_i \beta_{i'} s_{jj'}}$$

For example, when $p = 2$, the expression above becomes

$$B/W = \frac{\dfrac{n_1 n_2}{n_1 + n_2} (\beta_1^2 d_1^2 + \beta_2^2 d_2^2 + 2\beta_1 \beta_2 d_1 d_2)}{(n_1 + n_2 - 2)(\beta_1^2 s_{11} + \beta_2^2 s_{22} + 2\beta_1 \beta_2 s_{12})}$$

Expressed in this way, it is clear that the value of B/W remains unchanged if, for some nonzero constant c, β_1 is replaced by $c\beta_1$, β_2 by $c\beta_2, \ldots$, and β_p by $c\beta_p$. In practical terms, this means that if the set of values $\{b_1, b_2, \ldots, b_p\}$ maximizes B/W, so does the set $\{cb_1, cb_2, \ldots, cb_p\}$. As we shall see later, this "scale-invariance" property relates to the fact that it is only the relative sizes (i.e., ratios) of the coefficients that really matters when using a discriminant function for classifying individuals into groups.

With these remarks in mind, the particular set $\{b_1, b_2, \ldots, b_p\}$ of solutions which we choose to present explicitly is that set which is commonly printed out by means of standard discriminant-analysis computer programs (e.g., those in the BMD, SPSS, and SAS series). In particular, if

$$\mathbf{S}^{-1} = \begin{bmatrix} s^{11} & s^{12} & \cdots & s^{1p} \\ s^{21} & s^{22} & \cdots & s^{2p} \\ \vdots & \vdots & \ddots & \vdots \\ s^{p1} & s^{p2} & \cdots & s^{pp} \end{bmatrix} \tag{22.3}$$

is the inverse matrix[1] associated with the matrix $\mathbf{S}$ given earlier, the values $b_1, b_2, \ldots, b_p$ which maximize B/W are given as follows:

$$\begin{aligned} b_1 &= s^{11}d_1 + s^{12}d_2 + \cdots + s^{1p}d_p \\ b_2 &= s^{21}d_1 + s^{22}d_2 + \cdots + s^{2p}d_p \\ &\ \ \vdots \\ b_p &= s^{p1}d_1 + s^{p2}d_2 + \cdots + s^{pp}d_p \end{aligned} \tag{22.4}$$

The linear combination based on the b's we will call

$$\ell = b_1X_1 + b_2X_2 + \cdots + b_pX_p$$

and this linear combination maximizes the quantity B/W based on the sample at hand. Since we actually only have a *sample* from each population, the b's should be looked upon as estimates of the β's and ℓ can be considered to be an estimate of the optimal linear combination L, which could be determined if all population parameters (i.e., the true values of the means, variances, and covariances for the X's) were known.

It is worthwhile to examine the form of the expressions for the b's which have been given. First, notice that the b's are constructed as *linear combinations* of the *differences* (22.1) between the variable means in the two groups. Second, the coefficients of these mean differences are functions of *pooled* sums of squares and cross-products (22.2), and the validity of such pooling is contingent on the assumption that the variances of and covariances among the p variables are the same in each of the

---

[1] The values s^{ii} $(i = 1, \ldots, p, j = 1, \ldots, p)$ of the inverse matrix are defined so as to satisfy the following general mathematical relationship:

$$s^{i1}s_{1j} + s^{i2}s_{2j} + \cdots + s^{ip}s_{pj} = \begin{cases} 1 & \text{if } i = j \\ 0 & \text{if } i \neq j \end{cases}$$

Further discussion of matrix mathematics, including the notion of an inverse of a matrix, can be found in Appendix B.

two groups. Such an assumption will almost certainly never hold exactly in actual practice, but moderate departures from homogeneity do not seem to affect the behavior of the discriminant function seriously [e.g., see Lachenbruch (1975)].

22.4 CALCULATION OF ℓ USING DUMMY-DEPENDENT-VARIABLE REGRESSION

There is another approach to that discussed in the preceding section for computing the discriminant function coefficients. This alternative procedure is based on the use of multiple regression techniques to fit a model for which the dependent variable is dichotomous.

To be more specific, suppose that we define a dummy dependent variable Y which takes the value $n_2/(n_1+n_2)$ for an individual in our sample of individuals coming from group 1 and which takes the value $-n_1/(n_1+n_2)$ if the individual is from group 2. We could, in fact, use any two distinct values for Y to designate the two groups, but the particular values we have specified make things rather nice computationally.[2] Now, consider fitting the regression model

$$Y = \beta_0 + \beta_1 X_1 + \beta_2 X_2 + \cdots + \beta_p X_p + E$$

where these X's are exactly the same as those appearing in the expression for ℓ given earlier.

Since we have arranged by choice of coding that the overall mean of the dependent variable ($\bar{Y}$, say) is zero, it follows that the fitted model can be written as

$$\hat{Y} = \hat{\beta}_0 + \hat{\beta}_1 X_1 + \hat{\beta}_2 X_2 + \cdots + \hat{\beta}_p X_p$$

where $\hat{\beta}_0 = -\sum_{j=1}^{p} \hat{\beta}_j \bar{x}_j$, $\hat{\beta}_1$, $\hat{\beta}_2, \ldots$, and $\hat{\beta}_p$ are the usual least-squares estimates of $\beta_0, \beta_1, \ldots, \beta_p$, and where $\bar{x}_j = \sum_{i=1}^{2} \sum_{k=1}^{n_i} x_{ijk}/(n_1+n_2)$ is the *overall mean* of the values on variable X_j for both groups.

The question of interest here is what is the relationship between the $\hat{\beta}$'s obtained by means of the multiple regression approach above and the b's whose computational formulas were presented earlier. Without going into the mathematical details, it can be shown that $\hat{\beta}_j = cb_j$ for every j, where c is some positive constant; in other words, the regression procedure above can be used to produce a linear combination which differs from ℓ only by a constant multiplier. For the particular coding of Y that we have advocated, this constant multiplier c can be shown to have the specific value

$$c^* = \frac{n_1 n_2}{n_1 + n_2} \bigg/ \left[(n_1 + n_2 - 2) + \frac{n_1 n_2}{n_1 + n_2} D^2 \right] \tag{22.5}$$

---

[2] In general, Y can be defined as follows:

$$Y = \begin{cases} k_1 & \text{for an individual in the sample from population 1} \\ k_2 & \text{for an individual in the sample from population 2} \end{cases}$$

The values of k_1 and k_2 most frequently used other than the values already mentioned are $k_1 = 1$ and $k_2 = 0$.

where

$$D^2 = \sum_{j=1}^{p} \sum_{j'=1}^{p} d_j d_{j'} s^{jj'} \tag{22.6}$$

In general, other choices of codings for Y (see footnote 2) will change the value of the constant multiplier above, and the estimate of β_0 will also depend (linearly) on the new value of $\bar{Y}$.

The quantity D^2 of (22.6) has special significance. It is called "Mahalanobis' D^2" (after a famous Indian statistician), and it represents a generalized measure of the "distance" between the two populations; for example, $D^2 = (\bar{x}_{11} - \bar{x}_{21})^2/s^{11}$ when $p = 1$. Now, the usual assumptions made in discriminant analysis are that (1) the X's have a multivariate normal distribution (which implies, among other things, that each X separately has a normal distribution and that any linear combination of the X's is normal) and (2) the variances of and covariances among the X's are the same in the two groups. If these assumptions hold, it can be shown that

$$F = \frac{n_1 n_2 (n_1 + n_2 - p - 1)}{(n_1 + n_2)(n_1 + n_2 - 2)p} D^2 \frown F_{p, n_1 + n_2 - p - 1} \tag{22.7}$$

which provides a test as to *whether there are significant differences between the group means for all variables considered together*.

The preceding test can be related to the test for the significance of the fitted regression model. In particular, the analysis-of-variance table based on the regression is given in Table 22.1.

It can be shown that the usual ratio of mean squares,

$$\frac{SSR}{p} \bigg/ \frac{SSE}{n_1 + n_2 - p - 1}$$

is identical to

$$\frac{n_1 n_2 (n_1 + n_2 - p - 1)}{(n_1 + n_2)(n_1 + n_2 - 2)p} D^2$$

in value, and (given the stated discriminant-analysis assumptions, which are an inversion of the usual regression assumptions of Y normal and the X's nonstochastic)

TABLE 22.1 *ANOVA table for the dummy-dependent-variable regression*

SOURCE	df	SS
Due to regression	p	$SSR = \dfrac{n_1 n_2}{n_1 + n_2} \sum_{j=1}^{p} \hat{\beta}_j d_j$
Deviations from regression	$n_1 + n_2 - p - 1$	$SSE = \dfrac{n_1 n_2}{n_1 + n_2} \left(1 - \sum_{j=1}^{p} \hat{\beta}_j d_j\right)$
Total	$n_1 + n_2 - 1$	$\dfrac{n_1 n_2}{n_1 + n_2}$

provides an F test of the regression null hypothesis

$$H_0: \beta_1 = \beta_2 = \cdots = \beta_p = 0$$

If H_0 is rejected, one can conclude that there is evidence of between-group differences, although the utility of the discriminant function for assigning individuals to groups is a separate issue (to be discussed in Section 22.6).

There is also an interesting relationship between R^2, the squared multiple correlation coefficient obtained by means of the regression analysis, and the quantity D^2. In particular, it can be shown that

$$D^2 = \frac{(n_1 + n_2)(n_1 + n_2 - 2)}{n_1 n_2} \frac{R^2}{1 - R^2} \tag{22.8}$$

or equivalently that

$$R^2 = c^* D^2$$

where c^* is given by (22.5).

22.5 NUMERICAL EXAMPLE

The methodology presented in the preceding sections will be illustrated with data collected in a longitudinal epidemiologic study [see Kleinbaum et al. (1971)] designed to investigate the joint effects of three factors on the risk of developing coronary heart disease (CHD). At the start of the study period, the age, the diastolic blood pressure, and the cholesterol level were recorded for each of 832 white males free of CHD. By the end of the study period, 71 of these males had developed CHD. The statistical question of import here is whether it is possible to "discriminate" between the 71 white males who developed CHD and the remaining 761 who did not on the basis of the available data on age (X_1), diastolic blood pressure (X_2), and cholesterol level (X_3). In the discriminant analysis to follow, we will consider the CHD group of $n_1 = 71$ white males to be a sample from the (conceptual) aggregate (population 1, say) of all white males who would develop CHD under circumstances like those in this study. Similarly, population 2 will denote the collection of all those who would remain free of CHD under such circumstances, and the NCHD group of $n_2 = 761$ individuals will be looked upon as a sample from that conceptual population.

The computational aspects of discriminant analysis as described earlier will be best illustrated by relating them to the output of a typical computer program designed to perform a discriminant analysis (e.g., like the BMD and SPSS programs). We shall first consider discriminant-analysis procedures associated with the function $\ell = b_1 X_1 + b_2 X_2 + b_3 X_3$, which involves all three variables. We then illustrate the use of a regression analysis program to arrive at the same results. Finally, we shall illustrate the use of stepwise discriminant analysis as a tool in choosing a discriminant function, although this variable-selection procedure may not necessarily lead us to a discriminant function involving all three variables.

22.5.1 OUTPUT FROM A TYPICAL DISCRIMINANT-ANALYSIS PROGRAM

Initial computer output typically consists of various summary statistics. These appear in Table 22.2 for our particular data set.

Next, the discriminant function coefficients are calculated. First, the inverse matrix S^{-1}, given by (22.3), is computed, which here has the form

$$S^{-1} = \begin{bmatrix} 0.00560 & -0.00153 & -0.00045 \\ -0.00153 & 0.00553 & -0.00037 \\ -0.00045 & -0.00037 & 0.00063 \end{bmatrix}$$

TABLE 22.2   *Summary statistics for CHD study data*

Means $\{\bar{x}_{ij}\}$

VARIABLE	GROUP		$d_j = \bar{x}_{1j} - \bar{x}_{2j}$
	NCHD	CHD	
X_1 (AGE)	44.81	56.86	12.05
X_2 (DBP)	86.99	95.62	8.63
X_3 (CHOL)	201.27	221.51	20.24

Standard deviations

VARIABLE	GROUP	
	NCHD	CHD
X_1	14.98	10.28
X_2	14.50	15.37
X_3	43.01	38.83

Within-groups covariance matrix (S)

	X_1	X_2	X_3
X_1	214.26	72.37	195.61
X_2	72.37	212.44	175.53
X_3	195.61	175.53	1820.61

Within-groups correlation matrix $(r_{ij'} = s_{ij'}/\sqrt{s_{ij}s_{i'i'}})$

	X_1	X_2	X_3
X_1	1.00	0.34	0.31
X_2	0.34	1.00	0.28
X_3	0.31	0.28	1.00

Then the coefficient b_j associated with X_j is calculated [using (22.4)] as

$$b_j = \sum_{j'=1}^{p} s^{jj'} d_{j'} = \sum_{j'=1}^{p} s^{jj'} \bar{x}_{1j'} - \sum_{j'=1}^{p} s^{jj'} \bar{x}_{2j'}$$

We have expressed b_j as a *difference* between two quantities for a particular reason. Some computer programs (e.g., those specifically designed for two-group discriminant analysis only) provide the b_j's directly; others (e.g., those with the capacity to work with more than two groups) print out the quantity $\sum_{j'=1}^{p} s^{jj'} \bar{x}_{ij'}$ as the score for the jth variable in the ith group, so that b_j must be obtained by the subtraction process indicated above. For the data we are considering, the computer produced the table of scores given as Table 22.3. We have generated the column of discriminant function coefficients based on these scores. The constant b_0, which we have not previously defined, is important when using the calculated discriminant function for classification purposes, and we shall discuss this in the next section. As an explicit example of the calculations resulting in these scores and coefficients, we obtain b_1 from the table of means and from the elements of the first row of the matrix $\mathbf{S}^{-1}$ as

$$b_1 = [0.00560(56.86) - 0.00153(95.62) - 0.00045(221.51)]$$

$$- [0.00560(44.81) - 0.00153(86.99) - 0.00045(201.27)]$$

$$= 0.072 - 0.027 = 0.045$$

Thus, the calculated discriminant function ℓ has the specific form

$$\ell = 0.045X_1 + 0.022X_2 + 0.004X_3$$

and this is the basic function that we shall be using for assignment purposes.

The typical discriminant-analysis program next prints out the F statistic given by (22.7), which in our case has $p = 3$ and $(n_1 + n_2 - p - 1) = 828$ degrees of freedom and which tests *whether there are significant differences between the groups with respect to X_1, X_2, and X_3*. For the data we are considering, $F = 17.605$, which is highly significant (probably due to the large value for the denominator degrees of freedom) but which says little about the fit of the discriminant model or about the utility of the model for classification purposes. Partial F statistics (based on 1 and $n_1 + n_2 - p - 1$ degrees of freedom) reflecting the *relative* contribution of each variable in the final model are also printed out in addition to the overall F statistic. As we will see, these F's are identical in

TABLE 22.3   *Discriminant function coefficients for CHD study data*

VARIABLE	GROUP		b_j
	NCHD	CHD	
X_1	0.027	0.072	$b_1 = $ 0.045
X_2	0.338	0.360	$b_2 = $ 0.022
X_3	0.075	0.079	$b_3 = $ 0.004
(constant)	−23.561	−28.726	$b_0 = -5.165$

value to those obtained by means of a multiple regression approach to the problem and hence can be interpreted in this light. For our data, these F values (with 1 and 828 df) are 22.657 for X_1, 5.282 for X_2, and 1.675 for X_3, which suggests that X_3 might be a superfluous variable. We shall explore this possibility later by means of stepwise discriminant-analysis procedures.

Some programs do not print out the value of D^2. However, this is no problem since, from (22.7),

$$D^2 = \frac{(n_1+n_2)(n_1+n_2-2)p}{n_1 n_2 (n_1+n_2-p-1)} F$$

In our case, then, we find that

$$D^2 = \frac{832(830)(3)}{71(761)(828)}(17.605) = 0.815$$

We shall use this value of D^2 to relate the output above to that obtained by means of a multiple regression approach to the problem, and also to estimate misclassification rates.

22.5.2 REGRESSION OUTPUT FOR THE DISCRIMINANT-ANALYSIS PROBLEM

The output of a typical multiple regression program provides much (but not all) of the information obtained by means of a discriminant analysis. In particular, no classification tables (to be discussed in Section 22.6) would be forthcoming from a multiple regression program; but the discriminant function coefficients can be obtained from the regression coefficients, and the F statistics generated by each program are completely comparable. To see all this, let us use a typical multiple regression program to fit the model

$$Y = \beta_0 + \beta_1 X_1 + \beta_2 X_2 + \beta_3 X_3 + E$$

where $Y = n_2/(n_1+n_2) = 761/832$ for an individual in the CHD group and $Y = -n_1/(n_1+n_2) = -71/832$ for an individual in the NCHD group, and where X_1, X_2, and X_3 are as defined earlier. Typical output would consist first of the usual type of ANOVA table for a regression analysis, which for our data would be as follows:

SOURCE	df	SS	MS	F
Regression	3	3.894	1.298	17.605
Residual	828	61.047	0.074	($R^2 = 0.06$)

As pointed out in Section 22.4, the F statistic in this ANOVA table, which tests the regression null hypothesis $H_0: \beta_1 = \beta_2 = \beta_3 = 0$, is identical in value to the overall discriminant analysis F statistic which tests for overall differences between the group means. Finally, the table of estimated regression coefficients, their standard errors, and

TABLE 22.4 *Regression coefficients for CHD study data*

VARIABLE	COEFFICIENT	STANDARD ERROR	F VALUE
X_1	$\hat{\beta}_1 = 0.00331$	0.00070	22.657
X_2	$\hat{\beta}_2 = 0.00161$	0.00070	5.282
X_3	$\hat{\beta}_3 = 0.00031$	0.00024	1.675
(constant)	$\hat{\beta}_0 = -0.355$		

the associated partial F statistics appear in Table 22.4. As mentioned earlier, these partial F's are always identical in value to those given by the discriminant analysis.

To transform to the b's from the $\hat{\beta}$'s, we first obtain the value of D^2 using (22.8); then we determine the value of c^* using (22.5) as

$$c^* = \frac{n_1 n_2}{n_1 + n_2} \bigg/ \left[(n_1 + n_2 - 2) + \frac{n_1 n_2}{n_1 + n_2} D^2 \right]$$

$$= \frac{71(761)}{832} \bigg/ \left[830 + \frac{71(761)}{832}(0.815) \right] = 0.0736$$

Then we obtain b_j as $\hat{\beta}_j / c^*$ for $j = 1, 2, 3$. For example,

$$b_1 = \frac{\hat{\beta}_1}{c^*} = \frac{0.00331}{0.0736} = 0.045$$

We also point out that the low value of R^2 obtained from the regression analysis (namely, 0.06) indicates that the fitted regression model is not a good predictor of the dichotomous response under consideration. This emphasizes the often-overlooked point that significant F values do not necessarily imply a high value of R^2.

22.6 CLASSIFICATION USING THE DISCRIMINANT FUNCTION

In this section we shall discuss how to use the calculated discriminant function for classification purposes and how to evaluate its performance in this regard. Let us suppose that we have calculated the linear discriminant function

$$\ell = b_1 X_1 + b_2 X_2 + \cdots + b_p X_p$$

by means of the methodology previously discussed in Section 22.3 and that we are now interested in using ℓ to assign individuals to one of the two groups. Since such an assignment procedure is a statistical decision-making process, it is imperative to have a measure of the goodness of the decision rule. An excellent statistic in this regard is the *error rate* or *misclassification rate* (i.e., the probability of assigning an individual to the wrong population); there are, of course, two such rates, one for each population. A discussion of assignment rules and their associated error rates will now follow.

22.6.1 CUTOFF POINTS

As before, it is helpful to think of associating with each individual a score that is the value of ℓ based on that particular individual's set of observed variable values. Then, in order to use ℓ for allocating individuals to one of the two groups, we would need to specify a critical score or *cutoff point* such that an individual is assigned to one group if his score exceeds the cutoff point and to the other group if it does not. If we begin with

$$\bar{\ell}_1 = b_1\bar{x}_{11} + b_2\bar{x}_{12} + \cdots + b_p\bar{x}_{1p}$$

and

$$\bar{\ell}_2 = b_1\bar{x}_{21} + b_2\bar{x}_{22} + \cdots + b_p\bar{x}_{2p}$$

which are the observed mean scores for our samples from groups 1 and 2, respectively, we can construct a completely symmetrical classification rule by using as the cutoff point the mean of $\bar{\ell}_1$ and $\bar{\ell}_2$,

$$\frac{1}{2}(\bar{\ell}_1 + \bar{\ell}_2) = b_1\frac{\bar{x}_{11} + \bar{x}_{21}}{2} + b_2\frac{\bar{x}_{12} + \bar{x}_{22}}{2} + \cdots + b_p\frac{\bar{x}_{1p} + \bar{x}_{2p}}{2}$$

Then, if $\bar{\ell}_1 > \bar{\ell}_2$, the allocation rule is: Assign an individual with observed variable values $x_1, x_2, \ldots, x_p$ (and score $\ell = b_1x_1 + b_2x_2 + \cdots + b_px_p$) to group 1 if $\ell > \frac{1}{2}(\bar{\ell}_1 + \bar{\ell}_2)$ and to group 2 if $\ell < \frac{1}{2}(\bar{\ell}_1 + \bar{\ell}_2)$. In discriminant-analysis terminology, the quantity $-\frac{1}{2}(\bar{\ell}_1 + \bar{\ell}_2)$ is designated b_0; equivalently, then, we would allocate to group 1 or 2 according as the function

$$\ell^* = b_0 + b_1x_1 + b_2x_2 + \cdots + b_px_p \tag{22.9}$$

is greater than or less than zero. And, since the sign of ℓ^* becomes the sole criterion for classification, multiplying ℓ^* by any positive constant does not alter the allocation rule. In fact, if $n_1 = n_2$ so that $\bar{x}_i = \frac{1}{2}(\bar{x}_{1i} + \bar{x}_{2i})$, the classification rule using ℓ^* above would be identical to one based on the sign of the fitted regression model discussed in Section 22.4. Using just the sign of ℓ^* would not be correct, of course, if the cutoff point for ℓ^* was some value other than zero.

In certain instances, important considerations may warrant changing the cutoff point to something other than the symmetrical one discussed above. For example, suppose it is known that an individual selected at random has (a priori) probability p_1 of being selected from population 1 and (a priori) probability $p_2 = 1 - p_1$ of coming from population 2, where "a priori" means "before the sample is taken." Then, one might consider using some cutoff point which reflects the relative sizes of p_1 and p_2 in such a way as to make it more likely to be assigned to the population with the larger a priori probability. For example, using (22.9), we might choose $\log_e(p_2/p_1)$ instead of zero as the cutoff point. Then, if $p_1 > p_2$ (so that a randomly selected individual is more likely to have come from population 1 than from population 2), $\log_e(p_2/p_1)$ will be strictly negative and we will assign an individual to population 1 as long as ℓ^* exceeds this negative value (thus allowing for more persons to be assigned to population 1 than would be assigned using zero as the cutoff point). Analogously, the condition $p_1 < p_2$ implies that ℓ^* must exceed some positive quantity to merit assignment to group 1. In general, p_1 will be unknown; however, if the samples are selected randomly from the two populations, $n_1/(n_1 + n_2)$ provides a good estimate of p_1, so that $\log_e (p_2/p_1)$ can sometimes be approximated by $\log_e (n_2/n_1)$.

A modification of the cutoff point is also warranted when the cost or seriousness of an incorrect assignment is group-dependent. For example, it would be a much more serious error to classify an individual with cancer as not having cancer on the basis of a series of diagnostic tests (often called a "false negative" finding) than it would be to treat a cancer-free individual unnecessarily (a "false positive"). More quantitatively, if c_{12} is the cost of misclassifying a member of group 1 as a member of group 2 and if c_{21} is the cost of incorrectly assigning an individual from group 2 to group 1, then $\log_e (c_{21}/c_{12})$ has properties completely analogous to those of $\log_e (p_2/p_1)$. The choice of the cost ratio c_{21}/c_{12} must very often be based on a subjective evaluation of the situation by the experimenter.

A generalized cutoff point which takes both the ratio of a priori probabilities and the cost ratio into account is

$$\log_e \frac{p_2}{p_1} + \log_e \frac{c_{21}}{c_{12}} = \log_e \frac{p_2}{p_1} \frac{c_{21}}{c_{12}}$$

and a purely mathematical (and not just heuristic) justification for its use can be found, for example, in Anderson (1958). Note that this generalized cutoff point has the value zero when $p_2/p_1 = c_{12}/c_{21}$, which includes the special case when $p_1 = p_2$ and $c_{12} = c_{21}$.

22.6.2 ERROR RATES

For any specified cutoff point, it is possible to count how many individuals in the sample have been incorrectly classified using ℓ^*. In particular, the following table can always be constructed:

ACTUAL GROUP	ASSIGN GROUP 1	ASSIGN GROUP 2	
1	n_{11}	n_{12}	n_1
2	n_{21}	n_{22}	n_2

The ratios n_{12}/n_1 and n_{21}/n_2 are often used to estimate the misclassification (or error) rates for the two groups, but these *apparent error rates* generally underestimate the true misclassification rates and so paint an optimistic picture of the true state of affairs. This is because the function ℓ^* has been determined using these two particular samples and so should be expected to perform better with them than with a new group of individuals. Better misclassification rate estimates are available, and we will present them shortly.

Incidentally, when groups 1 and 2 designate sets of diseased and nondiseased individuals, respectively, it is common practice to refer to n_{11}/n_1 as the *sensitivity* and n_{22}/n_2 as the *specificity*. Naturally, the sensitivity and specificity (and also the two apparent error rates) will vary inversely with one another as the cutoff point is altered; that is, if the sensitivity goes up with a change in the cutoff point, the specificity must go down, and vice versa.

Now, calculation of the apparent error rates does not require any distributional assumptions. However, under the usual discriminant-analysis assumptions mentioned earlier, it is possible to obtain more realistic misclassification rate estimates than those provided by the apparent error rates. In particular, if the decision rule is to assign an individual to group 1 if

$$\ell^* > \log_e \frac{p_2 \, c_{21}}{p_1 \, c_{12}} \tag{22.10}$$

and to group 2 otherwise, then an estimate (say, $\hat{P}_1$) of the probability of incorrectly assigning to population 2 an individual from population 1 can be shown to be

$$\hat{P}_1 = \Phi\left[\frac{\log_e (p_2/c_{21})(p_1/c_{12}) - D^2/2}{\sqrt{D^2}} \right] \tag{22.11}$$

where D^2 is the Mahalanobis' distance defined earlier and where Φ represents the standard normal cumulative distribution function [i.e., $\Phi(z_0) = \Pr(Z < z_0)$ when $Z \frown N(0, 1)$]. Similarly, $\hat{P}_2$, an estimate of the probability of incorrectly assigning an individual from population 2 to population 1, is given by the expression

$$\hat{P}_2 = \Phi\left[-\frac{\log_e (p_2/c_{21})(p_1/c_{12}) + D^2/2}{\sqrt{D^2}} \right] \tag{22.12}$$

Note that when $\log_e [(p_2/p_1)(c_{21}/c_{12})] = 0$, so that we are back to the symmetrical cutoff point introduced earlier, we have

$$\hat{P}_1 = \hat{P}_2 = \Phi(-\sqrt{D^2}/2)$$

In his discussion on error rates, Lachenbruch (1975) pointed out that these estimates themselves may be misleading for small sample sizes and, on the basis of sampling experiments, has suggested using

$$\frac{n_1 + n_2 - p - 3}{n_1 + n_2 - 2} D^2$$

in place of D^2 in the formulas for $\hat{P}_1$ and $\hat{P}_2$ given by (22.11) and (22.12).

22.6.3 CONTINUATION OF OUR NUMERICAL EXAMPLE

As an assessment of the utility of the previously calculated discriminant function

$$\ell^* = -5.165 + 0.045 X_1 + 0.022 X_2 + 0.004 X_3$$

in classifying individuals correctly, the usual discriminant-analysis program provides a table giving for each sample the frequency distribution of the number of individuals assigned to each group. This enables one to calculate the apparent error rates discussed in Section 22.6.2. Based on the rule that an individual is assigned to the CHD group or the NCHD group according as ℓ^* is positive or negative, the following classification table is obtained:

	NUMBER CLASSIFIED INTO CHD GROUP ($\ell^* > 0$)	NUMBER CLASSIFIED INTO NCHD GROUP ($\ell^* < 0$)	TOTAL
CHD sample	51	20	71
NCHD sample	272	489	761

Thus, $20/71 = 0.282$ is the apparent error rate for the CHD group, and $272/761 = 0.357$ is the apparent error rate associated with the NCHD group. The alternative error rate estimates $\hat{P}_1$ and $\hat{P}_2$ designed to overcome the drawbacks of the apparent error rates have the values

$$\hat{P}_1 = \hat{P}_2 = \Phi(-\sqrt{D^2}/2) = \Phi(-\sqrt{0.815}/2)$$

$$= \Phi(-0.451) = 0.326$$

Incidentally, Lachenbruch's correction to D^2 has the value $826/830 = 0.995$, which is so close to 1 that it can be ignored.

Some discriminant-analysis programs have the option of allowing the user to specify values for the a priori probabilities p_1 and $p_2 = 1 - p_1$, so that, for example, classification can be based on whether ℓ^* is greater than or less than $\log_e (p_2/p_1)$. From the way in which our data were gathered, it is reasonable to estimate $\log_e (p_2/p_1)$ as

$$\log_e \frac{n_2}{n_1} = \log_e \frac{761}{71} = 2.372$$

and the classification table based on this new cutoff point is as follows:

	NUMBER CLASSIFIED INTO CHD GROUP ($\ell^* > 2.372$)	NUMBER CLASSIFIED INTO NCHD GROUP ($\ell^* < 2.372$)	TOTAL
CHD sample	0	71	71
NCHD sample	0	761	761

The values of $\hat{P}_1$ and $\hat{P}_2$ based on this modified cutoff point are

$$\hat{P}_1 = \Phi\left(\frac{2.372 - 0.815/2}{\sqrt{0.815}}\right) = 0.985$$

and

$$\hat{P}_2 = \Phi\left(-\frac{2.372 + 0.815/2}{\sqrt{0.815}}\right) = 0.001$$

These error rates are so one-sided because the information in our sample suggests that it is much more likely for an individual to be a member of the NCHD group (761/832) than to be a member of the CHD group (71/832). In addition, the one-sidedness reflects the fact that the cutoff point $\log_e (p_2/p_1)$ assumes (unrealistically) equal costs of misclassification, whereas a cutoff point of zero can be interpreted as having arisen from the more realistic condition that $p_2/p_1 = c_{12}/c_{21}$. In any case, the example above illustrates dramatically the effect of a change in the value of the cutoff point on the misclassification rates, and it suggests that careful consideration should be given to the choice of an appropriate cutoff point.

22.7 STEPWISE DISCRIMINANT ANALYSIS

We shall now analyze these data using a *stepwise-discriminant-analysis procedure*. This procedure operates in principle like that of stepwise multiple regression in the sense that one variable is included in the discriminant function at each step, this variable being the one that results in the most significant F value after adjusting for variables already included in the model. This step-by-step procedure continues until no further significant gain in discrimination (as reflected by these partial F's) can be achieved via the addition of more variables to the discriminant function. Stepwise discriminant analysis allows for an examination at every step of the importance both of variables which have been included and which are candidates for inclusion in the discriminant function. This is quite important since a variable that may have appeared to be fairly important at an early stage may, at a later stage, become superfluous because of the relationship between it and other variables already in the model. In general, such redundancy among variables would often go unnoticed when these variables are "forced" into the discriminant function (as they were in the analysis just completed) and could, in fact, lead to a loss in discriminatory power.

At each step in a stepwise discriminant analysis, the computer output consists of the overall F value, the partial F for each variable already in the model (often called the "F-to-remove" since a variable with an insignificant value would be dropped from the discriminant function), the partial F for each variable still a candidate for inclusion (which is called the "F-to-enter"), the discriminant-function coefficients, and the usual classification table. At the final step when there are exactly p variables in the model, the overall F has p and $(n_1 + n_2 - p - 1)$ degrees of freedom, an F-to-remove has 1 and $(n_1 + n_2 - p - 1)$ df, and an F-to-enter for a particular variable has 1 and $(n_1 + n_2 - p - 2)$ df, since it is the partial F obtained when that variable is the next variable added to the model.

For our data, the initial F's-to-enter (with 1 and 830 df) are 43.969 for X_1, 22.768 for X_2, and 14.607 for X_3. Thus, the variable X_1 (age) is the first variable to be included in the discriminant function and the associated output for the first step is as follows:

Step number: 1

Variable entered: X_1

Variables included and F-to-remove—degrees of freedom (1,830): X_1: 43.969

Variables not included and F-to-enter—degrees of freedom (1,829): X_2: 6.820, X_3: 3.202

Overall F: $F_{1,830} = 43.969$

Discriminant-function scores

| | GROUP | | |
VARIABLE	NCHD	CHD	b_i
X_1	0.209	0.265	$b_1 = 0.056$
(constant)	−5.380	−8.238	$b_0 = -2.858$

Classification table (based on $\ell^* = -2.858 + 0.056X_1$)

	NUMBER CLASSIFIED INTO CHD GROUP ($\ell^* > 0$)	NUMBER CLASSIFIED INTO NCHD GROUP ($\ell^* < 0$)	TOTAL
CHD sample	48	23	71
NCHD sample	279	482	761

Examination of the F's-to-enter indicates that variable X_2(DBP) is to be included in the model at the second step.

Step number: 2
Variable entered: X_2
Variables included and F-to-remove—degrees of freedom (1,829): X_1: 27.599, X_2: 6.819
Variables not included and F-to-enter—degrees of freedom (1,828): X_3: 1.675
Overall F: $F_{2,829} = 25.548$

Discriminant-function scores

| | GROUP | | |
VARIABLE	NCHD	CHD	b_i
X_1	0.080	0.128	0.048
X_2	0.382	0.406	0.024
(constant)	−19.111	−23.768	−4.657

Classification table (based on $\ell^* = -4.657 + 0.048X_1 + 0.024X_2$)

	NUMBER CLASSIFIED INTO CHD GROUP ($\ell^* > 0$)	NUMBER CLASSIFIED INTO NCHD GROUP ($\ell^* < 0$)	TOTAL
CHD sample	53	18	71
NCHD sample	267	494	761

It is not necessary, of course, to present step 3, since the output would be identical to that obtained earlier when the full model was considered.

Examination of the classification tables suggests that the model containing X_1 and X_2 does a better job of classifying individuals than does the model with X_1 alone or even the model containing all three variables. The partial F statistics reflect this situation in the sense that X_1 and X_2 are always associated with significant F-to-enter and F-to-remove values, whereas X_3's F-to-enter value becomes nonsignificant once X_1 and X_2 are in the model. This example illustrates how a loss in discriminatory power can result from forcing an essentially superfluous variable into the discriminant function. Finally, in a manner completely analogous to that given earlier, a stepwise-multiple-regression program can be used to produce the same information at each step that was provided by the stepwise-discriminant-analysis program. Incidentally, the R^2 values associated with steps 1, 2, and 3 are 0.050, 0.058, and 0.060, respectively, indicating minimal predictive power.

22.8 MISCELLANEOUS REMARKS

In this final section some special points of interest will be briefly discussed. First, it is not necessary that the discriminant function be linear in the X's. In other words, as long as the discriminant function is linear in its coefficients, all the procedures that we have described can be used no matter how the X's are defined. As an example, for the data of the previous section, we could have considered a model containing an interaction term of the form $X_1 X_2$, or such terms as $\log_e X_3$ or $X_3^{1/2}$. The analogy to multiple regression should be apparent.

Second, in typical situations where the purpose is to discriminate between diseased and nondiseased groups of individuals, it is often desirable to use a function of ℓ^* as a measure of the probability (or risk) of developing the disease. In particular, the *multiple logistic function*

$$P(\ell^*) = \frac{1}{1 + (p_2/p_1)(c_{21}/c_{12})e^{-\ell^*}}$$

is often used for this purpose. Note that as ℓ^* ranges from $-\infty$ to $+\infty$ the value of $P(\ell^*)$ increases monotonically from 0 to 1. The above form of the multiple logistic function is based on a cutoff point of $\log_e (p_2/p_1)(c_{21}/c_{12})$, so a value of ℓ^* exactly equal to $\log_e (p_2/p_1)(c_{21}/c_{12})$ implies a 50–50 chance of developing the disease. For a cutoff point of zero, then, $P(\ell^*) = 1/(1 + e^{-\ell^*})$ and so $P(0) = 1/2$. Based on the assumptions of multivariate normality and of variance and covariance homogeneity in the two populations, it is possible to argue on theoretical grounds alone in favor of the function $P(\ell^*)$. Although we will not present the argument here, it is given, for example, by Cornfield (1962), and applications based on the use of the multiple logistic function can be found in Cornfield (1962), Truett et al. (1967), and Kleinbaum et al. (1971).

Finally, we have not explicitly discussed how to estimate the variance of one of the discriminant-function coefficients b_i, $i = 1, 2, \ldots, p$. A large-sample expression for this estimated variance which has been suggested in the literature is

$$s^{ii}\left(\frac{1}{n_1} + \frac{1}{n_2}\right)$$

For the example we have been considering, the estimated variance of b_1 based on the formula above is

$$0.00560\left(\frac{1}{71}\times\frac{1}{761}\right)=8.62\times10^{-5}$$

A possible alternative method for obtaining an estimate of the variance of b_i can be based on the fact that $b_i = \hat{\beta}_i/c^*$, so that

$$\text{Var}(b_i)=\frac{\text{Var}(\hat{\beta}_i)}{c^{*2}}$$

Using this expression, we estimate the variance of b_1 as

$$\frac{(0.00070)^2}{(0.0736)^2}=9.05\times10^{-5}$$

which is close to the large-sample value above.

PROBLEMS

1. A retrospective study of risk factors for coronary heart disease (CHD) involved a sample of $n_1 = 15$ persons free of CHD and a second sample of $n_2 = 10$ recent CHD cases selected from the employment files for workers with at least 10 years of continuous service. For each person sampled, values of systolic blood pressure (SBP), cholesterol (CHOL), and age (AGE) were obtained from the company's previous medical records of 10 years ago. The goal of the study was to determine the extent to which the variables SBP, CHOL, and AGE can help to discriminate future cases from noncases. The data are as follows for noncases:

NONCASE NO.:	1	2	3	4	5	6	7
SBP	135	122	130	148	146	129	162
CHOL	227	228	219	245	223	215	245
AGE	45	41	49	52	54	47	60

NONCASE NO.:	8	9	10	11	12	13	14	15
SBP	160	144	166	138	152	138	140	134
CHOL	262	230	255	222	250	264	271	220
AGE	48	44	64	59	51	54	56	50

For cases the data are:

CASE NO.:	1	2	3	4	5	6	7	8	9	10
SBP	145	142	135	149	180	150	161	170	152	164
CHOL	238	232	225	230	255	240	253	280	271	260
AGE	60	64	54	48	43	43	63	63	62	65

(a) Consider the accompanying summary table of sample means and standard deviations for each variable in each group. What standard statistical procedure might you use, given the information in the table, to evaluate whether each variable individually discriminates cases from noncases? Carry out this procedure for SBP, CHOL, and AGE separately, and comment on your findings.

| | GROUP MEANS AND STANDARD DEVIATIONS | | | |
| | NONCASES | | CASES | |
VARIABLE	$\bar{X}$	S	$\bar{X}$	S
SBP	142.93	12.83	154.80	13.77
CHOL	238.40	18.61	248.40	18.39
AGE	51.60	6.37	56.50	8.81

(b) In assessing the combined ability of SBP, CHOL, and AGE to discriminate cases from noncases, how would you define the dummy dependent variable (for sample sizes $n_1 = 15$, $n_2 = 10$) if you were using multiple regression to carry out the discriminant-analysis computations?

(c) Using (22.5), determine for the regression model containing SBP, CHOL, and AGE that constant c^* which, when divided into the regression coefficients, yields the corresponding linear discriminant function coefficients. [*Note*: The quantity D^2 can be computed directly from the regression printout using the F statistic (with p and $n_1 + n_2 - p - 1$ df) for testing for significant overall regression by means of the following formula

$$D^2 = \frac{(n_1 + n_2)(n_1 + n_2 - 2)p}{n_1 n_2 (n_1 + n_2 - p - 1)} F$$

This F statistic for this data set has the value 1.871.]

(d) Given the multiple regression solution

$$\hat{Y} = 2.23602 - 0.01406(\text{SBP}) + 0.00242(\text{CHOL}) - 0.01393(\text{AGE})$$

compute the linear discriminant function using your answer to part (c), and then use this function to define two distinct classification rules according to the following specifications of a priori probabilities and costs of misclassification:

(1) $p_1 = 0.6$, $p_2 = 0.4$, $c_{12} = c_{21}$.

(2) $p_2 / p_1 = c_{12} / c_{21}$.

[*Note*: Each rule should be of the following general form: if $\ell^* > C$, assign to group 1 (NCHD); if $\ell^* < C$, assign to group 2 (CHD).]

(e) For classification rule (2) in part (d), determine the appropriate group assignment for noncase 1 and case 10 based on this rule.

(f) Consider the following classification tables based on the two classification rules specified in part (d):

(1)

ACTUAL GROUP	ASSIGNED GROUP	
	NCHD	CHD
NCHD	13	2
CHD	5	5

(2)

ACTUAL GROUP	ASSIGNED GROUP	
	NCHD	CHD
NCHD	11	4
CHD	3	7

Which table is preferable and why?

(g) For classification table (2) in part (f), determine the "apparent error rates" and the more precise estimated probabilities of misclassification discussed in the chapter.

(h) Using the regression ANOVA table, test whether or not there are significant differences between the two groups when all variables are considered together.

SOURCE	df	SS
SBP	1	1.0441
AGE|SBP	1	0.1956
CHOL|SBP, AGE	1	0.0259
Residual	21	4.7345

(i) The order of variables presented in the ANOVA table is precisely that order obtained from forward-stepwise-regression analysis. Based on this information, how would you modify the linear discriminant function to eliminate variables not making a significant contribution to discrimination?

(j) Given the fitted regression model

$$\hat{Y} = 2.1656 - 0.01466(SBP)$$

determine an appropriate decision rule for discriminating cases from non-cases based on this model. How might one evaluate whether or not this "reduced" discriminant function does a better job of discriminating than does the function containing all three variables?

(k) What is the multiple logistic function corresponding to the discriminant function found in part (j)? Using this function, what is the estimated risk of developing CHD for noncase 1 and for case 10.

2. An anthropologist, interested in developing a procedure for classifying skulls as coming from one of two Egyptian historical periods (I and II), applied discriminant analysis to the measurement data given here, which were obtained on a sample of five skulls from each period.

PERIOD	MAXIMUM BREADTH	BASI-ALVEOLAR LENGTH	NASAL HEIGHT	BASI-BREGMATIC HEIGHT
I	134.825	98.703	50.538	133.110
I	134.203	97.667	52.380	134.201
I	133.714	98.976	52.100	134.628
I	135.901	99.203	50.875	133.765
I	133.976	99.521	49.615	132.861
II	135.603	95.400	52.903	131.646
II	135.422	97.832	52.200	130.876
II	136.721	95.212	51.875	133.888
II	134.750	95.861	53.769	130.773
II	137.100	96.500	54.010	129.998

PERIOD	GROUP MEANS			
I	134.5238	98.8140	51.1016	133.7130
II	135.9192	96.1610	52.9514	131.4362

(a) Determine the linear discriminant function ℓ involving all four variables utilizing the following computer results of scores for each variable in each group:

	PERIOD I	PERIOD II
Maximum breadth	178.8834	180.6715
Basi-alveolar length	417.4353	410.6387
Nasal height	224.4495	223.8206
Basi-bregmatic height	284.1536	279.4690
(constant)	−57,389.2539	−56,314.6914

(b) Determine the value of b_0, and then define a classification rule assuming that the ratio of prior probabilities is inversely related to the ratio of costs of misclassifications, i.e., $p_2/p_1 = c_{12}/c_{21}$.

(c) Given that the F statistic of (22.7) has the value 11.730 for the model involving all four variables, test whether there are significant differences between the two groups based on this model.

(d) Use the classification rule developed in part (b) to assign the first skull in each group to one of the two periods.

(e) The classification table based on the rule in part (b) is as shown. Comment on the discrimination achieved.

ACTUAL GROUP	ASSIGNED GROUP	
	I	II
I	5	0
II	0	5

(f) The use of a stepwise-discriminant-function program on these data turned up only two variables, basi-alveolar length and basi-bregmatic height, as making a significant contribution to discrimination based on partial F tests. The scores based on this two-variable model are given in the table. Form the discriminant function, the classification rule (based on the cutoff point b_0), and the classification table for this model. Comment on the difference, if any, in discrimination achieved using only these two variables as opposed to using all four variables.

	PERIOD I	PERIOD II
Basi-alveolar length	273.3677	267.0010
Basi-bregmatic height	211.6374	207.3149
(constant)	−27,656.2852	−26,462.5547

3. A methodological study was undertaken in a large metropolitan area to describe the organizational climate of public high schools in the area. In this study, several variables describing various components of a school's organizational personality were derived by factor analysis and were then used in a two-group discriminant analysis to classify schools into one of two categories of organizational climate: *A*, satisfactory; and *B*, unsatisfactory. A discriminant function involving four variables was developed from data given.

CLIMATE TYPE	DIS*	HIN	MOR	ALO
A	2.58	1.86	1.77	1.86
A	3.12	2.97	1.28	1.71
A	1.87	1.99	0.72	1.89
A	1.65	1.92	1.86	0.98
A	1.42	2.31	1.25	0.85
A	2.41	3.12	0.82	1.28
A	1.25	1.78	1.06	2.01
A	−2.78	2.21	0.25	1.72
A	1.25	0.51	0.56	−0.67
A	−1.28	3.02	0.65	−0.97
B	0.85	1.21	0.48	1.65
B	0.96	1.35	0.38	1.57
B	−2.16	0.18	−0.17	−1.28
B	−1.85	−2.61	0.76	−2.01
B	−1.66	−1.89	0.85	−3.10

* DIS, disengagement (i.e., anomie) of teachers; HIN, hindrance from principal as perceived by teachers; MOR, morale; and ALO, aloofness (formality) exhibited by principal.

(a) To use multiple regression to carry out the discriminant analysis computations, how would you define the dummy dependent variable?

(b) The multiple regression solution involving all four variables is as follows:

$$\hat{Y} = -0.59047 - 0.03022(\text{DIS}) + 0.23192(\text{HIN}) + 0.38134(\text{MOR}) - 0.04191(\text{ALO})$$

The discriminant function scores for each group obtained via the use of a standard stepwise-discriminant-analysis program are given in the table. Form the linear discriminant function using the stepwise-discriminant-analysis computer output, and verify that the discriminant function coefficients are constant multiples of the corresponding regression coefficients.

	CLIMATE A	CLIMATE B
DIS	−1.22922	−0.87242
HIN	3.28807	0.54974
MOR	8.29556	3.79302
ALO	−0.58024	−0.08542
(constant)	−7.48262	−1.83262

(c) The F statistic of (22.7) has the value 5.0687 for this data set. Carry out the appropriate test of hypothesis using this statistic, and then state your conclusions clearly.

(d) Assuming that $\log_e [(p_2/p_1)(c_{21}/c_{12})] = 0$, define an appropriate decision rule for classifying schools as being either climate type A or climate type B.

(e) Use the rule developed in part (d) to classify the first school listed as climate type A in the data set and the first school listed as climate type B.

(f) Use the classification table given, which is based on the rule in part (d), to estimate the probabilities of misclassification.

ACTUAL GROUP	ASSIGNED GROUP	
	A	B
A	9	1
B	0	5

4. The data given were used in a discriminant analysis to describe the relationship between pregnancy characteristics of the mother and the prematurity status of her baby. Use the (SPSS) regression computer printout given to carry out a discriminant analysis of this data set using a discriminant function model involving only prepregnant maternal weight and also using a model with both variables included.

PREMATURITY STATUS*:	NP	NP	NP	NP	NP	NP	NP	NP	NP	NP
Prepregnant maternal weight	105	110	115	120	125	130	135	140	125	115
Maternal age at delivery	22	25	27	30	33	35	37	30	25	25

* P, premature delivery; NP, nonpremature delivery.

PREMATURITY STATUS	NP	NP	NP	NP	NP	P	P	P	P	P
Prepregnant maternal weight	135	110	120	105	115	145	125	130	140	135
Maternal age at delivery	22	26	32	28	24	35	30	32	40	37

Make sure to complete the following steps:

(a) Determine the linear discriminant function from the regression output.

(b) Determine the cutoff point b_0, and define your classification rule assuming that $p_2/p_1 = c_{12}/c_{21}$.

(c) Test for significant differences between the groups based on the fitted models.

(d) Assign each individual to one of the two groups based on the rule in part (b).

(e) Estimate the probabilities of misclassification.

(f) Draw some overall conclusions about the quality of discrimination achieved.

Computer printout for Problem 4 (SPSS)

```
DEPENDENT VARIABLE.. PREM      (Y = .25 if NP,  Y = -.75 if P)

VARIABLE(S) ENTERED ON STEP NUMBER 1..   MW  (Pre-pregnant Maternal Weight)

MULTIPLE R            0.61551   ANALYSIS OF VARIANCE  D.F.  SUM OF SQUARES   MEAN SQUARE      F
R SQUARE              0.37885   REGRESSION             1.      1.42068       1.42068     10.97842
ADJUSTED R SQUARE     0.37885   RESIDUAL              18.      2.32932       0.12941
STANDARD ERROR        0.35973
```

——————————— VARIABLES IN THE EQUATION ———————————

VARIABLE	B	BETA	STD ERROR B	F
MW	-0.00029	-0.61551	0.00009	10.978
(CONSTANT)	1.07384			

```
VARIABLE(S) ENTERED ON STEP NUMBER 2..   MAGE  (Maternal Age at Delivery)

MULTIPLE R            0.62360   ANALYSIS OF VARIANCE  D.F.  SUM OF SQUARES   MEAN SQUARE      F
R SQUARE              0.38888   REGRESSION             2.      1.45831       0.72915      5.40892
ADJUSTED R SQUARE     0.35493   RESIDUAL              17.      2.29169       0.13481
STANDARD ERROR        0.36716
```

——————————— VARIABLES IN THE EQUATION ———————————

VARIABLE	B	BETA	STD ERROR B	F
MW	-0.00045	-0.97191	0.00033	1.924
MAGE	0.03109	0.37021	0.05885	0.279
(CONSTANT)	0.77074			

5. A discriminant analysis was carried out on the data given in the table to determine the extent to which a person's views on political issues could be used to predict that person's political party affiliation. The independent variables represent indices derived from questionnaire responses regarding certain political issues. Use the computer printout given to perform a discriminant analysis of this data set using a discriminant function model containing all four variables. Complete the same steps as outlined in Problem 4.

Computer printout for Problem 5 (BMD excerpt)*

APPROXIMATE F   2.66922   DEGREES OF FREEDOM 4   10.00

F MATRIX - DEGREES OF FREEDOM   4   10

	GROUP PLUS
GROUP MINUS	2.66922

$\leftarrow$ F-statistic to test for significant differences between groups

F LEVEL INSUFFICIENT FOR FURTHER COMPUTATION

VARIABLE		FUNCTION PLUS	MINUS
DEF	1	0.96135	0.24029
ENV	2	2.62234	4.33101
RACE	3	0.63664	1.30335
ECON	4	0.27346	0.45406
CONSTANT		-3.52708	-7.67388

MINUS $\leftarrow$ PLUS = Republican, MINUS = Democrat

$\leftarrow$ Discriminant scores used to form linear discriminant function

GROUP WITH LARGEST PROB.	SQUARE OF DISTANCE FROM AND POSTERIOR PROBABILITY FOR GROUP

Square of distance = D^2, Posterior probability $= (1 + e^{|J^*|})^{-1}$ for group with largest probability

GROUP PLUS

CASE		PLUS		MINUS	
1	PLUS	6.450	0.974	13.719	0.026
2	PLUS	1.197	0.901	5.605	0.099
3	PLUS	2.395	0.986	10.918	0.014
4	MINUS	3.038	0.312	1.455	0.688
5	PLUS	4.496	0.751	6.701	0.249

GROUP MINUS

CASE		PLUS		MINUS	
1	PLUS	5.663	0.559	6.140	0.441
2	MINUS	7.066	0.358	5.901	0.642
3	MINUS	9.451	0.054	3.707	0.946
4	MINUS	5.779	0.118	1.764	0.882
5	MINUS	5.022	0.193	2.157	0.807
6	MINUS	8.818	0.040	2.480	0.960
7	MINUS	13.026	0.007	3.150	0.993
8	MINUS	14.026	0.006	3.927	0.994
9	PLUS	2.285	0.772	4.721	0.228
10	MINUS	4.927	0.097	0.476	0.903

* Supported by the Health Sciences Computing Facility, University of California, Los Angeles, California 90024, under NIH Grant RR-00003.

NUMBER OF CASES CLASSIFIED INTO GROUP

GROUP	PLUS	MINUS
PLUS	4	1
MINUS	2	8

$\leftarrow$ Classification table for which $\log_e \dfrac{p_2}{p_1} \dfrac{c_{21}}{c_{12}} = 0$

POLITICAL PARTY	DEFENSE SPENDING	ENVIRONMENT	RACE	ECONOMY
R	2.42	1.82	−0.65	−2.32
R	2.25	1.76	0.42	1.54
R	2.84	1.32	−0.93	1.65
D	1.79	2.25	1.72	1.15
D	1.01	2.86	0.10	2.79
D	−1.85	2.75	1.78	−1.11
D	−1.72	2.21	1.65	0.21
R	−1.01	1.86	1.41	0.51
D	−0.84	1.92	2.32	0.81
D	−0.77	3.17	1.39	2.31
D	−0.71	3.85	2.55	1.63
D	−0.80	3.65	3.10	1.75
D	−0.45	1.01	1.02	1.10
R	−0.17	0.92	1.25	2.86
D	−1.21	2.35	1.82	1.50

6. For the data in the table, use the computer printout to carry out a discriminant analysis to describe how the four symptom factors jointly can be used to classify psychiatric patients as either depressive or schizophrenic. Complete the same steps as outlined in Problem 4.

PSYCHOPATHOLOGICAL GROUP*	THINKING DISTURBANCE (TD)	WITHDRAWAL-RETARDATION (WR)	HOSTILE-SUSPICIOUSNESS (HS)	ANXIOUS-DEPRESSION (AD)
D	5.0	6.4	6.6	10.2
D	2.9	5.3	2.5	9.2
D	2.7	5.0	2.5	10.3
D	2.5	4.7	3.5	8.6
D	1.9	4.3	4.5	11.1
S	7.8	6.9	3.8	6.0
S	7.6	5.6	4.2	7.3
S	7.4	5.1	5.4	4.9
S	7.2	6.0	6.0	3.6
S	3.5	5.6	3.0	11.8

* D, depressive patient; S, schizophrenic patient.

Computer printout for Problem 6 (BMD excerpt)*

APPROXIMATE F   3.34015   DEGREES OF FREEDOM 4   5.00

F MATRIX - DEGREES OF FREEDOM   4   10

* Supported by the Health Sciences Computing Facility, University of California, Los Angeles, California 90024, under NIH Grant RR-00003.

```
            GROUP
            PLUS

GROUP
MINUS    3.34014  ◄──  F-statistic to test for significant
                       differences between groups
```

F LEVEL INSUFFICIENT FOR FURTHER COMPUTATION

Computer printout for Problem 6 (BMD excerpt)—*continued*

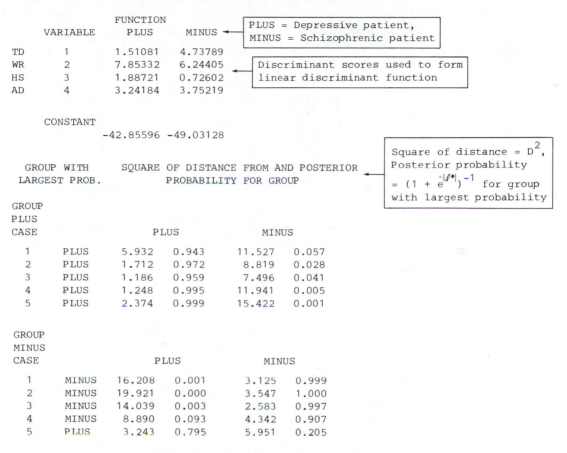

VARIABLE		FUNCTION PLUS	MINUS
TD	1	1.51081	4.73789
WR	2	7.85332	6.24405
HS	3	1.88721	0.72602
AD	4	3.24184	3.75219

PLUS = Depressive patient,
MINUS = Schizophrenic patient

Discriminant scores used to form linear discriminant function

CONSTANT
-42.85596 -49.03128

Square of distance = D^2, Posterior probability = $(1 + e^{-|l*|})^{-1}$ for group with largest probability

GROUP WITH LARGEST PROB.   SQUARE OF DISTANCE FROM AND POSTERIOR PROBABILITY FOR GROUP

GROUP
PLUS

CASE		PLUS		MINUS	
1	PLUS	5.932	0.943	11.527	0.057
2	PLUS	1.712	0.972	8.819	0.028
3	PLUS	1.186	0.959	7.496	0.041
4	PLUS	1.248	0.995	11.941	0.005
5	PLUS	2.374	0.999	15.422	0.001

GROUP
MINUS

CASE		PLUS		MINUS	
1	MINUS	16.208	0.001	3.125	0.999
2	MINUS	19.921	0.000	3.547	1.000
3	MINUS	14.039	0.003	2.583	0.997
4	MINUS	8.890	0.093	4.342	0.907
5	PLUS	3.243	0.795	5.951	0.205

NUMBER OF CASES CLASSIFIED INTO GROUP

GROUP	PLUS	MINUS
PLUS	5	0
MINUS	1	4

Classification table for which $\log_e \dfrac{p_2\, c_{21}}{p_1\, c_{12}} = 0$

7. A botanist was interested in classifying hybrid roses as either of the florabunda type or the multaflora type by using numerical characteristics of the leaves from a sample of bushes of each parental type to determine a classification index.[3] She selected 15 leaves from each of 10 bushes of each parental type and measured each leaf for two characteristics: the number of stomata on the outer face (X_1) and the number of stomata on the inner face (X_2). She then determined the means of the measurements for each bush and obtained the results given in the table.

[3] Adapted from a study by Mergen (1958).

	FLORABUNDA		MULTAFLORA	
BUSH NO.	X_1	X_2	X_1	X_2
1	128.5	133.8	95.2	88.4
2	133.7	133.6	91.0	95.7
3	126.5	131.1	100.0	103.9
4	131.3	130.5	110.3	112.6
5	119.4	120.6	88.7	93.8
6	123.8	133.6	110.6	108.7
7	132.1	122.3	115.5	116.0
8	134.3	134.6	102.6	101.8
9	129.8	133.5	94.8	98.6
10	130.1	128.7	114.8	115.9
Mean	128.95	130.23	102.35	103.54

(a) Determine the linear discriminant function ℓ involving X_1 and X_2 by utilizing the table of computer results of scores for each variable in each group.

	FLORABUNDA	MULTAFLORA
X_1	1.04420	0.82099
X_2	1.34694	1.07863
(constant)	−155.72366	−98.54755

(b) Determine the value of b_0 and then define a rule to classify hybrid roses as either of the florabunda type or multaflora type, assuming equal costs of misclassification and equal prior probabilities.

(c) If you were using multiple regression to carry out your discriminant-analysis computations, how would you define the dummy dependent variable?

(d) Using (22.5) and the note given in part (c) of Problem 1, determine the constant c^* which, when divided into the regression coefficients, yields the corresponding linear discriminant function coefficients. (The F statistic for testing for significant overall regression for these data has the value 30.9277.)

(e) Given the multiple regression solution

$$\hat{Y} = -3.4240 + 0.01337X_1 + 0.01607X_2$$

compute the linear discriminant function ℓ using your answer to part (d), and then check that the coefficients are the same as those obtained in part (a).

(f) Use your classification rule defined in part (b) to classify the hybrids in the table as either of the florabunda or the multaflora type.

HYBRID NO.	X_1	X_2
1	119.8	117.6
2	121.9	122.4
3	117.2	114.9
4	112.8	119.6

8. To discriminate between two species of clover that are quite difficult to distinguish from one another, measurements were taken on three characteristics for samples of each type,[4] with the results shown in the table.

SAMPLE NO.	SPECIES 1			SAMPLE NO.	SPECIES 2		
	X_1	X_2	X_3		X_1	X_2	X_3
1	0.547	0.696	0.217	1	0.062	0.540	0.090
2	0.695	0.624	0.158	2	0.155	0.485	0.120
3	0.321	0.620	0.149	3	0.100	0.550	0.012
4	0.307	0.542	0.126	4	0.141	0.420	0.046
5	0.429	0.562	0.130	5	0.126	0.550	0.006
6	0.210	0.877	0.228	6	0.051	0.511	0.062
7	0.592	0.571	0.148	7	0.131	0.604	0.026
8	0.565	0.612	0.103	8	-0.016	0.510	0.002
9	0.627	0.703	0.115	9	0.061	0.612	0.008
10	0.275	0.614	0.099	10	0.188	0.549	0.064
11	0.702	0.608	0.087	11	0.128	0.544	0.011
12	0.389	0.815	0.190	12	0.112	0.570	-0.008
Mean	0.472	0.654	0.146	Mean	0.103	0.537	0.037

(a) In using a multiple regression program to determine a discriminant function involving all three variables, how would you define the dummy dependent variable?

(b) Using the stepwise regression results given, assess the order of importance and significance of each variable. Based on these results, which variables would you use in a discriminant function for discriminating between species?

SOURCE	df	SS	FITTED EQUATIONS
X_1	1	4.1771	$\hat{Y}$ (only X_1 in model) $= -0.5432 + 1.89007X_1$
$X_3\|X_1$	1	0.7955	$\hat{Y}$ (only X_1 and X_3 in model) $=$
			$\quad -0.6668 + 1.25118X_1 + 3.36758X_3$
$X_2\|X_1, X_3$	1	0.0828	$\hat{Y}(X_1, X_2$ and X_3 in model) $=$
			$\quad -1.0726 + 1.25614X_1 + 0.79186X_2 + 2.63219X_3$
Residual	20	0.9446	

(c) Using the estimated multiple regression coefficients $\hat{\beta}_1$, $\hat{\beta}_2$, and $\hat{\beta}_3$ for the model containing all three independent variables, determine the cutoff point,

$$\frac{1}{2}(\bar{\ell}_1 + \bar{\ell}_2)c^* = \hat{\beta}_1\left(\frac{\bar{X}_{11} + \bar{X}_{21}}{2}\right) + \hat{\beta}_2\left(\frac{\bar{X}_{12} + \bar{X}_{22}}{2}\right) + \hat{\beta}_3\left(\frac{\bar{X}_{13} + \bar{X}_{23}}{2}\right)$$

that can be used to define a classification rule based on the assumption of equal a priori probabilities and equal costs of misclassification.

[4] Adapted from a study by Whitehead (1954).

(d) Give the appropriate classification rule based on the cutoff point obtained in part (c). Why is it not necessary to actually compute c^* in defining this rule? When would it be necessary to compute c^*?

(e) Determine an appropriate cutoff point and classification rule when only those variables determined to be significant discriminators in part (b) are considered, again assuming equal a priori probabilities and equal costs of misclassification.

(f) For the model in part (e), test whether there are significant differences between groups using the appropriate F test.

(g) The classification table obtained for both the classification rule involving all three variables and the rule involving only X_1 and X_3 is the same, as is shown. What does this imply regarding which model is more appropriate?

SPEC	ASSIGN TO SPEC 1	ASSIGN TO SPEC 2
1	12	0
2	0	12

CHAPTER 23 ANALYSIS OF CATEGORICAL DATA

23.1 PREVIEW

In this chapter we describe a method for analyzing multivariable data which are in the form of counts or frequencies corresponding to categories or combinations of categories of nominally treated variables. A simple form of such a categorical data layout is the standard (2×2) two-way contingency table of the form shown in Table 23.1. Here a, b, c, and d are the frequencies of observations (i.e., the observed cell counts) in each of the four cells. As we shall discuss in Section 23.2, the two basic kinds of questions generally asked about such two-way layouts concern issues regarding *independence* or *homogeneity*, and the usual method of analysis in either case involves the use of the chi-square statistic of the general form

$$\chi^2 = \sum \frac{(O-E)^2}{E}$$

where the O's refer to observed cell counts, and the E's refer to expected cell counts specified under a given null hypothesis of interest.

We shall describe how the analysis of such a simple (2×2) contingency table can be considered a special case of a more general method of analysis of categorical data which utilizes a *linear-models approach* similar to that of regression analysis. We

TABLE 23.1 *A (2×2) contingency table*

FACTOR A	FACTOR B	
	LEVEL 1	LEVEL 2
Level 1	a	b
Level 2	c	d

mentioned this methodology briefly in Section 2.3 as a method for describing the relationship between a nominal dependent variable and several nominally or ordinally treated independent variables. Our intention at that early point in the text was to distinguish this method from the more frequently used methods of regression analysis and discriminant analysis in terms of the manner of classification of the variables under study. In this regard, categorical data analysis involves nominally treated dependent variables, whereas regression analysis generally considers continuous dependent variables. Also, categorical data analysis treats the independent variables as nominal or ordinal, whereas discriminant analysis allows the independent variables to be continuous. In this chapter we shall go much further in our discussion of this method for analyzing categorical data to describe its general applicability (even to situations involving several dependent variables), the data configurations considered, the assumptions required, and the general methodology used.

The need for multivariable analysis of categorical data techniques has increased steadily in recent years, since present-day studies in the health, social science, and other fields now typically generate data on *several* qualitative (i.e., nominal or categorical) variables. Simple χ^2 statistics based on looking at such variables "two at a time" do not do justice to the multivariable nature of such data. Thus, general and powerful methods must be considered for analyzing multidimensional contingency tables (i.e., tables arising from cross-classifications of several qualitative variables), methods that provide the flexibility to fit multivariable models and to test appropriate hypotheses. Also, the qualitative nature of the response variable in a categorical data analysis precludes the use of standard multivariable techniques (e.g., analysis-of-variance and dummy-variable regression procedures) which rely on the assumptions of normality and homogeneity of variance for testing purposes.

As an example of a categorical data set which would benefit from the type of sophisticated analysis we are advocating, we will describe a set of data obtained from a large-scale study (Hogue et al., 1974) on postabortion fertility control in several countries conducted by the International Fertility Research Program (IFRP). The dependent variable in this study is dichotomous: an individual either *accepts* or *does not accept* postabortion fertility control (PAFC) at the time of follow-up. The independent variables of interest are *age* (*A*), number of living *children* (*C*), years of *education* completed (*E*), and country or *location* (*L*); each of these demographic variables has been categorized into two groups. Cross-classification according to these four factors then results in $2^4 = 16$ cells, and data on 1,052 cases distributed as in Table 23.2.

The IFRP personnel were concerned with studying the relationship between the four independent variables *A*, *C*, *E*, and *L* and the binomial response PAFC, with specific attention being directed at assessing the main effects of and the interactions among the independent variables. Their analysis objective would be best met by finding a simple predictive model with as few parameters as possible which would efficiently represent the multivariable relationship between the four demographic factors and the dichotomous response. In this regard, Grizzle et al. (1969) have described how to use linear regression techniques and weighted-least-squares analysis to build such models when the data are categorical in nature, and it is their approach that we shall adopt in this chapter. Their methodology involves the use of test statistics which have χ^2 distributions in large samples when the corresponding null hypotheses of interest are true. The theoretical justification for their approach can be found in papers by Wald (1943) and Neyman (1949).

TABLE 23.2 *Classification of cases accepting or not accepting postabortion fertility control by age (A), number of living children (C), years of education completed (E), and location (L)*

A	C	E	L	NUMBER ACCEPTING FERTILITY CONTROL	NUMBER NOT ACCEPTING FERTILITY CONTROL	PROPORTION ACCEPTING FERTILITY CONTROL
<25	0–1	0–6	India	4	8	0.333
25+	0–1	0–6	India	2	1	0.667
<25	0–1	7+	India	14	7	0.667
25+	0–1	7+	India	7	3	0.700
<25	2+	0–6	India	17	1	0.944
25+	2+	0–6	India	175	9	0.951
<25	2+	7+	India	12	2	0.857
25+	2+	7+	India	90	5	0.947
<25	0–1	0–6	Yugoslavia	12	21	0.364
25+	0–1	0–6	Yugoslavia	13	9	0.591
<25	0–1	7+	Yugoslavia	112	155	0.419
25+	0–1	7+	Yugoslavia	40	39	0.506
<25	2+	0–6	Yugoslavia	8	3	0.727
25+	2+	0–6	Yugoslavia	63	36	0.636
<25	2+	7+	Yugoslavia	13	9	0.591
25+	2+	7+	Yugoslavia	96	66	0.593
				678	374	0.588

[Two alternative approaches to the analysis of categorical data are that based on *maximum likelihood* as formulated by Bishop (1969) and Goodman (1970) and that based on *minimum discrimination information* as formulated by Ku et al. (1971). In large-sample situations, all three of these methods are essentially equivalent. In most applications, the choice of which method to use is a matter of practical and computational convenience, although we prefer the Grizzle–Starmer–Koch approach because it is operationally so similar to that of multiple regression.]

In Section 23.2 we shall provide a general discussion concerning categorical data (at which time we will establish notation with regard to data representation), and we shall examine a particular data set in some detail. In Sections 23.3 and 23.4 we shall describe the Grizzle–Starmer–Koch (GSK) linear-models approach to the analysis of categorical data. In Section 23.5 we shall illustrate the generality and flexibility of this methodology with some examples, one of which will be based on the data given in Table 23.2. The main goal in our working through several diverse examples will be to illustrate how to specify an underlying model, how to formulate (in terms of this model) hypotheses to be tested, and how to interpret (with respect to the model) the results of these statistical tests.

It is not expected that the reader will become an expert in analyzing complex categorical data sets after working through this chapter. Indeed, the purpose of the chapter is simply to introduce the reader to a general and powerful approach for analyzing such data sets, and this will be accomplished via a general (nonmathematical) discussion of the methodology and a consideration of some illustrative applications.

23.2 REPRESENTATION OF CATEGORICAL DATA

In any experimental situation, data of two general types are collected on each subject or unit under study:

1. Data on the group or subpopulation to which the subject belongs or on the set of experimental conditions imposed on the subject.
2. Data on what happens to the subject during the course of the experiment.

In the case of categorical data, we shall use the term "factor" when referring to type 1 ways of classification; in other words, a *factor* in categorical data analysis is analogous to an *independent variable* in regression analysis. When referring to type 2 ways of classification, we shall use the term *response*, which is akin to the term *dependent variable* in regression analysis. Hence, the data in any multidimensional table are simply the observed frequencies with which subjects belonging to the same combination of factor categories (i.e., the same subpopulation) give the same combination of responses (i.e., fall in the same response category).

To firm up these concepts, consider the hypothetical data set shown in Table 23.3. Here n_{ij} is the observed number (or frequency) of individuals in the ith subpopulation giving the jth category of response, $i = 1, 2, \ldots, s, j = 1, 2, \ldots r$, where $n_{i.} = \sum_{j=1}^{r} n_{ij}$ is the total number of individuals in the ith subpopulation. At this time we note that all marginal frequencies that depend on factor combinations only (i.e., the $n_{i.}$'s) are considered *fixed* numbers known prior to the performance of the study, while all other frequencies are random variables.

The "factor-independent variable/response-dependent variable" analogy that we alluded to above leads to a method for classifying the types of tables one may encounter according to the manner in which the data have been gathered, and/or according to the general types of questions one should be asking of the data. There are four principal types of multidimensional tables of usual interest: "multiresponse, no-factor" tables; "multiresponse, one-factor" tables; "one-response, multifactor" tables; and "multiresponse, multifactor" tables.

In multiresponse, no-factor tables (for which $s = 1$ in Table 23.3), it is considered that only one subpopulation is sampled and that two or more responses (i.e.,

TABLE 23.3 *Hypothetical set of categorical data*

SUBPOPULATIONS (factor combinations)	CATEGORIES OF RESPONSE (response combinations)				TOTAL
	1	2	$\cdots$	r	
1	n_{11}	n_{12}	$\cdots$	n_{1r}	$n_{1.}$
2	n_{21}	n_{22}	$\cdots$	n_{2r}	$n_{2.}$
$\vdots$	$\vdots$	$\vdots$		$\vdots$	$\vdots$
s	n_{s1}	n_{s2}	$\cdots$	n_{sr}	$n_{s.}$
					$n_{..} = \sum_{i=1}^{s} n_{i.}$

dependent variables) are measured on each sampling unit, each response being categorized to yield r combinations of response categories in total. For such a table, the researcher is interested in the relationships among the different responses, and the analysis problems here are analogous to those concerning questions of independence and correlation among several continuous dependent variables, $Y_1, Y_2, \ldots, Y_k$, say.

If the table is of the multiresponse, one-factor type, samples are considered to have been taken from each of $s(\geq 2)$ subpopulations corresponding to the categories of a single factor. The interest of the investigator in this case is to study the effects of the factor on the various interrelationships among the responses, as, for example, in the "multivariate" regression of several continuous dependent variables ($Y_1, Y_2, \ldots, Y_k$, say) on a single independent variable X (an issue we have not dealt with in this text).

With one-response, multifactor tables, samples are considered to have come from each of $s(\geq 2)$ subpopulations corresponding to combinations of categories of two or more factors, and one response with $r(\geq 2)$ categories is measured in each subpopulation. Here one would be interested in the way in which the factors jointly influence the response, and the analysis problems here are analogous to those arising in regression analysis involving one continuous dependent variable Y and several independent variables, $X_1, X_2, \ldots, X_p$, say.

Finally, for multiresponse, multifactor tables, samples are considered to have come from each of $s(\geq 2)$ subpopulations corresponding to combinations of categories of *several* factors, and *several* responses are measured in each subpopulation. For such tables, both the relationships among the responses and the ways in which the factors combine to affect these relationships are of interest, as they are when performing a multivariate regression of several continuous dependent variables ($Y_1, Y_2, \ldots, Y_k$) on several independent variables ($X_1, X_2, \ldots, X_p$).

As an example, Table 23.2 is best looked upon (for analysis purposes) as a one-response, multifactor table. There are four factors (A, C, E, and L), each at two levels, which results in $s = 2^4 = 16$ subpopulations (or factor combinations); the one response PAFC is *binomial*, giving $r = 2$ response categories.

[It is important to note that, based on the sampling scheme actually used to generate these data, a multiresponse, no-factor classification appears more technically correct. Thus, as this example illustrates, it may turn out that the sampling scheme actually used does not correspond to the sampling method appropriate for answering the primary research questions of interest. For example, for the IFRP data (Table 23.2), it is desirable to assume that the marginal totals for each row (i.e., subpopulation) were known (or fixed) in advance of data collection, even though, in actuality, only the total sample size of 1,052 was fixed. In such a case it is necessary to consider that the subpopulation (conditional) distributions based on the condition of *assuming* fixed marginal totals a priori adequately mimic the unconditional distributions based on actually sampling with fixed marginals a priori in order to justify the desired analysis. Thus, for the IFRP data, we must appeal to such a conditional argument in order to justify a one-response, multifactor analysis. In most applications, such an argument is usually quite tenable.]

In particular, if Π_{i1} is the true (but unknown) probability of an individual in the ith subpopulation accepting postabortion fertility control [so that $\Pi_{i2} (= 1 - \Pi_{i1})$ is the true (but unknown) probability of that individual *not* accepting postabortion fertility control], then the parameter of interest for the ith subpopulation is Π_{i1}.[1]

In the general case when $r > 2$, there are r probability parameters $\Pi_{i1}, \Pi_{i2}, \ldots, \Pi_{ir}$ associated with the ith subpopulation, and the (joint) distribution of the observed cell frequencies $n_{i1}, n_{i2}, \ldots, n_{ir}$ is *multinomial*,[2] where $\sum_{j=1}^{r} n_{ij} = n_{i.}$ and $\sum_{j=1}^{r} \Pi_{ij} = 1$.

In Table 23.4 the data layout of Table 23.3 is given in terms of these unknown multinomial cell probability parameters.

The analysis of data of the general form given in Table 23.3 usually revolves around the estimation of and the testing of hypotheses concerning functions of the unknown cell probabilities given in Table 23.4. In fact, as will be illustrated in the next section, most analysis questions can be handled by considering either linear functions or log-linear functions of the Π_{ij}'s. The GSK approach to be discussed in Section 23.3 allows for the fitting of such functions to a linear model, for evaluating the goodness of fit of the model, and for the testing of hypotheses about the parameters of the fitted model.

Finally, before proceeding to an in-depth discussion of the GSK methodology, we will present another example. This example will help to illustrate some of the concepts presented in this section and, in addition, will serve the purpose of emphasizing the fact that the conditions under which an experiment is conducted (e.g., the sampling scheme used to collect the data) determine whether particular dimensions in a contingency table should be considered as factors or responses (which would affect the subsequent data analysis and interpretations thereof).

EXAMPLE Consider Table 23.5, which summarizes (hypothetical) data from a political science study concerning party opinion in a particular city about a new governmental policy.

TABLE 23.4 *Cell probabilities for hypothetical data set in Table 23.3*

SUBPOPULATIONS (factor combinations)	CATEGORIES OF RESPONSE (response combinations)				TOTAL
	1	2	$\cdots$	r	
1	Π_{11}	Π_{12}	$\cdots$	Π_{1r}	1
2	Π_{21}	Π_{22}	$\cdots$	Π_{2r}	1
$\vdots$	$\vdots$	$\vdots$		$\vdots$	$\vdots$
s	Π_{s1}	Π_{s2}	$\cdots$	Π_{sr}	1

[1] The (joint) probability distribution of the responses for the binomial distribution in the ith subpopulation is given by

$$P(n_{i1}, n_{i2}) = \frac{n_{i.}!}{n_{i1}! n_{i2}!} \Pi_{i1}^{n_{i1}} \Pi_{i2}^{n_{i2}}$$

[2] The (joint) probability distribution of the responses for the multinomial distribution in the ith subpopulation is given by

$$P(n_{i1}, n_{i2}, \ldots, n_{ir}) = \frac{n_{i.}!}{n_{i1}! n_{i2}! \cdots n_{ir}!} \Pi_{i1}^{n_{i1}} \Pi_{i2}^{n_{i2}} \cdots \Pi_{ir}^{n_{ir}}$$

TABLE 23.5   *Hypothetical data from a political science study*

	FAVOR POLICY	DO *NOT* FAVOR POLICY	NO OPINION	TOTAL
Democrats	200	200	100	500
Republicans	250	175	75	500
Total	450	375	175	1,000

We shall consider two distinct sampling procedures that could have produced the data array in Table 23.5:

1. A random sample of 1,000 individuals in the city is selected and each individual is asked his or her party affiliation (for simplicity, we assume that "Democrat" and "Republican" are the only two possible responses) and his or her opinion concerning the new policy (favor, do not favor, or no opinion).

2. A random sample of 500 Democrats is selected from a list of registered Democrats in the city and each Democrat is asked his or her opinion concerning the new policy (favor, do not favor, or no opinion), and a completely analogous procedure is then used to obtain the opinions of 500 Republicans.

Sampling scheme 1 elicits *two responses* from each individual (political party and policy opinion), and no marginal frequencies are fixed other than the total sample size of 1,000. Under this sampling scheme, then, we are considering a two-response, no-factor table with $s = 1$ subpopulation and $r = 2 \times 3 = 6$ response categories. In the format of Tables 23.3 and 23.4, the data of Table 23.5 would look like that in Table 23.6. Here we note that $\sum_{j=1}^{6} \Pi_{1j} = 1$, since we are dealing with only *one* multinomial population.

Sampling scheme 2 results in *one response* (policy opinion) from each individual, and both row marginal totals are fixed a priori. Under this sampling scheme we are considering a one-resonse, one-factor table with $s = 2$ subpopulations and $r = 3$ response categories. In the format of Tables 23.3 and 23.4, we would have the

TABLE 23.6   *Representation of data in Table 23.5 based on a one-multinomial-population model*

	SIX RESPONSE CATEGORIES						TOTAL
	DEMOCRAT AND FAVOR	DEMOCRAT AND DO NOT FAVOR	DEMOCRAT AND NO OPINION	REPUBLICAN AND FAVOR	REPUBLICAN AND DO NOT FAVOR	REPUBLICAN AND NO OPINION	
One sub-population	$n_{11} = 200$	$n_{12} = 200$	$n_{13} = 100$	$n_{14} = 250$	$n_{15} = 175$	$n_{16} = 75$	$n_{1.} = 1,000$
	Π_{11}	Π_{12}	Π_{13}	Π_{14}	Π_{15}	Π_{16}	1

representation given in Table 23.7. Here we note that $\sum_{j=1}^{3} \Pi_{1j} = 1$ and $\sum_{j=1}^{3} \Pi_{2j} = 1$, since we are dealing with *two* multinomial populations.

To analyze the data in Table 23.5, a person with a knowledge of basic statistics would naturally think of using the standard χ^2 test involving a comparison of observed (O) versus expected (E) cell frequencies. In particular, for Table 23.5, suppose that we let O_{ij} be the observed cell frequency in the ith row ($i = 1$ for Democrats, $i = 2$, for Republicans) and jth column ($j = 1$ for "favor policy," $j = 2$ for "do not favor policy," $j = 3$ for "no opinion"). Further, suppose that we let $O_{i\cdot} = \sum_{j=1}^{3} O_{ij}$ be the ith row marginal total and $O_{\cdot j} = \sum_{i=1}^{2} O_{ij}$ be the jth column marginal total; then it can be seen from Table 23.5 that $O_{11} = 200$, $O_{12} = 200$, $O_{13} = 100$, $O_{21} = 250$, $O_{22} = 175$, $O_{23} = 75$, $O_{1\cdot} = O_{2\cdot} = 500$, $O_{\cdot 1} = 450$, $O_{\cdot 2} = 375$, and $O_{\cdot 3} = 175$.

Then the "expected" cell frequency in the ith row and jth column under the null hypothesis of "no association between the two methods of classification" is computed as

$$E_{ij} = \frac{O_{i\cdot} O_{\cdot j}}{O_{\cdot\cdot}}$$

where $O_{\cdot\cdot} = \sum_{j=1}^{3} O_{\cdot j} = \sum_{i=1}^{2} O_{i\cdot} = n_{\cdot\cdot}$ ($= 1{,}000$ in this example). For the data in Table 23.5 one can check that

$$E_{11} = E_{21} = \frac{500(450)}{1{,}000} = 225$$

$$E_{12} = E_{22} = \frac{500(375)}{1{,}000} = 187.5$$

$$E_{13} = E_{23} = \frac{500(175)}{1{,}000} = 87.5$$

Finally, the test statistic to be used here is of the form

$$\chi^2 = \sum_{i=1}^{2} \sum_{j=1}^{3} \frac{(O_{ij} - E_{ij})^2}{E_{ij}}$$

TABLE 23.7   *Representation of data in Table 23.5 based on a two-multinomial-population model*

TWO SUBPOPULATIONS	THREE RESPONSE CATEGORIES			TOTAL
	FAVOR POLICY	DO NOT FAVOR POLICY	NO OPINION	
Democrats (subpopulation 1)	$n_{11} = 200$	$n_{12} = 200$	$n_{13} = 100$	$n_{1\cdot} = 500$
	Π_{11}	Π_{12}	Π_{13}	1
Republicans (subpopulation 2)	$n_{21} = 250$	$n_{22} = 175$	$n_{23} = 75$	$n_{2\cdot} = 500$
	Π_{21}	Π_{22}	Π_{23}	1

which, for our example, has the value

$$\chi^2 = \frac{(200-225)^2}{225} + \frac{(200-187.5)^2}{187.5} + \frac{(100-87.5)^2}{87.5} + \frac{(250-225)^2}{225}$$

$$+ \frac{(175-187.5)^2}{187.5} + \frac{(75-87.5)^2}{87.5} = 2.78 + 0.83 + 1.79 + 2.78$$

$$+ 0.83 + 1.79 = 10.80$$

This computed value of χ^2 is significant at the 1% level since $\chi^2_{2,0.01} = 9.21$, the degrees of freedom for χ^2 being determined as the (number of rows -1) $\times$ (number of columns -1) $= (2-1) \times (3-1) = 1 \times 2 = 2$.

Thus, the null hypothesis of "no association between the two methods of classification" is rejected, but what does this statement really mean? What do we mean by the phrase "no association"? In other words, what does the (vague) statement "no association" mean in terms of the unknown parameters in the problem, the cell probabilities (the Π_{ij}'s)? The answers to these questions will depend on what sampling scheme was used to generate the data. In particular, even though the "$(O-E)^2/E$" chi-square statistic given above happens to be computationally the same for both sampling schemes described earlier, the particular null hypotheses being tested via this statistic are markedly different in the two instances.

For sampling scheme 1, the null hypothesis of no association is really one of *independence* (i.e., we wish to investigate the possibility of a "contingency" between the two methods of classification and hence the term "contingency table"). More specifically, as when two events A and B are said to be independent in a probabilistic sense if $P(A \text{ and } B) = P(A)P(B)$, so here the two responses (or methods of classification) are said to be independent when the probability of an individual being in a particular response category (i.e., a cell in the original contingency table, Table 23.5) is the product of the corresponding marginal probabilities associated with that response category. This is why $E_{ij} = O_i.O_{.j}/O_{..}$, since E_{ij} is expressible as

$$E_{ij} = O_{..} \left(\frac{O_{i.}}{O_{..}} \right) \left(\frac{O_{.j}}{O_{..}} \right)$$

where $O_{i.}/O_{..}$ and $O_{.j}/O_{..}$ can be looked upon as the (estimated) marginal probabilities of being in the ith row and jth column, respectively.

Let us pursue this point a little further. From Table 23.6 it follows that

$\Pi_D(= \Pi_{11} + \Pi_{12} + \Pi_{13})$ is the true (marginal) probability of being a *Democrat*,

$\Pi_R(= \Pi_{14} + \Pi_{15} + \Pi_{16})$ is the true (marginal) probability of being a *Republican*,

$\Pi_F(= \Pi_{11} + \Pi_{14})$ is the true (marginal) probability of *favoring* the policy,

$\Pi_{NF}(= \Pi_{12} + \Pi_{15})$ is the true (marginal) probability of *not favoring* the policy,

$\Pi_{NO}(= \Pi_{13} + \Pi_{16})$

is the true (marginal) probability of having *no opinion* concerning the policy.

Then the null hypothesis H_I of independence specifies that the unknown true cell probabilities have a particular structure, namely that they are formed as products of the above marginal probabilities as follows:

H_I: $\Pi_{11} = \Pi_D \cdot \Pi_F$

$\Pi_{12} = \Pi_D \cdot \Pi_{NF}$

$\Pi_{13} = \Pi_D \cdot \Pi_{NO}$

$\Pi_{14} = \Pi_R \cdot \Pi_F$

$\Pi_{15} = \Pi_R \cdot \Pi_{NF}$

$\Pi_{16} = \Pi_R \cdot \Pi_{NO}$

For sampling scheme 2, the null hypothesis of no association is really one of *homogeneity* (i.e., the two multinomial populations involved are identical in structure and hence the term "homogeneous"). More specifically, the null hypothesis H_H of homogeneity (or, equivalently, the null hypothesis of identicalness of the two multinomial distributions) means that the corresponding cell probabilities in Table 23.7 are equal:

H_H: $\Pi_{11} = \Pi_{21}$

$\Pi_{12} = \Pi_{22}$

$\Pi_{13} = \Pi_{23}$

This is why $E_{ij} = O_{i.}O_{.j}/O_{..}$ is expressible as

$$E_{ij} = O_{i.}\left(\frac{O_{1j} + O_{2j}}{O_{1.} + O_{2.}}\right)$$

where $(O_{1j} + O_{2j})/(O_{1.} + O_{2.})$ is the *pooled* estimate of the probability of giving the jth response under the null hypothesis that $\Pi_{1j} = \Pi_{2j}$, $j = 1, 2, 3$.

Of course, since $(\Pi_{11} + \Pi_{12} + \Pi_{13}) = 1$ and $(\Pi_{21} + \Pi_{22} + \Pi_{23}) = 1$, it follows that the conditions $\Pi_{11} = \Pi_{21}$ and $\Pi_{12} = \Pi_{22}$ together imply that $\Pi_{13} = \Pi_{23}$, so that the hypothesis H_H above can be more compactly stated as

H'_H: $\Pi_{11} = \Pi_{21}$

$\Pi_{12} = \Pi_{22}$

In this regard it is important in what follows to express any null hypothesis of interest in a way which involves a minimum number of parameters. An equivalent but more rigorous way to say this is that the constraints on the parameters imposed by the null hypothesis are to be mathematically independent of one another (e.g., the constraints given by H_H are linearly dependent, while those given by H'_H are not). Such mathematical independence is required under the GSK formulation.

Thus, it is clear that the null hypotheses of *independence* and *homogeneity* are quite different, even though the usual χ^2 test is computed exactly the same way for these two hypotheses.

There is one other important point to be made with reference to this particular example. As mentioned earlier, the GSK approach to the analysis of categorical data involves estimating and testing appropriate hypotheses concerning linear and/or log-linear functions of the unknown cell probabilities, since many categorical data-analysis questions involve consideration of such functions. Hence, it is of interest here to see just how the null hypotheses of independence and homogeneity discussed earlier can be represented in terms of such functions of the cell probabilities.

Now, the null hypothesis H_I of independence given earlier can be shown [e.g., see Bhapkar and Koch (1968)] to be equivalent to the null hypothesis

$$H_I': \quad \Delta_{11} = 1$$

$$\Delta_{12} = 1$$

where

$$\Delta_{11} = \frac{\Pi_{11}\Pi_{16}}{\Pi_{14}\Pi_{13}} \quad \text{and} \quad \Delta_{12} = \frac{\Pi_{12}\Pi_{16}}{\Pi_{15}\Pi_{13}}$$

Note that under the null hypothesis H_I,

$$\Delta_{11} = \frac{(\Pi_D \cdot \Pi_F)(\Pi_R \cdot \Pi_{NO})}{(\Pi_R \cdot \Pi_F)(\Pi_D \cdot \Pi_{NO})} = 1$$

and

$$\Delta_{12} = \frac{(\Pi_D \cdot \Pi_{NF})(\Pi_R \cdot \Pi_{NO})}{(\Pi_R \cdot \Pi_{NF})(\Pi_D \cdot \Pi_{NO})} = 1$$

It is then immediate that H_I' is equivalent to

$$H_I'': \quad \log_e \Delta_{11} = 0$$

$$\log_e \Delta_{12} = 0$$

where

$$\log_e \Delta_{11} = \log_e \Pi_{11} + \log_e \Pi_{16} - \log_e \Pi_{14} - \log_e \Pi_{13}$$

and

$$\log_e \Delta_{12} = \log_e \Pi_{12} + \log_e \Pi_{16} - \log_e \Pi_{15} - \log_e \Pi_{13}$$

Thus, H_I can be converted to an equivalent null hypothesis H_I'' involving *log-linear* functions of the unknown cell probabilities. By a log-linear function of the Π_{ij}'s, we mean that the function is of the form $\sum_{i=1}^{s} \sum_{j=1}^{r} k_{ij} \log_e \Pi_{ij}$ for suitable choices of the constants k_{ij} (for the previous example, each k_{ij} is either 1, -1, or 0, depending on the values of i and j). As we will see, the GSK approach would involve estimating the log functions in H_I'' and then testing H_I'' using these estimates (which would, of course, require variance estimates as well).

[There is a more general form of log-linear function considered in the GSK methodology; this is a function that is "linear in the logs of linear functions of the Π_{ij}," that is, a function of the general form

$$\sum_{\gamma=1}^{u} k_\gamma \log_e \left(\sum_{i=1}^{s} \sum_{j=1}^{r} a_{\gamma,ij} \Pi_{ij} \right)$$

In the next section we shall illustrate the utility of such a function with some particular examples.]

As for the hypothesis H_H of homogeneity, it is easy to see that

H_H': $\Pi_{11} = \Pi_{21}$

 $\Pi_{12} = \Pi_{22}$

can be equivalently written in the form

H_H'': $\Pi_{11} - \Pi_{21} = 0$

 $\Pi_{12} - \Pi_{22} = 0$

which involves *linear* functions of the unknown cell probabilities.

To summarize, then, we have indicated how the hypotheses H_I and H_H of independence and homogeneity can be equivalently expressed as hypotheses H_I'' and H_H'' stating that certain linear and/or log-linear functions of the unknown cell probabilities are equal to zero. The GSK approach can be used to test these and even more complex hypotheses (e.g., those concerning regression coefficients produced by means of a weighted-least-squares fitting of a linear model to the cell probabilities or functions thereof).

23.3 GSK LINEAR MODELS APPROACH TO THE ANALYSIS OF CATEGORICAL DATA

As we mentioned previously, the Grizzle–Starmer–Koch (GSK) linear-models approach to the analysis of categorical data is based on the application of general weighted-least-squares regression techniques to estimates of appropriate (linear and/or log-linear) functions of the cell proportions in complex categorical data layouts. From these results, statistical tests of particular hypotheses of interest are directly produced. For instance, the relationship between the dichotomous dependent variable PAFC and the independent variables A, C, E, and L in the example of Section 23.1 can be investigated with this methodology, and questions regarding variable selection, model appropriateness, and interaction can be pursued in the same spirit as that used in analysis of variance and stepwise-regression procedures for quantitative data.

The basic requirement here for any test of hypothesis is that the sample size be sufficiently large. By this we mean that the observed frequencies (i.e., the n_{ij}'s) in Table 23.3 should each be at least 10 in value (and preferably 25 or more).[3] The only other requirement is that there be either a sampling or observational basis for arguing that the inherent variability in the set of observed frequencies for any subpopulation can be characterized in terms of a multinomial distribution (as discussed previously).[4]

We shall now describe the steps involved in using the GSK methodology, inserting examples at each step where necessary for clarification:

Step 1: Formulate the appropriate multidimensional contingency table (in the form of Table 23.3) *for your problem*, specifying *s* (the number of subpopulations) and *r* (the number of response categories). The n_{ij} of this table are the data to be used in the analysis.

Step 2: Letting the Π_{ij} denote the unknown cell probabilities (see Table 23.4), *specify the functions of the Π_{ij} that you wish to study*. These functions will either be defined as:

(a) *u* functions $a_1(\Pi), a_2(\Pi), \ldots, a_u(\Pi)$, each of which is a *linear function of the Π_{ij}*, that is, a function of the form

$$a_\gamma(\Pi) = \sum_{i=1}^{s} \sum_{j=1}^{r} a_{\gamma,ij}\Pi_{ij}, \qquad \gamma = 1, 2, \ldots, u$$

for suitable choices of the $a_{\gamma,ij}$; or,

(b) *t* functions $f_1(\Pi), f_2(\Pi), \ldots, f_t(\Pi)$, each of which is a *log-linear function of the Π_{ij}*, that is, a function of the form

$$f_\alpha(\Pi) = \sum_{\gamma=1}^{u} k_{\alpha\gamma} \log_e a_\gamma(\Pi), \qquad \alpha = 1, 2, \ldots, t$$

for suitable choices of the $k_{\alpha\gamma}$, where the $a_\gamma(\Pi)$ are the linear functions defined in (a).

EXAMPLES

1. For the IFRP study data of Table 23.2, $s = 16$, $r = 2$, and the $u = 16$ *linear* functions of interest are $a_\gamma(\Pi) = \Pi_{\gamma 1}$, $\gamma = 1, \ldots, 16$, which are the true probabilities of fertility control acceptance for the 16 subpopulations. The $a_{\gamma,ij}$ for each linear function are given by

$$a_{\gamma,ij} = \begin{cases} 1 & \text{if } i = \gamma, j = 1 \\ 0 & \text{otherwise} \end{cases}$$

2. For the political science data of Table 23.7, $s = 2$, $r = 3$, and the homogeneity null hypothesis H_H'' involves the $u = 2$ *linear* functions $a_1(\Pi) = (\Pi_{11} - \Pi_{21})$ and $a_2(\Pi) = (\Pi_{12} - \Pi_{22})$. Each function is of the form $a_\gamma(\Pi) = \sum_{i=1}^{2} \sum_{j=1}^{3} a_{\gamma,ij}\Pi_{ij}$, where the $a_{\gamma,ij}$ are

---

[3] The GSK weighted-least-squares approach to estimation and testing suffers when several of the $n_{ij} = 0$. The effects of replacing such zero values for observed cell counts by some small positive value, as has been advocated by Grizzle, Starmer, and Koch, has not been definitively investigated.

[4] Under the multinomial structure, the research questions of interest are based entirely on the study of selected functions of estimates of the Π_{ij}, where the estimate of Π_{ij} is given by the relative frequency $p_{ij} = n_{ij}/n_i$. and the variances and covariances of these functions of the p_{ij} are themselves functions of the variances and covariances of the p_{ij}.

given by $a_{1,11} = 1$, $a_{1,21} = -1$, $a_{1,12} = a_{1,13} = a_{1,22} = a_{1,23} = 0$ for $a_1(\Pi)$, and $a_{2,12} = 1$, $a_{2,22} = -1$, $a_{2,11} = a_{2,13} = a_{2,21} = a_{2,23} = 0$ for $a_2(\Pi)$.

3. For the political science data of Table 23.6, $s = 1$, $r = 6$, and the independence null hypothesis H''_I involves the $t = 2$ *log-linear functions*

$$f_1(\Pi) = \log_e \Pi_{11} + \log_e \Pi_{16} - \log_e \Pi_{14} - \log_e \Pi_{13}$$

and

$$f_2(\Pi) = \log_e \Pi_{12} + \log_e \Pi_{16} - \log_e \Pi_{15} - \log_e \Pi_{13}.$$

Each of these functions is of the general form

$$f_\alpha(\Pi) = \sum_{\gamma=1}^{u} k_{\alpha\gamma} \log_e a_\gamma(\Pi)$$

where $u = 6$, $a_\gamma(\Pi) = \sum_{j=1}^{6} a_{\gamma,1j} \Pi_{1j} = \Pi_{1\gamma}$, so

$$a_{\gamma,1j} = \begin{cases} 1 & \text{if } j = \gamma \\ 0 & \text{otherwise} \end{cases}$$

and the $k_{\alpha\gamma}$ are

$$\begin{cases} \text{for } f_1(\Pi): & k_{11} = 1, k_{12} = 0, k_{13} = -1, k_{14} = -1, k_{15} = 0, k_{16} = 1 \\ \text{for } f_2(\Pi): & k_{21} = 0, k_{22} = 1, k_{23} = -1, k_{24} = 0, k_{25} = -1, k_{26} = 1 \end{cases}$$

Step 3: Determine the point estimates of the functions $a_\gamma(\Pi)$ *or* $f_\alpha(\Pi)$ *defined in* step 2 by substituting $p_{ij} = n_{ij}/n_{i.}$ for Π_{ij} everywhere it appears in each function. (These point estimates are printed out by the GSK computer program GENCAT, which we shall discuss in the next section.)

Step 4 (optional)[5]: When considering a single or multifactor problem, postulate a linear model for the parameters of interest (i.e., the functions of the Π_{ij} defined in step 2) *in terms of factor effects*, as follows:

Let $g_1(\Pi)$, $g_2(\Pi)$, ..., $g_m(\Pi)$ denote the set of functions specified in step 2 (i.e., $m = u$ or t, depending on whether the functions are *linear* or *log-linear*). Define your model by specifying v (the number of parameters in your model) and the $x_{\xi l}$ ($\xi = 1, 2, \ldots, m$; $l = 1, 2, \ldots, v$), so that the model is of the form

$$g_\xi(\Pi) = x_{\xi 1}\beta_1 + x_{\xi 2}\beta_2 + \cdots + x_{\xi v}\beta_v, \qquad \xi = 1, 2, \ldots, m$$

Weighted-least-squares procedures are used in the GSK approach to estimate the unknown β's in the above model. Even the multiresponse, no-factor situation can be couched in this linear-model framework by considering the trivial case where $v = 0$ or all $x_{\xi l}$ are 0. The $x_{\xi l}$'s are constants whose specification depends on the nature of the

---

[5] This step may be skipped, for example, if the situation being considered is a "multiresponse, no-factor" one (e.g., like a test for independence), since it is only of interest in such a case to test whether the functions of the Π_{ij} defined in step 2 are equal to zero.

data set, on the structure of the g_ε's, and on the particular research questions of interest. The weighted-least-squares analysis produces estimates $b_1, b_2, \ldots, b_v$ of $\beta_1, \beta_2, \ldots, \beta_v$ by fitting the model above to the observed responses $g_\varepsilon(p)$, which are obtained by replacing Π_{ij} by p_{ij} in each $g_\varepsilon(\Pi)$.

[We have briefly discussed weighted least squares in its most elementary form (straight-line regression) in Chapter 16. For the more mathematically sophisticated, we point out that the general form of the weighted-least-squares solution referred to above can be expressed in matrix notation as

$$\mathbf{b} = (\mathbf{X}'\hat{\mathbf{\Sigma}}^{-1}\mathbf{X})^{-1}\mathbf{X}'\hat{\mathbf{\Sigma}}^{-1}\mathbf{g}(p)$$

where $\mathbf{b} = (b_1, \ldots, b_v)'$ is the vector of estimated coefficients, $\mathbf{X}$ is the $(m \times v)$ matrix of the $x_{\varepsilon l}$, $\hat{\mathbf{\Sigma}}$ is the matrix of estimated variances and covariances of the estimated functions $g_\varepsilon(p)$, and $\mathbf{g(p)} = (g_1(p), \ldots, g_m(p))'$. For a brief introduction to matrix mathematics, the reader is referred to Appendix B.]

The *predicted value* $\hat{g}_\varepsilon(p)$ using the fitted model is then given by

$$\hat{g}_\varepsilon(p) = x_{\varepsilon 1}b_1 + x_{\varepsilon 2}b_2 + \cdots + x_{\varepsilon v}b_v$$

EXAMPLE   For the IFRP data (see Table 23.2), one postulated model for the Π_{i1} would contain an *overall mean effect* (μ), *the four main effects* ($\alpha_A, \alpha_C, \alpha_E, \alpha_L$), and the six *two-factor interaction effects* ($\gamma_{AC}, \gamma_{AE}, \gamma_{AL}, \gamma_{CE}, \gamma_{CL}, \gamma_{EL}$). In equation form, this model would be written

$$\Pi_{i1} = x_0\mu + x_A\alpha_A + x_C\alpha_C + x_E\alpha_E + x_L\alpha_L + x_Ax_C\gamma_{AC} + x_Ax_E\gamma_{AE}$$

$$+ x_Ax_L\gamma_{AL} + x_Cx_E\gamma_{CE} + x_Cx_L\gamma_{CL} + x_Ex_L\gamma_{EL}, \qquad i = 1, 2, \ldots, 16$$

where

$$x_0 \equiv 1, \qquad x_A = \begin{cases} 1 & \text{if } A = 25+ \\ 0 & \text{if } A = <25 \end{cases} \qquad x_C = \begin{cases} 1 & \text{if } C = 2+ \\ 0 & \text{if } C = 0\text{--}1 \end{cases}$$

$$x_E = \begin{cases} 1 & \text{if } E = 7+ \\ 0 & \text{if } E = 0\text{--}6 \end{cases} \qquad x_L = \begin{cases} 1 & \text{if } L = \text{Yugoslavia} \\ 0 & \text{if } L = \text{India} \end{cases}$$

Note that the x's are dummy variables, defined as in regression analysis. Analogously, then, for a factor having k categories, $(k - 1)$ dummy variables would be needed.

Step 5: State the null hypotheses that you wish to test in terms of the m functions $g_1(\Pi), g_2(\Pi), \ldots, g_m(\Pi)$, or in terms of the β's if you have postulated a linear model in step 4, and then perform the appropriate tests using the GSK methodology.

(a) If you hypothesize $H_0: g_1(\Pi) = g_2(\Pi) = \cdots = g_m(\Pi) = 0$, the test of this hypothesis uses a large sample χ^2 statistic with m degrees of freedom (which is printed out by the GSK computer program).

(b) If you have postulated a linear model in step 4:

(1) A test of the *goodness of fit* (GOF) of this model can be carried out using a large sample χ^2 statistic (printed out by the program) with $m - v$ degrees of freedom, where v is the number of parameters in the model. The null hypothesis for this test is

"H_0: the model provides a good fit to the data"; if the model does actually describe the data well, the differences between the observed $g_\varepsilon(p)$'s and the predicted values $\hat{g}_\varepsilon(p)$'s based on the fitted model will be small, and the GOF χ^2 statistic will be nonsignificant. If the GOF statistic is significant, however, the discrepancies between observed and predicted values (i.e., the residuals) are sufficiently large to warrant fitting an alternative model.

(2) A test of any *linear* null hypothesis about the β's in the postulated model may be carried out using a χ^2 statistic with d degrees of freedom when the null hypothesis is specified in terms of d (linearly independent) functions of the β's of the following form:

$$H_0: \quad c_{11}\beta_1 + c_{12}\beta_2 + \cdots + c_{1v}\beta_v = 0$$

$$c_{21}\beta_1 + c_{22}\beta_2 + \cdots + c_{2v}\beta_v = 0$$

$$\vdots$$

$$c_{d1}\beta_1 + c_{d2}\beta_2 + \cdots + c_{dv}\beta_v = 0$$

where the c's are constants whose values can be specified to form particular null hypotheses of interest. For example, the null hypothesis $H_0: \beta_1 = 0$ can be specified by letting $d = 1$, $c_{11} = 1$, and all the other c's equal 0; and the null hypothesis $H_0: \beta_1 = \beta_2 = \beta_3$ (or, equivalently, $H_0: \beta_1 - \beta_2 = 0$, $\beta_1 - \beta_3 = 0$) can be specified by letting $d = 2$, $c_{11} = 1$, $c_{12} = -1$, $c_{21} = 1$, $c_{23} = -1$, and all the other c's equal to zero.

Thus, if the linear model is of the usual ANOVA regression model form, separate χ^2 tests for main effects and interactions can be made by specifying the appropriate subsets of the β's that represent each effect and then assigning appropriate values for the c's. Furthermore, the results of each test may be summarized in the form of an ANOVA table (Table 23.8), as with regression analysis (although it should be recognized that the separate χ^2 values in such a table are not additive).

TABLE 23.8 *Typical ANOVA table for categorical data analysis*

SOURCE	df	χ^2
Main effect A (I_A levels)	$I_A - 1$	$\chi^2(A)$
Main effect B (I_B levels)	$I_B - 1$	$\chi^2(B)$
.	.	.
.	.	.
.	.	.
Interaction effect AB	$(I_A - 1)(I_B - 1)$	$\chi^2(AB)$
.	.	.
.	.	.
.	.	.
Residual (goodness of fit)	$m - v$	χ^2 (goodness of fit)

EXAMPLES   1.   For the political science data of Table 23.5, a test of the null hypothesis of homogeneity $H''_H: \Pi_{11} - \Pi_{21} = 0$ and $\Pi_{12} - \Pi_{22} = 0$ using the GSK methodology yields a χ^2 statistic (with $u = 2$ degrees of freedom) having the value 10.91. A GSK test of the null hypothesis of independence

H_I'': $\log_e \Pi_{11} + \log_e \Pi_{16} - \log_e \Pi_{14} - \log_e \Pi_{13} = 0$

$\log_e \Pi_{12} + \log_e \Pi_{16} - \log_e \Pi_{15} - \log_e \Pi_{13} = 0$

yields a χ^2 statistic (with $t = 2$ degrees of freedom) having the value 10.74. Note that the "$(O-E)^2/E$" χ^2 statistic described in Section 23.2 had a value 10.80 based on 2 degrees of freedom.

[The reader is no doubt wondering at this time why the two GSK statistics and the $(O-E)^2/E$ statistic all have slightly different values in this example. Without being too mathematical, we will try to explain why. First, the GSK statistics are actually based on the use of an $(O-E)^2/O$ criterion, which was suggested by Neyman (1949) as a modification of the usual $(O-E)^2/E$ statistic. This modified statistic has a χ^2 distribution for large samples if the postulated model is true. Bhapkar (1966) has pointed out the relationship between the modified χ^2 statistic and a Wald statistic (1943) calculated using a weighted-least-squares regression algorithm (as employed in the GSK approach). When the constraints on the cell probabilities imposed under a particular null hypothesis are *linear* (e.g., as they are under H_H''), the Neyman modified χ^2 statistic is algebraically the same as the Wald statistic. On the other hand, when the constraints on the cell probabilities imposed under a particular null hypothesis are *nonlinear* (e.g., as with the log-linear functions given under H_I''), the Wald statistic is algebraically equivalent to the Neyman statistic based on estimates of these nonlinear functions using a linear *approximation* to these constraints.]

2. For the IFRP data (Table 23.2) and the model postulated in the example given after step 4, separate tests for main effects and interactions can be carried out using the GSK approach to yield the summary table given as Table 23.9.

From this table it may be concluded that only the main effect of variable C is significant ($\chi^2 = 21.86$ with 1 df), and that the GOF test is not significant ($\chi^2 = 3.92$ with 5 df).[6]

Step 6: Estimate parameters of interest by computing appropriate confidence intervals:

(a) If you have tested H_0: $g_1(\Pi) = g_2(\Pi) = \cdots = g_m(\Pi) = 0$ in step 5 and also wish to estimate one or more of the m parameters $g_1(\Pi), \ldots, g_m(\Pi)$, the following general confidence-interval formula may be used:

$$g_\xi(p) \pm z_{1-\alpha/2} S_{g_\xi(p)}, \qquad \xi = 1, 2, \ldots, m$$

where $g_\xi(p)$ is the estimate of $g_\xi(\Pi)$ obtained by replacing Π_{ij} by $p_{ij} = n_{ij}/n_{i\cdot}$, and where $S_{g_\xi(p)}$ is the estimated standard error of $g_\xi(p)$ (the value of which is printed out by the GSK program GENCAT).

(b) If you have postulated a linear model in step 4, then GENCAT will print out (weighted-least-squares) estimates $b_1, \ldots, b_v$ of $\beta_1, \ldots, \beta_v$. Confidence intervals for

---

[6] As with the unequal cell number ANOVA situation (see footnote 3 in Chapter 20), care must be taken in interpreting main effect test results when interaction effects are included in the fitted model. In particular, it is recommended (as before) that the proper approach to model building involve examining the highest order effects in the model at each stage and then reducing the model to simpler form only when such effects are nonsignificant.

TABLE 23.9   *ANOVA table for the data in Table 23.2*

SOURCE		df	χ^2	P VALUE
Main effects	Age (H_0: $\alpha_A = 0$)	1	2.59	0.11
	Children (H_0: $\alpha_C = 0$)	1	21.86**	0.00
	Education (H_0: $\alpha_E = 0$)	1	1.76	0.19
	Location (H_0: $\alpha_L = 0$)	1	0.94	0.33
Two-factor interactions	AC (H_0: $\gamma_{AC} = 0$)	1	2.32	0.13
	AE (H_0: $\gamma_{AE} = 0$)	1	0.00	1.00
	AL (H_0: $\gamma_{AL} = 0$)	1	0.28	0.60
	CE (H_0: $\gamma_{CE} = 0$)	1	1.58	0.21
	CL (H_0: $\gamma_{CL} = 0$)	1	2.71	0.10
	EL (H_0: $\gamma_{EL} = 0$)	1	1.49	0.22
All two-factor interactions		6	10.58	0.10
Residual (goodness of fit)		5	3.92	0.56

the β's may be obtained from the b's using the formula

$$b_\ell \pm z_{1-\alpha/2} S_{b_\ell}$$

where S_{b_ℓ} denotes the estimated standard error of b_ℓ.

Also, a comparison of the observed values $g_\xi(p)$ with the predicted values $\hat{g}_\xi(p)$ can be made using the residuals

$$g_\xi(p) - \hat{g}_\xi(p), \qquad \xi = 1, 2, \ldots, m$$

23.4 GSK COMPUTER PROGRAM

The GSK linear-models approach to the analysis of categorical data, as described above in conceptual terms, can be implemented in practice by employing a computer program called GENCAT, written and used extensively in the Biostatistics Department at the University of North Carolina.[7] The user of this program must first enter the categorical data set under study into the system according to the layout in Table 23.3. Then, to generate estimates and tests of hypotheses concerning appropriate functions of interest, the user has to specify values for the various sets of constants described in Section 23.3. In particular, the examination of linear functions [the $a_\gamma(\Pi)$'s] requires specification of the $\{a_{\gamma,ij}\}$, while consideration of log-linear functions [the $f_\alpha(\Pi)$'s] involves giving values to the $\{k_{\alpha\gamma}\}$ as well; and the fitting of a linear model to either the $a_\gamma(p)$'s or the $f_\alpha(p)$'s means assigning values to the $\{x_{\xi\ell}\}$, and then the subsequent testing of hypotheses about the β's necessitates choosing appropriate values for the c's.

Once the user has appropriately specified these sets of constants, the computer will perform the necessary computations and will then print out the desired estimates

---

[7] This computer program can be obtained from the Program Librarian, Department of Biostatistics, School of Public Health, University of North Carolina, Chapel Hill, N.C. 27514. For a program description, see Landis et al. (1976).

and χ^2 test statistics directly. Of course, the proper specification by the user of these sets of constants requires some sophistication, which can only be cultivated through understanding and experience. Indeed, to perform an in-depth analysis of a complex set of categorical data using the GSK methodology requires considerable expertise, and one hope that we have in illustrating by example some of these analysis techniques is to be able to impress upon the reader the need for care in the interpretation of test results.

EXAMPLE 1   The purpose here is to briefly summarize the analysis of the political science data in Table 23.5, which has already been discussed in considerable detail. Specifically, the null hypothesis of *independence*,

$$H_I'': \quad f_1(\Pi) = \log_e \Delta_{11} = 0, \quad f_2(\Pi) = \log_e \Delta_{12} = 0$$

and the null hypothesis of *homogeneity*,

$$H_H'': \quad a_1(\Pi) = (\Pi_{11} - \Pi_{21}) = 0, \quad a_2(\Pi) = (\Pi_{12} - \Pi_{22}) = 0$$

involve log-linear and linear functions of the Π_{ij}'s, respectively, and the specification of the sets of constants $\{a_{\gamma, ij}\}$ and $\{k_{\alpha\gamma}\}$ needed to form these particular functions has been dealt with earlier. The GSK methodology produces χ^2 statistics (each with 2 degrees of freedom) having the values 10.74 and 10.91 for testing H_I'' and H_H'', respectively; the commonly used $(O - E)^2/E$ statistic has the value 10.80. We have already discussed why these three statistics differ slightly in value.

EXAMPLE 2   In epidemiology, a *case-control study* is an inquiry involving groups of individuals selected on the basis of whether they do (the "cases") or do not (the "controls") have a particular disease of interest; the two groups of cases and controls are then compared with respect to an existing or a past "exposure" to a particular characteristic judged to be of possible relevance to the etiology of the disease. Commonly, in a case-control study, a specific hypothesis is being tested—for example, that a connection exists between lung cancer and prior cigarette smoking habits, or between congenital malformation and maternal rubella during pregnancy.

The data from a case-control study can best be represented via a two-way layout, as indicated in Table 23.10. This is a one-response, one-factor table, where the dichotomous response ($r = 2$) is either "exposed" or "not exposed" and the factor has two levels corresponding to the ($s = 2$) subpopulations of cases and controls. Here, we

TABLE 23.10   *Data layout for a case-control study*

TWO SUBPOPULATIONS	TWO RESPONSE CATEGORIES		TOTALS	
	EXPOSED	NOT EXPOSED		
Cases	n_{11}	n_{12}	$n_{1\cdot}$	
	Π_{11}	Π_{12}		1
Controls	n_{21}	n_{22}	$n_{2\cdot}$	
	Π_{21}	Π_{22}		1

have random samples of $n_1.$ cases and $n_2.$ controls; n_{11} of the $n_1.$ cases have the exposure characteristic, while n_{21} of the $n_2.$ controls have been exposed. Note that $\Pi_{11} + \Pi_{12} = 1$ and that $\Pi_{21} + \Pi_{22} = 1$, in accord with our two (binomial)-population model.

Now, a parameter of more than just casual interest to epidemiologists involved in a case-control study is the *odds ratio* [see Cornfield (1956)], which is defined as

$$OR = \frac{\Pi_{11}/\Pi_{12}}{\Pi_{21}/\Pi_{22}} = \frac{\Pi_{11}\Pi_{22}}{\Pi_{12}\Pi_{21}}$$

As its name implies, OR is seen to be the ratio of the odds for a case being exposed as opposed to not being exposed (Π_{11}/Π_{12}) to the corresponding odds for a control (Π_{21}/Π_{22}).

The specific null hypothesis to be tested here is

$$H_0: \quad OR = \frac{\Pi_{11}\Pi_{22}}{\Pi_{12}\Pi_{21}} = 1$$

versus the alternative

$$H_A: \quad OR > 1$$

By taking natural logarithms, we can write H_0 as

$$H_0': \quad \log_e OR = \log_e \Pi_{11} + \log_e \Pi_{22} - \log_e \Pi_{12} - \log_e \Pi_{21} = 0$$

which is in a form amenable to investigation by the GSK approach.

In particular, H_0' involves exactly one log-linear function of the Π_{ij}'s, which (for $s = 2$, $r = 2$, and $u = 4$) is of the general form

$$f_1(\Pi) = \sum_{\gamma=1}^{4} k_{1\gamma} \log_e a_\gamma(\Pi)$$

where

$$a_\gamma(\Pi) = \sum_{i=1}^{2} \sum_{j=1}^{2} a_{\gamma,ij}\Pi_{ij}$$

$$= a_{\gamma,11}\Pi_{11} + a_{\gamma,12}\Pi_{12} + a_{\gamma,21}\Pi_{21} + a_{\gamma,22}\Pi_{22}$$

for $\gamma = 1, 2, 3, 4$. Now $a_1(\Pi) = \Pi_{11}$ if $a_{1,11} = 1$ and $a_{1,12} = a_{1,21} = a_{1,22} = 0$, and we can similarly arrange things so that $a_2(\Pi) = \Pi_{12}$, $a_3(\Pi) = \Pi_{21}$, and $a_4(\Pi) = \Pi_{22}$. So $f_1(\Pi) = \log_e$ OR if we then take $k_{11} = 1$, $k_{12} = -1$, $k_{13} = -1$, and $k_{14} = 1$. Based on these specifications, the GSK methodology would produce a χ^2 statistic based on 1 degree of freedom for testing H_0'. Also, the point estimate $\widehat{\log_e OR}$ of $\log_e$ OR would be given, along with its estimated standard error, so that a confidence interval for OR can be developed. As we know, this point estimate would be obtained by using $p_{ij} = n_{ij}/n_i.$ for Π_{ij} ($i = 1, 2$ and $j = 1, 2$) in the expression for $\log_e$ OR:

$$f_1(p) = \widehat{\log_e OR} = \log_e p_{11} + \log_e p_{22} - \log_e p_{12} - \log_e p_{21}$$

It can also be shown that an *explicit* expression for the estimated standard error of $\widehat{\log_e OR}$ as computed by the GSK approach is

$$SE(\widehat{\log_e OR}) = \sqrt{\frac{1}{n_{11}} + \frac{1}{n_{12}} + \frac{1}{n_{21}} + \frac{1}{n_{22}}}$$

which is a large-sample approximation to the actual standard error. Then, based on the reasonable assumption that $\widehat{\log_e OR}$ is approximately normally distributed for large samples, it follows that an approximate $100(1-\alpha)\%$ confidence interval for $\log_e OR$ is given by

$$P(L \le \log_e OR \le U) = 1 - \alpha$$

where

$$L = \widehat{\log_e OR} - z_{1-\alpha/2} \sqrt{\frac{1}{n_{11}} + \frac{1}{n_{12}} + \frac{1}{n_{21}} + \frac{1}{n_{22}}}$$

and

$$U = \widehat{\log_e OR} + z_{1-\alpha/2} \sqrt{\frac{1}{n_{11}} + \frac{1}{n_{12}} + \frac{1}{n_{21}} + \frac{1}{n_{22}}}$$

so that

$$P(e^L \le OR \le e^U) = 1 - \alpha$$

gives the corresponding large sample $100(1-\alpha)\%$ confidence interval for OR. We shall now consider a specific numerical example.

Table 23.11 records the frequency of past tonsillectomy (the *exposure* variable) among *cases* of Hodgkin's disease and their siblings (the *controls*).[8] Interest is in determining whether there is evidence that having had a tonsillectomy increases the risk of developing Hodgkin's disease.

TABLE 23.11   *Frequency of past tonsillectomy among cases of Hodgkin's disease and their siblings*

TWO SUBPOPULATIONS	HISTORY OF TONSILLECTOMY		TOTAL
	YES	NO	
Cases	$n_{11} = 90$	$n_{12} = 84$	$n_{1.} = 174$
Controls	$n_{21} = 165$	$n_{22} = 307$	$n_{2.} = 472$

[8] These data were taken from Johnson and Johnson (1972). Since the controls were actually siblings of the cases, a matched analysis [see Fleiss (1973)] would likely be more appropriate than an analysis which assumes that all observations in the table are independent. Nevertheless, we have proceeded under this assumption for illustrative purposes.

The GSK χ^2 statistic for testing H_0' has the value 14.72 with one degree of freedom, which is highly significant and thus provides strong evidence that the true value of the odds ratio is greater than 1. The point estimate of $\log_e$ OR as provided by the GSK methodology is

$$\log_e \text{OR} = \log_e \frac{90}{174} + \log_e \frac{307}{472} - \log_e \frac{84}{174} - \log_e \frac{165}{472} = 0.688$$

so the point estimate of OR would be taken to be

$$e^{0.688} = \frac{90(307)}{84(165)} = 1.99$$

and the estimated standard error of $\log_e$ OR is

$$\sqrt{\frac{1}{90} + \frac{1}{84} + \frac{1}{165} + \frac{1}{307}} = 0.180$$

Then, for an approximate 95% confidence interval for OR, we have

$$e^L = 1.99 e^{-1.96(0.180)} = 1.99 e^{-0.353} = 1.40$$

and

$$e^U = 1.99 e^{1.96(0.180)} = 1.99 e^{0.353} = 2.83$$

so that (approximately)

$$1.40 \le \text{OR} \le 2.83$$

Before leaving this example, one final concept deserves some attention. Our reason for working with H_0', which involves consideration of a log-linear function, is that we obtain directly as output the value of $\log_e$ OR along with an estimate of its standard error, thus enabling us to construct a confidence interval for OR as indicated above. However, if we are really only interested in testing the null hypothesis H_0 and are not concerned specifically with making inferences regarding the odds ratio parameter OR, we can utilize an alternative way of testing H_0 which would involve consideration of a linear function of the cell probabilities. This alternative approach is to be preferred because no approximations are required when dealing with linear functions as they are with log-linear functions. In particular, since OR = 1 is equivalent to $\Pi_{11}\Pi_{22} = \Pi_{12}\Pi_{21}$, or $\Pi_{11}(1-\Pi_{21}) = (1-\Pi_{11})\Pi_{21}$, it follows that testing H_0: OR = 1 is equivalent to testing the null hypothesis H_0'': $\Pi_{11} = \Pi_{21}$. Thus, H_0'' is easily seen to be a null hypothesis of *homogeneity* of the binomial populations for cases and controls, and the GSK χ^2 statistic for testing H_0'' has the value 14.67 with 1 degree of freedom. Again, because of the reasons given previously, this value is slightly different from both the GSK χ^2 statistic for testing H_0' (which has the value 14.72) and from the "$(O-E)^2/E$" χ^2 statistic (which has the value 14.96).

EXAMPLE 3 This example is one taken from the 1969 paper of Grizzle et al. The data in Table 23.12 are a tabulation of the severity of the "dumping syndrome," an undesirable consequence of surgery for duodenal ulcer. The four surgical procedures being considered are:

A: drainage and vagotomy

B: 25% resection (antrectomy) and vagotomy

C: 50% resection (hemigastrectomy) and vagotomy

D: 75% resection

Table 23.12 describes a one-response, two-factor situation, the factors being *hospitals* (designated 1, 2, 3, 4) and *surgical procedures* (designated A, B, C, D), and the response being the *clinical evaluation of the severity of the dumping syndrome* (with the $r = 3$ levels "none," "slight," and "moderate"). Thus, we have a total of $s = 4 \times 4 = 16$ trinomial populations, corresponding to the four hospitals performing each of the four surgical procedures.

The purpose of the analysis is to assess the relationship between the two factors and the response, as one would strive to do via analysis-of-variance or regression procedures applied to quantitative data. To proceed further we need to convert the three responses for each hospital–surgical procedure combination to a single score. If we assign the response categories none, slight, and moderate the weights 1, 2, and 3, respectively, a reasonable score for the γth "hospital–surgical procedure" combination ($\gamma = 1, 2, \ldots, 16$) would be the linear function

$$a_\gamma(p) = 1p_{\gamma 1} + 2p_{\gamma 2} + 3p_{\gamma 3}$$

where $p_{\gamma 1}$ is the estimated probability of the response category "none," $p_{\gamma 2}$ is the estimated probability of the response category "slight," and $p_{\gamma 3}$ is the estimated probability of the response category "moderate."[9] Here $a_\gamma(p)$ is estimating the parametric function $a_\gamma(\Pi) = 1\Pi_{\gamma 1} + 2\Pi_{\gamma 2} + 3\Pi_{\gamma 3}$.

TABLE 23.12   *Severity of dumping syndrome*

CLINICAL EVALUATION OF SEVERITY OF DUMPING SYNDROME	HOSPITAL															
	1				2				3				4			
	SURGICAL PROCEDURE				SURGICAL PROCEDURE				SURGICAL PROCEDURE				SURGICAL PROCEDURE			
	A	B	C	D	A	B	C	D	A	B	C	D	A	B	C	D
None	23	23	20	24	18	18	13	9	8	12	11	7	12	15	14	13
Slight	7	10	13	10	6	6	13	15	6	4	6	7	9	3	8	6
Moderate	2	5	5	6	1	2	2	2	3	4	2	4	1	2	3	4
Total	32	38	38	40	25	26	28	26	17	20	19	18	22	20	25	23
Score $(1p_{\gamma 1} + 2p_{\gamma 2} + 3p_{\gamma 3})$	1.3	1.5	1.6	1.6	1.3	1.4	1.6	1.7	1.7	1.6	1.5	1.8	1.5	1.4	1.6	1.6

[9] We recognize that using equally spaced numbers to describe the response categories may perhaps be too crude a scoring scheme to accurately compare these categories on an ordinal scale. Other scoring procedures might also be considered, one of which is called *ridit analysis* [see Bross (1958)]. Also, the moderate and slight categories could possibly be pooled together and consideration given to a simple binomial response.

The particular form for $a_\gamma(p)$ is obtained from the general expression

$$a_\gamma(p) = \sum_{i=1}^{16} \sum_{j=1}^{3} a_{\gamma,ij} p_{ij}$$

given earlier by taking

$$a_{\gamma,ij} = \begin{cases} 1 & \text{when } \gamma = i, j = 1 \\ 2 & \text{when } \gamma = i, j = 2 \\ 3 & \text{when } \gamma = i, j = 3 \end{cases}$$

and setting $a_{\gamma,ij} = 0$ if $\gamma \neq i$. For example, when $\gamma = 1$, then $a_{1,11} = 1$, $a_{1,12} = 2$, $a_{1,13} = 3$, and $a_{1,21} = a_{1,22} = a_{1,23} = \cdots = a_{1,16,1} = a_{1,16,2} = a_{1,16,3} = 0$, so

$$a_1(p) = 1p_{11} + 2p_{12} + 3p_{13}$$

which has the specific value

$$a_1(p) = 1\left(\frac{23}{32}\right) + 2\left(\frac{7}{32}\right) + 3\left(\frac{2}{32}\right) = 1.3$$

The numerical values for the $u = 16$ linear functions $a_1(p), a_2(p), \ldots, a_{16}(p)$ are given in the last line of Table 23.12. We shall attempt next to describe the variation in these 16 scores by means of a regression model involving the "hospital" and "surgical procedure" factors.

The regression model that we propose to fit by weighted least squares to the $a_\gamma(p)$'s is of the general form

$$a_\gamma(\Pi) = x_{\gamma 0}\mu + x_{\gamma 1}\alpha_1 + x_{\gamma 2}\alpha_2 + x_{\gamma 3}\alpha_3 + x_{\gamma A}\tau_A + x_{\gamma B}\tau_B + x_{\gamma C}\tau_C$$

where μ = overall mean effect; α_1, α_2, and α_3 are the differential effects of hospitals 1, 2, and 3, respectively; and τ_A, τ_B, and τ_C are the differential effects of surgical procedures A, B, and C, respectively.

An explanation as to how the x's are defined is best given after seeing Table 23.13, which lists the values of the x's associated with each of the 16 scores. From this tabulation one can see that the regression model we are considering is completely equivalent to a two-way ANOVA model with no interaction, the two ways of classification being by "hospitals" and by "surgical procedures." For example, the true score for hospital 1 and surgical procedure A, $a_1(\Pi) = 1\Pi_{11} + 2\Pi_{12} + 3\Pi_{13}$, is modeled as

$$a_1(\Pi) = (1)\mu + (1)\alpha_1 + (0)\alpha_2 + (0)\alpha_3 + (1)\tau_A + (0)\tau_B + (0)\tau_C$$

$$= \mu + \alpha_1 + \tau_A$$

and, similarly, $a_{11}(\Pi) = \mu + \alpha_3 + \tau_C$.

Now, what about the "-1" values appearing in the table? These values arise because of the restrictions

$$\alpha_1 + \alpha_2 + \alpha_3 + \alpha_4 = 0 \quad \text{and} \quad \tau_A + \tau_B + \tau_C + \tau_D = 0$$

TABLE 23.13 *Dependent and independent variable values for data in Table 23.12*
23.12

HOSPITAL–SURGICAL PROCEDURE COMBINATION	SCORE $a_\gamma(p)$	INDEPENDENT VARIABLE VALUES						
		$x_{\gamma 0}$	$x_{\gamma 1}$	$x_{\gamma 2}$	$x_{\gamma 3}$	$x_{\gamma A}$	$x_{\gamma B}$	$x_{\gamma C}$
1,A	$a_1(p) = 1.3$	1	1	0	0	1	0	0
1,B	$a_2(p) = 1.5$	1	1	0	0	0	1	0
1,C	$a_3(p) = 1.6$	1	1	0	0	0	0	1
1,D	$a_4(p) = 1.6$	1	1	0	0	-1	-1	-1
2,A	$a_5(p) = 1.3$	1	0	1	0	1	0	0
2,B	$a_6(p) = 1.4$	1	0	1	0	0	1	0
2,C	$a_7(p) = 1.6$	1	0	1	0	0	0	1
2,D	$a_8(p) = 1.7$	1	0	1	0	-1	-1	-1
3,A	$a_9(p) = 1.7$	1	0	0	1	1	0	0
3,B	$a_{10}(p) = 1.6$	1	0	0	1	0	1	0
3,C	$a_{11}(p) = 1.5$	1	0	0	1	0	0	1
3,D	$a_{12}(p) = 1.8$	1	0	0	1	-1	-1	-1
4,A	$a_{13}(p) = 1.5$	1	-1	-1	-1	1	0	0
4,B	$a_{14}(p) = 1.4$	1	-1	-1	-1	0	1	0
4,C	$a_{15}(p) = 1.6$	1	-1	-1	-1	0	0	1
4,D	$a_{16}(p) = 1.6$	1	-1	-1	-1	-1	-1	-1

or, equivalently,

$$\alpha_4 = -\alpha_1 - \alpha_2 - \alpha_3 \quad \text{and} \quad \tau_D = -\tau_A - \tau_B - \tau_C$$

The reader will remember from our earlier discussions of analysis of variance that these standard linear restrictions are imposed because there are only 3 degrees of freedom for hospitals and 3 degrees of freedom for surgical procedures, so there can be only three α effects (in our case, α_1, α_2, and α_3) and only three τ effects (in our case, τ_A, τ_B, and τ_C) appearing *explicitly* in the model, the remaining effects (in our case, α_4 and τ_D) being (linear) functions of the effects in the model. For example, from Table 23.13, the true score $a_{16}(\Pi)$ for hospital 4 and surgical procedure D is given by

$$a_{16}(\Pi) = (1)\mu + (-1)\alpha_1 + (-1)\alpha_2 + (-1)\alpha_3 + (-1)\tau_A + (-1)\tau_B + (-1)\tau_C$$

$$= \mu + (-\alpha_1 - \alpha_2 - \alpha_3) + (-\tau_A - \tau_B - \tau_C) = \mu + \alpha_4 + \tau_D$$

A fitting of this model by the GSK weighted-least-squares methodology produces the following estimates:

$$\hat{\mu} = 1.54, \quad \hat{\alpha}_1 = -0.04, \quad \hat{\alpha}_2 = -0.04, \quad \hat{\alpha}_3 = 0.11,$$

$$\hat{\alpha}_4 = -\hat{\alpha}_1 - \hat{\alpha}_2 - \hat{\alpha}_3 = -0.03, \quad \hat{\tau}_A = -0.11, \quad \hat{\tau}_B = -0.07,$$

$$\hat{\tau}_C = 0.05, \quad \hat{\tau}_D = -\hat{\tau}_A - \hat{\tau}_B - \hat{\tau}_C = 0.13$$

Since (in the notation of Section 23.3) there are $m = u = 16$ responses [the $a_\gamma(p)$'s] and $v = 7$ parameters in the model, the χ^2 goodness-of-fit test for assessing how well the linear model describes the variation in the $a_\gamma(p)$'s has $m - v = 16 - 7 = 9$ degrees of freedom. The value of this error (or residual) sum of squares for these data is 6.32, which does not approach statistical significance; this can be interpreted to mean that an additive model having only mean, hospital, and surgical procedure effects fits the data adequately.

[It is important to remember that a test of significance concerning one or more effects in a model is adjusted for the presence of other effects in the model, and so an ANOVA table presented under these conditions should *not* be read as describing *orthogonal* effects but rather as providing a summary of several tests on different *nonorthogonal* effects. In particular, such an ANOVA table for the data we are presently considering would contain the "sum of squares for hospitals *adjusted* for surgical procedure effects" and the "sum of squares for surgical procedures *adjusted* for hospital effects," as well as the "goodness-of-fit sum of squares."]

Next, the testing of hypotheses concerning the hospital and surgical procedure effects in the model requires specification of the c's in the general linear hypothesis

$$H_0: \quad c_{10}\mu + c_{11}\alpha_1 + c_{12}\alpha_2 + c_{13}\alpha_3 + c_{1A}\tau_A + c_{1B}\tau_B + c_{1C}\tau_C = 0$$

$$c_{20}\mu + c_{21}\alpha_1 + c_{22}\alpha_2 + c_{23}\alpha_3 + c_{2A}\tau_A + c_{2B}\tau_B + c_{2C}\tau_C = 0$$

$$\vdots$$

$$c_{d0}\mu + c_{d1}\alpha_1 + c_{d2}\alpha_2 + c_{d3}\alpha_3 + c_{dA}\tau_A + c_{dB}\tau_B + c_{dC}\tau_C = 0$$

Thus, to test

$$H_0: \quad \alpha_1 = 0$$
$$\alpha_2 = 0$$
$$\alpha_3 = 0$$

we simply take $c_{11} = c_{22} = c_{33} = 1$ and set the rest of the c's equal to zero; similarly, to test

$$H_0: \quad \tau_A = 0$$
$$\tau_B = 0$$
$$\tau_C = 0$$

we set $c_{1A} = c_{2B} = c_{3C} = 1$ and set the rest of the c's equal to zero. The GSK chi-square statistics for testing these two hypotheses each have $d = 3$ degrees of freedom. There are no significant hospital effects, since the computed χ^2 value is 2.33. However, the surgical procedure effects are significant at less than the 0.05 level, the χ^2 value being 8.90. These results can be summarized in an ANOVA table (Table 23.14).

Finally, it is of interest to relate the severity of the dumping syndrome to the amount of stomach removed; since the fractions of stomach removed are approximately $0, \frac{1}{4}, \frac{1}{2},$ and $\frac{3}{4}$ for surgical procedures $A, B, C,$ and D, respectively, we may ask if the severity is *linearly* related to the fraction of stomach removed. This question can be answered by testing

$$H_0: \quad -3\tau_A - \tau_B + \tau_C + 3\tau_D = 0$$

TABLE 23.14   *ANOVA table for data in Table 23.12*

SOURCE OF VARIATION	df	SS (or χ^2 values)
Hospitals (adjusted for "surgical procedures")	3	2.33
Surgical procedures (adjusted for "hospitals")	3	8.90
Residual (goodness of fit)	9	6.32

due to the *equal spacings* in the fractions of stomach removed.[10] Expressed in terms of τ_A, τ_B, and τ_C, this hypothesis is of the form $-3\tau_A - \tau_B + \tau_C + 3(-\tau_A - \tau_B - \tau_C) = -6\tau_A - 4\tau_B - 2\tau_C = 0$, which is generated by taking $c_{1A} = -6$, $c_{1B} = -4$, $c_{1C} = -2$ and setting all other c's equal to zero. The resulting χ^2 statistic is 8.74 based on $d = 1$ degree of freedom, which is highly significant. Therefore, we reject the null hypothesis of no linear trend and conclude that the severity of the dumping syndrome tends to increase approximately linearly with the fraction of stomach removed.

EXAMPLE 4   This example will deal with the IFRP study introduced in Section 23.1, which focuses on the problem of describing the determinants of postabortion fertility control (PAFC) using the demographic variables age (A), number of living children (C), education (E), and location (L). The basic data has been given in Table 23.2, columns 5 and 6, listing the $r = 2$ observed cell counts for each of the $s = 16$ binomial populations (i.e., n_{i1} and n_{i2}, $i = 1, 2, \ldots, 16$) corresponding to the 2^4 combinations of the four (dichotomous) factors. The last column gives the estimated probability $p_{i1} = n_{i1}/(n_{i1} + n_{i2})$ of accepting postabortion fertility control for each of the 16 populations. As mentioned earlier, we are concerned here with a one-response, four-factor table, and the specific analysis objective will be to find a simple predictive model with as few parameters as possible which best describes the relationship between the four demographic variables (A, C, E, and L) and the dependent variable p_{i1}, the observed proportion accepting postabortion fertility control. Variable-selection and model-simplification techniques will be given special attention in this analysis.

The first model to be fitted to the data is a complete (or "saturated") model containing all 16 effects associated with a complete 2^4 factorial design. That is, the model contains an *overall mean effect* (μ), all four *main effects* (α_A, α_C, α_E, α_L), all six *two-factor interactions* (γ_{AC}, γ_{AE}, γ_{AL}, γ_{CE}, γ_{CL}, γ_{EL}), all four *three-factor interactions* (τ_{ACE}, τ_{ACL}, τ_{AEL}, τ_{CEL}), and the one *four-factor interaction* (δ_{AECL}). In equation form this model is as follows:

$$\Pi_{i1} = x_0\mu + x_A\alpha_A + x_C\alpha_C + x_E\alpha_E + x_L\alpha_L + x_Ax_C\gamma_{AC}$$

$$+ x_Ax_E\gamma_{AE} + x_Ax_L\gamma_{AL} + x_Cx_E\gamma_{CE} + x_Cx_L\gamma_{CL} + x_Ex_L\gamma_{EL}$$

$$+ x_Ax_Cx_E\tau_{ACE} + x_Ax_Cx_L\tau_{ACL} + x_Ax_Ex_L\tau_{AEL} + x_Cx_Ex_L\tau_{CEL}$$

$$+ x_Ax_Cx_Ex_L\delta_{ACEL} \tag{23.1}$$

---

[10] As mentioned in Section 17.8, the coefficients -3, -1, 1, and 3 in the contrast ($-3\tau_A -\tau_B + \tau_C + 3\tau_D$) are *orthogonal polynomial* coefficients [see Armitage (1971) for a further discussion.]

where

$$x_0 \equiv 1$$

$$x_A = \begin{cases} 1 & \text{if population } i \text{ has factor } A = 25+ \\ 0 & \text{if population } i \text{ has factor } A = <25 \end{cases}$$

$$x_C = \begin{cases} 1 & \text{if population } i \text{ has factor } C = 2+ \\ 0 & \text{if population } i \text{ has factor } C = 0\text{--}1 \end{cases}$$

$$x_E = \begin{cases} 1 & \text{if population } i \text{ has factor } E = 7+ \\ 0 & \text{if population } i \text{ has factor } E = 0\text{--}6 \end{cases}$$

$$x_L = \begin{cases} 1 & \text{if population } i \text{ has factor } L = \text{Yugoslavia} \\ 0 & \text{if population } i \text{ has factor } L = \text{India} \end{cases}$$

For example, for row (population) $i = 10$ of Table 23.2, $A = 25+$, $C = 0\text{--}1$, $E = 0\text{--}6$, and $L = $ Yugoslavia, so

$$\Pi_{10,1} = (1)\mu + (1)\alpha_A + (0)\alpha_C + (0)\alpha_E + (1)\alpha_L + (1)(0)\gamma_{AC}$$
$$+ (1)(0)\gamma_{AE} + (1)(1)\gamma_{AL} + (0)(0)\gamma_{CE} + (0)(1)\gamma_{CL}$$
$$+ (0)(1)\gamma_{EL} + (1)(0)(0)\tau_{ACE} + (1)(0)(1)\tau_{ACL} + (1)(0)(1)\tau_{AEL}$$
$$+ (0)(0)(1)\tau_{CEL} + (1)(0)(0)(1)\delta_{ACEL}$$
$$= \mu + \alpha_A + \alpha_L + \gamma_{AL}$$

The value of the x's for each of the 16 populations as listed in Table 23.2 are given in Table 23.15.

TABLE 23.15 *Values of x's for the model (23.1) based on the data of Table 23.2*

POPULATION NO.	VALUES OF INDEPENDENT VARIABLES															
	$x_0,$	$x_A,$	$x_C,$	$x_E,$	$x_L,$	$x_Ax_C,$	$x_Ax_E,$	$x_Ax_L,$	$x_Cx_E,$	$x_Cx_L,$	$x_Ex_L,$	$x_Ax_Cx_E,$	$x_Ax_Cx_L,$	$x_Ax_Ex_L,$	$x_Cx_Ex_L,$	$x_Ax_Cx_Ex_L$
1	1	0	0	0	0	0	0	0	0	0	0	0	0	0	0	0
2	1	1	0	0	0	0	0	0	0	0	0	0	0	0	0	0
3	1	0	0	1	0	0	0	0	0	0	0	0	0	0	0	0
4	1	1	0	1	0	0	1	0	0	0	0	0	0	0	0	0
5	1	0	1	0	0	0	0	0	0	0	0	0	0	0	0	0
6	1	1	1	0	0	1	0	0	0	0	0	0	0	0	0	0
7	1	0	1	1	0	0	0	0	1	0	0	0	0	0	0	0
8	1	1	1	1	0	1	1	0	1	0	0	1	0	0	0	0
9	1	0	0	0	1	0	0	0	0	0	0	0	0	0	0	0
10	1	1	0	0	1	0	0	1	0	0	0	0	0	0	0	0
11	1	0	0	1	1	0	0	0	0	0	1	0	0	0	0	0
12	1	1	0	1	1	0	1	1	0	0	1	0	0	1	0	0
13	1	0	1	0	1	0	0	0	0	1	0	0	0	0	0	0
14	1	1	1	0	1	1	0	1	0	1	0	0	1	0	0	0
15	1	0	1	1	1	0	0	0	1	1	1	0	0	0	1	0
16	1	1	1	1	1	1	1	1	1	1	1	1	1	1	1	1

A listing of the results of the significance tests concerning the various individual effects in model (23.1) (except for μ) is given in Table 23.16, together with the results of global tests concerning the joint effects of the second-order and third-order interactions. (It is assumed that the reader, after having worked through the previous examples, will be able to specify the values of the c's required to generate the appropriate hypotheses considered in Table 23.16.)

From Table 23.16 it is clear that the main effect of factor C (adjusted for the other effects in the model) is highly significant (P value < 0.001) and that the CE interaction effect is also significant (P value < 0.05). However, none of the three-factor interactions or the four-factor interactions are significant. This suggests that a simpler model than the saturated one just considered could be found to describe more *efficiently* the relationship between the four factors and the response.

The most obvious choice for a simpler model would be one *not* containing three- and four-factor interaction terms, that is, one containing only the main effects and the two-factor interactions corresponding to the first eleven columns of Table 23.15. A summary of the results of the significance tests associated with the fitting of this reduced model (which we presented in Section 23.3) is given in Table 23.17.

A goodness-of-fit χ^2 test can be performed concerning this 11-term model; this residual sum of squares has the value 3.92 with $16 - 11 = 5$ degrees of freedom, which is not significant. The conclusions here are somewhat different from those implied by Table 23.16. As with the complete model, the main effect of factor C is highly significant, but now no other effects are significant at the 0.05 level. The two effects "closest" to significance are the A and CL effects, whereas the CE interaction, which was significant for the complete model, is now clearly nonsignificant.

From the estimated parameters (i.e., $\hat{\mu}$, the $\hat{\alpha}$'s, and the $\hat{\gamma}$'s) obtained from fitting this reduced model, predicted (or adjusted) values for the postabortion fertility-control acceptance rates can be calculated.

TABLE 23.16   *Summary of tests of hypotheses concerning parameters in model (23.1)*

SOURCE OF VARIATION	df	χ^2 VALUE	P VALUE
A	1	1.20	0.27
C	1	17.42**	0.00
E	1	3.82	0.05
L	1	0.04	0.85
AC	1	1.11	0.29
AE	1	0.72	0.40
AL	1	0.10	0.75
CE	1	4.34*	0.04
CL	1	1.32	0.25
EL	1	2.08	0.15
ACE	1	1.08	0.30
ACL	1	0.00	1.00
AEL	1	0.17	0.68
CEL	1	0.67	0.41
$ACEL$	1	0.12	0.73
All two-factor interactions	6	6.65	0.35
All three-factor interactions	4	2.90	0.58

TABLE 23.17   *Summary of tests of hypotheses for a model containing the main effects and two-factor interactions corresponding to the first 11 columns of Table 23.15*

SOURCE OF VARIATION	df	χ^2 VALUE	P VALUE
A	1	2.59	0.11
C	1	21.86**	0.00
E	1	1.76	0.19
L	1	0.94	0.33
AC	1	2.32	0.13
AE	1	0.00	1.00
AL	1	0.28	0.60
CE	1	1.58	0.21
CL	1	2.71	0.10
EL	1	1.49	0.22
All two-factor interactions	6	10.58	0.10
Residual (goodness of fit)	5	3.92	0.56

[It is of importance to discuss the reason for preferring predicted or adjusted rates to the crude rates obtained for each population. If all 16 effects in the complete model were significant, the crude rates would be appropriate to use. However, if it can be established that one or more of the 16 effects is not significant and so should not be included in the model, the original crude rates would contain a certain amount of noise due to the presence of these nonsignificant effects, while the predicted (or adjusted) rates would be based on a more efficient model not containing these noise producers and so would be "smoothed" with respect to such "random variation."]

The $\hat{p}_{i1}$'s are calculated using the general expression

$$\hat{p}_{i1} = \hat{\mu} + x_A \hat{\alpha}_A + x_C \hat{\alpha}_C + x_E \hat{\alpha}_E + x_L \hat{\alpha}_L + x_A x_C \hat{\gamma}_{AC} + x_A x_E \hat{\gamma}_{AE}$$

$$+ x_A x_L \hat{\gamma}_{AL} + x_C x_E \hat{\gamma}_{CE} + x_C x_L \hat{\gamma}_{CL} + x_E x_L \hat{\gamma}_{EL} \tag{23.2}$$

Table 23.18 gives the observed p_{i1}, the predicted $\hat{p}_{i1}$, and the residual $(p_{i1} - \hat{p}_{i1})$ for each of the 16 populations based on model (23.2).

It is to be noted from Table 23.18 that fairly large residuals with absolute values exceeding 0.05 in value arise for populations 1, 3, 7, and 10. Also, Johnson and Koch (1971) recommend that, regardless of the number of degrees of freedom associated with the goodness-of-fit test, the critical value of χ^2 for the test should be 3.84 (the $\alpha = 0.05$ value of the χ^2 distribution with 1 degree of freedom) to ensure that the residual sum of squares does not contain any hidden but individually significant component, and we see that this conservative critical value of 3.84 is exceeded (although just barely) for the 11-term model by the goodness-of-fit χ^2 value of 3.92. Thus, even though this model fits the data fairly well, it is good strategy to consider some other possible candidates, and this is what we shall do.

At this point it becomes important to realize that the search for a best model in a categorical data-analysis setting is philosophically similar in many respects to that conducted in the usual regression analysis situation, although there unfortunately

TABLE 23.18   *Values of p_{i1}, $\hat{p}_{i1}$, and $p_{i1} - \hat{p}_{i1}$ calculated using model (23.2)*

POPULATION NO.	1	2	3	4	5	6	7	8
p_{i1}	0.333	0.667	0.667	0.700	0.944	0.951	0.857	0.947
$\hat{p}_{i1}$	0.465	0.628	0.578	0.737	0.925	0.951	0.926	0.947
$p_{i1} - \hat{p}_{i1}$	−0.132	0.039	0.089	−0.037	0.019	0.000	−0.069	0.000

POPULATION NO.	9	10	11	12	13	14	15	16
p_{i1}	0.364	0.591	0.419	0.506	0.727	0.636	0.591	0.593
$\hat{p}_{i1}$	0.379	0.495	0.419	0.530	0.679	0.658	0.607	0.581
$p_{i1} - \hat{p}_{i1}$	−0.015	0.096	0.000	−0.024	0.048	−0.022	−0.016	0.012

does not exist a well-documented algorithm for categorical data which operates like stepwise regression to incorporate or delete terms one step at a time according to a specified criterion (e.g., a partial F test). Nevertheless, through a comparison of several reasonable candidate models using the goodness-of-fit test, a careful examination of residual values, and χ^2 tests of significance concerning the effects in the model, the best one or two models can usually be identified.

With this in mind, 14 candidate models were fitted to the data in Table 23.2 and Table 23.19 presents a summary of the tests of hypotheses concerning the parameters

TABLE 23.19   *P values for tests of hypotheses concerning effects in several candidate models fitted to the data in Table 23.2*

SOURCE OF VARIATION	MODEL NO. 1	2	3	4	5	6	7	8	9	10	11	12	13	14
A	0.27	0.11	0.11	0.03	0.11	—	0.00	—	—	—	—	0.11	0.03	0.33
C	0.00	0.00	0.00	0.00	0.00	0.00	—	0.00	—	0.00	0.00	0.00	0.00	0.00
E	0.05	0.19	0.63	0.01	—	0.64	0.23	—	—	—	—	—	—	—
L	0.85	0.33	0.00	—	0.00	0.00	0.00	0.00	0.00	—	0.04	0.05	0.05	0.13
AC	0.29	0.13	—	—	—	—	—	—	—	—	—	—	0.13	0.45
AE	0.40	1.00	—	—	—	—	—	—	—	—	—	—	—	—
AL	0.75	0.60	—	—	—	—	—	—	—	—	—	—	—	0.83
CE	0.04	0.21	—	—	—	—	—	—	—	—	—	—	—	—
CL	0.25	0.10	—	—	—	—	—	—	—	—	0.02	0.02	0.02	0.25
EL	0.15	0.22	—	—	—	—	—	—	—	—	—	—	—	—
ACE	0.30	—	—	—	—	—	—	—	—	—	—	—	—	—
ACL	1.00	—	—	—	—	—	—	—	—	—	—	—	—	0.90
AEL	0.68	—	—	—	—	—	—	—	—	—	—	—	—	—
CEL	0.41	—	—	—	—	—	—	—	—	—	—	—	—	—
ACEL	0.73	—	—	—	—	—	—	—	—	—	—	—	—	—
Residual (goodness of fit)	1.00	0.56	0.21	0.00	0.26	0.14	0.00	0.19	0.00	0.00	0.46	0.60	0.73	0.59

associated with each of these fitted models. The entries in Table 23.19 are P values; values greater than 0.05 indicate nonsignificant effects or, in the case of the goodness-of-fit test, that the model provides a reasonable fit. Dashes in the table indicate effects that were *not* included in the particular model fitted.

From Table 23.19, models 4, 7, 9, and 10 can be eliminated from consideration because they do not provide a reasonable fit to the data. Note that each of these four models omits either the C-factor main effect or the L-factor main effect, suggesting that both of these two effects should be part of any predictive model finally selected.

Finally, by examining the remaining 10 models with respect to goodness-of-fit statistics and residual patterns, and by adopting the policy that each effect in a given model should be significant at the 5% level, we would select models 11, 12, and 13 as the three most qualified candidates. A choice among these three models would now have to be based on general grounds of interpretability and relevance.

PROBLEMS 1. The accompanying table displays information gathered from a demographic study concerning regional opinion about family planning.

REGION	APPROVE	DISAPPROVE	NO OPINION	TOTAL
East	500	350	150	1,000
South	400	300	300	1,000
Total	900	650	450	2,000

(a) Describe the two kinds of sampling schemes that could have produced the table and identify which scheme is associated with a two-response, no-factor representation of the data and which scheme is associated with a one-response, one-factor representation. Which of the two sampling schemes do you think was actually used?

(b) Which of the two null hypotheses of *independence* and *homogeneity* is appropriate for each of the sampling schemes identified?

(c) Construct a table corresponding to the two-response, no-factor representation of the data. What are r (the number of response categories) and s (the number of subpopulations)?

(d) For the two-response, no-factor case, specify the structure of each of the unknown true cell probabilities under the appropriate null hypothesis [i.e., describe the form of Π_{ij} under H_0 for each (i, j) pair].

(e) For the two-response, no-factor case, express the appropriate null hypothesis in terms of linear and/or log-linear functions of the unknown cell probabilities. Also, specify the constants $\{a_{\gamma,ij}\}$ and $\{k_{a\gamma}\}$ needed to form these functions. [*Hint*: Use the fact that H_0 can be written $(\Pi_{11}\Pi_{16}/\Pi_{14}\Pi_{13} = 1$ and $\Pi_{12}\Pi_{16}/\Pi_{15}\Pi_{13} = 1)$ to define your two functions.]

(f) Construct a table corresponding to the one-response, one-factor representation of the data.

(g) For the one-response, one-factor case, specify the structure of the unknown true cell probabilities under the appropriate null hypothesis.

(h) For the one-response, one-factor case, express the appropriate null hypothesis in terms of linear and/or log-linear functions of the unknown cell probabilities. Also, specify the constants needed to form these functions.

(i) Carry out the usual χ^2 test for the data of this problem. (Use $\alpha = 0.05$.)

(j) Compare the value of the usual χ^2 to the GSK modified χ^2 values for independence and homogeneity ($\chi^2_{INDEP} = 62.89$ and $\chi^2_{HOM} = 67.14$).

2. The data below represent a cross-classification of 6,099 males, aged 20–30, according to the grip strength of each hand.

RIGHT HAND	LEFT HAND				TOTAL
	HIGHEST GRADE	SECOND GRADE	THIRD GRADE	LOWEST GRADE	
Highest grade	1,275	159	103	42	1,579
Second grade	206	1,233	361	49	1,849
Third grade	98	312	1,521	139	2,070
Lowest grade	23	60	120	398	601
Total	1,602	1,764	2,105	628	6,099

(a) Describe the sampling scheme that produces a two-response, no-factor representation for this data set and construct the corresponding table of unknown cell probabilities. (For notational simplicity, let Π_{ij} denote the cell probability associated with the ith row and jth column in the table, where $i = 1$, 2, 3, 4 and $j = 1, 2, 3, 4$.)

(b) Suppose that it is of interest to test whether the distribution of persons according to right-hand grip-strength grade is the same as the corresponding distribution for left-hand grades. How can this null hypothesis be expressed in terms of homogeneity of marginal distributions? That is, specify H_0 in terms of the marginal probabilities $\Pi_{i.}$ and $\Pi_{.j}$ for all i and j, where $\Pi_{i.} = \sum_{j=1}^{4} \Pi_{ij}$ and $\Pi_{.j} = \sum_{i=1}^{4} \Pi_{ij}$.

(c) Express the null hypothesis given in part (b) in terms of $u = 3$ linear functions $a_1(\Pi)$, $a_2(\Pi)$, and $a_3(\Pi)$ of the unknown cell probabilities. Also, specify the constants $\{a_{\gamma,ij}\}$ needed to form these linear functions.

(d) Carry out the test of the null hypothesis in part (c) using the GSK methodology (computed $\chi^2 = 7.155$).

(e) Determine the point estimates of the functions $a_\gamma(\Pi)$, $\gamma = 1, 2, 3$.

3. The data given in the table come from a study of burglary–larceny arrests in Mecklenburg County, North Carolina [Clarke and Koch (1975)]. The investigators wished to assess the relationship of prior arrest history and type of offense to the probability that an arrest results in a prison sentence.

PRIOR ARREST HISTORY	TYPE OF OFFENSE	OUTCOME	
		NO PRISON SENTENCE	PRISON SENTENCE
None	Nonresidential burglary	17	38
None	Other	21	244
Some	Nonresidential burglary	42	67
Some	Other	67	302

(a) Construct the table of unknown cell probabilities (Π_{ij}) associated with a one-response, two-factor representation of the data set. What kind of sampling scheme would produce such a table? Do the data suggest that such a sampling scheme was actually used? If not, how can the one-response, two-factor representation be justified?

(b) Consider the following null hypothesis:

$$H_0: \Pi_{12} = \Pi_{22} = \Pi_{32} = \Pi_{42}$$

What does this null hypothesis mean? Express this hypothesis in terms of three linear functions of the unknown cell probabilities appropriate for the GSK methodology.

(c) What does the null hypothesis

$$H_0: \quad \Pi_{12} = \Pi_{22}, \quad \Pi_{32} = \Pi_{42}$$

mean? Express this null hypothesis in terms of two linear functions of the unknown cell probabilities.

(d) Answer the same questions as in part (c) for the null hypothesis

$$H_0: \quad \Pi_{12} = \Pi_{32}, \quad \Pi_{22} = \Pi_{42}$$

(e) · The GSK chi-square statistics obtained for the null hypotheses specified in parts (b)–(d) are given by

$$\chi^2(b) = 52.53$$
$$\chi^2(c) = 28.83$$
$$\chi^2(d) = 16.40$$

Using these results, carry out the corresponding statistical tests, making sure to specify in each case the appropriate degrees of freedom.

(f) Letting $a_\gamma(\Pi) = \Pi_{\gamma 2}$, $\gamma = 1, \ldots, 4$, postulate a linear regression model for these linear functions in terms of factor effects. Make sure to specify v (the number of parameters), m (the number of linear functions), and the $x_{\xi l}$ ($\xi = 1, 2, \ldots, m$; $l = 1, 2, \ldots, v$), using the general model form:

$$g_\xi(\Pi) = x_{\xi 1}\beta_1 + x_{\xi 2}\beta_2 + \cdots + x_{\xi v}\beta_v, \qquad \xi = 1, \ldots, m$$

(*Note*: Postulate a model that contains an overall effect, an effect due to prior arrest history, an effect due to type of offense, and an interaction effect.)

(g) Fill in the blanks in the summary table of χ^2 tests concerning the effects in the model defined in part (f).

SOURCE	df	χ^2	P VALUE
Type of offense		12.705	
Prior arrest history		15.442	
Interaction		0.101	

(h) For each χ^2 statistic given in the table in part (g), specify the appropriate null hypothesis in terms of d linearly independent linear functions of the regression coefficients.

(i) Based on your test results in part (g), how might you simplify the regression model to more efficiently describe the data?

(j) Why is there no goodness-of-fit test presented in the table given in part (g)?

(k) Compute a 95% confidence interval for each of the effects in the model postulated in part (f) by using the point estimates and standard errors printed out by the GSK computer program, as listed in the table.

	OVERALL EFFECT	TYPE OF OFFENSE EFFECT	PRIOR ARREST HISTORY EFFECT	INTERACTION EFFECT
Coefficient	0.0792	0.2299	0.1023	−0.0261
Standard error	0.0166	0.0645	0.0260	0.0821

4. A questionnaire developed for the Family Nurse Practitioner Study at the University of North Carolina (Wagner et al., 1976) was mailed to three types of physicians for the purpose of developing criteria for appropriate medical care. The information in the table allows for an assessment of the extent to which certain factors are related to whether or not a physician responded to the questionnaire.

	MD TYPE					
	INFECTIOUS-DISEASE PEDIATRICIANS		GENERAL PEDIATRICIANS		FAMILY PHYSICIANS	
YEAR OF GRADUATION	RESPONDENTS	NON-RESPONDENTS	RESPONDENTS	NON-RESPONDENTS	RESPONDENTS	NON-RESPONDENTS
Before 1940	5	5	38	33	56	86
1940–1950	9	10	85	58	42	51
1951–1960	23	14	102	60	81	61
After 1960	8	2	49	21	27	35
Total	45	31	274	172	206	233

(a) Construct the table of unknown cell probabilities corresponding to a one-response, two-factor representation of this data set.

(b) Given that it is of interest to describe the response rate for physicians receiving the questionnaire in terms of the factors "MD type" and "year of graduation," specify the appropriate linear functions of the Π_{ij}'s that define the functions of interest for this study.

(c) Determine the point estimates of the linear functions specified in part (b).

(d) Postulate a linear model that describes the response rate in terms of the main effects and two-factor interaction effects for the factors "year of graduation" and "MD type"; that is, specify v, m, and the $x_{\xi i}$'s for a model of the general form

$$g_\xi(\Pi) = x_{\xi 1}\beta_1 + x_{\xi 2}\beta_2 + \cdots + x_{\xi v}\beta_v, \qquad \xi = 1, \ldots, m$$

(e) Use the ANOVA table to carry out the appropriate significance tests regarding main effects and interactions. What do you conclude from these tests?

SOURCE	df	χ^2	P VALUE
MD type		12.630	
Year of graduation		9.669	
Interaction		6.732	

(f) For each χ^2 statistic given in the table in part (e), specify the appropriate null hypothesis in terms of d linearly independent linear functions of the regression coefficients.

(g) Based on the test results in part (e), how would you modify the model postulated in part (d) to achieve possibly better prediction of response rate?

5. The data in the table were obtained from an occupational health study (Higgins and Koch, 1977) concerning factors related to the development of byssinosis in workers employed in a certain textile industry.

WORKPLACE	LENGTH OF EMPLOYMENT (years)	SMOKING	RESPONSE (BYSSINOTIC OR NOT)	
			YES	NO
Dusty	<10	No	7	119
		Yes	30	203
	10–20	No	3	17
		Yes	16	51
	>20	No	8	64
		Yes	41	110
Not dusty	<10	No	12	1004
		Yes	14	1340
	10–20	No	2	209
		Yes	5	409
	>20	No	8	777
		Yes	19	951

(a) Identify the response-factor representation that is most appropriate for analyzing this data set, and construct the corresponding table of unknown cell probabilities.

(b) Specify 12 linear functions $a_\gamma(\Pi)$ of the Π_{ij} that identify the functions of interest for this data set.

(c) What are the point estimates of the linear functions specified in part (b)?

(d) Postulate a linear model appropriate for this data set which corresponds to the summary table.

SOURCE	df	χ^2
Workplace (W)	1	7.356
Length of employment (E)	2	0.156
Smoking (S)	1	2.705
$W \times E$ interaction	?	2.832
$W \times S$ interaction	?	8.402
$E \times S$ interaction	?	2.285
$W \times E \times S$ interaction	?	1.611

(e) Carry out the appropriate test associated with each χ^2 statistic presented in the table in part (d). What are your conclusions about the extent to which the selected factors are related to the development of byssinosis?

(f) Specify an appropriate linear model corresponding to the following modified summary table of χ^2 tests [which omits certain interaction effects found to be nonsignificant in part (e)].

SOURCE	df	χ^2
W	1	20.619
E	2	1.017
S	1	0.625
$W \times E$ interaction	2	12.758
$W \times S$ interaction	1	13.324
Goodness of fit	?	4.414

(g) Carry out the appropriate test associated with each χ^2 statistic given in the table in part (f) and draw appropriate conclusions.

(h) The predicted response, $\hat{g}_\varepsilon(p)$, obtained for the modified model in part (f), is given by the expression:

$$\hat{g}_\varepsilon(p) = 0.0128 + 0.1319x_1 - 0.0035x_2$$
$$- 0.0038x_3 + 0.0025x_4 - 0.0956x_1x_2$$
$$+ 0.0045x_1x_3 + 0.0921x_1x_4,$$

where the variables in this model are defined as follows:

$$x_1 = \begin{cases} 1 & \text{if dusty} \\ 0 & \text{if not dusty} \end{cases} \qquad x_2 = \begin{cases} 1 & \text{if} < 10 \text{ years of} \\ & \text{employment} \\ 0 & \text{otherwise} \end{cases}$$

$$x_3 = \begin{cases} 1 & \text{if } 10\text{–}20 \text{ years of} \\ & \text{employment} \\ 0 & \text{otherwise} \end{cases} \qquad x_4 = \begin{cases} 1 & \text{if smoker} \\ 0 & \text{if nonsmoker} \end{cases}$$

Use this result to determine the value of the predicted response $\hat{g}_\varepsilon(p)$ for each subpopulation and compare the observed values $g_\varepsilon(p)$ with these predicted values. Fill in the table and evaluate the results.

	SUBPOPULATION NO.											
	1	2	3	4	5	6	7	8	9	10	11	12
$g_\varepsilon(p) = p_{i1}$												
$\hat{g}_\varepsilon(p) = \hat{p}_{i1}$												
$p_{i1} - \hat{p}_{i1}$												

(i) Based on your results in parts (g) and/or (h), how would you, if at all, modify the linear model to obtain possibly better prediction?

6. In an experiment reported by Kastenbaum and Lamphiear (1959), litters of mice were treated in either one of two ways, and the distribution of the number of depletions per litter before weaning was observed to be as given in the table.

(a) Identify the response-factor representation that characterizes this data set and construct the corresponding table of unknown cell probabilities. What are the values of r and s?

(b) One function of the unknown cell probabilities for each subpopulation of interest has the form

$$a_\gamma(\Pi) = \Pi_{\gamma 1} + \Pi_{\gamma 2}, \qquad \gamma = 1, 2, \ldots, 10$$

LITTER SIZE	TREATMENT	NUMBER OF DEPLETIONS			TOTAL NUMBER OF LITTERS
		0	1	2+	
7	A	58	11	5	74
	B	75	19	7	101
8	A	49	14	10	73
	B	58	17	8	83
9	A	33	18	15	66
	B	45	22	10	77
10	A	15	13	15	43
	B	39	22	18	79
11	A	4	12	17	33
	B	5	15	8	28

which is the probability of one or more depletions in the γth subpopulation. Express this function as a linear function of the Π_{ij}'s by determining appropriate values for $a_{\gamma, ij}$ in the general formula

$$a_\gamma(\Pi) = \sum_{i=1}^{s} \sum_{j=1}^{r} a_{\gamma, ij} \, \Pi_{ij}$$

(c) Using the data given, fill in the blanks in the accompanying table to provide point estimates of the functions $a_\gamma(\Pi)$.

LITTER SIZE	TREATMENT	$\hat{a}_\gamma(\Pi)$
7	A	0.216
	B	0.257
8	A	0.329
	B	0.301
9	A	0.500
	B	0.416
10	A	
	B	
11	A	
	B	

(d) Based on the point estimates obtained in part (c), does there appear to be a treatment effect? A litter effect? Explain.

(e) Write down a linear model that describes $a_\gamma(\Pi)$ as a function of the main effects of treatment and litter size (with no interaction terms).

(f) Use the ANOVA table to carry out the appropriate significance tests regarding the effects in the model given in part (e). Make sure to state the appropriate null hypothesis for each test.

SOURCE	df	χ^2	P VALUE
Treatments		1.2166	
Litters		141.7631	
Residual (goodness of fit)		3.2246	

(g) Based on the results in part (f), what conclusions do you draw?

REFERENCES

Allen, D. M., and Grizzle, J. E. (1969). "Analysis of Growth and Dose Response Curves," *Biometrics*, *25*: 357–382.

Anderson, T. W. (1958). *An Introduction to Multivariate Statistical Analysis*. New York: John Wiley & Sons, Inc.

Anscombe, F. J. (1960). "Rejection of Outliers," *Technometrics*, *2*: 123–147.

Anscombe, F. J., and Tukey, J. W. (1963). "The Examination and Analysis of Residuals," *Technometrics*, *5*: 141–160.

Appelbaum, M. I., and Cramer, E. M. (1974). "Some Problems in the Non-Orthogonal Analysis of Variance," *Psych. Bull.*, *81*(6): 335–343.

Arkin, N. D. (1976). "Diagnosis for Suspected Cervical Cancer," Master's thesis, Department of Epidemiology, University of North Carolina, Chapel Hill, N.C.

Armitage, P. (1971). *Statistical Methods in Medical Research*. Oxford: Blackwell Scientific Publications.

Bartlett, M. S. (1947). "The Use of Transformations," *Biometrics*, *3*: 39–52.

Bhapkar, V. P. (1966). "A Note on the Equivalence of Two Test Criteria for Hypotheses in Categorical Data," *J. Amer. Statist. Assoc.*, *61*: 228–235.

Bhapkar, V. P., and Koch, G. G. (1968). "Hypotheses of No Interaction in Multi-Dimensional Contingency Tables," *Technometrics*, *10*: 107–123.

Bishop, Y. M. M. (1969). "Full Contingency Tables, Logits, and Split Contingency Tables," *Biometrics*, *25*: 383–399.

Blalock, H. M., Jr. (1964). *Causal Inferences in Non-experimental Research*. Chapel Hill, N.C.: University of North Carolina Press.

Blalock, H. M., Jr., ed. (1971). *Causal Models in the Social Sciences*. Chicago: Aldine Publishing Company.

Bliss, C. I. (1936). "The Size Factor in the Action of Arsenic upon Silkworms' Larvae," *J. Exptl. Biol.*, *13*: 95–110.

Bliss, C. I. (1967, 1970). *Statistics in Biology*, Vols. 1 and 2. New York: McGraw-Hill Book Company.

Bross, I. D. (1958). "How To Use Ridit Analysis," *Biometrics*, *14*: 18–38.

Clark, M. F., Lechyeka, M., and Cook, C. A. (1940). "The Biological Assay of Riboflavia," *J. Nutr.*, *20*: 133–144.

Clarke, S. H., and Koch, G. G. (1975). "Who Goes to Prison? The Likelihood of Receiving an Active Sentence," *Popular Government, 41*(2): 25–37.

Cornfield, J. (1956). "A Statistical Problem Arising from Retrospective Studies," *Proc. Third Berkeley Symp., 4*: 135–148.

Cornfield, J. (1962). "Joint Dependence of Risk of Coronary Heart Disease on Serum Cholesterol and Systolic Blood Pressure: A Discriminant Function Analysis," *Fed. Proc., 21*: 58–61.

Daly, M. B. (1973). "The Effect of Neighborhood Racial Characteristics on the Attitudes, Social Behavior, and Health of Low Income Housing Residents," Ph.D. dissertation, Department of Epidemiology, University of North Carolina, Chapel Hill, N.C.

Davis, E. A., Jr. (1955). "Seasonal Changes in the Energy Balance of the English Sparrow," *Auk., 72*(4): 385–411.

Draper, N. R., and Smith, H. (1966). *Applied Regression Analysis.* New York: John Wiley & Sons, Inc.

Fisher, R. A. (1921). "On the 'Probable Error' of a Coefficient of Correlation Deduced from a Small Sample," *Metron, 1*: 3–23.

Fisher, R. A. (1936). "The Use of Multiple Measurement in Taxonomic Problems," *Ann. Eugenics, 7*: 179–188.

Fleiss, J. L. (1973). *Statistical Methods for Rates and Proportions.* New York: John Wiley & Sons, Inc.

Goodman, J. A. (1970). "The Multivariate Analysis of Qualitative Data: Interactions Among Multiple Classifications," *J. Amer. Statist. Assoc., 65*: 226–256.

Green, L. W. (1970). "Manual for Scoring Socioeconomic Status for Research on Health Behaviors," *Public Health Rept., 85*: 815–827.

Grizzle, J. E., Starmer, C. F., and Koch, G. G. (1969). "Analysis of Categorical Data by Linear Models," *Biometrics, 25*: 489–504.

Gruber, F. J. (1970). "Industrialization and Health," Ph.D. dissertation, Department of Epidemiology, University of North Carolina, Chapel Hill, N.C.

Guenther, W. C. (1964). *Analysis of Variance.* Englewood Cliffs, N.J.: Prentice-Hall, Inc.

Hamilton, M. (1971). "Sudden Death and Water Hardness in North Carolina Counties in 1956–1964," Master's thesis, Department of Epidemiology, University of North Carolina, Chapel Hill, N.C.

Hamilton, M., Pickering, G. W., Roberts, J. A. F., and Sowry, G. S. C. (1954). "The Aetiology of Essential Hypertension; the Arterial Pressure in the General Population," *Clin. Sci., 13*: 11.

Harburg, E., Erfurt, J. C., Chapel, C., et al. (1973). "Socioecological Stressor Areas and Black–White Blood Pressure," *Detroit J. Chron. Dis., 26*: 595–611.

Harman, H. H. (1960). *Modern Factor Analysis.* Chicago: University of Chicago Press.

Higgins, J. E., and Koch, G. G. (1977). "Variable Selection and Generalized Chi-Square Analysis of Categorical Data Applied to a Large Cross-Sectional Occupational Health Survey," *Internat. Statist. Rev., 45*: 51–62.

Hocking, R. R. (1973). "A Discussion of the Two-Way Mixed Model," *Amer. Statistician, 27*(4): 148–152.

Hogue, C. J., Kleinbaum, D. G., Omran, A. R., Gruber, F. J., and Freeman, D. H., Jr. (1974). "The Impact of Personal Characteristics on Post-Abortion Contraceptive Acceptance," paper presented to 102nd Annual Meeting of the American Public Health Association, New Orleans.

Hulka, B. S., Kupper, L. L., Cassel, J. C., and Thompson, S. J. (1971). "A Method for Measuring Physicians' Awareness of Patients' Concerns," *HSMHA Health Rept., 86*: 741–751.

Hulka, B. S., Kupper, L. L., and Cassel, J. C. (1972). "Determinants of Physician Utilization: Approach to a Service-Oriented Classification of Symptoms," *Medical Care, 10*(4): 300–309.

James, S. A. (1973). "The Effects of the Race of Experimenter and Race of Comparison Norm on Social Influence in Same Race and Biracial Problem-Solving Dyads," Ph.D. dissertation, Department of Clinical Psychology, Washington University, St. Louis, Mo.

James, S. A., and Kleinbaum, D. G. (1976). "Socioecologic Stress and Hypertension-Related Mortality Rates in North Carolina," *Amer. J. Public Health, 66*(4): 354–358.

Jerne, N. K. (1951). "A Study of Avidity Based on Rabbit Skin Responses to Diphtheria Antitoxin Mixtures," *Acta. Path. et Microb. Scand.*, Suppl. 87.

Johnson, S. K., and Johnson, R. E. (1972). "Tonsillectomy History in Hodgkin's Disease," *New England Journal of Medicine, 287:* 1122–1125.

Johnson, W. D., and Koch, G. G. (1971). "A Note on the Weighted Least Squares Analysis of the Ries–Smith Contingency Table Data," *Technometrics, 13:* 438–447.

Kastenbaum, M. A., and Lamphiear, D. E. (1959). "Calculation of Chi-Square To Test the No-Three-Factor-Interaction Hypothesis," *Biometrics, 15:* 107–115.

Kleinbaum, D. G., and Kleinbaum, A. (1976a). "A Team Approach for Systematic Design and Evaluation of Visually Oriented Modules," *Modular Instruction in Statistics—Report of ASA Study.* J. R. O'Fallon and J. Service, eds., American Statistical Association, Washington, D. C., pp. 115–121.

Kleinbaum, D. G., and Kleinbaum, A. (1976b). "Measuring the Attitudes of Students Taking Their First Course in Statistics," unpublished report, Department of Biostatistics, University of North Carolina, Chapel Hill, N.C.

Kleinbaum, D. G., Kupper, L. L., Cassel, J. C., and Tyroler, H. A. (1971). "Multivariate Analysis of Risk of Coronary Heart Disease in Evans County, Georgia," *Arch. Intern. Med., 128:* 943–948.

Ku, H. H., Varner, R., and Kullback, S. (1971). "Analysis of Multidimensional Contingency Tables," *J. Amer. Statist. Assoc., 66:* 55–64.

Kupper, L. L., Stewart, J. R., and Williams, K. A. (1976). "A Note on Controlling Significance Levels in Stepwise Regression," *Amer. J. Epid., 103*(1): 13–15.

Lachenbruch, P. A. (1975). *Discriminant Analysis.* New York: Hafner Press.

Landis, J. R., Stanish, W. M., Freeman, J. L., and Koch, G. G. (1976). "A Computer Program for the Generalized Chi-Square Analysis of Categorical Data Using Weighted Least Squares (GENCAT)," *Computer Programs in Biomedicine, 6*(4): 196–231.

Lehmann, E. L. (1975). *Non-parametrics: Statistical Methods Based on Ranks.* San Francisco, Calif.: Holden-Day, Inc.

Lindman, H. R. (1974). *Analysis of Variance in Complex Experimental Designs.* San Francisco: W. H. Freeman and Co.

Marquardt, D. W. (1963). "An Algorithm for Least-Squares Estimation of Non-Linear Parameters," *J. Soc. Ind. Appl. Math., 11*(2): 431–441.

Marquardt, D. W., and Snee, R. D. (1975). "Ridge Regression in Practice," *Amer. Statistician, 29*(1): 3–20.

McMichael, A. J., Spirtas, R., Kupper, L. L., and Gamble, J. F. (1975). "Solvent Exposure and Leukemia Among Rubber Workers: An Epidemiologic Study," *J. Occup. Med., 17*(4): 234–239.

Mendenhall, W. (1968). *Introduction to Linear Models and the Design and Analysis of Experiments*. Belmont, Calif.: Wadsworth Publishing Company, Inc.

Mergen, F. (1958). "Genetic Variation in Needle Characteristics of Slash Pine and in Some of Its Hybrids," *Silvae Genetica, 7*: 1–9.

Miller, R. G., Jr. (1966). *Simultaneous Statistical Inference*. New York: McGraw-Hill Book Company.

Morrison, D. F. (1967). *Multivariate Statistical Methods*. New York: McGraw-Hill Book Company.

Nagasawa, S., Osano, S., and Kondo, K. (1964). "An Analytical Method for Evaluating the Susceptibility of Fish Species to an Agricultural Chemical," *Jap. J. Appl. Ent. Zool., 8*: 118–122.

Neter, J., and Wasserman, W. (1974). *Applied Linear Statistical Models*. Homewood, Ill.: Richard D. Irwin, Inc.

Neyman, J. (1949). "Contributions to the Theory of the χ^2 Test," *Proc. Berkeley Symp. Math. Statist. Prob.*, 239–273.

Ostle, B. (1963). *Statistics in Research*, 2nd ed. Ames, Iowa: Iowa State University Press.

Overall, J. E., and Klett, C. J. (1972). *Applied Multivariate Analysis*. New York: McGraw-Hill Book Company.

Packer, P. E. (1951). "An Approach to Watershed Protection Criteria," *J. Forestry, 49*: 638–644.

Patrick, R., Cassel, J. C., Tyroler, H. A., Stanley, L., and Wild, J. (1974). "The Ponape Study of the Health Effects of Cultural Change," paper presented to Annual Meeting, Society for Epidemiologic Research, Berkeley, Calif.

Peng, K. C. (1967). *Design and Analysis of Scientific Experiments*. Reading, Mass.: Addison-Wesley Publishing Company, Inc.

Pope, P. T., and Webster, J. T. (1972). "The Use of an *F*-Statistic in Stepwise Regression Procedures," *Technometrics, 14*(2): 327–340.

Price, R. D. (1954). "The Survival of Bacterium Tularense in Lice and Louse Feces," *Amer. J. Trop. Med. Hyg., 3*: 179–186.

Rummel, R. J. (1970). *Applied Factor Analysis*. Evanston, Ill.: Northwestern University Press.

Schønheyder, F. (1936). "The Quantitative Determination of Vitamin K," *Biochem. J., 30*: 890–896.

Schreiner, H. R., Gregoine, R. C., and Lawrie, J. A. (1962). "New Biological Effect of the Gases of the Helium Group," *Science, 136*: 653–654.

Searle, S. R. (1971). *Linear Models*. New York: John Wiley & Sons, Inc.

Siegel, S. (1956). *Nonparametric Statistics for the Behavioral Sciences*. New York: McGraw-Hill Book Company.

Snedecor, G. W., and Cochran, W. G. (1967). *Statistical Methods*, 6th ed. Ames, Iowa: Iowa State University Press.

Sokal, R. R., and Karlen, I. (1964). "Competition Among Genotypes in *Tribolium castaneum* at Varying Densities and Gene Frequencies (the Black Locus)," *Genetics, 49*: 195–211.

Sokal, R. R., and Rohlf, F. J. (1969). *Biometry*. San Francisco: W. H. Freeman and Company.

Sokal, R. R., and Thomas, P. A. (1965). "Geographic Variation of *Pemphigus populitransversus* in Eastern North America: Stem Mothers and New Data on Alates," *Univ. Kansas Sci. Bull., 46*: 201–252.

Speed, F. M., and Hocking, R. R. (1976). "The Use of the *R*( )-Notation with Unbalanced Data," *Amer. Statistician, 30*(1): 30–33.

Stamler, J. (1967). *Lectures in Preventive Cardiology.* New York: Grune & Stratton.

Steele, R. G. D., and Torrie, J. H. (1960). *Principles and Procedures of Statistics, with Special Reference to Biological Sciences.* New York: McGraw-Hill Book Company.

Thompson, S. J. (1972). "The Doctor–Patient Relationship and Outcomes of Pregnancy," Ph.D. dissertation. Department of Epidemiology, University of North Carolina, Chapel Hill, N.C.

Truett, J., Cornfield, J., and Kannel, W. (1967). "A Multivariate Analysis of the Risk of Coronary Heart Disease in Framingham," *J. Chron. Dis., 20*: 511–524.

Tukey, J. W. (1949). "One Degree of Freedom for Non-additivity," *Biometrics, 5*: 232.

Wagner, E. H., Greenberg, R. A., Imrey, P. B., Williams, C. A., Wolf, S. H., and Ibrahim, M. A. (1976). "Influence of Training and Experience on Selecting Criteria To Evaluate Medical Care," *New England J. Med., 294*: 871–876.

Wald, A. (1943). "Tests of Statistical Hypotheses Concerning Several Parameters When the Number of Observations Is Large," *Trans. Amer. Math. Soc., 54*: 426–482.

Whitehead, F. H. (1954). "An Example of Taxonomic Discrimination by Biometric Methods," *New Phytologist, 53*: 496–510.

Yoshida, M. (1961). "Ecological and Physiological Researches on the Wireworm, *Melanotus caudex* Lewis. Iwata," *Shizuoka Pref.*, Japan.

APPENDIX A

TABLES

TABLE A-1   *Standard normal cumulative probabilities**

z	0.00	0.01	0.02	0.03	0.04	0.05	0.06	0.07	0.08	0.09
-3.8	0.0001	0.0001	0.0001	0.0001	0.0001	0.0001	0.0001	0.0001	0.0001	0.0001
-3.7	0.0001	0.0001	0.0001	0.0001	0.0001	0.0001	0.0001	0.0001	0.0001	0.0001
-3.6	0.0002	0.0002	0.0001	0.0001	0.0001	0.0001	0.0001	0.0001	0.0001	0.0001
-3.5	0.0002	0.0002	0.0002	0.0002	0.0002	0.0002	0.0002	0.0002	0.0002	0.0002
-3.4	0.0003	0.0003	0.0003	0.0003	0.0003	0.0003	0.0003	0.0003	0.0003	0.0002
-3.3	0.0005	0.0005	0.0005	0.0004	0.0004	0.0004	0.0004	0.0004	0.0004	0.0003
-3.2	0.0007	0.0007	0.0006	0.0006	0.0006	0.0006	0.0006	0.0005	0.0005	0.0005
-3.1	0.0010	0.0009	0.0009	0.0009	0.0008	0.0008	0.0008	0.0008	0.0007	0.0007
-3.0	0.0014	0.0013	0.0013	0.0012	0.0012	0.0011	0.0011	0.0011	0.0010	0.0010
-2.9	0.0019	0.0018	0.0018	0.0017	0.0016	0.0016	0.0015	0.0015	0.0014	0.0014
-2.8	0.0026	0.0025	0.0024	0.0023	0.0023	0.0022	0.0021	0.0021	0.0020	0.0019
-2.7	0.0035	0.0034	0.0033	0.0032	0.0031	0.0030	0.0029	0.0028	0.0027	0.0026
-2.6	0.0047	0.0045	0.0044	0.0043	0.0041	0.0040	0.0039	0.0038	0.0037	0.0036
-2.5	0.0062	0.0060	0.0059	0.0057	0.0055	0.0054	0.0052	0.0051	0.0049	0.0048
-2.4	0.0082	0.0080	0.0078	0.0076	0.0073	0.0071	0.0069	0.0068	0.0066	0.0064
-2.3	0.0107	0.0104	0.0102	0.0099	0.0096	0.0094	0.0091	0.0089	0.0087	0.0084
-2.2	0.0139	0.0136	0.0132	0.0129	0.0125	0.0122	0.0119	0.0116	0.0113	0.0110
-2.1	0.0179	0.0174	0.0170	0.0166	0.0162	0.0158	0.0154	0.0150	0.0146	0.0143
-2.0	0.0228	0.0222	0.0217	0.0212	0.0207	0.0202	0.0197	0.0192	0.0188	0.0183
-1.9	0.0287	0.0281	0.0274	0.0268	0.0262	0.0256	0.0250	0.0244	0.0239	0.0233
-1.8	0.0359	0.0351	0.0344	0.0336	0.0329	0.0322	0.0314	0.0307	0.0301	0.0294
-1.7	0.0446	0.0436	0.0427	0.0418	0.0409	0.0401	0.0392	0.0384	0.0375	0.0367
-1.6	0.0548	0.0537	0.0526	0.0516	0.0505	0.0495	0.0485	0.0475	0.0465	0.0455
-1.5	0.0668	0.0655	0.0643	0.0630	0.0618	0.0606	0.0594	0.0582	0.0571	0.0559
-1.4	0.0808	0.0793	0.0778	0.0764	0.0749	0.0735	0.0721	0.0708	0.0694	0.0681
-1.3	0.0968	0.0951	0.0934	0.0918	0.0901	0.0885	0.0869	0.0853	0.0838	0.0823
-1.2	0.1151	0.1131	0.1112	0.1093	0.1075	0.1057	0.1038	0.1020	0.1003	0.0985
-1.1	0.1357	0.1335	0.1314	0.1292	0.1271	0.1251	0.1230	0.1210	0.1190	0.1170
-1.0	0.1587	0.1562	0.1539	0.1515	0.1492	0.1469	0.1446	0.1423	0.1401	0.1379
-0.9	0.1841	0.1814	0.1788	0.1762	0.1736	0.1711	0.1685	0.1660	0.1635	0.1611
-0.8	0.2119	0.2090	0.2061	0.2033	0.2005	0.1977	0.1949	0.1922	0.1894	0.1867
-0.7	0.2420	0.2389	0.2358	0.2327	0.2297	0.2266	0.2236	0.2206	0.2177	0.2148
-0.6	0.2743	0.2709	0.2676	0.2643	0.2611	0.2578	0.2546	0.2514	0.2483	0.2451
-0.5	0.3085	0.3050	0.3015	0.2981	0.2946	0.2912	0.2877	0.2843	0.2810	0.2776
-0.4	0.3446	0.3409	0.3372	0.3336	0.3300	0.3264	0.3228	0.3192	0.3156	0.3121
-0.3	0.3821	0.3783	0.3745	0.3707	0.3669	0.3632	0.3594	0.3557	0.3520	0.3483
-0.2	0.4207	0.4168	0.4129	0.4090	0.4052	0.4013	0.3974	0.3936	0.3897	0.3859
-0.1	0.4602	0.4562	0.4522	0.4483	0.4443	0.4404	0.4364	0.4325	0.4286	0.4247
-0.0	0.5000	0.4960	0.4920	0.4880	0.4840	0.4801	0.4761	0.4721	0.4681	0.4641

z	0.00	0.01	0.02	0.03	0.04	0.05	0.06	0.07	0.08	0.09
0.0	0.5000	0.5040	0.5080	0.5120	0.5160	0.5199	0.5239	0.5279	0.5319	0.5359
0.1	0.5398	0.5438	0.5478	0.5517	0.5557	0.5596	0.5636	0.5675	0.5714	0.5753
0.2	0.5793	0.5832	0.5871	0.5910	0.5948	0.5987	0.6026	0.6064	0.6103	0.6141
0.3	0.6179	0.6217	0.6255	0.6293	0.6331	0.6368	0.6406	0.6443	0.6480	0.6517
0.4	0.6554	0.6591	0.6628	0.6664	0.6700	0.6736	0.6772	0.6808	0.6844	0.6879
0.5	0.6915	0.6950	0.6985	0.7019	0.7054	0.7088	0.7123	0.7157	0.7190	0.7224
0.6	0.7257	0.7291	0.7324	0.7357	0.7389	0.7422	0.7454	0.7486	0.7517	0.7549
0.7	0.7580	0.7611	0.7642	0.7673	0.7703	0.7734	0.7764	0.7794	0.7823	0.7852
0.8	0.7881	0.7910	0.7939	0.7967	0.7995	0.8023	0.8051	0.8078	0.8106	0.8133
0.9	0.8159	0.8186	0.8212	0.8238	0.8264	0.8289	0.8315	0.8340	0.8365	0.8389
1.0	0.8413	0.8438	0.8461	0.8485	0.8508	0.8531	0.8554	0.8577	0.8599	0.8621
1.1	0.8643	0.8665	0.8686	0.8708	0.8729	0.8749	0.8770	0.8790	0.8810	0.8830
1.2	0.8849	0.8869	0.8888	0.8907	0.8925	0.8943	0.8962	0.8980	0.8997	0.9015
1.3	0.9032	0.9049	0.9066	0.9082	0.9099	0.9115	0.9131	0.9147	0.9162	0.9177
1.4	0.9192	0.9207	0.9222	0.9236	0.9251	0.9265	0.9279	0.9292	0.9306	0.9319
1.5	0.9332	0.9345	0.9357	0.9370	0.9382	0.9394	0.9406	0.9418	0.9429	0.9441
1.6	0.9452	0.9463	0.9474	0.9484	0.9495	0.9505	0.9515	0.9525	0.9535	0.9545
1.7	0.9554	0.9564	0.9573	0.9582	0.9591	0.9599	0.9608	0.9616	0.9625	0.9633
1.8	0.9641	0.9649	0.9656	0.9664	0.9671	0.9678	0.9686	0.9693	0.9699	0.9706
1.9	0.9713	0.9719	0.9726	0.9732	0.9738	0.9744	0.9750	0.9756	0.9761	0.9767
2.0	0.9772	0.9778	0.9783	0.9788	0.9793	0.9798	0.9803	0.9808	0.9812	0.9817
2.1	0.9821	0.9826	0.9830	0.9834	0.9838	0.9842	0.9846	0.9850	0.9854	0.9857
2.2	0.9861	0.9864	0.9868	0.9871	0.9875	0.9878	0.9881	0.9884	0.9887	0.9890
2.3	0.9893	0.9896	0.9898	0.9901	0.9904	0.9906	0.9909	0.9911	0.9913	0.9916
2.4	0.9918	0.9920	0.9922	0.9924	0.9927	0.9929	0.9931	0.9932	0.9934	0.9936
2.5	0.9938	0.9940	0.9941	0.9943	0.9945	0.9946	0.9948	0.9949	0.9951	0.9952
2.6	0.9953	0.9955	0.9956	0.9957	0.9959	0.9960	0.9961	0.9962	0.9963	0.9964
2.7	0.9965	0.9966	0.9967	0.9968	0.9969	0.9970	0.9971	0.9972	0.9973	0.9974
2.8	0.9974	0.9975	0.9976	0.9977	0.9977	0.9978	0.9979	0.9979	0.9980	0.9981
2.9	0.9981	0.9982	0.9982	0.9983	0.9984	0.9984	0.9985	0.9985	0.9986	0.9986
3.0	0.9986	0.9987	0.9987	0.9988	0.9988	0.9989	0.9989	0.9989	0.9990	0.9990
3.1	0.9990	0.9991	0.9991	0.9991	0.9992	0.9992	0.9992	0.9992	0.9993	0.9993
3.2	0.9993	0.9993	0.9994	0.9994	0.9994	0.9994	0.9994	0.9995	0.9995	0.9995
3.3	0.9995	0.9995	0.9995	0.9996	0.9996	0.9996	0.9996	0.9996	0.9996	0.9997
3.4	0.9997	0.9997	0.9997	0.9997	0.9997	0.9997	0.9997	0.9997	0.9997	0.9998
3.5	0.9998	0.9998	0.9998	0.9998	0.9998	0.9998	0.9998	0.9998	0.9998	0.9998
3.6	0.9998	0.9998	0.9999	0.9999	0.9999	0.9999	0.9999	0.9999	0.9999	0.9999
3.7	0.9999	0.9999	0.9999	0.9999	0.9999	0.9999	0.9999	0.9999	0.9999	0.9999
3.8	0.9999	0.9999	0.9999	0.9999	0.9999	0.9999	0.9999	0.9999	0.9999	0.9999
3.9	1.0000									

z	P (Z < z)
-4.265	0.00001
-3.891	0.00005
-3.719	0.0001
-3.291	0.0005
-3.090	0.001
-2.576	0.005
-2.326	0.01
-2.054	0.02
-1.960	0.025
-1.881	0.03
-1.751	0.04
-1.645	0.05
-1.555	0.06
-1.476	0.07
-1.405	0.08
-1.341	0.09
-1.282	0.10
-1.036	0.15
-0.842	0.20
-0.674	0.25
-0.524	0.30
-0.385	0.35
-0.253	0.40
-0.126	0.45
0	0.50

z	P (Z < z)
0	0.50
0.126	0.55
0.253	0.60
0.385	0.65
0.524	0.70
0.674	0.75
0.842	0.80
1.036	0.85
1.282	0.90
1.341	0.91
1.405	0.92
1.476	0.93
1.555	0.94
1.645	0.95
1.751	0.96
1.881	0.97
1.960	0.975
2.054	0.98
2.326	0.99
2.576	0.995
3.090	0.999
3.291	0.9995
3.719	0.9999
3.891	0.99995
4.265	0.99999

*Table entry is the area under the standard normal curve to the left of the indicated z value, thus giving $P(Z < z)$.

TABLE A-2 Percentiles of the t distribution

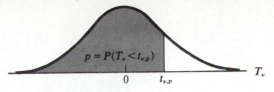

$p = P(T_\nu < t_{\nu,p})$

(a) *Student's t distribution*

% df	55	65	75	85	90	95	97.5	99	99.5	99.95
1	0.158	0.510	1.000	1.963	3.078	6.314	12.706	31.821	63.657	636.619
2	0.142	0.445	0.816	1.386	1.886	2.920	4.303	6.965	9.925	31.599
3	0.137	0.424	0.765	1.250	1.638	2.353	3.182	4.541	5.841	12.924
4	0.134	0.414	0.741	1.190	1.533	2.132	2.776	3.747	4.604	8.610
5	0.132	0.408	0.727	1.156	1.476	2.015	2.571	3.365	4.032	6.869
6	0.131	0.404	0.718	1.134	1.440	1.943	2.447	3.143	3.707	5.959
7	0.130	0.402	0.711	1.119	1.415	1.895	2.365	2.998	3.499	5.408
8	0.130	0.399	0.706	1.108	1.397	1.860	2.306	2.896	3.355	5.041
9	0.129	0.398	0.703	1.100	1.383	1.833	2.262	2.821	3.250	4.781
10	0.129	0.397	0.700	1.093	1.372	1.812	2.228	2.764	3.169	4.587
11	0.129	0.396	0.697	1.088	1.363	1.796	2.201	2.718	3.106	4.437
12	0.128	0.395	0.695	1.083	1.356	1.782	2.179	2.681	3.055	4.318
13	0.128	0.394	0.694	1.079	1.350	1.771	2.160	2.650	3.012	4.221
14	0.128	0.393	0.692	1.076	1.345	1.761	2.145	2.624	2.977	4.140
15	0.128	0.393	0.691	1.074	1.341	1.753	2.131	2.602	2.947	4.073
16	0.128	0.392	0.690	1.071	1.337	1.746	2.120	2.583	2.921	4.015
17	0.128	0.392	0.689	1.069	1.333	1.740	2.110	2.567	2.898	3.965
18	0.127	0.392	0.688	1.067	1.330	1.734	2.101	2.552	2.878	3.922
19	0.127	0.391	0.688	1.066	1.328	1.729	2.093	2.539	2.861	3.883
20	0.127	0.391	0.687	1.064	1.325	1.725	2.086	2.528	2.845	3.850
21	0.127	0.391	0.686	1.063	1.323	1.721	2.080	2.518	2.831	3.819
22	0.127	0.390	0.686	1.061	1.321	1.717	2.074	2.508	2.819	3.792
23	0.127	0.390	0.685	1.060	1.319	1.714	2.069	2.500	2.807	3.768
24	0.127	0.390	0.685	1.059	1.318	1.711	2.064	2.492	2.797	3.745
25	0.127	0.390	0.684	1.058	1.316	1.708	2.060	2.485	2.787	3.725
26	0.127	0.390	0.684	1.058	1.315	1.706	2.056	2.479	2.779	3.707
27	0.127	0.389	0.684	1.057	1.314	1.703	2.052	2.473	2.771	3.690
28	0.127	0.389	0.683	1.056	1.313	1.701	2.048	2.467	2.763	3.674
29	0.127	0.389	0.683	1.055	1.311	1.699	2.045	2.462	2.756	3.659
30	0.127	0.389	0.683	1.055	1.310	1.697	2.042	2.457	2.750	3.646
35	0.127	0.388	0.682	1.052	1.306	1.690	2.030	2.438	2.724	3.591
40	0.126	0.388	0.681	1.050	1.303	1.684	2.021	2.423	2.704	3.551
45	0.126	0.388	0.680	1.049	1.301	1.679	2.014	2.412	2.690	3.520
50	0.126	0.388	0.679	1.047	1.299	1.676	2.009	2.403	2.678	3.496
60	0.126	0.387	0.679	1.045	1.296	1.671	2.000	2.390	2.660	3.460
70	0.126	0.387	0.678	1.044	1.294	1.667	1.994	2.381	2.648	3.435
80	0.126	0.387	0.678	1.043	1.292	1.664	1.990	2.374	2.639	3.416
90	0.126	0.387	0.677	1.042	1.291	1.662	1.987	2.368	2.632	3.402
100	0.126	0.386	0.677	1.042	1.290	1.660	1.984	2.364	2.626	3.390
120	0.126	0.386	0.677	1.041	1.289	1.658	1.980	2.358	2.617	3.373
140	0.126	0.386	0.676	1.040	1.288	1.656	1.977	2.353	2.611	3.361
160	0.126	0.386	0.676	1.040	1.287	1.654	1.975	2.350	2.607	3.352
180	0.126	0.386	0.676	1.039	1.286	1.653	1.973	2.547	2.603	3.345
200	0.126	0.386	0.676	1.039	1.286	1.653	1.972	2.345	2.601	3.340
∞	0.126	0.385	0.674	1.036	1.282	1.645	1.960	2.326	2.576	3.291

TABLE A-3   Percentiles of the chi-square distribution

$$p = P(\chi_\nu^2 < \chi_{\nu,p}^2)$$

(b) χ^2 distribution

df \ %	0.5	1	2.5	5	10	20	30	40	50	60	70	80	90	95	97.5	99	99.5	99.95
1	0.0001	0.0002	0.001	0.004	0.016	0.064	0.148	0.275	0.455	0.708	1.074	1.642	2.706	3.841	5.024	6.635	7.879	12.116
2	0.010	0.020	0.051	0.103	0.211	0.446	0.713	1.022	1.386	1.833	2.408	3.219	4.605	5.991	7.378	9.210	10.597	15.202
3	0.072	0.115	0.216	0.352	0.584	1.005	1.424	1.869	2.366	2.946	3.665	4.642	6.251	7.815	9.348	11.345	12.838	17.730
4	0.207	0.297	0.484	0.711	1.064	1.649	2.195	2.753	3.357	4.045	4.878	5.989	7.779	9.488	11.143	13.277	14.860	19.997
5	0.412	0.554	0.831	1.145	1.610	2.343	3.000	3.655	4.351	5.132	6.064	7.289	9.236	11.070	12.833	15.086	16.750	22.105
6	0.676	0.872	1.237	1.635	2.204	3.070	3.828	4.570	5.348	6.211	7.231	8.558	10.645	12.592	14.449	16.812	18.548	24.103
7	0.989	1.239	1.690	2.167	2.833	3.822	4.671	5.493	6.346	7.283	8.383	9.803	12.017	14.067	16.013	18.475	20.278	26.018
8	1.344	1.646	2.180	2.733	3.490	4.594	5.527	6.423	7.344	8.351	9.524	11.030	13.362	15.507	17.535	20.090	21.955	27.868
9	1.735	2.088	2.700	3.325	4.168	5.380	6.393	7.357	8.343	9.414	10.656	12.242	14.684	16.919	19.023	21.666	23.589	29.666
10	2.156	2.558	3.247	3.940	4.865	6.179	7.267	8.295	9.342	10.473	11.781	13.442	15.987	18.307	20.483	23.209	25.188	31.420
11	2.603	3.053	3.816	4.575	5.578	6.989	8.148	9.237	10.341	11.530	12.899	14.631	17.275	19.675	21.920	24.725	26.757	33.137
12	3.074	3.571	4.404	5.226	6.304	7.807	9.034	10.182	11.340	12.584	14.011	15.812	18.549	21.026	23.337	26.217	28.300	34.821
13	3.565	4.107	5.009	5.892	7.042	8.634	9.926	11.129	12.340	13.636	15.119	16.985	19.812	22.362	24.736	27.688	29.819	36.478
14	4.075	4.660	5.629	6.571	7.790	9.467	10.821	12.078	13.339	14.685	16.222	18.151	21.064	23.685	26.119	29.141	31.319	38.109
15	4.601	5.229	6.262	7.261	8.547	10.307	11.721	13.030	14.339	15.733	17.322	19.311	22.307	24.996	27.488	30.578	32.801	39.719
16	5.142	5.812	6.908	7.962	9.312	11.152	12.624	13.983	15.338	16.780	18.418	20.465	23.542	26.296	28.845	32.000	34.267	41.308
17	5.697	6.408	7.564	8.672	10.085	12.002	13.531	14.937	16.338	17.824	19.511	21.615	24.769	27.587	30.191	33.409	35.718	42.879
18	6.265	7.015	8.231	9.390	10.865	12.857	14.440	15.893	17.338	18.868	20.601	22.760	25.989	28.869	31.526	34.805	37.156	44.434
19	6.844	7.633	8.907	10.117	11.651	13.716	15.352	16.850	18.338	19.910	21.689	23.900	27.204	30.144	32.852	36.191	38.582	45.973
20	7.434	8.260	9.591	10.851	12.443	14.578	16.266	17.809	19.337	20.951	22.775	25.038	28.412	31.410	34.170	37.566	39.997	47.498
21	8.034	8.897	10.283	11.591	13.240	15.445	17.182	18.768	20.337	21.991	23.858	26.171	29.615	32.671	35.479	38.932	41.401	49.011
22	8.643	9.542	10.982	12.338	14.041	16.314	18.101	19.729	21.337	23.031	24.939	27.301	30.813	33.924	36.781	40.289	42.796	50.511
23	9.260	10.196	11.689	13.091	14.848	17.187	19.021	20.690	22.337	24.069	26.018	28.429	32.007	35.172	38.076	41.638	44.181	52.000
24	9.886	10.856	12.401	13.848	15.659	18.062	19.943	21.752	23.337	25.106	27.096	29.553	33.196	36.415	39.364	42.980	45.559	53.479
25	10.520	11.524	13.120	14.611	16.473	18.940	20.867	22.616	24.337	26.143	28.172	30.675	34.382	37.652	40.646	44.314	46.928	54.947
26	11.160	12.198	13.844	15.379	17.292	19.820	21.792	23.579	25.336	27.179	29.246	31.795	35.563	38.885	41.923	45.642	48.290	56.407
27	11.808	12.879	14.573	16.151	18.114	20.703	22.719	24.544	26.336	28.214	30.319	32.912	36.741	40.113	43.195	46.963	49.645	57.858
28	12.461	13.565	15.308	16.928	18.939	21.588	23.647	25.509	27.336	29.249	31.391	34.027	37.916	41.337	44.461	48.278	50.993	59.300
29	13.121	14.256	16.047	17.708	19.768	22.475	24.577	26.475	28.336	30.283	32.461	35.139	39.087	42.557	45.722	49.588	52.336	60.735
30	13.787	14.953	16.791	18.493	20.599	23.364	25.508	27.442	29.336	31.316	33.530	36.250	40.256	43.773	46.979	50.892	53.672	62.162
35	17.192	18.509	20.569	22.465	24.797	27.836	30.178	32.282	34.336	36.475	38.859	41.778	46.059	49.802	53.203	57.342	60.275	69.199
40	20.707	22.164	24.433	26.509	29.051	32.345	34.872	37.134	39.335	41.622	44.165	47.269	51.805	55.758	59.342	63.691	66.766	76.095
45	24.311	25.901	28.366	30.612	33.350	36.884	39.585	41.995	44.335	46.761	49.452	52.729	57.505	61.656	65.410	69.957	73.166	82.876
50	27.991	29.707	32.357	34.764	37.689	41.449	44.313	46.864	49.335	51.892	54.723	58.164	63.167	67.505	71.420	76.154	79.490	89.561
60	35.534	37.485	40.482	43.188	46.459	50.641	53.809	56.620	59.335	62.135	65.227	68.972	74.397	79.082	83.298	88.379	91.952	102.695
70	43.275	45.442	48.758	51.739	55.329	59.898	63.346	66.396	69.334	72.358	75.689	79.715	85.527	90.531	95.023	100.425	104.215	115.578
80	51.172	53.540	57.153	60.391	64.278	69.207	72.915	76.188	79.334	82.566	86.120	90.405	96.578	101.879	106.629	112.329	116.321	128.261
90	59.196	61.754	65.647	69.126	73.291	78.558	82.511	85.993	89.334	92.761	96.524	101.054	107.565	113.145	118.136	124.116	128.299	140.782
100	67.328	70.065	74.222	77.929	82.358	87.945	92.129	95.808	99.334	102.946	106.906	111.667	118.498	124.342	129.561	135.807	140.169	153.167
120	83.852	86.923	91.573	95.705	100.624	106.806	111.419	115.465	119.334	123.289	127.616	132.806	140.233	146.567	152.211	158.950	163.648	177.603
140	100.655	104.034	109.137	113.659	119.029	125.758	130.766	135.149	139.334	143.604	148.269	153.854	161.827	168.613	174.648	181.840	186.847	201.683
160	117.679	121.346	126.870	131.756	137.546	144.783	150.158	154.856	159.334	163.898	168.876	174.828	183.311	190.516	196.915	204.530	209.824	225.481
180	134.884	138.820	144.741	149.969	156.153	163.868	169.588	174.580	179.334	184.173	189.446	195.743	204.704	212.304	219.044	227.056	232.620	249.048
200	152.241	156.432	162.728	168.279	174.835	183.003	189.049	194.319	199.334	204.434	209.985	216.609	226.021	233.994	241.058	249.445	255.264	272.423

TABLE A-4 Percentiles of the F distribution

Upper 25% point of the F distribution

$$p = P(F_{\nu_1,\nu_2} < F_{\nu_1,\nu_2;p})$$

(c) F distribution

DEGREES OF FREEDOM FOR NUMERATOR

ν₂＼ν₁	1	2	3	4	5	6	7	8	9	10	11	12	13	14	15	16	17	18	19	20	25	30	40	50	100	150	200
1	5.83	7.50	8.20	8.58	8.82	8.98	9.10	9.19	9.26	9.32	9.37	9.41	9.44	9.47	9.49	9.52	9.53	9.55	9.57	9.58	9.63	9.67	9.71	9.74	9.80	9.81	9.82
2	2.57	3.00	3.15	3.23	3.28	3.31	3.34	3.35	3.37	3.38	3.39	3.39	3.40	3.41	3.41	3.41	3.42	3.42	3.42	3.43	3.44	3.44	3.45	3.46	3.47	3.47	3.47
3	2.02	2.28	2.36	2.39	2.41	2.42	2.43	2.44	2.44	2.44	2.45	2.45	2.45	2.46	2.46	2.46	2.46	2.46	2.46	2.46	2.46	2.47	2.47	2.47	2.47	2.47	2.47
4	1.81	2.00	2.05	2.06	2.07	2.08	2.08	2.08	2.08	2.08	2.08	2.08	2.08	2.08	2.08	2.08	2.08	2.08	2.08	2.08	2.08	2.08	2.08	2.08	2.08	2.08	2.08
5	1.69	1.85	1.88	1.89	1.89	1.89	1.89	1.89	1.89	1.89	1.89	1.89	1.89	1.88	1.88	1.88	1.88	1.88	1.88	1.88	1.88	1.88	1.88	1.88	1.87	1.87	1.87
6	1.62	1.76	1.78	1.79	1.79	1.78	1.78	1.78	1.77	1.77	1.77	1.77	1.77	1.76	1.76	1.76	1.76	1.76	1.76	1.76	1.75	1.75	1.75	1.75	1.74	1.74	1.74
7	1.57	1.70	1.72	1.72	1.71	1.71	1.70	1.70	1.69	1.69	1.69	1.68	1.68	1.68	1.68	1.68	1.67	1.67	1.67	1.67	1.67	1.66	1.66	1.66	1.65	1.65	1.65
8	1.54	1.66	1.67	1.66	1.66	1.65	1.64	1.64	1.63	1.63	1.63	1.62	1.62	1.62	1.62	1.62	1.61	1.61	1.61	1.61	1.60	1.60	1.59	1.59	1.58	1.58	1.58
9	1.51	1.62	1.63	1.63	1.62	1.61	1.60	1.60	1.59	1.59	1.58	1.58	1.58	1.57	1.57	1.57	1.57	1.56	1.56	1.56	1.55	1.55	1.54	1.54	1.53	1.53	1.53
10	1.49	1.60	1.60	1.59	1.59	1.58	1.57	1.56	1.56	1.55	1.55	1.54	1.54	1.54	1.53	1.53	1.53	1.53	1.53	1.52	1.52	1.51	1.51	1.50	1.49	1.49	1.49
11	1.47	1.58	1.58	1.57	1.56	1.55	1.54	1.53	1.53	1.52	1.52	1.51	1.51	1.51	1.50	1.50	1.50	1.50	1.49	1.49	1.49	1.48	1.47	1.47	1.46	1.46	1.46
12	1.46	1.56	1.56	1.55	1.54	1.53	1.52	1.51	1.51	1.50	1.50	1.49	1.49	1.49	1.48	1.47	1.47	1.47	1.47	1.47	1.46	1.45	1.45	1.44	1.43	1.43	1.43
13	1.45	1.55	1.55	1.53	1.52	1.51	1.50	1.49	1.49	1.48	1.47	1.47	1.47	1.46	1.46	1.46	1.45	1.45	1.45	1.45	1.44	1.43	1.42	1.42	1.41	1.41	1.40
14	1.44	1.53	1.53	1.52	1.51	1.50	1.49	1.48	1.47	1.46	1.46	1.45	1.45	1.44	1.44	1.44	1.44	1.43	1.43	1.43	1.41	1.41	1.41	1.40	1.39	1.39	1.38
15	1.43	1.52	1.52	1.51	1.49	1.48	1.47	1.46	1.46	1.45	1.44	1.44	1.43	1.43	1.43	1.42	1.42	1.42	1.41	1.41	1.40	1.40	1.39	1.38	1.37	1.37	1.37
16	1.42	1.51	1.51	1.50	1.48	1.47	1.46	1.45	1.44	1.44	1.43	1.43	1.42	1.42	1.41	1.41	1.41	1.40	1.40	1.40	1.39	1.38	1.37	1.37	1.36	1.35	1.35
17	1.42	1.51	1.50	1.49	1.47	1.46	1.45	1.44	1.43	1.43	1.42	1.41	1.41	1.41	1.40	1.40	1.39	1.39	1.39	1.39	1.38	1.37	1.36	1.36	1.34	1.34	1.34
18	1.41	1.50	1.49	1.48	1.46	1.45	1.44	1.43	1.42	1.42	1.41	1.40	1.40	1.40	1.39	1.39	1.38	1.38	1.38	1.38	1.37	1.36	1.35	1.35	1.33	1.33	1.32
19	1.41	1.49	1.49	1.47	1.46	1.44	1.43	1.42	1.41	1.41	1.40	1.40	1.39	1.39	1.38	1.38	1.37	1.37	1.37	1.37	1.35	1.35	1.34	1.33	1.32	1.31	1.31
20	1.40	1.49	1.48	1.46	1.45	1.44	1.43	1.42	1.41	1.40	1.39	1.39	1.38	1.38	1.37	1.37	1.37	1.36	1.36	1.36	1.34	1.34	1.33	1.32	1.31	1.30	1.30
21	1.40	1.48	1.48	1.46	1.44	1.43	1.42	1.41	1.40	1.39	1.39	1.38	1.37	1.37	1.37	1.36	1.36	1.35	1.35	1.35	1.34	1.33	1.32	1.32	1.30	1.30	1.29
22	1.40	1.48	1.47	1.45	1.44	1.42	1.41	1.40	1.39	1.39	1.38	1.37	1.37	1.36	1.36	1.36	1.35	1.35	1.35	1.34	1.33	1.32	1.31	1.31	1.29	1.29	1.28
23	1.39	1.47	1.47	1.45	1.43	1.42	1.41	1.40	1.39	1.38	1.37	1.37	1.36	1.36	1.35	1.35	1.35	1.34	1.34	1.34	1.33	1.32	1.31	1.31	1.29	1.28	1.28
24	1.39	1.47	1.46	1.44	1.43	1.41	1.40	1.39	1.38	1.38	1.37	1.36	1.36	1.35	1.35	1.34	1.34	1.34	1.33	1.33	1.32	1.31	1.30	1.30	1.28	1.28	1.27
25	1.39	1.47	1.46	1.44	1.42	1.41	1.40	1.39	1.38	1.37	1.36	1.36	1.35	1.35	1.34	1.34	1.33	1.33	1.33	1.33	1.31	1.31	1.30	1.29	1.27	1.27	1.26
26	1.38	1.46	1.45	1.44	1.42	1.41	1.39	1.38	1.37	1.37	1.36	1.35	1.35	1.34	1.34	1.33	1.33	1.33	1.32	1.32	1.31	1.30	1.29	1.28	1.27	1.26	1.26
27	1.38	1.46	1.45	1.43	1.42	1.40	1.39	1.38	1.37	1.36	1.35	1.35	1.34	1.34	1.33	1.33	1.33	1.32	1.32	1.32	1.30	1.30	1.28	1.28	1.26	1.25	1.25
28	1.38	1.46	1.45	1.43	1.41	1.40	1.39	1.38	1.37	1.36	1.35	1.34	1.34	1.33	1.33	1.32	1.32	1.32	1.31	1.31	1.30	1.29	1.28	1.27	1.25	1.25	1.25
29	1.38	1.45	1.45	1.43	1.41	1.40	1.38	1.37	1.36	1.35	1.35	1.34	1.33	1.33	1.32	1.32	1.32	1.31	1.31	1.31	1.29	1.29	1.27	1.26	1.25	1.24	1.24
30	1.38	1.45	1.44	1.42	1.41	1.39	1.38	1.37	1.36	1.35	1.34	1.34	1.33	1.33	1.32	1.32	1.31	1.31	1.31	1.30	1.29	1.28	1.27	1.26	1.24	1.24	1.24
32	1.37	1.45	1.44	1.42	1.40	1.39	1.38	1.37	1.36	1.35	1.34	1.33	1.33	1.32	1.32	1.31	1.31	1.30	1.30	1.30	1.29	1.28	1.26	1.26	1.24	1.23	1.23
34	1.37	1.44	1.43	1.42	1.40	1.38	1.37	1.36	1.35	1.34	1.34	1.33	1.32	1.32	1.31	1.31	1.30	1.30	1.29	1.29	1.28	1.27	1.26	1.25	1.23	1.22	1.22
36	1.37	1.44	1.43	1.41	1.40	1.38	1.37	1.36	1.35	1.34	1.33	1.33	1.32	1.31	1.31	1.30	1.30	1.30	1.29	1.29	1.28	1.26	1.25	1.24	1.22	1.22	1.21
38	1.36	1.44	1.43	1.41	1.39	1.38	1.37	1.35	1.35	1.34	1.33	1.32	1.31	1.31	1.30	1.30	1.29	1.29	1.29	1.28	1.27	1.26	1.24	1.24	1.22	1.21	1.21
40	1.36	1.44	1.42	1.40	1.39	1.37	1.36	1.35	1.34	1.33	1.32	1.31	1.31	1.30	1.30	1.29	1.29	1.28	1.28	1.28	1.26	1.25	1.24	1.23	1.21	1.21	1.20
42	1.36	1.43	1.42	1.40	1.38	1.37	1.35	1.34	1.33	1.32	1.32	1.31	1.30	1.30	1.29	1.29	1.28	1.28	1.28	1.27	1.26	1.25	1.23	1.23	1.21	1.20	1.20
44	1.36	1.43	1.42	1.40	1.38	1.36	1.35	1.34	1.33	1.32	1.31	1.31	1.30	1.29	1.29	1.28	1.28	1.27	1.27	1.27	1.26	1.24	1.23	1.22	1.20	1.19	1.19
46	1.36	1.43	1.42	1.40	1.38	1.36	1.35	1.34	1.33	1.32	1.31	1.30	1.30	1.29	1.28	1.28	1.28	1.27	1.27	1.26	1.25	1.24	1.22	1.22	1.20	1.19	1.18
48	1.36	1.43	1.42	1.39	1.38	1.36	1.35	1.34	1.33	1.32	1.31	1.30	1.29	1.29	1.28	1.28	1.27	1.27	1.27	1.26	1.25	1.24	1.22	1.21	1.19	1.19	1.18
50	1.35	1.43	1.41	1.39	1.37	1.36	1.34	1.33	1.32	1.31	1.30	1.30	1.29	1.28	1.28	1.27	1.27	1.27	1.26	1.26	1.25	1.23	1.22	1.21	1.19	1.18	1.18
60	1.35	1.42	1.41	1.38	1.37	1.35	1.33	1.32	1.31	1.30	1.29	1.29	1.28	1.27	1.27	1.26	1.26	1.25	1.25	1.25	1.23	1.22	1.21	1.20	1.18	1.17	1.16
70	1.35	1.41	1.40	1.38	1.36	1.34	1.33	1.32	1.31	1.30	1.29	1.28	1.27	1.27	1.26	1.26	1.25	1.25	1.24	1.24	1.23	1.21	1.20	1.19	1.16	1.16	1.15
80	1.34	1.41	1.40	1.38	1.36	1.34	1.33	1.32	1.30	1.30	1.28	1.28	1.27	1.26	1.26	1.26	1.25	1.24	1.24	1.24	1.23	1.21	1.19	1.18	1.16	1.15	1.15
90	1.34	1.41	1.39	1.37	1.35	1.34	1.32	1.31	1.30	1.29	1.28	1.27	1.26	1.26	1.25	1.25	1.24	1.24	1.23	1.23	1.22	1.20	1.19	1.18	1.15	1.15	1.14
100	1.34	1.41	1.39	1.37	1.35	1.33	1.32	1.30	1.29	1.28	1.28	1.27	1.26	1.25	1.25	1.24	1.24	1.23	1.23	1.23	1.21	1.20	1.18	1.17	1.14	1.14	1.13
125	1.34	1.40	1.39	1.38	1.35	1.33	1.32	1.30	1.29	1.28	1.27	1.26	1.25	1.25	1.24	1.24	1.23	1.23	1.22	1.22	1.20	1.19	1.17	1.16	1.14	1.13	1.12
150	1.33	1.40	1.38	1.36	1.34	1.33	1.31	1.30	1.29	1.27	1.27	1.26	1.25	1.24	1.24	1.23	1.23	1.22	1.22	1.21	1.20	1.19	1.17	1.16	1.13	1.12	1.11
200	1.33	1.40	1.38	1.36	1.33	1.32	1.30	1.29	1.28	1.26	1.26	1.25	1.24	1.24	1.23	1.22	1.22	1.21	1.21	1.20	1.19	1.18	1.16	1.15	1.12	1.11	1.10
300	1.33	1.39	1.38	1.35	1.33	1.32	1.30	1.29	1.28	1.27	1.26	1.25	1.24	1.23	1.23	1.22	1.22	1.21	1.21	1.20	1.19	1.18	1.15	1.14	1.11	1.10	1.09
500	1.33	1.39	1.37	1.35	1.33	1.31	1.30	1.29	1.28	1.26	1.25	1.24	1.24	1.23	1.22	1.22	1.21	1.21	1.20	1.20	1.18	1.17	1.15	1.14	1.11	1.10	1.08
1000	1.32	1.39	1.37	1.35	1.33	1.31	1.29	1.28	1.27	1.26	1.25	1.24	1.23	1.23	1.22	1.21	1.21	1.20	1.20	1.19	1.18	1.16	1.14	1.13	1.10	1.08	1.07

DEGREES OF FREEDOM FOR DENOMINATOR

TABLE A-4 Percentiles of the F distribution (continued)

Upper 10% point of the F distribution

DEGREES OF FREEDOM FOR NUMERATOR

den	1	2	3	4	5	6	7	8	9	10	11	12	13	14	15	16	17	18	19	20	25	30	40	50	100	150	200
1	39.9	49.5	53.6	55.8	57.2	58.2	58.9	59.4	59.9	60.2	60.5	60.7	60.9	61.1	61.2	61.3	61.5	61.6	61.7	61.7	62.1	62.3	62.5	62.7	63.0	63.1	63.2
2	8.53	9.00	9.16	9.24	9.29	9.33	9.35	9.37	9.38	9.39	9.40	9.41	9.41	9.42	9.42	9.43	9.43	9.44	9.44	9.44	9.45	9.46	9.47	9.47	9.48	9.48	9.49
3	5.54	5.46	5.39	5.34	5.31	5.28	5.27	5.25	5.24	5.23	5.22	5.22	5.21	5.20	5.20	5.20	5.19	5.19	5.19	5.18	5.17	5.17	5.16	5.15	5.14	5.14	5.14
4	4.54	4.32	4.19	4.11	4.05	4.01	3.98	3.95	3.94	3.92	3.91	3.90	3.89	3.88	3.87	3.86	3.86	3.85	3.85	3.84	3.83	3.82	3.80	3.80	3.78	3.77	3.77
5	4.06	3.78	3.62	3.52	3.45	3.40	3.37	3.34	3.32	3.30	3.28	3.27	3.26	3.25	3.24	3.23	3.22	3.22	3.21	3.21	3.19	3.17	3.16	3.15	3.13	3.12	3.12
6	3.78	3.46	3.29	3.18	3.11	3.05	3.01	2.98	2.96	2.94	2.92	2.90	2.89	2.88	2.87	2.86	2.85	2.85	2.84	2.84	2.81	2.80	2.78	2.77	2.75	2.74	2.73
7	3.59	3.26	3.07	2.96	2.88	2.83	2.78	2.75	2.72	2.70	2.68	2.67	2.65	2.64	2.63	2.62	2.61	2.61	2.60	2.59	2.57	2.56	2.54	2.52	2.50	2.49	2.48
8	3.46	3.11	2.92	2.81	2.73	2.67	2.62	2.59	2.56	2.54	2.52	2.50	2.49	2.48	2.46	2.45	2.45	2.44	2.43	2.42	2.40	2.38	2.36	2.35	2.32	2.31	2.31
9	3.36	3.01	2.81	2.69	2.61	2.55	2.51	2.47	2.44	2.42	2.40	2.38	2.36	2.35	2.34	2.33	2.32	2.31	2.30	2.30	2.27	2.25	2.23	2.22	2.19	2.18	2.17
10	3.29	2.92	2.73	2.61	2.52	2.46	2.41	2.38	2.35	2.32	2.30	2.28	2.27	2.26	2.24	2.23	2.22	2.22	2.21	2.20	2.17	2.16	2.13	2.12	2.09	2.08	2.07
11	3.23	2.86	2.66	2.54	2.45	2.39	2.34	2.30	2.27	2.25	2.23	2.21	2.19	2.18	2.17	2.16	2.15	2.14	2.13	2.12	2.10	2.08	2.05	2.04	2.01	1.99	1.99
12	3.18	2.81	2.61	2.48	2.39	2.33	2.28	2.24	2.21	2.19	2.17	2.15	2.13	2.12	2.10	2.09	2.08	2.08	2.07	2.06	2.03	2.01	1.99	1.97	1.94	1.93	1.92
13	3.14	2.76	2.56	2.43	2.35	2.28	2.23	2.20	2.16	2.14	2.12	2.10	2.08	2.07	2.05	2.04	2.03	2.02	2.01	2.01	1.98	1.96	1.93	1.92	1.88	1.87	1.86
14	3.10	2.73	2.52	2.39	2.31	2.24	2.19	2.15	2.12	2.10	2.07	2.05	2.04	2.02	2.01	2.00	1.99	1.98	1.97	1.96	1.93	1.91	1.89	1.87	1.83	1.82	1.82
15	3.07	2.70	2.49	2.36	2.27	2.21	2.16	2.12	2.09	2.06	2.04	2.02	2.00	1.99	1.97	1.96	1.95	1.94	1.93	1.92	1.89	1.87	1.85	1.83	1.79	1.78	1.77
16	3.05	2.67	2.46	2.33	2.24	2.18	2.13	2.09	2.06	2.03	2.01	1.99	1.97	1.95	1.94	1.93	1.92	1.91	1.90	1.89	1.86	1.84	1.81	1.79	1.76	1.74	1.74
17	3.03	2.64	2.44	2.31	2.22	2.15	2.10	2.06	2.03	2.00	1.98	1.96	1.94	1.93	1.91	1.90	1.89	1.88	1.87	1.86	1.83	1.81	1.78	1.76	1.73	1.71	1.71
18	3.01	2.62	2.42	2.29	2.20	2.13	2.08	2.04	2.00	1.98	1.95	1.93	1.92	1.90	1.89	1.89	1.86	1.85	1.84	1.84	1.80	1.78	1.75	1.74	1.70	1.68	1.68
19	2.99	2.61	2.40	2.27	2.18	2.11	2.06	2.02	1.98	1.96	1.93	1.91	1.89	1.88	1.86	1.85	1.84	1.83	1.82	1.81	1.78	1.76	1.73	1.71	1.67	1.66	1.65
20	2.97	2.59	2.38	2.25	2.16	2.09	2.04	2.00	1.96	1.94	1.91	1.89	1.87	1.86	1.84	1.83	1.82	1.81	1.80	1.79	1.76	1.74	1.71	1.69	1.65	1.64	1.63
21	2.96	2.57	2.36	2.23	2.14	2.08	2.02	1.98	1.95	1.92	1.90	1.87	1.86	1.84	1.83	1.81	1.80	1.79	1.78	1.78	1.74	1.72	1.69	1.67	1.63	1.62	1.61
22	2.95	2.56	2.35	2.22	2.13	2.06	2.01	1.97	1.93	1.90	1.88	1.86	1.84	1.83	1.81	1.80	1.79	1.78	1.77	1.76	1.73	1.70	1.67	1.65	1.61	1.60	1.59
23	2.94	2.55	2.34	2.21	2.11	2.05	1.99	1.95	1.92	1.89	1.87	1.84	1.83	1.81	1.80	1.78	1.77	1.76	1.75	1.74	1.71	1.69	1.66	1.64	1.59	1.58	1.57
24	2.93	2.54	2.33	2.19	2.10	2.04	1.98	1.94	1.91	1.88	1.85	1.83	1.81	1.80	1.78	1.77	1.76	1.75	1.74	1.73	1.70	1.67	1.64	1.62	1.58	1.56	1.56
25	2.92	2.53	2.32	2.18	2.09	2.02	1.97	1.93	1.89	1.87	1.84	1.82	1.80	1.79	1.77	1.76	1.75	1.74	1.73	1.72	1.68	1.66	1.63	1.61	1.56	1.55	1.54
26	2.91	2.52	2.31	2.17	2.08	2.01	1.96	1.92	1.88	1.86	1.83	1.81	1.79	1.77	1.76	1.75	1.73	1.72	1.71	1.71	1.67	1.65	1.61	1.59	1.55	1.54	1.53
27	2.90	2.51	2.30	2.17	2.07	2.00	1.95	1.91	1.87	1.85	1.82	1.80	1.78	1.76	1.75	1.74	1.72	1.71	1.70	1.70	1.66	1.64	1.60	1.58	1.54	1.52	1.52
28	2.89	2.50	2.29	2.16	2.06	2.00	1.94	1.90	1.87	1.84	1.81	1.79	1.77	1.75	1.74	1.73	1.71	1.70	1.69	1.69	1.65	1.63	1.59	1.57	1.53	1.51	1.50
29	2.89	2.50	2.28	2.15	2.06	1.99	1.93	1.89	1.86	1.83	1.80	1.78	1.76	1.75	1.73	1.72	1.71	1.69	1.68	1.68	1.64	1.62	1.58	1.56	1.52	1.50	1.49
30	2.88	2.49	2.28	2.14	2.05	1.98	1.93	1.88	1.85	1.82	1.79	1.77	1.75	1.74	1.72	1.71	1.70	1.69	1.68	1.67	1.63	1.61	1.57	1.55	1.51	1.49	1.48
32	2.87	2.48	2.26	2.13	2.04	1.97	1.91	1.87	1.83	1.81	1.78	1.76	1.74	1.72	1.71	1.69	1.68	1.67	1.66	1.65	1.62	1.59	1.56	1.53	1.49	1.47	1.46
34	2.86	2.47	2.25	2.12	2.02	1.96	1.90	1.86	1.82	1.79	1.77	1.75	1.73	1.71	1.70	1.68	1.67	1.66	1.65	1.64	1.60	1.58	1.54	1.52	1.47	1.46	1.45
36	2.85	2.46	2.24	2.11	2.01	1.94	1.89	1.85	1.81	1.78	1.76	1.73	1.71	1.70	1.68	1.67	1.66	1.65	1.64	1.63	1.59	1.56	1.53	1.51	1.46	1.44	1.43
38	2.84	2.45	2.23	2.10	2.01	1.94	1.88	1.84	1.80	1.77	1.75	1.72	1.70	1.69	1.67	1.66	1.65	1.63	1.62	1.61	1.58	1.55	1.52	1.49	1.45	1.43	1.42
40	2.84	2.44	2.23	2.09	2.00	1.93	1.87	1.83	1.79	1.76	1.74	1.71	1.70	1.68	1.66	1.65	1.64	1.62	1.61	1.61	1.57	1.54	1.51	1.48	1.43	1.42	1.41
42	2.83	2.43	2.22	2.08	1.99	1.92	1.86	1.82	1.78	1.75	1.73	1.71	1.69	1.67	1.65	1.64	1.63	1.62	1.61	1.60	1.56	1.53	1.50	1.47	1.42	1.41	1.40
44	2.82	2.43	2.21	2.08	1.98	1.91	1.86	1.81	1.78	1.75	1.72	1.70	1.68	1.66	1.65	1.63	1.62	1.61	1.60	1.59	1.55	1.52	1.49	1.46	1.41	1.39	1.39
46	2.82	2.42	2.21	2.07	1.98	1.91	1.85	1.81	1.77	1.74	1.71	1.69	1.67	1.65	1.64	1.62	1.61	1.60	1.59	1.58	1.54	1.52	1.48	1.46	1.40	1.39	1.38
48	2.81	2.42	2.20	2.07	1.97	1.90	1.85	1.80	1.77	1.73	1.71	1.69	1.67	1.65	1.63	1.62	1.61	1.59	1.58	1.57	1.54	1.51	1.47	1.45	1.40	1.38	1.37
50	2.81	2.41	2.20	2.06	1.97	1.90	1.84	1.80	1.76	1.73	1.70	1.68	1.66	1.64	1.63	1.61	1.60	1.59	1.58	1.57	1.53	1.50	1.46	1.44	1.39	1.37	1.36
60	2.79	2.39	2.18	2.04	1.95	1.87	1.82	1.77	1.74	1.71	1.68	1.66	1.64	1.62	1.60	1.59	1.58	1.56	1.55	1.54	1.50	1.48	1.44	1.41	1.36	1.34	1.33
70	2.78	2.38	2.16	2.03	1.93	1.86	1.80	1.76	1.72	1.69	1.66	1.64	1.62	1.60	1.59	1.57	1.56	1.55	1.54	1.53	1.49	1.46	1.42	1.39	1.34	1.31	1.30
80	2.77	2.37	2.15	2.02	1.92	1.85	1.79	1.75	1.71	1.68	1.65	1.63	1.61	1.59	1.57	1.56	1.55	1.53	1.52	1.51	1.47	1.44	1.40	1.38	1.32	1.30	1.28
90	2.76	2.36	2.15	2.01	1.91	1.84	1.78	1.74	1.70	1.67	1.64	1.62	1.60	1.58	1.56	1.55	1.54	1.52	1.51	1.50	1.46	1.43	1.39	1.36	1.30	1.28	1.27
100	2.76	2.36	2.14	2.00	1.91	1.83	1.78	1.73	1.69	1.66	1.64	1.61	1.59	1.57	1.56	1.54	1.53	1.52	1.50	1.49	1.45	1.42	1.38	1.35	1.29	1.27	1.26
125	2.75	2.35	2.13	1.99	1.89	1.82	1.77	1.72	1.68	1.65	1.62	1.60	1.58	1.56	1.54	1.53	1.51	1.50	1.49	1.48	1.44	1.41	1.36	1.34	1.27	1.25	1.23
150	2.74	2.34	2.12	1.98	1.89	1.81	1.76	1.71	1.67	1.64	1.61	1.59	1.57	1.55	1.53	1.52	1.50	1.49	1.48	1.47	1.43	1.40	1.35	1.33	1.26	1.23	1.22
200	2.73	2.33	2.11	1.97	1.88	1.80	1.75	1.70	1.66	1.63	1.60	1.58	1.56	1.54	1.52	1.51	1.49	1.48	1.47	1.46	1.41	1.38	1.34	1.31	1.24	1.21	1.20
300	2.72	2.32	2.10	1.96	1.87	1.79	1.74	1.69	1.65	1.62	1.59	1.57	1.55	1.53	1.51	1.50	1.48	1.47	1.46	1.45	1.40	1.37	1.32	1.30	1.22	1.19	1.18
500	2.72	2.31	2.09	1.96	1.86	1.79	1.73	1.68	1.64	1.61	1.58	1.56	1.54	1.52	1.50	1.49	1.47	1.46	1.45	1.44	1.39	1.36	1.31	1.28	1.21	1.18	1.16
1000	2.71	2.31	2.09	1.95	1.85	1.78	1.72	1.68	1.64	1.61	1.58	1.55	1.53	1.51	1.49	1.48	1.46	1.45	1.44	1.43	1.38	1.35	1.30	1.27	1.20	1.16	1.15

DEGREES OF FREEDOM FOR DENOMINATOR

TABLE A-4 Percentiles of the F distribution (continued)

Upper 5% point of the F distribution

DEGREES OF FREEDOM FOR NUMERATOR

denom \ num	1	2	3	4	5	6	7	8	9	10	11	12	13	14	15	16	17	18	19	20	25	30	40	50	100	150	200
1	161	200	216	225	230	234	237	239	241	242	243	244	245	245	246	246	247	247	248	248	249	250	251	252	253	253	254
2	18.5	19.0	19.2	19.2	19.3	19.3	19.4	19.4	19.4	19.4	19.4	19.4	19.4	19.4	19.4	19.4	19.4	19.4	19.4	19.4	19.4	19.5	19.5	19.5	19.5	19.5	19.5
3	10.1	9.55	9.28	9.12	9.01	8.94	8.89	8.85	8.81	8.79	8.76	8.74	8.73	8.71	8.70	8.69	8.68	8.67	8.67	8.66	8.63	8.62	8.59	8.58	8.55	8.54	8.54
4	7.71	6.94	6.59	6.39	6.26	6.16	6.09	6.04	6.00	5.96	5.94	5.91	5.89	5.87	5.86	5.84	5.83	5.82	5.81	5.80	5.77	5.75	5.72	5.70	5.66	5.65	5.65
5	6.61	5.79	5.41	5.19	5.05	4.95	4.88	4.82	4.77	4.74	4.70	4.68	4.66	4.64	4.62	4.60	4.59	4.58	4.57	4.56	4.52	4.50	4.46	4.44	4.41	4.39	4.39
6	5.99	5.14	4.76	4.53	4.39	4.28	4.21	4.15	4.10	4.06	4.03	4.00	3.98	3.96	3.94	3.92	3.91	3.90	3.88	3.87	3.83	3.81	3.77	3.75	3.71	3.70	3.69
7	5.59	4.74	4.35	4.12	3.97	3.87	3.79	3.73	3.68	3.64	3.60	3.57	3.55	3.53	3.51	3.49	3.48	3.47	3.46	3.44	3.40	3.38	3.34	3.32	3.27	3.26	3.25
8	5.32	4.46	4.07	3.84	3.69	3.58	3.50	3.44	3.39	3.35	3.31	3.28	3.26	3.24	3.22	3.20	3.19	3.17	3.16	3.15	3.11	3.08	3.04	3.02	2.97	2.96	2.95
9	5.12	4.26	3.86	3.63	3.48	3.37	3.29	3.23	3.18	3.14	3.10	3.07	3.05	3.03	3.01	2.99	2.97	2.96	2.95	2.94	2.89	2.86	2.83	2.80	2.76	2.74	2.73
10	4.96	4.10	3.71	3.48	3.33	3.22	3.14	3.07	3.02	2.98	2.94	2.91	2.89	2.86	2.85	2.83	2.81	2.80	2.79	2.77	2.73	2.70	2.66	2.64	2.59	2.57	2.56
11	4.84	3.98	3.59	3.36	3.20	3.09	3.01	2.95	2.90	2.85	2.82	2.79	2.76	2.74	2.72	2.70	2.69	2.67	2.66	2.65	2.60	2.57	2.53	2.51	2.46	2.44	2.43
12	4.75	3.89	3.49	3.26	3.11	3.00	2.91	2.85	2.80	2.75	2.72	2.69	2.66	2.64	2.62	2.60	2.58	2.57	2.56	2.54	2.50	2.47	2.43	2.40	2.35	2.33	2.32
13	4.67	3.81	3.41	3.18	3.03	2.92	2.83	2.77	2.71	2.67	2.63	2.60	2.58	2.55	2.53	2.51	2.50	2.48	2.47	2.46	2.41	2.38	2.34	2.31	2.26	2.24	2.23
14	4.60	3.74	3.34	3.11	2.96	2.85	2.76	2.70	2.65	2.60	2.57	2.53	2.51	2.48	2.46	2.44	2.43	2.41	2.40	2.39	2.34	2.31	2.27	2.24	2.19	2.17	2.16
15	4.54	3.68	3.29	3.05	2.90	2.79	2.71	2.64	2.59	2.54	2.51	2.48	2.45	2.42	2.40	2.38	2.37	2.35	2.34	2.33	2.28	2.25	2.20	2.18	2.12	2.10	2.10
16	4.49	3.63	3.24	3.01	2.85	2.74	2.66	2.59	2.54	2.49	2.46	2.42	2.40	2.37	2.35	2.33	2.32	2.30	2.29	2.28	2.23	2.19	2.15	2.12	2.07	2.05	2.04
17	4.45	3.59	3.20	2.96	2.81	2.70	2.61	2.55	2.49	2.45	2.41	2.38	2.35	2.33	2.31	2.29	2.27	2.26	2.24	2.23	2.18	2.15	2.10	2.08	2.02	2.00	1.99
18	4.41	3.55	3.16	2.93	2.77	2.66	2.58	2.51	2.46	2.41	2.37	2.34	2.31	2.29	2.27	2.25	2.23	2.22	2.20	2.19	2.14	2.11	2.06	2.04	1.98	1.96	1.95
19	4.38	3.52	3.13	2.90	2.74	2.63	2.54	2.48	2.42	2.38	2.34	2.31	2.28	2.26	2.23	2.21	2.20	2.18	2.17	2.16	2.11	2.07	2.03	2.00	1.94	1.92	1.91
20	4.35	3.49	3.10	2.87	2.71	2.60	2.51	2.45	2.39	2.35	2.31	2.28	2.25	2.22	2.20	2.18	2.17	2.15	2.14	2.12	2.07	2.04	1.99	1.97	1.91	1.89	1.88
21	4.32	3.47	3.07	2.84	2.68	2.57	2.49	2.42	2.37	2.32	2.28	2.25	2.22	2.20	2.18	2.16	2.14	2.12	2.11	2.10	2.05	2.01	1.96	1.94	1.88	1.86	1.84
22	4.30	3.44	3.05	2.82	2.66	2.55	2.46	2.40	2.34	2.30	2.26	2.23	2.20	2.17	2.15	2.13	2.11	2.10	2.08	2.07	2.02	1.98	1.94	1.91	1.85	1.83	1.82
23	4.28	3.42	3.03	2.80	2.64	2.53	2.44	2.37	2.32	2.27	2.24	2.20	2.18	2.15	2.13	2.11	2.09	2.08	2.06	2.05	2.00	1.96	1.91	1.88	1.82	1.80	1.79
24	4.26	3.40	3.01	2.78	2.62	2.51	2.42	2.36	2.30	2.25	2.22	2.18	2.15	2.13	2.11	2.09	2.07	2.05	2.04	2.03	1.97	1.94	1.89	1.86	1.80	1.78	1.77
25	4.24	3.39	2.99	2.76	2.60	2.49	2.40	2.34	2.28	2.24	2.20	2.16	2.14	2.11	2.09	2.07	2.05	2.04	2.02	2.01	1.96	1.92	1.87	1.84	1.78	1.76	1.75
26	4.23	3.37	2.98	2.74	2.59	2.47	2.39	2.32	2.27	2.22	2.18	2.15	2.12	2.09	2.07	2.05	2.03	2.02	2.00	1.99	1.94	1.90	1.85	1.82	1.76	1.74	1.73
27	4.21	3.35	2.96	2.73	2.57	2.46	2.37	2.31	2.25	2.20	2.17	2.13	2.10	2.08	2.06	2.04	2.02	2.00	1.99	1.97	1.92	1.88	1.84	1.81	1.74	1.72	1.71
28	4.20	3.34	2.95	2.71	2.56	2.45	2.36	2.29	2.24	2.19	2.15	2.12	2.09	2.06	2.04	2.02	2.00	1.99	1.97	1.96	1.91	1.87	1.82	1.79	1.73	1.70	1.69
29	4.18	3.33	2.93	2.70	2.55	2.43	2.35	2.28	2.22	2.18	2.14	2.10	2.08	2.05	2.03	2.01	1.99	1.97	1.96	1.94	1.89	1.85	1.81	1.77	1.71	1.69	1.67
30	4.17	3.32	2.92	2.69	2.53	2.42	2.33	2.27	2.21	2.16	2.13	2.09	2.06	2.04	2.01	1.99	1.98	1.96	1.95	1.93	1.88	1.84	1.79	1.76	1.70	1.67	1.66
32	4.15	3.29	2.90	2.67	2.51	2.40	2.31	2.24	2.19	2.14	2.10	2.07	2.04	2.01	1.99	1.97	1.95	1.94	1.92	1.91	1.85	1.82	1.77	1.74	1.67	1.64	1.63
34	4.13	3.28	2.88	2.65	2.49	2.38	2.29	2.23	2.17	2.12	2.08	2.05	2.02	1.99	1.97	1.95	1.93	1.92	1.90	1.89	1.83	1.80	1.75	1.71	1.65	1.62	1.61
36	4.11	3.26	2.87	2.63	2.48	2.36	2.28	2.21	2.15	2.11	2.07	2.03	2.00	1.98	1.95	1.93	1.92	1.90	1.88	1.87	1.81	1.78	1.73	1.69	1.62	1.60	1.59
38	4.10	3.24	2.85	2.62	2.46	2.35	2.26	2.19	2.14	2.09	2.05	2.02	1.99	1.96	1.94	1.92	1.90	1.88	1.87	1.85	1.80	1.76	1.71	1.68	1.61	1.58	1.57
40	4.08	3.23	2.84	2.61	2.45	2.34	2.25	2.18	2.12	2.08	2.04	2.00	1.97	1.95	1.92	1.90	1.89	1.87	1.85	1.84	1.78	1.74	1.69	1.66	1.59	1.56	1.55
42	4.07	3.22	2.83	2.59	2.44	2.32	2.24	2.17	2.11	2.06	2.03	1.99	1.96	1.94	1.91	1.89	1.87	1.86	1.84	1.83	1.77	1.73	1.68	1.65	1.57	1.55	1.53
44	4.06	3.21	2.82	2.58	2.43	2.31	2.23	2.16	2.10	2.05	2.01	1.98	1.95	1.92	1.90	1.88	1.86	1.84	1.83	1.81	1.76	1.72	1.67	1.63	1.56	1.53	1.52
46	4.05	3.20	2.81	2.57	2.42	2.30	2.22	2.15	2.09	2.04	2.00	1.97	1.94	1.91	1.89	1.87	1.85	1.83	1.82	1.80	1.75	1.71	1.65	1.62	1.55	1.52	1.51
48	4.04	3.19	2.80	2.57	2.41	2.29	2.21	2.14	2.08	2.03	1.99	1.96	1.93	1.90	1.88	1.86	1.84	1.82	1.81	1.79	1.74	1.70	1.64	1.61	1.54	1.51	1.49
50	4.03	3.18	2.79	2.56	2.40	2.29	2.20	2.13	2.07	2.03	1.99	1.95	1.92	1.89	1.87	1.85	1.83	1.81	1.80	1.78	1.73	1.69	1.63	1.60	1.52	1.50	1.48
60	4.00	3.15	2.76	2.53	2.37	2.25	2.17	2.10	2.04	1.99	1.95	1.92	1.89	1.86	1.84	1.82	1.80	1.78	1.76	1.75	1.69	1.65	1.59	1.56	1.48	1.45	1.44
70	3.98	3.13	2.74	2.50	2.35	2.23	2.14	2.07	2.02	1.97	1.93	1.89	1.86	1.84	1.81	1.79	1.77	1.75	1.74	1.72	1.66	1.62	1.57	1.53	1.45	1.42	1.40
80	3.96	3.11	2.72	2.49	2.33	2.21	2.13	2.06	2.00	1.95	1.91	1.88	1.84	1.82	1.79	1.77	1.75	1.73	1.72	1.70	1.64	1.60	1.54	1.51	1.43	1.39	1.38
90	3.95	3.10	2.71	2.47	2.32	2.20	2.11	2.04	1.99	1.94	1.90	1.86	1.83	1.80	1.78	1.76	1.74	1.72	1.70	1.69	1.63	1.59	1.53	1.49	1.41	1.38	1.36
100	3.94	3.09	2.70	2.46	2.31	2.19	2.10	2.03	1.97	1.93	1.89	1.85	1.82	1.79	1.77	1.75	1.73	1.71	1.69	1.68	1.62	1.57	1.52	1.48	1.39	1.36	1.34
125	3.92	3.07	2.68	2.44	2.29	2.17	2.08	2.01	1.96	1.91	1.87	1.83	1.80	1.77	1.75	1.73	1.71	1.69	1.67	1.66	1.60	1.55	1.49	1.45	1.36	1.33	1.31
150	3.90	3.06	2.66	2.43	2.27	2.16	2.07	2.00	1.94	1.89	1.85	1.82	1.79	1.76	1.73	1.71	1.69	1.67	1.66	1.64	1.58	1.54	1.48	1.44	1.34	1.31	1.29
200	3.89	3.04	2.65	2.42	2.26	2.14	2.06	1.98	1.93	1.88	1.84	1.80	1.77	1.74	1.72	1.69	1.67	1.66	1.64	1.62	1.56	1.52	1.46	1.41	1.32	1.28	1.26
300	3.87	3.03	2.63	2.40	2.24	2.13	2.04	1.97	1.91	1.86	1.82	1.78	1.75	1.72	1.70	1.68	1.66	1.64	1.62	1.61	1.54	1.50	1.43	1.39	1.30	1.26	1.23
500	3.86	3.01	2.62	2.39	2.23	2.12	2.03	1.96	1.90	1.85	1.81	1.77	1.74	1.71	1.69	1.66	1.64	1.62	1.61	1.59	1.53	1.48	1.42	1.38	1.28	1.23	1.21
1000	3.85	3.00	2.61	2.38	2.22	2.11	2.02	1.95	1.89	1.84	1.80	1.76	1.73	1.70	1.68	1.65	1.63	1.61	1.60	1.58	1.52	1.47	1.41	1.36	1.26	1.22	1.19

DEGREES OF FREEDOM FOR DENOMINATOR

TABLE A-4 Percentiles of the F distribution (continued)

Upper 2.5% point of the F distribution

DEGREES OF FREEDOM FOR DENOMINATOR (rows) × DEGREES OF FREEDOM FOR NUMERATOR (columns)

den\num	1	2	3	4	5	6	7	8	9	10	11	12	13	14	15	16	17	18	19	20	25	30	40	50	100	150	200
1	648	800	864	900	922	937	948	957	963	969	973	977	980	983	985	987	989	990	992	993	998	1001	1006	1008	1013	1015	1016
2	38.5	39.0	39.2	39.2	39.3	39.3	39.4	39.4	39.4	39.4	39.4	39.4	39.4	39.4	39.4	39.4	39.4	39.4	39.4	39.4	39.5	39.5	39.5	39.5	39.5	39.5	39.5
3	17.4	16.0	15.4	15.1	14.9	14.7	14.6	14.5	14.5	14.4	14.4	14.3	14.3	14.3	14.3	14.2	14.2	14.2	14.2	14.2	14.1	14.1	14.0	14.0	14.0	13.9	13.9
4	12.2	10.6	9.98	9.60	9.36	9.20	9.07	8.98	8.90	8.84	8.79	8.75	8.71	8.68	8.66	8.63	8.61	8.59	8.58	8.56	8.50	8.46	8.41	8.38	8.32	8.30	8.29
5	10.0	8.43	7.76	7.39	7.15	6.98	6.85	6.76	6.68	6.62	6.57	6.52	6.49	6.46	6.43	6.40	6.38	6.36	6.34	6.33	6.27	6.23	6.18	6.14	6.08	6.06	6.05
6	8.81	7.26	6.60	6.23	5.99	5.82	5.70	5.60	5.52	5.46	5.41	5.37	5.33	5.30	5.27	5.24	5.22	5.20	5.18	5.17	5.11	5.07	5.01	4.98	4.92	4.89	4.88
7	8.07	6.54	5.89	5.52	5.29	5.12	4.99	4.90	4.82	4.76	4.71	4.67	4.63	4.60	4.57	4.54	4.52	4.50	4.48	4.47	4.40	4.36	4.31	4.28	4.21	4.19	4.18
8	7.57	6.06	5.42	5.05	4.82	4.65	4.53	4.43	4.36	4.30	4.24	4.20	4.16	4.13	4.10	4.08	4.05	4.03	4.02	4.00	3.94	3.89	3.84	3.81	3.74	3.72	3.70
9	7.21	5.71	5.08	4.72	4.48	4.32	4.20	4.10	4.03	3.96	3.91	3.87	3.83	3.80	3.77	3.74	3.72	3.70	3.68	3.67	3.60	3.56	3.51	3.47	3.40	3.38	3.37
10	6.94	5.46	4.83	4.47	4.24	4.07	3.95	3.85	3.78	3.72	3.66	3.62	3.58	3.55	3.52	3.50	3.47	3.45	3.44	3.42	3.35	3.31	3.26	3.22	3.15	3.13	3.12
11	6.72	5.26	4.63	4.28	4.04	3.88	3.76	3.66	3.59	3.53	3.47	3.43	3.39	3.36	3.33	3.30	3.28	3.26	3.24	3.23	3.16	3.12	3.06	3.03	2.96	2.93	2.92
12	6.55	5.10	4.47	4.12	3.89	3.73	3.61	3.51	3.44	3.37	3.32	3.28	3.24	3.21	3.18	3.15	3.13	3.11	3.09	3.07	3.01	2.96	2.91	2.87	2.80	2.78	2.76
13	6.41	4.97	4.35	4.00	3.77	3.60	3.48	3.39	3.31	3.25	3.20	3.15	3.12	3.08	3.05	3.03	3.00	2.98	2.96	2.95	2.88	2.84	2.78	2.74	2.67	2.65	2.63
14	6.30	4.86	4.24	3.89	3.66	3.50	3.38	3.29	3.21	3.15	3.09	3.05	3.01	2.98	2.95	2.92	2.90	2.88	2.86	2.84	2.78	2.73	2.67	2.64	2.56	2.54	2.53
15	6.20	4.77	4.15	3.80	3.58	3.41	3.29	3.20	3.12	3.06	3.01	2.96	2.92	2.89	2.86	2.84	2.81	2.79	2.77	2.76	2.69	2.64	2.59	2.55	2.47	2.45	2.44
16	6.12	4.69	4.08	3.73	3.50	3.34	3.22	3.12	3.05	2.99	2.93	2.89	2.85	2.82	2.79	2.76	2.74	2.72	2.70	2.68	2.61	2.57	2.51	2.47	2.40	2.37	2.36
17	6.04	4.62	4.01	3.66	3.44	3.28	3.16	3.06	2.98	2.92	2.87	2.82	2.79	2.75	2.72	2.70	2.67	2.65	2.63	2.62	2.55	2.50	2.44	2.41	2.33	2.30	2.29
18	5.98	4.56	3.95	3.61	3.38	3.22	3.10	3.01	2.93	2.87	2.81	2.77	2.73	2.70	2.67	2.64	2.62	2.60	2.58	2.56	2.49	2.44	2.38	2.35	2.27	2.24	2.23
19	5.92	4.51	3.90	3.56	3.33	3.17	3.05	2.96	2.88	2.82	2.76	2.72	2.68	2.65	2.62	2.59	2.57	2.55	2.53	2.51	2.44	2.39	2.33	2.30	2.22	2.19	2.18
20	5.87	4.46	3.86	3.51	3.29	3.13	3.01	2.91	2.84	2.77	2.72	2.68	2.64	2.60	2.57	2.55	2.52	2.50	2.48	2.46	2.40	2.35	2.29	2.25	2.17	2.14	2.13
21	5.83	4.42	3.82	3.48	3.25	3.09	2.97	2.87	2.80	2.73	2.68	2.64	2.60	2.56	2.53	2.51	2.48	2.46	2.44	2.42	2.36	2.31	2.25	2.21	2.13	2.10	2.09
22	5.79	4.38	3.78	3.44	3.22	3.05	2.93	2.84	2.76	2.70	2.65	2.60	2.56	2.53	2.50	2.47	2.45	2.43	2.41	2.39	2.32	2.27	2.21	2.17	2.09	2.06	2.05
23	5.75	4.35	3.75	3.41	3.18	3.02	2.90	2.81	2.73	2.67	2.62	2.57	2.53	2.50	2.47	2.44	2.42	2.39	2.37	2.36	2.29	2.24	2.18	2.14	2.06	2.03	2.01
24	5.72	4.32	3.72	3.38	3.15	2.99	2.87	2.78	2.70	2.64	2.59	2.54	2.50	2.47	2.44	2.41	2.39	2.36	2.35	2.33	2.26	2.21	2.15	2.11	2.02	2.00	1.98
25	5.69	4.29	3.69	3.35	3.13	2.97	2.85	2.75	2.68	2.61	2.56	2.51	2.48	2.44	2.41	2.38	2.36	2.34	2.32	2.30	2.23	2.18	2.12	2.08	2.00	1.97	1.95
26	5.66	4.27	3.67	3.33	3.10	2.94	2.82	2.73	2.65	2.59	2.54	2.49	2.45	2.42	2.39	2.36	2.34	2.31	2.29	2.28	2.21	2.16	2.09	2.05	1.97	1.94	1.92
27	5.63	4.24	3.65	3.31	3.08	2.92	2.80	2.71	2.63	2.57	2.51	2.47	2.43	2.39	2.36	2.34	2.31	2.29	2.27	2.25	2.18	2.13	2.07	2.03	1.94	1.91	1.90
28	5.61	4.22	3.63	3.29	3.06	2.90	2.78	2.69	2.61	2.55	2.49	2.45	2.41	2.37	2.34	2.32	2.29	2.27	2.25	2.23	2.16	2.11	2.05	2.01	1.92	1.89	1.88
29	5.59	4.20	3.61	3.27	3.04	2.88	2.76	2.67	2.59	2.53	2.48	2.43	2.39	2.36	2.32	2.30	2.27	2.25	2.23	2.21	2.14	2.09	2.03	1.99	1.90	1.87	1.86
30	5.57	4.18	3.59	3.25	3.03	2.87	2.75	2.65	2.57	2.51	2.46	2.41	2.37	2.34	2.31	2.28	2.26	2.23	2.21	2.20	2.12	2.07	2.01	1.97	1.88	1.85	1.84
32	5.53	4.15	3.56	3.22	3.00	2.84	2.71	2.62	2.54	2.48	2.43	2.38	2.34	2.31	2.28	2.25	2.22	2.20	2.18	2.16	2.09	2.04	1.98	1.93	1.85	1.82	1.80
34	5.50	4.12	3.53	3.19	2.97	2.81	2.69	2.59	2.52	2.45	2.40	2.35	2.31	2.28	2.25	2.22	2.20	2.17	2.15	2.13	2.06	2.01	1.95	1.90	1.82	1.78	1.77
36	5.47	4.09	3.50	3.17	2.94	2.78	2.66	2.57	2.49	2.43	2.37	2.33	2.29	2.25	2.22	2.20	2.17	2.15	2.13	2.11	2.04	1.99	1.92	1.88	1.79	1.76	1.74
38	5.45	4.07	3.48	3.15	2.92	2.76	2.64	2.55	2.47	2.41	2.35	2.31	2.27	2.23	2.20	2.17	2.15	2.13	2.11	2.09	2.01	1.96	1.90	1.85	1.76	1.73	1.71
40	5.42	4.05	3.46	3.13	2.90	2.74	2.62	2.53	2.45	2.39	2.33	2.29	2.25	2.21	2.18	2.15	2.13	2.11	2.09	2.07	1.99	1.94	1.88	1.83	1.74	1.71	1.69
42	5.40	4.03	3.45	3.11	2.89	2.73	2.61	2.51	2.43	2.37	2.32	2.27	2.23	2.20	2.16	2.14	2.11	2.09	2.07	2.05	1.98	1.92	1.86	1.81	1.72	1.69	1.67
44	5.39	4.02	3.43	3.09	2.87	2.71	2.59	2.50	2.42	2.36	2.30	2.26	2.22	2.18	2.15	2.11	2.10	2.07	2.05	2.03	1.96	1.91	1.84	1.80	1.70	1.67	1.65
46	5.37	4.00	3.42	3.08	2.86	2.70	2.58	2.48	2.41	2.34	2.29	2.24	2.20	2.17	2.13	2.11	2.08	2.06	2.04	2.02	1.94	1.89	1.82	1.78	1.69	1.65	1.63
48	5.35	3.99	3.40	3.07	2.84	2.69	2.56	2.47	2.39	2.33	2.27	2.23	2.19	2.15	2.12	2.09	2.07	2.05	2.02	2.01	1.93	1.88	1.81	1.77	1.67	1.64	1.62
50	5.34	3.97	3.39	3.05	2.83	2.67	2.55	2.46	2.38	2.32	2.26	2.22	2.18	2.14	2.11	2.08	2.06	2.03	2.01	1.99	1.92	1.87	1.80	1.75	1.66	1.62	1.60
60	5.29	3.93	3.34	3.01	2.79	2.63	2.51	2.41	2.33	2.27	2.22	2.17	2.13	2.09	2.06	2.03	2.01	1.98	1.96	1.94	1.87	1.82	1.74	1.70	1.60	1.56	1.54
70	5.25	3.89	3.31	2.97	2.75	2.59	2.47	2.38	2.30	2.24	2.18	2.14	2.10	2.06	2.03	2.00	1.97	1.95	1.93	1.91	1.83	1.78	1.71	1.66	1.56	1.52	1.50
80	5.22	3.86	3.28	2.95	2.73	2.57	2.45	2.35	2.28	2.21	2.16	2.11	2.07	2.03	2.00	1.97	1.95	1.92	1.90	1.88	1.81	1.75	1.68	1.63	1.53	1.49	1.47
90	5.20	3.84	3.26	2.93	2.71	2.55	2.43	2.34	2.26	2.19	2.14	2.09	2.05	2.02	1.98	1.95	1.93	1.91	1.88	1.86	1.79	1.73	1.66	1.61	1.50	1.46	1.44
100	5.18	3.83	3.25	2.92	2.70	2.54	2.42	2.32	2.24	2.18	2.12	2.08	2.04	2.00	1.97	1.94	1.91	1.89	1.87	1.85	1.77	1.71	1.64	1.59	1.48	1.44	1.42
125	5.15	3.80	3.22	2.89	2.67	2.51	2.39	2.30	2.22	2.15	2.10	2.05	2.01	1.97	1.94	1.91	1.89	1.86	1.84	1.82	1.74	1.68	1.61	1.56	1.45	1.40	1.38
150	5.13	3.78	3.20	2.87	2.65	2.49	2.37	2.28	2.20	2.13	2.08	2.03	1.99	1.95	1.92	1.89	1.87	1.84	1.82	1.80	1.72	1.67	1.59	1.54	1.42	1.38	1.35
200	5.10	3.76	3.18	2.85	2.63	2.47	2.35	2.26	2.18	2.11	2.06	2.01	1.97	1.93	1.90	1.87	1.84	1.82	1.80	1.78	1.70	1.64	1.56	1.51	1.39	1.35	1.32
300	5.07	3.73	3.16	2.83	2.61	2.45	2.33	2.23	2.16	2.09	2.04	1.99	1.95	1.91	1.88	1.85	1.82	1.80	1.77	1.75	1.67	1.62	1.54	1.48	1.36	1.31	1.28
500	5.05	3.72	3.14	2.81	2.59	2.43	2.31	2.22	2.14	2.07	2.02	1.97	1.93	1.89	1.86	1.83	1.80	1.78	1.76	1.74	1.65	1.60	1.52	1.46	1.34	1.28	1.25
1000	5.04	3.70	3.13	2.80	2.58	2.42	2.30	2.20	2.13	2.06	2.01	1.96	1.92	1.88	1.85	1.82	1.79	1.77	1.74	1.72	1.64	1.58	1.50	1.45	1.32	1.26	1.23

TABLE A-4 Percentiles of the F distribution (continued)

Upper 1% point of the F distribution

											DEGREES OF FREEDOM FOR NUMERATOR																	
	1	2	3	4	5	6	7	8	9	10	11	12	13	14	15	16	17	18	19	20	25	30	40	50	100	150	200
1	4052	5000	5403	5625	5764	5859	5928	5981	6022	6056	6083	6106	6126	6143	6157	6170	6181	6192	6201	6209	6240	6261	6287	6303	6334	6345	6350
2	98.5	99.2	99.2	99.2	99.3	99.3	99.4	99.4	99.4	99.4	99.4	99.4	99.4	99.4	99.4	99.4	99.4	99.4	99.4	99.4	99.5	99.5	99.5	99.5	99.5	99.5	99.5
3	34.1	30.8	29.5	28.7	28.2	27.9	27.7	27.5	27.3	27.2	27.1	27.1	27.0	26.9	26.9	26.8	26.8	26.8	26.7	26.7	26.6	26.5	26.4	26.4	26.2	26.2	26.2
4	21.2	18.0	16.7	16.0	15.5	15.2	15.0	14.8	14.7	14.5	14.5	14.4	14.3	14.2	14.2	14.2	14.1	14.1	14.0	14.0	13.9	13.8	13.7	13.7	13.6	13.6	13.5
5	16.3	13.3	12.1	11.4	11.0	10.7	10.5	10.3	10.2	10.1	9.96	9.89	9.82	9.77	9.72	9.68	9.64	9.61	9.58	9.55	9.45	9.38	9.29	9.24	9.13	9.09	9.08
6	13.7	10.9	9.78	9.15	8.75	8.47	8.26	8.10	7.98	7.87	7.79	7.72	7.66	7.60	7.56	7.52	7.48	7.45	7.42	7.40	7.30	7.23	7.14	7.09	6.99	6.95	6.93
7	12.2	9.55	8.45	7.85	7.46	7.19	6.99	6.84	6.72	6.62	6.54	6.47	6.41	6.36	6.31	6.28	6.24	6.21	6.18	6.16	6.06	5.99	5.91	5.86	5.75	5.72	5.70
8	11.3	8.65	7.59	7.01	6.63	6.37	6.18	6.03	5.91	5.81	5.73	5.67	5.61	5.56	5.52	5.48	5.44	5.41	5.38	5.36	5.26	5.20	5.12	5.07	4.96	4.93	4.91
9	10.6	8.02	6.99	6.42	6.06	5.80	5.61	5.47	5.35	5.26	5.18	5.11	5.05	5.01	4.96	4.92	4.89	4.86	4.83	4.81	4.71	4.65	4.57	4.52	4.41	4.38	4.36
10	10.0	7.56	6.55	5.99	5.64	5.39	5.20	5.06	4.94	4.85	4.77	4.71	4.65	4.60	4.56	4.52	4.49	4.46	4.43	4.41	4.31	4.25	4.17	4.12	4.01	3.98	3.96
11	9.65	7.21	6.22	5.67	5.32	5.07	4.89	4.74	4.63	4.54	4.46	4.40	4.34	4.29	4.25	4.21	4.18	4.15	4.12	4.10	4.01	3.94	3.86	3.81	3.71	3.67	3.66
12	9.33	6.93	5.95	5.41	5.06	4.82	4.64	4.50	4.39	4.30	4.22	4.16	4.10	4.05	4.01	3.97	3.94	3.91	3.88	3.86	3.76	3.70	3.62	3.57	3.47	3.43	3.41
13	9.07	6.70	5.74	5.21	4.86	4.62	4.44	4.30	4.19	4.10	4.02	3.96	3.91	3.86	3.82	3.78	3.75	3.72	3.69	3.66	3.57	3.51	3.43	3.38	3.27	3.24	3.22
14	8.88	6.51	5.56	5.04	4.69	4.46	4.28	4.14	4.03	3.94	3.86	3.80	3.75	3.70	3.66	3.62	3.59	3.56	3.53	3.51	3.41	3.35	3.27	3.22	3.11	3.08	3.06
15	8.68	6.36	5.42	4.89	4.56	4.32	4.14	4.00	3.89	3.80	3.73	3.67	3.61	3.56	3.52	3.49	3.45	3.42	3.40	3.37	3.28	3.21	3.13	3.08	2.98	2.94	2.92
16	8.53	6.23	5.29	4.77	4.44	4.20	4.03	3.89	3.78	3.69	3.62	3.55	3.50	3.45	3.41	3.37	3.34	3.31	3.28	3.26	3.16	3.10	3.02	2.97	2.86	2.83	2.81
17	8.40	6.11	5.19	4.67	4.34	4.10	3.93	3.79	3.68	3.59	3.52	3.46	3.40	3.35	3.31	3.27	3.24	3.21	3.19	3.16	3.07	3.00	2.92	2.87	2.76	2.73	2.71
18	8.29	6.01	5.09	4.58	4.25	4.01	3.84	3.71	3.60	3.51	3.43	3.37	3.32	3.27	3.23	3.19	3.16	3.13	3.10	3.08	2.98	2.92	2.84	2.78	2.68	2.64	2.62
19	8.18	5.93	5.01	4.50	4.17	3.94	3.77	3.63	3.52	3.43	3.36	3.30	3.24	3.19	3.15	3.12	3.08	3.05	3.03	3.00	2.91	2.84	2.76	2.71	2.60	2.57	2.55
20	8.10	5.85	4.94	4.43	4.10	3.87	3.70	3.56	3.46	3.37	3.29	3.23	3.18	3.13	3.09	3.05	3.02	2.99	2.96	2.94	2.84	2.78	2.69	2.64	2.54	2.50	2.48
21	8.02	5.78	4.87	4.37	4.04	3.81	3.64	3.51	3.40	3.31	3.24	3.17	3.12	3.07	3.03	2.99	2.96	2.93	2.90	2.88	2.79	2.72	2.64	2.58	2.48	2.44	2.42
22	7.95	5.72	4.82	4.31	3.99	3.76	3.59	3.45	3.35	3.26	3.18	3.12	3.07	3.02	2.98	2.94	2.91	2.88	2.85	2.83	2.73	2.67	2.58	2.53	2.42	2.38	2.36
23	7.88	5.66	4.76	4.26	3.94	3.71	3.54	3.41	3.30	3.21	3.14	3.07	3.02	2.97	2.93	2.89	2.86	2.83	2.80	2.78	2.69	2.62	2.54	2.48	2.37	2.34	2.32
24	7.82	5.61	4.72	4.22	3.90	3.67	3.50	3.36	3.26	3.17	3.09	3.03	2.98	2.93	2.89	2.85	2.82	2.79	2.76	2.74	2.64	2.58	2.49	2.44	2.33	2.29	2.27
25	7.77	5.57	4.68	4.18	3.85	3.63	3.46	3.32	3.22	3.13	3.06	2.99	2.94	2.89	2.85	2.81	2.78	2.75	2.72	2.70	2.60	2.54	2.45	2.40	2.29	2.25	2.23
26	7.72	5.53	4.64	4.14	3.82	3.59	3.42	3.29	3.18	3.09	3.02	2.96	2.90	2.86	2.81	2.78	2.75	2.72	2.69	2.66	2.57	2.50	2.42	2.36	2.25	2.21	2.19
27	7.68	5.49	4.60	4.11	3.78	3.56	3.39	3.26	3.15	3.06	2.99	2.93	2.87	2.82	2.78	2.75	2.71	2.68	2.66	2.63	2.54	2.47	2.38	2.33	2.22	2.18	2.16
28	7.64	5.45	4.57	4.07	3.75	3.53	3.36	3.23	3.12	3.03	2.96	2.90	2.84	2.79	2.75	2.72	2.68	2.65	2.63	2.60	2.51	2.44	2.35	2.30	2.19	2.15	2.13
29	7.60	5.42	4.54	4.04	3.73	3.50	3.33	3.20	3.09	3.00	2.93	2.87	2.81	2.77	2.73	2.69	2.66	2.63	2.60	2.57	2.48	2.41	2.33	2.27	2.16	2.12	2.10
30	7.56	5.39	4.51	4.02	3.70	3.47	3.30	3.17	3.07	2.98	2.91	2.84	2.79	2.74	2.70	2.66	2.63	2.60	2.57	2.55	2.45	2.39	2.30	2.25	2.13	2.09	2.07
32	7.50	5.34	4.46	3.97	3.65	3.43	3.26	3.13	3.02	2.93	2.86	2.80	2.74	2.70	2.65	2.62	2.58	2.55	2.53	2.50	2.41	2.34	2.25	2.20	2.08	2.04	2.02
34	7.44	5.29	4.42	3.93	3.61	3.39	3.22	3.09	2.98	2.89	2.82	2.76	2.70	2.66	2.61	2.58	2.54	2.51	2.49	2.46	2.37	2.30	2.21	2.16	2.04	2.00	1.98
36	7.40	5.25	4.38	3.89	3.57	3.35	3.18	3.05	2.95	2.86	2.79	2.72	2.67	2.62	2.58	2.54	2.51	2.48	2.45	2.43	2.33	2.26	2.18	2.12	2.00	1.96	1.94
38	7.35	5.21	4.34	3.86	3.54	3.32	3.15	3.02	2.92	2.83	2.75	2.69	2.64	2.59	2.55	2.51	2.48	2.45	2.42	2.40	2.30	2.23	2.14	2.09	1.97	1.93	1.90
40	7.31	5.18	4.31	3.83	3.51	3.29	3.12	2.99	2.89	2.80	2.73	2.66	2.61	2.56	2.52	2.48	2.45	2.42	2.39	2.37	2.27	2.20	2.11	2.06	1.94	1.90	1.87
42	7.28	5.15	4.29	3.80	3.49	3.27	3.10	2.97	2.86	2.78	2.70	2.64	2.59	2.54	2.50	2.46	2.43	2.40	2.37	2.34	2.25	2.18	2.09	2.03	1.91	1.87	1.85
44	7.25	5.12	4.26	3.78	3.47	3.24	3.08	2.95	2.84	2.75	2.68	2.62	2.56	2.52	2.47	2.44	2.40	2.37	2.35	2.32	2.22	2.15	2.07	2.01	1.89	1.84	1.82
46	7.22	5.10	4.24	3.76	3.44	3.22	3.06	2.93	2.82	2.73	2.66	2.60	2.54	2.50	2.45	2.42	2.38	2.35	2.33	2.30	2.20	2.13	2.04	1.99	1.86	1.82	1.80
48	7.19	5.08	4.22	3.74	3.43	3.20	3.04	2.91	2.80	2.71	2.64	2.58	2.53	2.48	2.44	2.40	2.37	2.33	2.31	2.28	2.18	2.12	2.02	1.97	1.84	1.80	1.78
50	7.17	5.06	4.20	3.72	3.41	3.19	3.02	2.89	2.78	2.70	2.63	2.56	2.51	2.46	2.42	2.38	2.35	2.32	2.29	2.27	2.17	2.10	2.01	1.95	1.82	1.78	1.76
60	7.08	4.98	4.13	3.65	3.34	3.12	2.95	2.82	2.72	2.63	2.56	2.50	2.44	2.39	2.35	2.31	2.28	2.25	2.22	2.20	2.10	2.03	1.94	1.88	1.75	1.70	1.68
70	7.01	4.92	4.07	3.60	3.29	3.07	2.91	2.78	2.67	2.59	2.51	2.45	2.40	2.35	2.31	2.27	2.23	2.20	2.18	2.15	2.05	1.98	1.89	1.83	1.70	1.65	1.62
80	6.96	4.88	4.04	3.56	3.26	3.04	2.87	2.74	2.64	2.55	2.48	2.42	2.36	2.31	2.27	2.23	2.20	2.17	2.14	2.12	2.01	1.94	1.85	1.79	1.65	1.61	1.58
90	6.93	4.85	4.01	3.53	3.23	3.01	2.84	2.72	2.61	2.52	2.45	2.39	2.33	2.29	2.24	2.21	2.17	2.14	2.11	2.09	1.99	1.92	1.82	1.76	1.62	1.57	1.55
100	6.90	4.82	3.98	3.51	3.21	2.99	2.82	2.69	2.59	2.50	2.43	2.37	2.31	2.27	2.22	2.19	2.15	2.12	2.09	2.07	1.97	1.89	1.80	1.74	1.60	1.55	1.52
125	6.84	4.78	3.94	3.47	3.17	2.95	2.79	2.66	2.55	2.47	2.39	2.33	2.28	2.23	2.19	2.15	2.11	2.08	2.05	2.03	1.93	1.85	1.76	1.69	1.55	1.50	1.47
150	6.81	4.75	3.91	3.45	3.14	2.92	2.76	2.63	2.53	2.44	2.37	2.31	2.25	2.20	2.16	2.12	2.09	2.06	2.03	2.00	1.90	1.83	1.73	1.66	1.52	1.46	1.43
200	6.76	4.71	3.88	3.41	3.11	2.89	2.73	2.60	2.50	2.41	2.34	2.27	2.22	2.17	2.13	2.09	2.06	2.03	2.00	1.97	1.87	1.79	1.69	1.63	1.48	1.42	1.39
300	6.72	4.68	3.85	3.38	3.08	2.86	2.70	2.57	2.47	2.38	2.31	2.24	2.19	2.14	2.10	2.06	2.03	1.99	1.97	1.94	1.84	1.76	1.66	1.59	1.44	1.38	1.35
500	6.69	4.65	3.82	3.36	3.05	2.84	2.68	2.55	2.44	2.36	2.28	2.22	2.17	2.12	2.07	2.04	2.00	1.97	1.94	1.92	1.81	1.74	1.63	1.57	1.41	1.34	1.31
1000	6.66	4.63	3.80	3.34	3.04	2.82	2.66	2.53	2.43	2.34	2.27	2.20	2.15	2.10	2.06	2.02	1.98	1.95	1.92	1.90	1.79	1.72	1.61	1.54	1.38	1.32	1.28

DEGREES OF FREEDOM FOR DENOMINATOR

TABLE A-4 Percentiles of the F distribution (continued)

Upper 0.5% point of the F distribution

DEGREES OF FREEDOM FOR NUMERATOR

df₂	1	2	3	4	5	6	7	8	9	10	11	12	13	14	15	16	17	18	19	20	25	30	40	50	100	150	200
1	****	****	****	****	****	****	****	****	****	****	****	****	****	****	****	****	****	****	****	****	****	****	****	****	****	****	****
2	199	199	199	199	199	199	199	199	199	199	199	199	199	199	199	199	199	199	199	199	199	199	199	199	199	199	199
3	55.6	49.8	47.5	46.2	45.4	44.8	44.4	44.1	43.9	43.7	43.5	43.4	43.3	43.2	43.1	43.0	42.9	42.9	42.8	42.8	42.6	42.5	42.3	42.2	42.0	42.0	41.9
4	31.3	26.3	24.3	23.2	22.5	21.9	21.6	21.4	21.1	21.0	20.8	20.7	20.6	20.5	20.4	20.4	20.3	20.3	20.2	20.2	20.0	19.9	19.8	19.7	19.5	19.4	19.4
5	22.8	18.3	16.5	15.6	14.9	14.5	14.2	14.0	13.8	13.6	13.5	13.4	13.3	13.2	13.1	13.1	13.0	13.0	12.9	12.9	12.8	12.7	12.5	12.5	12.3	12.2	12.2
6	18.6	14.5	12.9	12.0	11.5	11.1	10.8	10.6	10.4	10.3	10.1	10.0	9.95	9.88	9.81	9.76	9.71	9.66	9.62	9.59	9.45	9.36	9.24	9.17	9.03	8.98	8.95
7	16.2	12.4	10.9	10.1	9.52	9.16	8.89	8.68	8.51	8.38	8.27	8.18	8.10	8.03	7.97	7.91	7.87	7.83	7.79	7.75	7.62	7.53	7.42	7.35	7.22	7.17	7.15
8	14.7	11.0	9.60	8.81	8.30	7.95	7.69	7.50	7.34	7.21	7.10	7.01	6.94	6.87	6.81	6.76	6.72	6.68	6.64	6.61	6.48	6.40	6.29	6.22	6.09	6.04	6.02
9	13.6	10.1	8.72	7.96	7.47	7.13	6.88	6.69	6.54	6.42	6.31	6.23	6.15	6.09	6.03	5.98	5.94	5.90	5.86	5.83	5.71	5.62	5.52	5.45	5.32	5.28	5.26
10	12.8	9.43	8.08	7.34	6.87	6.54	6.30	6.12	5.97	5.85	5.75	5.66	5.59	5.53	5.47	5.42	5.38	5.34	5.31	5.27	5.15	5.07	4.97	4.90	4.77	4.73	4.71
11	12.2	8.91	7.60	6.88	6.42	6.10	5.86	5.68	5.54	5.42	5.32	5.24	5.16	5.10	5.05	5.00	4.96	4.92	4.89	4.86	4.74	4.65	4.55	4.49	4.36	4.31	4.29
12	11.8	8.51	7.23	6.52	6.07	5.76	5.52	5.35	5.20	5.09	4.99	4.91	4.84	4.77	4.72	4.67	4.63	4.59	4.56	4.53	4.41	4.33	4.23	4.17	4.04	3.99	3.97
13	11.4	8.19	6.93	6.23	5.79	5.48	5.25	5.08	4.94	4.82	4.72	4.64	4.57	4.51	4.46	4.41	4.37	4.33	4.30	4.27	4.15	4.07	3.97	3.91	3.78	3.74	3.71
14	11.1	7.92	6.68	6.00	5.56	5.26	5.03	4.86	4.72	4.60	4.51	4.43	4.36	4.30	4.25	4.20	4.16	4.12	4.09	4.06	3.94	3.86	3.76	3.70	3.57	3.53	3.50
15	10.8	7.70	6.48	5.80	5.37	5.07	4.85	4.67	4.54	4.42	4.33	4.25	4.18	4.12	4.07	4.02	3.98	3.95	3.91	3.88	3.77	3.69	3.58	3.52	3.39	3.35	3.33
16	10.6	7.51	6.30	5.64	5.21	4.91	4.69	4.52	4.38	4.27	4.18	4.10	4.03	3.97	3.92	3.87	3.83	3.80	3.76	3.73	3.62	3.54	3.44	3.37	3.25	3.20	3.18
17	10.4	7.35	6.16	5.50	5.07	4.78	4.56	4.39	4.25	4.14	4.05	3.97	3.90	3.84	3.79	3.75	3.71	3.67	3.64	3.61	3.49	3.41	3.31	3.25	3.12	3.07	3.05
18	10.2	7.21	6.03	5.37	4.96	4.66	4.44	4.28	4.14	4.03	3.94	3.86	3.79	3.73	3.68	3.64	3.60	3.56	3.53	3.50	3.38	3.30	3.20	3.14	3.01	2.96	2.94
19	10.1	7.09	5.92	5.27	4.85	4.56	4.34	4.18	4.04	3.93	3.84	3.76	3.70	3.64	3.59	3.54	3.50	3.46	3.43	3.40	3.29	3.21	3.11	3.04	2.91	2.87	2.85
20	9.94	6.99	5.82	5.17	4.76	4.47	4.26	4.09	3.96	3.85	3.76	3.68	3.61	3.55	3.50	3.46	3.42	3.38	3.35	3.32	3.20	3.12	3.02	2.96	2.83	2.78	2.76
21	9.83	6.89	5.73	5.09	4.68	4.39	4.18	4.01	3.88	3.77	3.68	3.60	3.54	3.48	3.43	3.38	3.34	3.31	3.27	3.24	3.13	3.05	2.95	2.88	2.75	2.71	2.68
22	9.73	6.81	5.65	5.02	4.61	4.32	4.11	3.94	3.81	3.70	3.61	3.54	3.47	3.41	3.36	3.31	3.27	3.24	3.21	3.18	3.06	2.98	2.88	2.82	2.69	2.64	2.62
23	9.63	6.73	5.58	4.95	4.54	4.26	4.05	3.88	3.75	3.64	3.55	3.47	3.41	3.35	3.30	3.25	3.21	3.18	3.15	3.12	3.00	2.92	2.82	2.76	2.62	2.58	2.56
24	9.55	6.66	5.52	4.89	4.49	4.20	3.99	3.83	3.69	3.59	3.50	3.42	3.35	3.30	3.25	3.20	3.16	3.12	3.09	3.06	2.95	2.87	2.77	2.70	2.57	2.52	2.50
25	9.48	6.60	5.46	4.84	4.43	4.15	3.94	3.78	3.64	3.54	3.45	3.37	3.30	3.25	3.20	3.15	3.11	3.08	3.04	3.01	2.90	2.82	2.72	2.65	2.52	2.47	2.45
26	9.41	6.54	5.41	4.79	4.38	4.10	3.89	3.73	3.60	3.49	3.40	3.33	3.26	3.20	3.15	3.11	3.07	3.03	3.00	2.97	2.85	2.77	2.67	2.61	2.47	2.43	2.40
27	9.34	6.49	5.36	4.74	4.34	4.06	3.85	3.69	3.56	3.45	3.36	3.28	3.22	3.16	3.11	3.07	3.03	2.99	2.96	2.93	2.81	2.73	2.63	2.57	2.43	2.38	2.36
28	9.28	6.44	5.32	4.70	4.30	4.02	3.81	3.65	3.52	3.41	3.32	3.25	3.18	3.12	3.07	3.03	2.99	2.95	2.92	2.89	2.77	2.69	2.59	2.53	2.39	2.35	2.32
29	9.23	6.40	5.28	4.66	4.26	3.98	3.77	3.61	3.48	3.38	3.29	3.21	3.15	3.09	3.04	2.99	2.95	2.92	2.88	2.86	2.74	2.66	2.56	2.49	2.36	2.31	2.29
30	9.18	6.35	5.24	4.62	4.23	3.95	3.74	3.58	3.45	3.34	3.25	3.18	3.11	3.06	3.01	2.96	2.92	2.89	2.85	2.82	2.71	2.63	2.52	2.46	2.32	2.28	2.25
32	9.09	6.28	5.17	4.56	4.17	3.89	3.68	3.52	3.39	3.29	3.20	3.12	3.06	3.00	2.95	2.90	2.86	2.83	2.80	2.77	2.65	2.57	2.47	2.40	2.26	2.22	2.19
34	9.01	6.22	5.11	4.50	4.11	3.84	3.63	3.47	3.34	3.24	3.15	3.07	3.01	2.95	2.90	2.85	2.81	2.78	2.75	2.72	2.60	2.52	2.42	2.35	2.21	2.16	2.14
36	8.94	6.16	5.06	4.46	4.06	3.79	3.58	3.42	3.30	3.19	3.10	3.03	2.96	2.90	2.85	2.81	2.77	2.73	2.70	2.67	2.56	2.48	2.37	2.30	2.17	2.12	2.09
38	8.88	6.11	5.02	4.41	4.02	3.75	3.54	3.39	3.26	3.15	3.06	2.99	2.92	2.87	2.82	2.77	2.73	2.70	2.66	2.63	2.52	2.44	2.33	2.27	2.12	2.08	2.05
40	8.83	6.07	4.98	4.37	3.99	3.71	3.51	3.35	3.22	3.12	3.03	2.95	2.89	2.83	2.78	2.74	2.70	2.66	2.63	2.60	2.48	2.40	2.30	2.23	2.09	2.04	2.01
42	8.78	6.03	4.94	4.34	3.95	3.68	3.48	3.32	3.19	3.09	3.00	2.92	2.86	2.80	2.75	2.71	2.67	2.63	2.60	2.57	2.45	2.37	2.26	2.20	2.06	2.00	1.98
44	8.74	5.99	4.91	4.31	3.92	3.65	3.45	3.29	3.16	3.06	2.97	2.89	2.83	2.77	2.72	2.68	2.64	2.60	2.57	2.54	2.42	2.34	2.24	2.17	2.03	1.97	1.95
46	8.70	5.96	4.88	4.28	3.90	3.62	3.42	3.26	3.14	3.03	2.94	2.87	2.80	2.75	2.70	2.65	2.61	2.58	2.54	2.51	2.40	2.32	2.21	2.14	2.00	1.95	1.92
48	8.66	5.93	4.85	4.25	3.87	3.60	3.40	3.24	3.11	3.01	2.92	2.85	2.78	2.72	2.67	2.63	2.59	2.55	2.52	2.49	2.37	2.29	2.19	2.12	1.97	1.92	1.90
50	8.63	5.90	4.83	4.23	3.85	3.58	3.38	3.22	3.09	2.99	2.90	2.82	2.76	2.70	2.65	2.61	2.57	2.53	2.50	2.47	2.35	2.27	2.16	2.10	1.95	1.90	1.87
60	8.49	5.79	4.73	4.14	3.76	3.49	3.29	3.13	3.01	2.90	2.82	2.74	2.68	2.62	2.57	2.53	2.49	2.45	2.42	2.39	2.27	2.19	2.08	2.01	1.86	1.81	1.78
70	8.40	5.72	4.66	4.08	3.70	3.43	3.23	3.08	2.95	2.85	2.76	2.68	2.62	2.56	2.51	2.47	2.43	2.39	2.36	2.33	2.21	2.13	2.02	1.95	1.80	1.74	1.71
80	8.33	5.67	4.61	4.03	3.65	3.39	3.19	3.03	2.91	2.80	2.72	2.64	2.58	2.52	2.47	2.43	2.39	2.35	2.32	2.29	2.17	2.08	1.97	1.90	1.75	1.69	1.66
90	8.28	5.62	4.57	3.99	3.62	3.35	3.15	3.00	2.87	2.77	2.68	2.61	2.54	2.49	2.44	2.39	2.35	2.32	2.28	2.25	2.13	2.05	1.94	1.87	1.71	1.65	1.62
100	8.24	5.59	4.54	3.96	3.59	3.33	3.13	2.97	2.85	2.74	2.66	2.58	2.52	2.46	2.41	2.37	2.33	2.29	2.26	2.23	2.11	2.02	1.91	1.84	1.68	1.62	1.59
125	8.17	5.53	4.49	3.91	3.54	3.28	3.08	2.93	2.80	2.70	2.61	2.54	2.47	2.42	2.37	2.32	2.28	2.24	2.21	2.18	2.06	1.98	1.86	1.79	1.63	1.56	1.53
150	8.12	5.49	4.45	3.88	3.51	3.25	3.05	2.89	2.77	2.67	2.58	2.51	2.44	2.38	2.33	2.29	2.25	2.21	2.18	2.15	2.03	1.94	1.83	1.76	1.59	1.53	1.49
200	8.06	5.44	4.41	3.84	3.47	3.21	3.01	2.86	2.73	2.63	2.54	2.47	2.40	2.35	2.30	2.25	2.21	2.18	2.14	2.11	1.99	1.91	1.79	1.71	1.54	1.48	1.44
300	8.00	5.39	4.36	3.80	3.43	3.17	2.97	2.82	2.69	2.59	2.51	2.43	2.37	2.31	2.26	2.21	2.17	2.14	2.10	2.07	1.95	1.87	1.75	1.67	1.50	1.43	1.39
500	7.95	5.35	4.33	3.76	3.40	3.14	2.94	2.79	2.66	2.56	2.48	2.40	2.34	2.28	2.23	2.19	2.14	2.11	2.07	2.04	1.92	1.84	1.72	1.64	1.46	1.39	1.35
1000	7.91	5.33	4.30	3.74	3.37	3.11	2.92	2.77	2.64	2.54	2.45	2.38	2.32	2.26	2.21	2.16	2.12	2.09	2.05	2.02	1.90	1.81	1.69	1.61	1.43	1.36	1.31

DEGREES OF FREEDOM FOR DENOMINATOR

TABLE A-4 Percentiles of the F distribution (continued)

Upper 0.1% point of the F distribution

DEGREES OF FREEDOM FOR NUMERATOR

df den.	1	2	3	4	5	6	7	8	9	10	11	12	13	14	15	16	17	18	19	20	25	30	40	50	100	150	200
1	****	****	****	****	****	****	****	****	****	****	****	****	****	****	****	****	****	****	****	****	****	****	****	****	****	****	****
2	999	999	999	999	999	999	999	999	999	999	999	999	999	999	999	999	999	999	999	999	999	999	999	999	999	999	999
3	167	148	141	137	135	133	132	131	130	129	129	128	128	128	128	127	127	127	127	126	126	125	125	125	124	124	124
4	74.1	61.2	56.2	53.4	51.7	50.5	49.7	49.0	48.5	48.1	47.7	47.4	47.2	46.9	46.8	46.6	46.5	46.3	46.2	46.1	45.7	45.4	45.1	44.9	44.5	44.3	44.3
5	47.2	37.1	33.2	31.1	29.8	28.8	28.2	27.6	27.2	26.9	26.6	26.4	26.2	26.1	25.9	25.8	25.7	25.6	25.5	25.4	25.1	24.9	24.6	24.4	24.1	24.0	24.0
6	35.5	27.0	23.7	21.9	20.8	20.0	19.5	19.0	18.7	18.4	18.2	18.0	17.8	17.7	17.6	17.4	17.4	17.3	17.2	17.1	16.9	16.7	16.4	16.3	16.0	15.9	15.9
7	29.2	21.7	18.8	17.2	16.2	15.5	15.0	14.6	14.3	14.1	13.9	13.7	13.6	13.4	13.3	13.2	13.1	13.1	13.0	12.9	12.7	12.5	12.3	12.0	12.0	11.9	11.8
8	25.4	18.5	15.8	14.4	13.5	12.9	12.4	12.0	11.8	11.5	11.4	11.2	11.1	10.9	10.8	10.8	10.7	10.6	10.5	10.5	10.3	10.1	9.92	9.80	9.57	9.49	9.45
9	22.9	16.4	13.9	12.6	11.7	11.1	10.7	10.4	10.1	9.89	9.72	9.57	9.44	9.33	9.24	9.15	9.08	9.01	8.95	8.90	8.69	8.55	8.37	8.26	8.04	7.96	7.93
10	21.0	14.9	12.6	11.3	10.5	9.92	9.52	9.20	8.96	8.75	8.59	8.45	8.32	8.22	8.13	8.05	7.98	7.91	7.86	7.80	7.60	7.47	7.30	7.19	6.98	6.91	6.87
11	19.7	13.8	11.6	10.3	9.58	9.05	8.66	8.35	8.12	7.92	7.76	7.63	7.51	7.41	7.32	7.24	7.17	7.11	7.06	7.01	6.81	6.68	6.52	6.42	6.21	6.14	6.10
12	18.6	13.0	10.8	9.63	8.89	8.38	8.00	7.71	7.48	7.29	7.14	7.00	6.89	6.79	6.71	6.63	6.57	6.51	6.45	6.40	6.22	6.09	5.93	5.83	5.63	5.56	5.52
13	17.8	12.3	10.2	9.07	8.35	7.86	7.49	7.21	6.98	6.80	6.65	6.52	6.41	6.31	6.23	6.16	6.09	6.03	5.98	5.93	5.75	5.63	5.47	5.37	5.17	5.10	5.07
14	17.1	11.8	9.73	8.62	7.92	7.44	7.08	6.80	6.58	6.40	6.26	6.13	6.02	5.93	5.85	5.78	5.71	5.66	5.60	5.56	5.38	5.25	5.10	5.00	4.81	4.74	4.71
15	16.6	11.3	9.34	8.25	7.57	7.09	6.74	6.47	6.26	6.08	5.94	5.81	5.71	5.62	5.54	5.46	5.40	5.35	5.29	5.25	5.07	4.95	4.80	4.70	4.51	4.44	4.41
16	16.1	11.0	9.01	7.94	7.27	6.80	6.46	6.19	5.98	5.81	5.67	5.55	5.44	5.35	5.27	5.20	5.14	5.09	5.04	4.99	4.82	4.70	4.54	4.45	4.26	4.19	4.16
17	15.7	10.7	8.73	7.68	7.02	6.56	6.22	5.96	5.75	5.58	5.44	5.32	5.22	5.13	5.05	4.99	4.92	4.87	4.82	4.78	4.60	4.48	4.33	4.24	4.05	3.98	3.95
18	15.4	10.4	8.49	7.46	6.81	6.35	6.02	5.76	5.56	5.39	5.25	5.13	5.03	4.94	4.87	4.80	4.74	4.68	4.63	4.59	4.42	4.30	4.15	4.06	3.87	3.80	3.77
19	15.1	10.2	8.28	7.27	6.62	6.18	5.85	5.59	5.39	5.22	5.08	4.97	4.87	4.78	4.70	4.64	4.58	4.52	4.47	4.43	4.26	4.14	3.99	3.90	3.71	3.65	3.61
20	14.9	9.95	8.10	7.10	6.46	6.02	5.69	5.44	5.24	5.08	4.94	4.82	4.72	4.64	4.56	4.49	4.44	4.38	4.33	4.29	4.12	4.00	3.86	3.77	3.58	3.51	3.48
21	14.6	9.77	7.94	6.95	6.32	5.88	5.56	5.31	5.11	4.95	4.81	4.70	4.60	4.51	4.44	4.37	4.31	4.26	4.21	4.17	4.00	3.88	3.74	3.64	3.46	3.39	3.36
22	14.4	9.61	7.80	6.81	6.19	5.76	5.44	5.19	4.99	4.83	4.70	4.58	4.49	4.40	4.33	4.26	4.20	4.15	4.10	4.06	3.89	3.78	3.63	3.54	3.35	3.28	3.25
23	14.2	9.47	7.67	6.70	6.08	5.65	5.33	5.09	4.89	4.73	4.60	4.48	4.39	4.30	4.23	4.16	4.10	4.05	4.00	3.96	3.79	3.68	3.53	3.44	3.25	3.19	3.16
24	14.0	9.34	7.55	6.59	5.98	5.55	5.23	4.99	4.80	4.64	4.51	4.39	4.30	4.21	4.14	4.07	4.02	3.96	3.92	3.87	3.71	3.59	3.45	3.36	3.17	3.10	3.07
25	13.9	9.22	7.45	6.49	5.89	5.46	5.15	4.91	4.71	4.56	4.42	4.31	4.22	4.13	4.06	3.99	3.94	3.88	3.84	3.79	3.63	3.52	3.37	3.28	3.09	3.03	2.99
26	13.7	9.12	7.36	6.41	5.80	5.38	5.07	4.83	4.64	4.48	4.35	4.24	4.14	4.06	3.99	3.92	3.86	3.81	3.77	3.72	3.56	3.44	3.30	3.21	3.02	2.95	2.92
27	13.6	9.02	7.27	6.33	5.73	5.31	5.00	4.76	4.57	4.41	4.28	4.17	4.08	3.99	3.92	3.86	3.80	3.75	3.70	3.66	3.49	3.38	3.23	3.14	2.96	2.89	2.86
28	13.5	8.93	7.19	6.25	5.66	5.24	4.93	4.69	4.50	4.35	4.22	4.11	4.01	3.93	3.86	3.80	3.74	3.69	3.64	3.60	3.43	3.32	3.18	3.09	2.90	2.83	2.80
29	13.4	8.85	7.12	6.19	5.59	5.18	4.87	4.64	4.45	4.29	4.16	4.05	3.96	3.88	3.80	3.74	3.68	3.63	3.59	3.54	3.38	3.27	3.12	3.03	2.84	2.78	2.74
30	13.3	8.77	7.05	6.12	5.53	5.12	4.82	4.58	4.39	4.24	4.11	4.00	3.91	3.82	3.75	3.69	3.63	3.58	3.53	3.49	3.33	3.22	3.07	2.98	2.79	2.73	2.69
32	13.1	8.64	6.94	6.01	5.43	5.02	4.72	4.48	4.30	4.14	4.02	3.91	3.81	3.73	3.66	3.60	3.54	3.49	3.44	3.40	3.24	3.13	2.98	2.89	2.70	2.64	2.60
34	13.0	8.52	6.83	5.92	5.34	4.93	4.63	4.40	4.22	4.06	3.94	3.83	3.74	3.65	3.58	3.52	3.46	3.41	3.37	3.33	3.16	3.05	2.91	2.82	2.63	2.56	2.52
36	12.8	8.42	6.74	5.84	5.26	4.86	4.56	4.33	4.14	3.99	3.87	3.76	3.67	3.59	3.51	3.45	3.40	3.34	3.30	3.26	3.10	2.98	2.84	2.75	2.56	2.49	2.46
38	12.7	8.33	6.66	5.76	5.19	4.79	4.49	4.26	4.08	3.93	3.80	3.70	3.60	3.52	3.45	3.39	3.34	3.28	3.24	3.20	3.04	2.92	2.78	2.69	2.50	2.43	2.40
40	12.6	8.25	6.59	5.70	5.13	4.73	4.44	4.21	4.02	3.87	3.75	3.64	3.55	3.47	3.40	3.34	3.28	3.23	3.18	3.14	2.98	2.87	2.73	2.64	2.44	2.38	2.34
42	12.5	8.18	6.53	5.64	5.07	4.68	4.38	4.16	3.97	3.83	3.70	3.59	3.50	3.42	3.35	3.29	3.23	3.18	3.14	3.10	2.94	2.83	2.68	2.59	2.40	2.33	2.29
44	12.4	8.12	6.48	5.59	5.02	4.63	4.34	4.11	3.93	3.78	3.66	3.55	3.46	3.38	3.31	3.25	3.19	3.14	3.10	3.06	2.89	2.78	2.64	2.55	2.35	2.28	2.25
46	12.4	8.06	6.42	5.54	4.98	4.59	4.30	4.07	3.89	3.74	3.62	3.51	3.42	3.34	3.27	3.21	3.15	3.10	3.06	3.02	2.86	2.74	2.60	2.51	2.31	2.24	2.21
48	12.3	8.00	6.38	5.50	4.94	4.55	4.26	4.03	3.85	3.70	3.58	3.48	3.38	3.31	3.24	3.17	3.12	3.07	3.02	2.98	2.82	2.71	2.56	2.47	2.28	2.21	2.17
50	12.2	7.96	6.34	5.46	4.90	4.51	4.22	4.00	3.82	3.67	3.55	3.44	3.35	3.27	3.20	3.14	3.09	3.04	2.99	2.95	2.79	2.68	2.53	2.44	2.25	2.18	2.14
60	12.0	7.77	6.17	5.31	4.76	4.37	4.09	3.86	3.69	3.54	3.42	3.32	3.23	3.15	3.08	3.02	2.96	2.91	2.87	2.83	2.67	2.55	2.41	2.32	2.12	2.05	2.01
70	11.8	7.64	6.06	5.20	4.66	4.28	3.99	3.77	3.60	3.45	3.33	3.23	3.14	3.06	2.99	2.93	2.88	2.83	2.78	2.74	2.58	2.47	2.32	2.23	2.03	1.95	1.92
80	11.7	7.54	5.97	5.12	4.58	4.20	3.92	3.70	3.53	3.39	3.27	3.16	3.07	3.00	2.93	2.87	2.81	2.76	2.72	2.68	2.52	2.41	2.26	2.16	1.96	1.89	1.85
90	11.6	7.47	5.91	5.06	4.53	4.15	3.87	3.65	3.48	3.34	3.22	3.11	3.02	2.95	2.88	2.82	2.76	2.71	2.67	2.63	2.47	2.36	2.21	2.11	1.91	1.83	1.79
100	11.5	7.41	5.86	5.02	4.48	4.11	3.83	3.61	3.44	3.30	3.18	3.07	2.99	2.91	2.84	2.78	2.73	2.68	2.63	2.59	2.43	2.32	2.17	2.08	1.87	1.79	1.75
125	11.4	7.30	5.77	4.93	4.40	4.03	3.75	3.54	3.37	3.23	3.11	3.00	2.92	2.84	2.77	2.71	2.66	2.61	2.56	2.52	2.36	2.25	2.10	2.01	1.79	1.71	1.67
150	11.3	7.24	5.71	4.88	4.35	3.98	3.71	3.49	3.32	3.18	3.06	2.96	2.87	2.80	2.73	2.67	2.61	2.56	2.52	2.48	2.32	2.21	2.06	1.96	1.74	1.66	1.62
200	11.2	7.15	5.63	4.81	4.29	3.92	3.65	3.43	3.26	3.12	3.00	2.90	2.82	2.74	2.67	2.61	2.56	2.51	2.46	2.42	2.26	2.15	2.00	1.90	1.68	1.60	1.55
300	11.0	7.07	5.56	4.75	4.22	3.86	3.59	3.38	3.21	3.07	2.95	2.85	2.76	2.69	2.62	2.56	2.50	2.46	2.41	2.37	2.21	2.10	1.94	1.85	1.62	1.53	1.48
500	11.0	7.00	5.51	4.69	4.18	3.81	3.54	3.33	3.16	3.02	2.91	2.81	2.72	2.64	2.58	2.52	2.46	2.41	2.37	2.33	2.17	2.05	1.90	1.80	1.57	1.48	1.43
1000	10.9	6.96	5.46	4.65	4.14	3.78	3.51	3.30	3.13	2.99	2.87	2.77	2.69	2.61	2.54	2.48	2.43	2.38	2.34	2.30	2.14	2.02	1.87	1.77	1.53	1.44	1.38

DEGREES OF FREEDOM FOR DENOMINATOR

Values of $\frac{1}{2}\log_e\frac{1+r}{1-r}$ *for given values of r*

	0.000	0.001	0.002	0.003	0.004	0.005	0.006	0.007	0.008	0.009
0.000	0.0000	0.0010	0.0020	0.0030	0.0040	0.0050	0.0060	0.0070	0.0080	0.0090
0.010	0.0100	0.0110	0.0120	0.0130	0.0140	0.0150	0.0160	0.0170	0.0180	0.0190
0.020	0.0200	0.0210	0.0220	0.0230	0.0240	0.0250	0.0260	0.0270	0.0280	0.0290
0.030	0.0300	0.0310	0.0320	0.0330	0.0340	0.0350	0.0360	0.0370	0.0380	0.0390
0.040	0.0400	0.0410	0.0420	0.0430	0.0440	0.0450	0.0460	0.0470	0.0480	0.0490
0.050	0.0501	0.0511	0.0521	0.0531	0.0541	0.0551	0.0561	0.0571	0.0581	0.0591
0.060	0.0601	0.0611	0.0621	0.0631	0.0641	0.0651	0.0661	0.0671	0.0681	0.0691
0.070	0.0701	0.0711	0.0721	0.0731	0.0741	0.0751	0.0761	0.0771	0.0782	0.0792
0.080	0.0802	0.0812	0.0822	0.0832	0.0842	0.0852	0.0862	0.0872	0.0882	0.0892
0.090	0.0902	0.0912	0.0922	0.0933	0.0943	0.0953	0.0963	0.0973	0.0983	0.0993
0.100	0.1003	0.1013	0.1024	0.1034	0.1044	0.1054	0.1064	0.1074	0.1084	0.1094
0.110	0.1105	0.1115	0.1125	0.1135	0.1145	0.1155	0.1165	0.1175	0.1185	0.1195
0.120	0.1206	0.1216	0.1226	0.1236	0.1246	0.1257	0.1267	0.1277	0.1287	0.1297
0.130	0.1308	0.1318	0.1328	0.1338	0.1348	0.1358	0.1368	0.1379	0.1389	0.1399
0.140	0.1409	0.1419	0.1430	0.1440	0.1450	0.1460	0.1470	0.1481	0.1491	0.1501
0.150	0.1511	0.1522	0.1532	0.1542	0.1552	0.1563	0.1573	0.1583	0.1593	0.1604
0.160	0.1614	0.1624	0.1634	0.1644	0.1655	0.1665	0.1676	0.1686	0.1696	0.1706
0.170	0.1717	0.1727	0.1737	0.1748	0.1758	0.1768	0.1779	0.1789	0.1799	0.1810
0.180	0.1820	0.1830	0.1841	0.1851	0.1861	0.1872	0.1882	0.1892	0.1903	0.1913
0.190	0.1923	0.1934	0.1944	0.1954	0.1965	0.1975	0.1986	0.1996	0.2007	0.2017
0.200	0.2027	0.2038	0.2048	0.2059	0.2069	0.2079	0.2090	0.2100	0.2111	0.2121
0.210	0.2132	0.2142	0.2153	0.2163	0.2174	0.2184	0.2194	0.2205	0.2215	0.2226
0.220	0.2237	0.2247	0.2258	0.2268	0.2279	0.2289	0.2300	0.2310	0.2321	0.2331
0.230	0.2342	0.2353	0.2363	0.2374	0.2384	0.2395	0.2405	0.2416	0.2427	0.2437
0.240	0.2448	0.2458	0.2469	0.2480	0.2490	0.2501	0.2511	0.2522	0.2533	0.2543
0.250	0.2554	0.2565	0.2575	0.2586	0.2597	0.2608	0.2618	0.2629	0.2640	0.2650
0.260	0.2661	0.2672	0.2682	0.2693	0.2704	0.2715	0.2726	0.2736	0.2747	0.2758
0.270	0.2769	0.2779	0.2790	0.2801	0.2812	0.2823	0.2833	0.2844	0.2855	0.2866
0.280	0.2877	0.2888	0.2898	0.2909	0.2920	0.2931	0.2942	0.2953	0.2964	0.2975
0.290	0.2986	0.2997	0.3008	0.3019	0.3029	0.3040	0.3051	0.3062	0.3073	0.3084
0.300	0.3095	0.3106	0.3117	0.3128	0.3139	0.3150	0.3161	0.3172	0.3183	0.3195
0.310	0.3206	0.3217	0.3228	0.3239	0.3250	0.3261	0.3272	0.3283	0.3294	0.3305
0.320	0.3317	0.3328	0.3339	0.3350	0.3361	0.3372	0.3384	0.3395	0.3406	0.3417
0.330	0.3428	0.3439	0.3451	0.3462	0.3473	0.3484	0.3496	0.3507	0.3518	0.3530
0.340	0.3541	0.3552	0.3564	0.3575	0.3586	0.3597	0.3609	0.3620	0.3632	0.3643
0.350	0.3654	0.3666	0.3677	0.3689	0.3700	0.3712	0.3723	0.3734	0.3746	0.3757
0.360	0.3769	0.3780	0.3792	0.3803	0.3815	0.3826	0.3838	0.3850	0.3861	0.3873
0.370	0.3884	0.3896	0.3907	0.3919	0.3931	0.3942	0.3954	0.3966	0.3977	0.3989
0.380	0.4001	0.4012	0.4024	0.4036	0.4047	0.4059	0.4071	0.4083	0.4094	0.4106
0.390	0.4118	0.4130	0.4142	0.4153	0.4165	0.4177	0.4189	0.4201	0.4213	0.4225
0.400	0.4236	0.4248	0.4260	0.4272	0.4284	0.4296	0.4308	0.4320	0.4332	0.4344
0.410	0.4356	0.4368	0.4380	0.4392	0.4404	0.4416	0.4429	0.4441	0.4453	0.4465
0.420	0.4477	0.4489	0.4501	0.4513	0.4526	0.4538	0.4550	0.4562	0.4574	0.4587
0.430	0.4599	0.4611	0.4623	0.4636	0.4648	0.4660	0.4673	0.4685	0.4697	0.4710
0.440	0.4722	0.4735	0.4747	0.4760	0.4772	0.4784	0.4797	0.4809	0.4822	0.4835
0.450	0.4847	0.4860	0.4872	0.4885	0.4897	0.4910	0.4923	0.4935	0.4948	0.4061
0.460	0.4973	0.4986	0.4999	0.5011	0.5024	0.5037	0.5049	0.5062	0.5075	0.5088
0.470	0.5101	0.5114	0.5126	0.5139	0.5152	0.5165	0.5178	0.5191	0.5204	0.5217
0.480	0.5230	0.5243	0.5256	0.5279	0.5282	0.5295	0.5308	0.5321	0.5334	0.5347
0.490	0.5361	0.5374	0.5387	0.5400	0.5413	0.5427	0.5440	0.5453	0.5466	0.5480

r	0.000	0.001	0.002	0.003	0.004	0.005	0.006	0.007	0.008	0.009
0.500	0.5493	0.5506	0.5520	0.5533	0.5547	0.5560	0.5573	0.5587	0.5600	0.5614
0.510	0.5627	0.5641	0.5654	0.5668	0.5681	0.5695	0.5709	0.5722	0.5736	0.5750
0.520	0.5763	0.5777	0.5791	0.5805	0.5818	0.5832	0.5846	0.5860	0.5874	0.5888
0.530	0.5901	0.5915	0.5929	0.5943	0.5957	0.5971	0.5985	0.5999	0.6013	0.6027
0.540	0.6042	0.6056	0.6070	0.6084	0.6098	0.6112	0.6127	0.6141	0.6155	0.6170
0.550	0.6184	0.6198	0.6213	0.6227	0.6241	0.6256	0.6270	0.6285	0.6299	0.6314
0.560	0.6328	0.6343	0.6358	0.6372	0.6387	0.6401	0.6416	0.6431	0.6446	0.6460
0.570	0.6475	0.6490	0.6505	0.6520	0.6535	0.6550	0.6565	0.6579	0.6594	0.6610
0.580	0.6625	0.6640	0.6655	0.6670	0.6685	0.6700	0.6715	0.6731	0.6746	0.6761
0.590	0.6777	0.6792	0.6807	0.6823	0.6838	0.6854	0.6869	0.6885	0.6900	0.6916
0.600	0.6931	0.6947	0.6963	0.6978	0.6994	0.7010	0.7026	0.7042	0.7057	0.7073
0.610	0.7089	0.7105	0.7121	0.7137	0.7153	0.7169	0.7185	0.7201	0.7218	0.7234
0.620	0.7250	0.7266	0.7283	0.7299	0.7315	0.7332	0.7348	0.7364	0.7381	0.7398
0.630	0.7414	0.7431	0.7447	0.7464	0.7481	0.7497	0.7514	0.7531	0.7548	0.7565
0.640	0.7582	0.7599	0.7616	0.7633	0.7650	0.7667	0.7684	0.7701	0.7718	0.7736
0.650	0.7753	0.7770	0.7788	0.7805	0.7823	0.7840	0.7858	0.7875	0.7893	0.7910
0.660	0.7928	0.7946	0.7964	0.7981	0.7999	0.8017	0.8035	0.8053	0.8071	0.8089
0.670	0.8107	0.8126	0.8144	0.8162	0.8180	0.8199	0.8217	0.8236	0.8254	0.8273
0.680	0.8291	0.8310	0.8328	0.8347	0.8366	0.8385	0.8404	0.8423	0.8442	0.8461
0.690	0.8480	0.8499	0.8518	0.8537	0.8556	0.8576	0.8595	0.8614	0.8634	0.8653
0.700	0.8673	0.8693	0.8712	0.8732	0.8752	0.8772	0.8792	0.8812	0.8832	0.8852
0.710	0.8872	0.8892	0.8912	0.8933	0.8953	0.8973	0.8994	0.9014	0.9035	0.9056
0.720	0.9076	0.9097	0.9118	0.9139	9.9160	0.9181	0.9202	0.9223	0.9245	0.9266
0.730	0.9287	0.9309	0.9330	0.9352	0.9373	0.9395	0.9417	0.9439	0.9461	0.9483
0.740	0.9505	0.9527	0.9549	0.9571	0.9594	0.9616	0.9639	0.9661	0.9684	0.9707
0.750	0.9730	0.9752	0.9775	0.9799	0.9822	0.9845	0.9868	0.9892	0.9915	0.9939
0.760	0.9962	0.9986	1.0010	1.0034	1.0058	1.0082	1.0106	1.0130	1.0154	1.0179
0.770	1.0203	1.0228	1.0253	1.0277	1.0302	1.0327	1.0352	1.0378	1.0403	1.0428
0.780	1.0454	1.0479	1.0505	1.0531	1.0557	1.0583	1.0609	1.0635	1.0661	1.0688
0.790	1.0714	1.0741	1.0768	1.0795	1.0822	1.0849	1.0876	1.0903	1.0931	1.0958
0.800	1.0986	1.1014	1.1041	1.1070	1.1098	1.1127	1.1155	1.1184	1.1212	1.1241
0.810	1.1270	1.1299	1.1329	1.1358	1.1388	1.1417	1.1447	1.1477	1.1507	1.1538
0.820	1.1568	1.1599	1.1630	1.1660	1.1692	1.1723	1.1754	1.1786	1.1817	1.1849
0.830	1.1870	1.1913	1.1946	1.1979	1.2011	1.2044	1.2077	1.2111	1.2144	1.2178
0.840	1.2212	1.2246	1.2280	1.2315	1.2349	1.2384	1.2419	1.2454	1.2490	1.2526
0.850	1.2561	1.2598	1.2634	1.2670	1.2708	1.2744	1.2782	1.2819	1.2857	1.2895
0.860	1.2934	1.2972	1.3011	1.3050	1.3089	1.3129	1.3168	1.3209	1.3249	1.3290
0.870	1.3331	1.3372	1.3414	1.3456	1.3498	1.3540	1.3583	1.3626	1.3670	1.3714
0.880	1.3758	1.3802	1.3847	1.3892	1.3938	1.3984	1.4030	1.4077	1.4124	1.4171
0.890	1.4219	1.4268	1.4316	1.4366	1.4415	1.4465	1.4516	1.4566	1.4618	1.4670
0.900	1.4722	1.4775	1.4828	1.4883	1.4937	1.4992	1.5047	1.5103	1.5160	1.5217
0.910	1.5275	1.5334	1.5393	1.5453	1.5513	1.5574	1.5636	1.5698	1.5762	1.5825
0.920	1.5890	1.5956	1.6022	1.6089	1.6157	1.6226	1.6296	1.6366	1.6438	1.6510
0.930	1.6584	1.6659	1.6734	1.6811	1.6888	1.6967	1.7047	1.7129	1.7211	1.7295
0.940	1.7380	1.7467	1.7555	1.7645	1.7736	1.7828	1.7923	1.8019	1.8117	1.8216
0.950	1.8318	1.8421	1.8527	1.8635	1.8745	1.8857	1.8972	1.9090	1.9210	1.9333
0.960	1.9459	1.9588	1.9721	1.9857	1.9996	2.0140	2.0287	2.0439	2.0595	2.0756
0.970	2.0923	2.1095	2.1273	2.1457	2.1649	2.1847	2.2054	2.2269	2.2494	2.2729
0.980	2.2976	2.3223	2.3507	2.3796	2.4101	2.4426	2.4774	2.5147	2.5550	2.5988
0.990	2.6467	2.6996	2.7587	2.8257	2.9031	2.9945	3.1063	3.2504	3.4534	3.8002

r	z
0.9999	4.95172
0.99999	6.10303

Source: Albert E. Waugh, *Statistical Tables and Problems,* McGraw-Hill Book Company, New York, 1952, Table A11, pp. 40-41, with the kind permission of the author and publisher.

Note: To obtain $\frac{1}{2}\log_e\frac{(1+r)}{(1-r)}$ when r is negative, use the negative of the value corresponding to the absolute value of r; e.g., if $r = -0.242$, $\frac{1}{2}\log_e\frac{1+0.242}{1-0.242} = -0.2469$

TABLE A-6 Upper α point of Studentized range,* $q_{k,\nu} = R/S$, k = sample size for range R, ν = number of degrees of freedom for S

$[enter = q_{k,\nu,1-\alpha}$ where $P(q_{k,\nu} > q_{k,\nu,1-\alpha}) = \alpha]$

$\alpha = 0.05$

ν \ k	2	3	4	5	6	7	8	9	10	11	12	13	14	15	16	17	18	19	20
1	18.0	27.0	32.8	37.1	40.4	43.1	45.4	47.4	49.1	50.6	52.0	53.2	54.3	55.4	56.3	57.2	58.0	58.8	59.6
2	6.08	8.33	9.80	10.9	11.7	12.4	13.0	13.5	14.0	14.4	14.7	15.1	15.4	15.7	15.9	16.1	16.4	16.6	16.8
3	4.50	5.91	6.82	7.50	8.04	8.48	8.85	9.18	9.46	9.72	9.95	10.2	10.3	10.5	10.7	10.8	11.0	11.1	11.2
4	3.93	5.04	5.76	6.29	6.71	7.05	7.35	7.60	7.83	8.03	8.21	8.37	8.52	8.66	8.79	8.91	9.03	9.13	9.23
5	3.64	4.60	5.22	5.67	6.03	6.33	6.58	6.80	6.99	7.17	7.32	7.47	7.60	7.72	7.83	7.93	8.03	8.12	8.21
6	3.46	4.34	4.90	5.30	5.63	5.90	6.12	6.32	6.49	6.65	6.79	6.92	7.03	7.14	7.24	7.34	7.43	7.51	7.59
7	3.34	4.16	4.68	5.06	5.36	5.61	5.82	6.00	6.16	6.30	6.43	6.55	6.66	6.76	6.85	6.94	7.02	7.10	7.17
8	3.26	4.04	4.53	4.89	5.17	5.40	5.60	5.77	5.92	6.05	6.18	6.29	6.39	6.48	6.57	6.65	6.73	6.80	6.87
9	3.20	3.95	4.41	4.76	5.02	5.24	5.43	5.59	5.74	5.87	5.98	6.09	6.19	6.28	6.36	6.44	6.51	6.58	6.64
10	3.15	3.88	4.33	4.65	4.91	5.12	5.30	5.46	5.60	5.72	5.83	5.93	6.03	6.11	6.19	6.27	6.34	6.40	6.47
11	3.11	3.82	4.26	4.57	4.82	5.03	5.20	5.35	5.49	5.61	5.71	5.81	5.90	5.98	6.06	6.13	6.20	6.27	6.33
12	3.08	3.77	4.20	4.51	4.75	4.95	5.12	5.27	5.39	5.51	5.61	5.71	5.80	5.88	5.95	6.02	6.09	6.15	6.21
13	3.06	3.73	4.15	4.45	4.69	4.88	5.05	5.19	5.32	5.43	5.53	5.63	5.71	5.79	5.86	5.93	5.99	6.05	6.11
14	3.03	3.70	4.11	4.41	4.64	4.83	4.99	5.13	5.25	5.36	5.46	5.55	5.64	5.71	5.79	5.85	5.91	5.97	6.03
15	3.01	3.67	4.08	4.37	4.59	4.78	4.94	5.08	5.20	5.31	5.40	5.49	5.57	5.65	5.72	5.78	5.85	5.90	5.96
16	3.00	3.65	4.05	4.33	4.56	4.74	4.90	5.03	5.15	5.26	5.35	5.44	5.52	5.59	5.66	5.73	5.79	5.84	5.90
17	2.98	3.63	4.02	4.30	4.52	4.70	4.86	4.99	5.11	5.21	5.31	5.39	5.47	5.54	5.61	5.67	5.73	5.79	5.84
18	2.97	3.61	4.00	4.28	4.49	4.67	4.82	4.96	5.07	5.17	5.27	5.35	5.43	5.50	5.57	5.63	5.69	5.74	5.79
19	2.96	3.59	3.98	4.25	4.47	4.65	4.79	4.92	5.04	5.14	5.23	5.31	5.39	5.46	5.53	5.59	5.65	5.70	5.75
20	2.95	3.58	3.96	4.23	4.45	4.62	4.77	4.90	5.01	5.11	5.20	5.28	5.36	5.43	5.49	5.55	5.61	5.66	5.71
24	2.92	3.53	3.90	4.17	4.37	4.54	4.68	4.81	4.92	5.01	5.10	5.18	5.26	5.32	5.38	5.44	5.49	5.55	5.59
30	2.89	3.49	3.85	4.10	4.30	4.46	4.60	4.72	4.82	4.92	5.00	5.08	5.15	5.21	5.27	5.33	5.38	5.43	5.47
40	2.86	3.44	3.79	4.04	4.23	4.39	4.52	4.63	4.73	4.82	4.90	4.98	5.04	5.11	5.16	5.22	5.27	5.31	5.36
60	2.83	3.40	3.74	3.98	4.16	4.31	4.44	4.55	4.65	4.73	4.81	4.88	4.94	5.00	5.06	5.11	5.15	5.20	5.24
120	2.80	3.36	3.68	3.92	4.10	4.24	4.36	4.47	4.56	4.64	4.71	4.78	4.84	4.90	4.95	5.00	5.04	5.09	5.13
∞	2.77	3.31	3.63	3.86	4.03	4.17	4.29	4.39	4.47	4.55	4.62	4.68	4.74	4.80	4.85	4.89	4.93	4.97	5.01

TABLE A-6 *Upper α point of Studentized range (continued)*

α = 0.01

ν \ k	2	3	4	5	6	7	8	9	10	11	12	13	14	15	16	17	18	19	20
1	90.0	135	164	186	202	216	227	237	246	253	260	266	272	277	282	286	290	294	298
2	14.0	19.0	22.3	24.7	26.6	28.2	29.5	30.7	31.7	32.6	33.4	34.1	34.8	35.4	36.0	36.5	37.0	37.5	37.9
3	8.26	10.6	12.2	13.3	14.2	15.0	15.6	16.2	16.7	17.1	17.5	17.9	18.2	18.5	18.8	19.1	19.3	19.5	19.8
4	6.51	8.12	9.17	9.96	10.6	11.1	11.5	11.9	12.3	12.6	12.8	13.1	13.3	13.5	13.7	13.9	14.1	14.2	14.4
5	5.70	6.97	7.80	8.42	8.91	9.32	9.67	9.97	10.2	10.5	10.7	10.9	11.1	11.2	11.4	11.6	11.7	11.8	11.9
6	5.24	6.33	7.03	7.56	7.97	8.32	8.61	8.87	9.10	9.30	9.49	9.65	9.81	9.95	10.1	10.2	10.3	10.4	10.5
7	4.95	5.92	6.54	7.01	7.37	7.68	7.94	8.17	8.37	8.55	8.71	8.86	9.00	9.12	9.24	9.35	9.46	9.55	9.65
8	4.74	5.63	6.20	6.63	6.96	7.24	7.47	7.68	7.87	8.03	8.18	8.31	8.44	8.55	8.66	8.76	8.85	8.94	9.03
9	4.60	5.43	5.96	6.35	6.66	6.91	7.13	7.32	7.49	7.65	7.78	7.91	8.03	8.13	8.23	8.32	8.41	8.49	8.57
10	4.48	5.27	5.77	6.14	6.43	6.67	6.87	7.05	7.21	7.36	7.48	7.60	7.71	7.81	7.91	7.99	8.07	8.15	8.22
11	4.39	5.14	5.62	5.97	6.25	6.48	6.67	6.84	6.99	7.13	7.25	7.36	7.46	7.56	7.65	7.73	7.81	7.88	7.95
12	4.32	5.04	5.50	5.84	6.10	6.32	6.51	6.67	6.81	6.94	7.06	7.17	7.26	7.36	7.44	7.52	7.59	7.66	7.73
13	4.26	4.96	5.40	5.73	5.98	6.19	6.37	6.53	6.67	6.79	6.90	7.01	7.10	7.19	7.27	7.34	7.42	7.48	7.55
14	4.21	4.89	5.32	5.63	5.88	6.08	6.26	6.41	6.54	6.66	6.77	6.87	6.96	7.05	7.12	7.20	7.27	7.33	7.39
15	4.17	4.83	5.25	5.56	5.80	5.99	6.16	6.31	6.44	6.55	6.66	6.76	6.84	6.93	7.00	7.07	7.14	7.20	7.26
16	4.13	4.78	5.19	5.49	5.72	5.92	6.08	6.22	6.35	6.46	6.56	6.66	6.74	6.82	6.90	6.97	7.03	7.09	7.15
17	4.10	4.74	5.14	5.43	5.66	5.85	6.01	6.15	6.27	6.38	6.48	6.57	6.66	6.73	6.80	6.87	6.94	7.00	7.05
18	4.07	4.70	5.09	5.38	5.60	5.79	5.94	6.08	6.20	6.31	6.41	6.50	6.58	6.65	6.72	6.79	6.85	6.91	6.96
19	4.05	4.67	5.05	5.33	5.55	5.73	5.89	6.02	6.14	6.25	6.34	6.43	6.51	6.58	6.65	6.72	6.78	6.84	6.89
20	4.02	4.64	5.02	5.29	5.51	5.69	5.84	5.97	6.09	6.19	6.29	6.37	6.45	6.52	6.59	6.65	6.71	6.76	6.82
24	3.96	4.54	4.91	5.17	5.37	5.54	5.69	5.81	5.92	6.02	6.11	6.19	6.26	6.33	6.39	6.45	6.51	6.56	6.61
30	3.89	4.45	4.80	5.05	5.24	5.40	5.54	5.65	5.76	5.85	5.93	6.01	6.08	6.14	6.20	6.26	6.31	6.36	6.41
40	3.82	4.37	4.70	4.93	5.11	5.27	5.39	5.50	5.60	5.69	5.77	5.84	5.90	5.96	6.02	6.07	6.12	6.17	6.21
60	3.76	4.28	4.60	4.82	4.99	5.13	5.25	5.36	5.45	5.53	5.60	5.67	5.73	5.79	5.84	5.89	5.93	5.98	6.02
120	3.70	4.20	4.50	4.71	4.87	5.01	5.12	5.21	5.30	5.38	5.44	5.51	5.56	5.61	5.66	5.71	5.75	5.79	5.83
∞	3.64	4.12	4.40	4.60	4.76	4.88	4.99	5.08	5.16	5.23	5.29	5.35	5.40	5.45	5.49	5.54	5.57	5.61	5.65

*From pp. 176-177 of *Biometrika Tables for Statisticians*, Vol. I, by E. S. Pearson and H. O. Hartley, published by the Biometrika Trustees, Cambridge University Press, Cambridge, 1954. Reproduced with the kind permission of the authors and the publisher. Corrections of ±1 in the last figure, supplied by James Pacheres, have been incorporated in 41 entries.

MATRICES AND THEIR RELATIONSHIP TO REGRESSION ANALYSIS

B.1 PREVIEW

Statisticians have found matrix mathematics to be a very useful vehicle for compactly presenting the concepts, methods, and formulas of regression analysis and other multivariable methods. Moreover, matrix formulation of such topics has had the important practical implication of permitting extremely efficient and accurate use of the electronic computer for carrying out multivariable analyses on large data sets.

This appendix will summarize some of the more elementary but important notions and manipulations of matrix algebra and will use this tool to describe the general least-squares procedures of multiple regression. Admittedly, the material in this appendix is somewhat more mathematical in nature than in the main body of the text and is not absolutely necessary knowledge for the applied user of the multivariable methods we describe. Nevertheless, the reader who is comfortable with the mathematical level used here should find the matrix formulation of regression analysis a powerful and unifying supplement which may facilitate the learning of more advanced multivariable methods.

B.2 DEFINITIONS

A *matrix* may be simply defined as a rectangular array of numbers. For example,

$$\mathbf{A} = \begin{bmatrix} 2 & 3 & 1 \\ 1 & 1 & 2 \end{bmatrix}, \quad \mathbf{B} = \begin{bmatrix} 2 & 1 \\ 3 & 2 \end{bmatrix}, \quad \mathbf{C} = \begin{bmatrix} 1 \\ 1 \\ 3 \end{bmatrix}$$

are all matrices.

The *dimensions* of a matrix are the number of rows and the number of columns that it has. For example, the matrix **A** above has *two* rows and *three* columns:

$$
\begin{array}{cccc}
 & \overset{\text{column}}{\underset{\downarrow}{\text{one}}} & \overset{\text{column}}{\underset{\downarrow}{\text{two}}} & \overset{\text{column}}{\underset{\downarrow}{\text{three}}} \\
\text{row one} \to & 2 & 3 & 1 \\
\text{row two} \to & 1 & 1 & 2
\end{array}
$$

It is customary to say that **A** is a $(2_{\text{number of rows}} \times 3_{\text{number of columns}})$ matrix or to write $\mathbf{A}_{2\times3}$. The dimensions of the matrices **B** and **C** above are (2×2) and (3×1), respectively. An example of a (1×4) matrix is any matrix with one row and four columns, such as $\mathbf{D}_{1\times4} = [-2, 3, 3, 0]$. Incidentally, matrices that contain only one row or only one column are often referred to as *vectors*, so that the matrices

$$
\mathbf{C} = \begin{bmatrix} 1 \\ 1 \\ 3 \end{bmatrix} \quad \text{and} \quad \mathbf{D} = [-2, 3, 3, 0]
$$

are examples of a column vector and a row vector, respectively. Also, the matrix **B** is called a *square matrix* because it has the *same* number of rows and columns.

The numbers forming the "rectangular array" of a matrix are called the *elements* of the matrix. If we let a_{ij} denote the element in the ith row and jth column of the matrix **A** above,

$$
a_{11} = 2, \quad a_{12} = 3, \quad a_{13} = 1
$$
$$
a_{21} = 1, \quad a_{22} = 1, \quad a_{23} = 2
$$

It is often informative to write

$$
\mathbf{A}_{2\times3} = ((a_{ij}))
$$

indicating that the matrix **A** (represented by a capital letter) with two rows and three columns has typical element a_{ij} (represented by the corresponding lowercase letter). Thus, if you were given that

$$
\mathbf{B}_{3\times2} = ((b_{ij}))
$$

where $b_{21} = 3$, $b_{31} = 2$, $b_{11} = -2$, $b_{12} = 6$, $b_{22} = 0$, $b_{32} = 1$, you should construct **B** to be

$$
\mathbf{B} = \begin{bmatrix} -2 & 6 \\ 3 & 0 \\ 2 & 1 \end{bmatrix}
$$

B.3 MATRICES IN REGRESSION ANALYSIS

Given any set of multivariable data suitable for a regression analysis, a number of key matrices can be defined which directly correspond to the basic components of the regression model being postulated. For example, consider the following $n = 12$ pairs of observations on $Y = $ WGT and $X = $ HGT:

CHILD NO.	1	2	3	4	5	6	7	8	9	10	11	12
Y (WGT)	64	71	53	67	55	58	77	57	56	51	76	68
X (HGT)	57	59	49	62	51	50	55	48	42	42	61	57

We can, in correspondence with the straight-line regression model $Y = \beta_0 + \beta_1 X + E$, define four matrices:

Y, the vector of observations on Y

X, the matrix of independent variables

β, the vector of parameters to be estimated

E, the vector of errors

For the data given, these matrices are defined as follows:

$$\mathbf{Y}_{12 \times 1} = \begin{bmatrix} 64 \\ 71 \\ 53 \\ 67 \\ 55 \\ 58 \\ 77 \\ 57 \\ 56 \\ 51 \\ 76 \\ 68 \end{bmatrix}, \quad \mathbf{X}_{12 \times 2} = \begin{bmatrix} 1 & 57 \\ 1 & 59 \\ 1 & 49 \\ 1 & 62 \\ 1 & 51 \\ 1 & 50 \\ 1 & 55 \\ 1 & 48 \\ 1 & 42 \\ 1 & 42 \\ 1 & 61 \\ 1 & 57 \end{bmatrix}, \quad \boldsymbol{\beta}_{2 \times 1} = \begin{bmatrix} \beta_0 \\ \beta_1 \end{bmatrix}, \quad \mathbf{E}_{12 \times 1} = \begin{bmatrix} E_1 \\ E_2 \\ E_3 \\ E_4 \\ E_5 \\ E_6 \\ E_7 \\ E_8 \\ E_9 \\ E_{10} \\ E_{11} \\ E_{12} \end{bmatrix}$$

Notice that the first column of the **X** matrix of independent variables contains only 1's. This is the general convention to be used for any regression model containing a constant term β_0; motivation for adopting this convention follows by imagining the β_0 term to be of the form $\beta_0 X_0$, where X_0 is a dummy variable always taking the value 1. The vector of errors **E** contains random (and unobservable) error values, one for each pair of observations, which represent the differences between the observed Y values and their (unknown) expected values under the given model.

B.4 TRANSPOSE OF A MATRIX

The transpose A' of a matrix A is defined to be that matrix whose (i, j)th element a'_{ij} is equal to the (j, i)th element of A. For example, if $A = \begin{bmatrix} 2 & 3 & 1 \\ 1 & 1 & 2 \end{bmatrix}$, then $A' = \begin{bmatrix} 2 & 1 \\ 3 & 1 \\ 1 & 2 \end{bmatrix}$, since $a'_{11} = a_{11} = 2$, $a'_{12} = a_{21} = 1$, $a'_{21} = a_{12} = 3$, $a'_{22} = a_{22} = 1$, $a'_{31} = a_{13} = 1$, $a'_{32} = a_{23} = 2$. Another way of looking at this is that the first column of A becomes the first row of A', the second column of A becomes the second row of A', and so on. Thus, if A is $(r \times c)$, then A' is $(c \times r)$.

As examples, the transposes of

$$A_{3 \times 2} = \begin{bmatrix} 1 & 2 \\ 3 & 1 \\ 1 & 1 \end{bmatrix}, \quad B_{3 \times 3} = \begin{bmatrix} 1 & 0 & 2 \\ 0 & 4 & -5 \\ 2 & -5 & 3 \end{bmatrix}, \quad C_{4 \times 1} = \begin{bmatrix} 0 \\ 1 \\ 2 \\ -2 \end{bmatrix}$$

are

$$A'_{2 \times 3} = \begin{bmatrix} 1 & 3 & 1 \\ 2 & 1 & 1 \end{bmatrix}, \quad B'_{3 \times 3} = \begin{bmatrix} 1 & 0 & 2 \\ 0 & 4 & -5 \\ 2 & -5 & 3 \end{bmatrix}, \quad C'_{1 \times 4} = [0 \quad 1 \quad 2 \quad -2]$$

Also, the transpose of the matrix $X_{12 \times 2}$ of the previous section is given by

$$X'_{2 \times 12} = \begin{bmatrix} 1 & 1 & 1 & 1 & 1 & 1 & 1 & 1 & 1 & 1 & 1 & 1 \\ 57 & 59 & 49 & 62 & 51 & 50 & 55 & 48 & 42 & 42 & 61 & 57 \end{bmatrix}$$

Note that in the examples above, the matrix B is such that $B = B'$ (equality here means that corresponding elements are equal). A matrix satisfying this condition is said to be a *symmetric matrix*. Note that a symmetric matrix A must always be square, since otherwise A and A' would have different dimensions and so could not possibly be equal. A necessary and sufficient condition for the square matrix $A = ((a_{ij}))$ to be symmetric is that $a_{ij} = a_{ji}$ for every $i \neq j$.

Correlation matrices such as

$$R_1 = \begin{bmatrix} 1 & r_{xy} \\ r_{xy} & 1 \end{bmatrix} \quad \text{or} \quad R_2 = \begin{bmatrix} 1 & r_{12} & r_{13} \\ r_{12} & 1 & r_{23} \\ r_{13} & r_{23} & 1 \end{bmatrix}$$

are always symmetric.

An important special case where the above condition for symmetry is satisfied is when $a_{ij} = a_{ji} = 0$ for every $i \neq j$. A square matrix having this property is said to be a *diagonal matrix*, the general form of which is given (for the 3×3 case) by

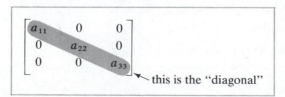

$$\begin{bmatrix} a_{11} & 0 & 0 \\ 0 & a_{22} & 0 \\ 0 & 0 & a_{33} \end{bmatrix}$$

this is the "diagonal"

The most often used diagonal matrix is the *identity matrix* **I**, which has 1's on the diagonal; for example, the (3×3) identity matrix is

$$\mathbf{I} = \begin{bmatrix} 1 & 0 & 0 \\ 0 & 1 & 0 \\ 0 & 0 & 1 \end{bmatrix}$$

We will see shortly that an identity matrix serves the same algebraic function for matrix multiplication that the number 1 serves for ordinary scalar multiplication.

B.5 MATRIX ADDITION

The sum of two matrices, say **A** and **B**, is obtained by adding together the corresponding elements of each matrix. Clearly, such addition can only be performed when the two matrices have the exact same dimensions. Thus, for example, we can add the matrices

$$\mathbf{A}_{2 \times 3} = \begin{bmatrix} 2 & 3 & 1 \\ 1 & 1 & 2 \end{bmatrix} \quad \text{and} \quad \mathbf{B}_{2 \times 3} = \begin{bmatrix} 4 & 1 & 5 \\ 1 & 3 & 1 \end{bmatrix}$$

(since they have the same dimensions) to get

$$\mathbf{A} + \mathbf{B}_{2 \times 3} = \begin{bmatrix} 2+4 & 3+1 & 1+5 \\ 1+1 & 1+3 & 2+1 \end{bmatrix} = \begin{bmatrix} 6 & 4 & 6 \\ 2 & 4 & 3 \end{bmatrix}$$

We could not, however, add the matrices **A** and **C**, where

$$\mathbf{C}_{3 \times 2} = \begin{bmatrix} 5 & 4 \\ 1 & 4 \\ 2 & 6 \end{bmatrix}$$

since the dimensions of these matrices are not the same.

An example of a more abstract use of matrix addition would be to sum the two (12×1) vectors

$$\mathbf{D}_{12 \times 1} = \begin{bmatrix} \beta_0 + 57\beta_1 \\ \beta_0 + 59\beta_1 \\ \beta_0 + 49\beta_1 \\ \vdots \\ \beta_0 + 57\beta_1 \end{bmatrix} \quad \text{and} \quad \mathbf{E}_{12 \times 1} = \begin{bmatrix} E_1 \\ E_2 \\ E_3 \\ \vdots \\ E_{12} \end{bmatrix}$$

to obtain

$$\underset{12\times1}{\mathbf{D}+\mathbf{E}} = \begin{bmatrix} \beta_0 + 57\beta_1 + E_1 \\ \beta_0 + 59\beta_1 + E_2 \\ \beta_0 + 49\beta_1 + E_3 \\ \cdot \\ \cdot \\ \cdot \\ \beta_0 + 57\beta_1 + E_{12} \end{bmatrix}$$

Actually, you may recognize that each element of the matrix $\mathbf{D}+\mathbf{E}$ is obtained by substituting for X in the right side of the straight-line regression equation

$$Y = \beta_0 + \beta_1 X + E$$

each of the twelve X (HGT) values given in the data set of Section B.3.

B.6   MATRIX MULTIPLICATION

Multiplication of two matrices is somewhat more complicated than addition. The first rule to remember is that the product $\mathbf{AB}$ of two matrices $\mathbf{A}$ and $\mathbf{B}$ can exist if and only if the *number of columns of* $\mathbf{A}$ *is equal to the number of rows of* $\mathbf{B}$. Thus, if $\mathbf{A}$ is (2×3) and $\mathbf{B}$ is (3×4), the product $\mathbf{AB}$ exists since $\mathbf{A}$ has 3 columns and $\mathbf{B}$ has 3 rows. However, the product $\mathbf{BA}$ does not exist, since the number of columns of $\mathbf{B}$ (i.e., 4) is not equal to the number of rows of $\mathbf{A}$ (i.e., 2). Notationally, therefore, a matrix product can exist only if the dimensions of the matrices can be represented as follows:

$$\mathbf{A}_{m\times n} \quad \times \quad \mathbf{B}_{n\times p} \quad = \mathbf{AB}_{m\times p}$$

equal

product dimensions

Note also from the expression above that the dimensions $(m \times p)$ of the product matrix $\mathbf{AB}$ are given by the number of rows of the *pre*multiplier (i.e., $\mathbf{A}$) and the number of columns of the *post*multiplier (i.e., $\mathbf{B}$). For example, if $\mathbf{A}$ is (2×3) and $\mathbf{B}$ is (3×4), the dimensions of the product $\mathbf{AB}$ are (2×4).

Now, to describe how to actually carry out matrix multiplication, consider the two matrices

$$\mathbf{A}_{2\times3} = \begin{bmatrix} 2 & 1 & 0 \\ 0 & 3 & 1 \end{bmatrix} \quad \text{and} \quad \mathbf{B}_{3\times2} = \begin{bmatrix} 1 & -2 \\ 1 & 0 \\ 3 & 2 \end{bmatrix}$$

Since $\mathbf{A}$ is (2×3) and $\mathbf{B}$ is (3×2), the product $\mathbf{AB}$ will be a (2×2) matrix. If we let the elements of $\mathbf{AB}$ be denoted by

$$\mathbf{AB}_{2\times 2} = ((c_{ij})) = \begin{bmatrix} c_{11} & c_{12} \\ c_{21} & c_{22} \end{bmatrix}$$

we can obtain the upper-left-hand-corner element c_{11} by working with the first row of $\mathbf{A}$ and the first column of $\mathbf{B}$, as follows:

$$
\begin{array}{ccc}
\mathbf{A} & \mathbf{B} & \mathbf{AB} \\
\text{row 1}\begin{bmatrix} 2 & 1 & 0 \\ 0 & 3 & 1 \end{bmatrix} & \begin{bmatrix} 1 & -2 \\ 1 & 0 \\ 3 & 2 \end{bmatrix} = & \begin{bmatrix} (2\times 1)+(1\times 1)+(0\times 3) & ? \\ = (3) & \\ ? & ? \end{bmatrix} \\
\text{row 2} & & \\
& \text{column 1}\quad\text{column 2} &
\end{array}
$$

What we have done here was calculate the product of each element in row 1 of $\mathbf{A}$ with the corresponding element in column 1 of $\mathbf{B}$, and then add up these three products to obtain $c_{11} = 3$:

$$\mathbf{B}:\quad \text{column 1}$$
$$c_{11} = (2\times 1)+(1\times 1)+(0\times 3) = 3$$
$$\mathbf{A}:\quad \text{row 1}$$

To find the element in the second row and first column of $\mathbf{AB}$ (i.e., c_{21}), we would work with the second row of $\mathbf{A}$ and the first column of $\mathbf{B}$, as follows:

$$
\begin{array}{ccc}
\mathbf{A} & \mathbf{B} & \mathbf{AB} \\
\text{row 1}\begin{bmatrix} 2 & 1 & 0 \\ 0 & 3 & 1 \end{bmatrix} & \begin{bmatrix} 1 & -2 \\ 1 & 0 \\ 3 & 2 \end{bmatrix} = & \begin{bmatrix} 3 & ? \\ (0\times 1)+(3\times 1)+(1\times 3) & \\ = (6) & ? \end{bmatrix} \\
\text{row 2} & & \\
& \text{column 1}\quad\text{column 2} &
\end{array}
$$

Thus, for the element c_{21}, we find

$$\mathbf{B}:\quad \text{column 1}$$
$$c_{21} = (0\times 1)+(3\times 1)+(1\times 3) = 6$$
$$\mathbf{A}:\quad \text{row 2}$$

Continuing this process, we will find

$$c_{12} = (2\times -2)+(1\times 0)+(0\times 2) = -4$$
$$c_{22} = (0\times -2)+(3\times 0)+(1\times 2) = 2$$

Thus,

$$\mathbf{AB}_{2\times2} = \begin{bmatrix} 2 & 1 & 0 \\ 0 & 3 & 1 \end{bmatrix} \begin{bmatrix} 1 & -2 \\ 1 & 0 \\ 3 & 2 \end{bmatrix} = \begin{bmatrix} 3 & -4 \\ 6 & 2 \end{bmatrix}$$

In general, if $\mathbf{A}_{m\times n} = ((a_{ij}))$ and $\mathbf{B}_{n\times p} = ((b_{ij}))$, the (i, j)th element c_{ij} of the product

$$\mathbf{AB}_{m\times p} = ((c_{ij}))$$

is defined to be

$$c_{ij} = \sum_{l=1}^{n} a_{il}b_{lj}, \qquad i = 1, 2, \ldots, m; \quad j = 1, 2, \ldots, p$$

Thus, as another example, if

$$\mathbf{A}_{2\times2} = \begin{bmatrix} -1 & 3 \\ 2 & 2 \end{bmatrix} \quad \text{and} \quad \mathbf{B}_{2\times1} = \begin{bmatrix} 0 \\ 1 \end{bmatrix}$$

then $m = 2$, $p = 1$, $n = 2$, and

$$c_{11} = \sum_{l=1}^{2} a_{1l}b_{l1} = (-1 \times 0) + (3 \times 1) = 3$$

$$c_{21} = \sum_{l=1}^{2} a_{2l}b_{l1} = (2 \times 0) + (2 \times 1) = 2$$

so that

$$\mathbf{AB}_{2\times1} = \begin{bmatrix} -1 & 3 \\ 2 & 2 \end{bmatrix} \begin{bmatrix} 0 \\ 1 \end{bmatrix} = \begin{bmatrix} 3 \\ 2 \end{bmatrix}$$

Here are a few other examples for additional practice:

1. Find $\mathbf{AI}$ and $\mathbf{IA}$, where

$$\mathbf{A}_{2\times2} = \begin{bmatrix} -1 & 3 \\ 2 & 2 \end{bmatrix} \quad \text{and} \quad \mathbf{I}_{2\times2} = \begin{bmatrix} 1 & 0 \\ 0 & 1 \end{bmatrix}$$

2. Find $\mathbf{X'X}$, where

$$\mathbf{X}_{3\times2} = \begin{bmatrix} 1 & 10 \\ 1 & 15 \\ 1 & 20 \end{bmatrix}$$

The answer to example 1 is

$$\mathbf{AI} = \mathbf{IA} = \mathbf{A}$$

which indicates why **I** is generally referred to as the *identity matrix*, since, like the scalar identity 1, the product of any square matrix (e.g., **A**) with an appropriate identity matrix (of the right dimensions) will always yield the original matrix **A**, whether **I** premultiplies or postmultiplies **A**.

The answer to example 2 is

$$\mathbf{X'X}_{2\times 2} = \begin{bmatrix} 1 & 1 & 1 \\ 10 & 15 & 20 \end{bmatrix} \begin{bmatrix} 1 & 10 \\ 1 & 15 \\ 1 & 20 \end{bmatrix} = \begin{bmatrix} 3 & 45 \\ 45 & 725 \end{bmatrix}$$

which is a symmetric matrix, as will be the case whenever any matrix is multiplied (pre or post) by its own transpose.

B.7 INVERSE OF A MATRIX

The definition of an *inverse matrix* parallels the basic property of the *reciprocal* of an ordinary (scalar) number. That is, for any nonzero number a, its reciprocal $1/a$ satisfies the equation

$$a \times \frac{1}{a} = \frac{1}{a} \times a = 1$$

In words, when a is either pre- or postmultiplied by its reciprocal, the result is the scalar 1. Analogously, we say that a matrix **A** has an inverse $\mathbf{A}^{-1}$ if and only if

$$\mathbf{AA}^{-1} = \mathbf{A}^{-1}\mathbf{A} = \mathbf{I}$$

That is, *the product of **A** by its inverse must be equal to an identity matrix*. We hasten to add at this point that *only square matrices can have inverses*. Thus, if **A** is $(n \times n)$, $\mathbf{A}^{-1}$ must also be $(n \times n)$.

We shall not describe in this appendix any of the many algorithms available for actually computing the inverse of a matrix [for further details, see Draper and Smith (1966) or Mendenhall (1968)]. For most practical applications, one can use standard computer packages for such computations. We wish to emphasize instead that one important attribute of inverses in statistical analyses is that their use permits the solution of matrix equations for a matrix of unknowns (e.g., the vector $\boldsymbol{\beta}$ of unknown regression parameters), in a manner similar to how "division" is used in ordinary algebra. More specifically, in regression analysis, the use of an inverse is crucial in that it provides for an efficient solution of the least-squares equations, as well as providing compact formulas for additional components of the analysis (e.g., the variances of and covariances among the estimated regression coefficients).

Some examples of matrix inverses are given as follows:
The inverse of

$$\mathbf{A} = \begin{bmatrix} 2 & 0 & 1 \\ 1 & -1 & 2 \\ 1 & 0 & 0 \end{bmatrix} \quad \text{is} \quad \mathbf{A}^{-1} = \begin{bmatrix} 0 & 0 & 1 \\ 2 & -1 & -3 \\ 1 & 0 & -2 \end{bmatrix}$$

(The reader can check this out by multiplying **A** and $\mathbf{A}^{-1}$ together to obtain $\mathbf{I}_{3\times 3}$.)

The inverse of

$$\mathbf{X'X} = \begin{bmatrix} n & \sum\limits_{i=1}^{n} X_i \\[2ex] \sum\limits_{i=1}^{n} X_i & \sum\limits_{i=1}^{n} X_i^2 \end{bmatrix}$$

is

$$(\mathbf{X'X})^{-1} = \begin{bmatrix} \dfrac{\sum\limits_{i=1}^{n} X_i^2}{n \sum\limits_{i=1}^{n} (X_i - \bar{X})^2} & \dfrac{-\bar{X}}{\sum\limits_{i=1}^{n} (X_i - \bar{X})^2} \\[4ex] \dfrac{-\bar{X}}{\sum\limits_{i=1}^{n} (X_i - \bar{X})^2} & \dfrac{1}{\sum\limits_{i=1}^{n} (X_i - \bar{X})^2} \end{bmatrix}$$

B.8 MATRIX FORMULATION OF REGRESSION ANALYSIS

We have previously seen in Section B.3 that when fitting a straight-line model

$$Y = \beta_0 + \beta_1 X + E$$

to a set of data consisting of n pairs of observations on the variables X and Y, we can define several matrices to characterize the regression problem under consideration:

$\mathbf{Y}$, the $(n \times 1)$ vector of observations on Y

$\mathbf{X}$, the $(n \times 2)$ matrix of independent variables

$\boldsymbol{\beta}$, the (2×1) vector of parameters

$\mathbf{E}$, the $(n \times 1)$ vector of random errors

In general, whenever we are considering a regression problem involving p independent variables using a model such as

$$Y = \beta_0 + \beta_1 X_1 + \beta_2 X_2 + \cdots + \beta_p X_p + E$$

we can analogously construct appropriate matrices based on the multivariable data set being considered. Thus, if the data on the ith individual consists of the $(p + 1)$ values

$$Y_i, X_{i1}, X_{i2}, \ldots, X_{ip}, \qquad i = 1, 2, \ldots, n$$

the following matrices can be constructed:

$$\mathbf{Y}_{n \times 1} = \begin{bmatrix} Y_1 \\ Y_2 \\ \cdot \\ \cdot \\ \cdot \\ Y_n \end{bmatrix}, \quad \text{the vector of observations on } Y$$

$$\mathbf{X}_{n \times (p+1)} = \begin{bmatrix} 1 & X_{11} & X_{12} & \cdots & X_{1p} \\ 1 & X_{21} & X_{22} & \cdots & X_{2p} \\ \vdots & \vdots & \vdots & & \vdots \\ 1 & X_{n1} & X_{n2} & \cdots & X_{np} \end{bmatrix}, \quad \text{the matrix of independent variables}$$

$$\boldsymbol{\beta}_{(p+1) \times 1} = \begin{bmatrix} \beta_0 \\ \beta_1 \\ \beta_2 \\ \vdots \\ \beta_p \end{bmatrix}, \quad \text{the vector of parameters}$$

$$\mathbf{E}_{n \times 1} = \begin{bmatrix} E_1 \\ E_2 \\ \vdots \\ E_n \end{bmatrix}, \quad \text{the vector of random errors}$$

Using the matrices above in conjunction with the notions of matrix addition and multiplication, we are then able to formulate the general regression model in matrix terms as follows:

$$\mathbf{Y}_{n \times 1} = \mathbf{X}_{n \times (p+1)} \boldsymbol{\beta}_{(p+1) \times 1} + \mathbf{E}_{n \times 1} \tag{B.1}$$

This compact equation (B.1) summarizes in a single statement the n equations

$$Y_i = \beta_0 + \beta_1 X_{i1} + \beta_2 X_{i2} + \cdots + \beta_p X_{ip} + E_i$$

$i = 1, 2, \ldots, n$. Note that this equivalence follows from the following matrix calculations:

$$\mathbf{X}\boldsymbol{\beta} + \mathbf{E} = \begin{bmatrix} 1 & X_{11} & X_{12} & \cdots & X_{1p} \\ 1 & X_{21} & X_{22} & \cdots & X_{2p} \\ \vdots & \vdots & & & \vdots \\ \vdots & \vdots & & & \vdots \\ 1 & X_{n1} & X_{n2} & \cdots & X_{np} \end{bmatrix} \begin{bmatrix} \beta_0 \\ \beta_1 \\ \vdots \\ \vdots \\ \beta_p \end{bmatrix} + \begin{bmatrix} E_1 \\ E_2 \\ \vdots \\ \vdots \\ E_n \end{bmatrix}$$

$$= \begin{bmatrix} \beta_0 + \beta_1 X_{11} + \cdots + \beta_p X_{1p} \\ \beta_0 + \beta_1 X_{21} + \cdots + \beta_p X_{2p} \\ \vdots \\ \beta_0 + \beta_1 X_{n1} + \cdots + \beta_p X_{np} \end{bmatrix} + \begin{bmatrix} E_1 \\ E_2 \\ \vdots \\ E_n \end{bmatrix}$$

$$= \begin{bmatrix} \beta_0 + \beta_1 X_{11} + \cdots + \beta_p X_{1p} + E_1 \\ \beta_0 + \beta_1 X_{21} + \cdots + \beta_p X_{2p} + E_2 \\ \vdots \\ \beta_0 + \beta_1 X_{n1} + \cdots + \beta_p X_{np} + E_n \end{bmatrix}$$

Based on the general matrix equation given by (B.1), a description of all the essential features of regression analysis can be expressed in matrix notation. In particular, the least-squares solution for the estimates of the regression coefficients in the parameter vector $\boldsymbol{\beta}$ can now be compactly written. This least-squares solution, in matrix terms, is that vector $\hat{\boldsymbol{\beta}}$ which minimizes the error sum of squares (given in matrix terms)

$$(\mathbf{Y} - \mathbf{X}\hat{\boldsymbol{\beta}})'(\mathbf{Y} - \mathbf{X}\hat{\boldsymbol{\beta}})$$

The solution to this minimization problem, which is obtained via the use of matrix calculus, yields the following easy-to-remember matrix formula:

$$\hat{\boldsymbol{\beta}} = (\mathbf{X}'\mathbf{X})^{-1}\mathbf{X}'\mathbf{Y}$$

where $\hat{\boldsymbol{\beta}}'_{1\times(p+1)} = [\hat{\beta}_0, \hat{\beta}_1, \ldots, \hat{\beta}_p]$ denotes the vector of estimated regression coefficients.

Thus, although we have pointed out in the text the futility of *explicitly* giving the solutions to the least-squares equations for models of more complexity than a straight line, we can at least *implicitly* express these solutions in matrix notation and, because of modern computer technology, can conveniently carry through with the computation of the least-squares solutions using this matrix representation.

At this point it is not our intention in this appendix to carry through completely with the matrix formulation of every other aspect of a regression analysis. The reader is referred to Draper and Smith (1966) for a fuller treatment of this matrix approach. However, some additional matrix results will be briefly summarized below to give more of a flavor for the utility of the matrix approach:

The *vector of predicted responses* is given by $\hat{\mathbf{Y}} = \mathbf{X}\hat{\boldsymbol{\beta}}$.

The *error sum of squares* SSE is given by

$$\mathrm{SSE} = \mathbf{Y}'\mathbf{Y} - \hat{\boldsymbol{\beta}}'\mathbf{X}'\mathbf{Y}$$

The *regression sum of squares* is given by $\hat{\beta}'X'Y - n\bar{Y}^2$.

The *variances of the regression coefficients*, that is, $\sigma_{\hat{\beta}_i}^2$, $i = 0, 1, 2, \ldots, p$, are given by the diagonal elements of the matrix

$$(X'X)^{-1}\sigma^2$$

Finally, although we do not present the details here, any test of a (linear) statistical hypothesis concerning some subset of the regression coefficients can be formulated and carried out entirely by means of matrix operations.

SOLUTIONS TO EXERCISES

Chapter 3

1. e   *2.* c   *3.* a   *4.* e   *5.* e   *6.* d   *7.* a   *8.* e   *9.* e   *10.* c
11. b   *12.* b   *13.* b   *14.* b   *15.* c   *16.* c   *17.* b   *18.* d
19. a   *20.* b

Chapter 5

1. (c) $\hat{\beta}_0(Y \text{ on } X) = -1.884$, $\hat{\beta}_1(Y \text{ on } X) = 0.235$
$\hat{\beta}_0(Z \text{ on } X) = -2.690$, $\hat{\beta}_1(Z \text{ on } X) = 0.196$

(d) 95% CI for $\beta_1(Y \text{ on } X)$: $0.131 < \beta_1 < 0.339$
95% CI for $\beta_1(Z \text{ on } X)$: $0.191 < \beta_1 < 0.201$
For $\alpha = 0.05$, reject H_0: $\beta_1 = 0$ for both regression equations in favor of H_A: $\beta_1 \neq 0$, since neither CI contains the value *zero*.

(e) For 90% *confidence bands*, use

$Y \text{ on } X$: $0.701 + 0.235(X_0 - 11) \pm 0.884\sqrt{0.091 + (X_0 - 11)^2/110}$

$Z \text{ on } X$: $-0.535 + 0.196(X_0 - 11) \pm 0.052\sqrt{0.091 + (X_0 - 11)^2/110}$

for various values of X_0.
For 90% *prediction bands*, use

$Y \text{ on } X$: $0.701 + 0.235(X_0 - 11) \pm 0.884\sqrt{1.091 + (X_0 - 11)^2/110}$

$Z \text{ on } X$: $-0.535 + 0.196(X_0 - 11) \pm 0.052\sqrt{1.091 + (X_0 - 11)^2/110}$

(f) The straight-line regression of Z on X produces a better fit, since the relationship between Y and X is apparently not linear.

2.3 (a) $\hat{\beta}_1 = 7.024$, $\hat{\beta}_0 = 140.800$

(b) (1) $\hat{\beta}_0$ and $\bar{Y}$(nonsmokers) have the same value.

(2) $(\hat{\beta}_0 + \hat{\beta}_1)$ and $\bar{Y}$(smokers) have the same value.

(3) The straight-line regression problem is, in this case, equivalent to the two-sample problem of comparing two population means. In fact, it can be shown by substituting $X = 0$ and $X = 1$ separately into the straight-line model that β_0 is equivalent to μ(nonsmokers) and $(\beta_0 + \beta_1)$ is equivalent to μ(smokers).

(c) Computed $T = 1.398$. Do not reject H_0 since $0.10 < P < 0.20$.

(d) Yes.

2.4 (a) $\hat{\beta}_1 = 21.492$, $\hat{\beta}_0 = 70.576$

(c) Computed $T = 6.062$. Reject H_0 since $P < 0.001$.

(d) 95% CI for $\mu_{Y|\bar{x}}$: $140.99 < \mu_{Y|\bar{x}} < 148.07$.

(e) For 95% prediction bands, use the following formula:

$$144.531 + 21.492(X_0 - 3.441) \pm 20.035\sqrt{1.031 + (X_0 - 3.441)^2/7.660}$$

(f) Yes, since H_0 is rejected in part (c), which agrees with one's visual impression of the data.

(g) Not obviously.

3. (f) The straight-line-model assumption.

(h) The data suggest that a (concave-downward) parabola would provide a better fit.

4. (c) The reliability of a predicted value $\hat{Y}$ is questionable for any X outside the range of X over which the fitted model is based.

(d) 95% CI for slope: $-0.494 < \beta_1 < -0.004$.

(f) The presence of the outlier appears to make a considerable difference, since the negative straight-line relationship between DI and IQ is much more pronounced and the fit of the line to the data improves considerably when the outlier is removed.

6. (g) Substitute various values of X_0 into the following formula:

$$519.304 - 14.041(X_0 - 9.058) \pm 28.948\sqrt{0.077 + (X_0 - 9.058)^2/92.424}$$

7. (d) Computed $T = -57.813$, which lies below $t_{17,0.005} = -2.898$. Thus, for a two-sided test, reject H_0 at $\alpha = 0.01$.

(f) Use for the construction of 95% confidence and prediction bands the following formulas:

95% *confidence bands*: $10.175 + 0.299(X_0 - 39.737)$

$$\pm 1.711\sqrt{0.053 + (X_0 - 39.737)^2/4{,}473.684}$$

95% *prediction bands*: $10.175 + 0.299(X_0 - 39.737)$

$$\pm 1.711\sqrt{1.053 + (X_0 - 39.737)^2/4{,}473.684}$$

9. (d) Use the following formula for construction of 99% confidence bands:

$$1.6108 - 1.785(X_0 - 0.7425) \pm 0.1762\sqrt{0.0833 + (X_0 - 0.7425)^2/1.1264}$$

(e) $\hat{Y}' = 862.978X'^{-1.785}$

(f) At maximum dose ($X_0 = 1.15$), the 99% CI for $\mu_{Y|X_0}$ is given by

$$0.7983 < \mu_{Y|X_0 = 1.15} < 0.9685$$

Raising each limit to a power of 10, we obtain the following 99% CI for $\mu_{Y'|X_0' = 14.125}$:

$$6.2849 < \mu_{Y'|X_0' = 14.125} < 9.3004$$

At minimum dose ($X_0 = 0.18$), the corresponding 99% CI's are

$$2.5080 < \mu_{Y|X_0 = 0.18} < 2.7218$$

and

$$322.107 < \mu_{Y'|X_0' = 1.514} < 526.987$$

(g) Plot the scatter diagram of the (X', Y') pairs and draw the fitted straight line of Y' on X' on the scatter. Then compare by eye whether using Y on X gives better fit to the (X, Y) scatter than does using Y' on X' to the (X', Y') scatter. In addition, using methods described in Chapter 7, you could perform a test for lack of fit of each model and compare results.

10. (c) 94% CI's for $\mu_{Y'|X}$ at $X = 5$, 4.5, and 4: $280.802 < \mu_{Y'|X=5} < 352.858$, $931.108 < \mu_{Y'|X=4.5} < 1{,}059.25$, and $2{,}795.12 < \mu_{Y'|X=4} < 3{,}512.37$.

11. (c) Growth-rate scores for each of the three tests for each gas would provide more information.

(d) 90% CI: $2.26 < \mu_{Y|100} < 2.69$.

(e) No observations were taken at $X = 200$ or more (i.e., the value 200 is outside the region of experimentation).

(f) The X values chosen did not uniformly cover the region of experimentation (notice the big gaps between $X = 39.9$, $X = 83.8$, and $X = 131.3$).

Chapter 6

1. (a) $r_{XY} = 0.862$, $r_{XZ} = 0.999$.

(b) $0.543 < \rho_{XY} < 0.960$, $0.992 < \rho_{XZ} < 0.9995$

(c) $r_{XY}^2 = 0.744 = (8.166 - 2.090)/8.166$, $r_{XZ}^2 = 0.998 = (4.233 - 0.0071)/4.233$

(d) The regression of log dry weight on age provides a better fit, since it explains a much greater proportion of the total variation in the dependent variable than does the regression of dry weight on age.

2. (d) The quantities $\hat{\beta}_1$ and r have meaning only in the context of an assumed *linear* relationship between X and Y, whereas the true model is a parabola containing no linear component.

3. (a) $r = 0.9798$

(c) Computed $T = 13.85$ (df $= 8$). Reject H_0 since $P < 0.001$.

(d) No observations have been taken between $X = 3$ and $X = 20$, thus making the point (20, 20) an outlier in the scatter diagram. It is dangerous to "extrapolate" to a point *far* outside the range of most of the data.

4. (b) 99% CI for ρ: $0.4435 < \rho_{SBP,QUET} < 0.8923$. Since this CI does not contain 0, reject H_0: $\rho = 0$ at $\alpha = 0.01$.

8. (a) Including outlier: $r_{DI,IQ} = -0.4747$, $r^2_{DI,IQ} = 0.2253$. Excluding outlier: $r_{DI,IQ} = -0.8069$, $r^2_{DI,IQ} = 0.6511$.
   (b) Excluding outlier: computed $T = -5.291$ (df $= 16$). Reject H_0: $\rho = 0$ since $P < 0.001$.

13. (b) 99% CI for ρ: $0.847 < \rho_{uv} < 0.997$.

Chapter 7

1. (a)

SOURCE	df	SS	MS	F
Regression	1	6.0748	6.0748	26.12
Residual	9	2.0936	0.2326	
Total	10	8.1684		

2. For the regression of SBP on QUET:
   (a)

SOURCE	df	SS	MS	F
Regression	1	3,537.95	3,537.95	36.75
Residual	30	2,888.02	96.27	
Total	31	6,425.97		

3. For the regression of QUET on AGE:
   (a)

SOURCE	df	SS	MS	F
Regression	1	4.9361	4.9361	54.36
Residual $\begin{cases} \text{lack of fit} \\ \text{pure error} \end{cases}$	$30 \begin{cases} 20 \\ 10 \end{cases}$	$2.7236 \begin{cases} 2.0745 \\ 0.6491 \end{cases}$	$0.0908 \begin{cases} 0.0649 \\ 0.0649 \end{cases}$	1.60
Total	31	7.6597		

   (b) Reject H_0 since $P < 0.001$.
   (c) Straight-line model is adequate.

12. (b) Fitted straight line: $\hat{Y} = -0.4624 + 0.0358X$

   (c)

SOURCE	df	SS	MS	F
Regression	1	12.8304	12.8304	632.04
Residual $\begin{cases} \text{lack of fit} \\ \text{pure error} \end{cases}$	$18 \begin{cases} 2 \\ 16 \end{cases}$	$0.3662 \begin{cases} 0.1514 \\ 0.2148 \end{cases}$	$0.0203 \begin{cases} 0.0757 \\ 0.0134 \end{cases}$	5.65

   (d) Computed $F = 632.04$ (1 and 18 df), $P < 0.001$. Conclude significant straightline regression.
   (e) Computed $F = 5.65$ (2 and 16 df), $0.01 < P < 0.025$. Conclude significant lack of fit at $\alpha = 0.025$ but not at $\alpha = 0.01$.

Chapter 8

1. (a) Smokers: $S\hat{B}P = 79.26 + 20.12QUET$
   Nonsmokers: $S\hat{B}P = 49.31 + 26.30QUET$

   (b) t test for equal slopes (one-sided): $T = 0.892$; $0.10 < P < 0.20$; do not reject H_0.

   (c) t test for equal intercepts (two-sided): $T = -1.248$; $0.20 < P < 0.40$; do not reject H_0.

   (d) Since the tests for slope and intercept did not lead to rejection, we conclude that the two lines are coincident (recognizing, nevertheless, that a more efficient test is possible using multiple regression analysis).

2. (e) Z test for equal correlations (two-sided): $Z = 0.594$; $P = 0.56$, do not reject H_0. The test for equality of slopes is not equivalent to the test for equality of correlations; that is, "equal slopes" does not imply "equal correlations."

6. (a) Males: $\hat{Y} = 13.767 + 14.966X$
   Females: $\hat{Y} = 15.656 + 12.735X$

   (b) t test for equal slopes (two-sided): $T = 0.82$, 28 df, $0.40 < P < 0.60$; do not reject H_0: lines are parallel.

   (c) 99% CI for $(\beta_{1M} - \beta_{1F})$: $-5.264 < (\beta_{1M} - \beta_{1F}) < 9.726$.

   (d) t test for equal intercepts (two-sided): $T = 0.69$, 28 df, $0.40 < P < 0.60$; do not reject H_0: equal intercepts. Since both the test for equal intercepts and the test for equal slopes are not rejected, conclude that there is no evidence that the lines are not coincident.

7. (d) 95% CI for $(\mu_{Y^A|15} - \mu_{Y^B|15})$: $4.201 < (\mu_{Y^A|15} - \mu_{Y^B|15}) < 4.799$.

Chapter 9

1. (c)

SOURCE		df	SS	MS	F
Regression		1	12.7050	12.7050	43.69
Residual {	lack of fit	4	4.4196	1.1049	57.03
	pure error	12	0.2325	0.0194	
Total		17	17.3571		

(d)

SOURCE		df	SS	MS	F
Regression {	degree 1 (X)	1	12.7050	12.7050	43.69
	degree 2 $(X^2|X)$	1	3.9051	3.9051	78.46
Residual {	lack of fit	3	0.5145	0.1715	8.85
	pure error	12	0.2325	0.0194	
Total		17	17.3571		

(e) $r_{XY}^2 = 0.732$, R^2 (quadratic regression of Y on X) $= 0.957$.

(f) Test for significance of straight-line regression of Y on X: $F = 12.705/0.2908 = 43.69$ (1 and 16 df); $P < 0.001$; reject H_0. Test for adequacy of straight-line model: $F = 1.1049/0.0194 = 57.03$ (4 and 12 df); $P < 0.001$; reject H_0; conclude straight-line model is not adequate.

(g) Test for significance of quadratic regression: $F = 8.3051/0.0498 = 166.77$ (2 and 15 df); $P < 0.001$; reject H_0. Test for addition of X^2 term: partial $F(X^2|X) = 3.9051/0.0498 = 78.42$ (1 and 15 df); $P < 0.001$; reject H_0. Test for adequacy of quadratic model: $F = 0.1715/0.0194 = 8.85$ (3 and 12 df); $0.001 < P < 0.005$; reject H_0; that is, conclude quadratic model is not adequate.

(h) Test for significance of straight-line regression of ln Y on X:

$$F = 4550.35; \ P \ll 0.001; \text{ reject } H_0.$$

Test for straight-line model adequacy of ln Y on X:

$$F = 0.6819 \ (4 \text{ and } 12 \text{ df}); \ P > 0.25; \text{ conclude that model is adequate.}$$

(i) R^2 (straight-line regression of ln Y on X) $= 0.996$. R^2 (quadratic regression of Y on X) $= 0.957$. The straight-line fit of ln Y on X provides a better fit to that data than the quadratic model of Y on X.

(j) (1) Homoscedasticity assumption appears to be much more reasonable when using ln Y on X than when using Y on X.

    (2) Straight-line regression of ln Y on X is preferred: gives higher R^2; adequate model; satisfies assumption of homoscedasticity; better graphical fit.

(k) Independence assumption is violated.

7. (b) Straight-line regression: $F = 8.218$ (1 and 10 df); $0.01 < P < 0.025$. Addition of X^2: $F = 2.250$ (1 and 9 df); $0.10 < P < 0.25$. Addition of X^3: $F = 12.451$ (1 and 8 df); $0.005 < P < 0.01$.

(c) The cubic model is best for several reasons: it appears to fit the scatter better, the addition of X^3 to the model is highly significant, and the change in R^2 is considerable when going from degree 1 ($R^2 = 0.451$) to degree 2 ($R^2 = 0.561$) to degree 3 ($R^2 = 0.828$).

Chapter 10

1. (a) (1) $\hat{Y}(\text{AGE} = 50, \text{SMK} = 1, \text{QUET} = 3.5) = 145.77$.

    (2) $\hat{Y}(\text{AGE} = 50, \text{SMK} = 0, \text{QUET} = 3.5) = 135.83$.

    (3) $\hat{Y}(\text{AGE} = 50, \text{SMK} = 1, \text{QUET} = 3.5) - \hat{Y}(\text{AGE} = 50, \text{SMK} = 1, \text{QUET} = 3.0) = 4.30$.

(b) $R^2(\text{model 1}) = 0.601$; $R^2(\text{model 2}) = 0.730$; $R^2(\text{model 3}) = 0.761$.

(c) Model 1: $H_0: \beta_1 = 0$ for the model $Y = \beta_0 + \beta_1 X_1 + E$; $F = 45.177$ (1 and 30 df); $P < 0.001$; reject H_0.

Model 2: $H_0: \beta_1 = \beta_2 = 0$ for the model $Y = \beta_0 + \beta_1 X_1 + \beta_2 X_2 + E$; $F = 39.164$ (2 and 29 df); $P < 0.001$; reject H_0.

Model 3: $H_0: \beta_1 = \beta_2 = \beta_3 = 0$ for the model $Y = \beta_0 + \beta_1 X_1 + \beta_2 X_2 + \beta_3 X_3 + E$; $F = 29.710$ (3 and 28 df); $P < 0.001$; reject H_0.

(d) (1) Partial $F(\text{SMK}|\text{AGE}) = 13.830$ (1 and 29 df); $P < 0.001$; reject H_0.

    (2) Partial $F(\text{QUET}|\text{AGE}, \text{SMK}) = 3.648$ (1 and 28 df); $0.05 < P < 0.10$; reject H_0 at $\alpha = 0.10$ but not at $\alpha = 0.05$.

(e) Best model would include the variables AGE and SMK if $\alpha \leqslant 0.05$, but all three variables if $\alpha = .10$.

4. (b) The model with X_2 and X_3 since it has the highest overall F value, even though the partial $F(X_2|X_3)$ is not significant at $\alpha = 0.01$.

   (d) Partial $F(X_1|X_2, X_3) = 1.37$ (1 and 16 df); $P > 0.25$; no significant improvement by adding X_1 to model.

   (e) Increase in R^2 is 0.016, which is very small.

   (f) Partial $F(X_4|X_2, X_3) = 0.004$ (1 and 16 df); $P > 0.25$.

   (g) X_2 and X_3 are the only two important variables, and X_3 is more important than X_2. Based on R^2 values, the same conclusion should be reached.

6. (d) X_1 should be retained; debatable for X_2 and X_3.

7. (d) Partial $F(X_2|X_1) = 25.302$ (1 and 12 df); $P < 0.001$.

   (e) X_1, because it has a higher (though negative) correlation with Y ($r = -0.843$) than does X_2 ($r = 0.807$). Equivalently, X_1 is more highly significant from part (b) than is X_2 from part (c).

   (f) Best model contains both X_1 and X_2.

   (g) $X_1 = 0.043$.

   (h) $X_1 = X_1^0 + (\hat{\beta}_2/\hat{\beta}_1)X_2^0$.

8. (c) $F(X_2) = 208.248$ (1 and 9 df); $P < 0.001$.
   Partial $F(X_3|X_2) = 2.022$ (1 and 8 df); $0.10 < P < 0.25$.
   Partial $F(X_1|X_2, X_3) = 3.019$ (1 and 7 df); $0.10 < P < 0.25$.

   (d) X_2, because it entered the model ahead of X_3 after X_1 was already in the model.

   (e) The addition of X_2 to a model containing X_1 is of significance, since partial $F(X_2|X_1) = 11.218$ (1 and 8 df), so that $0.01 < P < 0.025$. The addition of X_3 to a model containing both X_1 and X_2 is not significant, since partial $F(X_3|X_1, X_2) = 3.168$ (1 and 7 df), so that $0.10 < P < 0.25$. The computer information given for this problem is not sufficient to determine whether or not X_3 adds significantly to the prediction of Y after controlling for X_1.

   (f) X_1, which denotes slope gradient, might be forced in first if it was already generally accepted that slope gradient was an important factor in soil erosion and/or if the focus of the study was to see if any other independent variable was important after taking slope gradient into account.

   (g) Forcing in X_1 first, the best model (using $\alpha = 0.025$) involves X_1 and X_2. Not forcing in X_1, the best model involves only X_2.

Chapter 11

1. (a) AGE

   (b) $r_{SBP,SMK|AGE} = 0.568$, $r_{SBP,QUET|AGE} = 0.318$. The variable SMK should be considered next, since it has the higher partial correlation. However, this variable should not be included as an important predictor of Y unless it provides *significant* predictive power (as measured by a test of hypothesis) over that already achieved by AGE.

   (c) H_0: $\rho_{SBP,SMK|AGE} = 0$; partial $F(SMK|AGE) = 13.830$ (1 and 29 df); $P < 0.001$.

   (d) $r_{SBP,QUET|AGE,SMK} = 0.340$; partial $F(QUET|AGE, SMK) = 3.648$ (1 and 28 df); $0.05 < P < 0.10$.

   (e) AGE (most important), then SMK, and finally QUET. For $\alpha = 0.05$, only AGE and SMK would be considered important. For $\alpha = 0.01$, only AGE would be considered important enough to be included in the model.

   (f) $r^2_{SBP(QUET,SMK)|AGE} = 0.4009$; multiple–partial $F(SMK, QUET|AGE) = 9.371$ (2 and 28 df); $P < 0.001$.

(g) Beta(1) = 0.586, beta(2) = 0351, beta(3) = 0.297, which puts AGE first, SMK second, and QUET last, in agreement with the previous stepwise-regression order.

2. (a) For H_0: $\rho_{YX_3|X_1} = 0$, $T = 1.152$ (85 df); $P > 0.20$; do *not* reject H_0.
       For H_0: $\rho_{YX_3|X_1,X_2} = 0$, $T = 1.117$ (84 df); $P > 0.20$; do *not* reject H_0.

   (b)

SOURCE	df	SS	MS
X_1	1	0.8087	0.8087
$X_2|X_1$	1	0.0344	0.0344
$X_3|X_1, X_2$	1	0.2958	0.2958
Residual	84	19.9111	0.2370

   (c) H_0: $\rho_{Y(\text{interaction terms})|\text{main effects}} = 0$.
       Multiple–partial $F(\text{interaction terms}|\text{main effects}) = 0.7488$ (25 and 54 df); $P > 0.25$.

   (d) Overall $F(X_1) = 1.600$ (1 and 86 df); $0.10 < P < 0.25$.
       Overall $F(\text{main effects}) = 1.4435$ (8 and 79 df); $0.10 < P < 0.25$.
       Overall $F(\text{main effects} + \text{interactions}) = 0.8894$ (33 and 54 df); $P < 0.25$.

   (e) No evidence of a relationship has been found.

3. (a) Compute and interpret $r_{XY|Z}$, and test H_0: $\rho_{XY|Z} = 0$.

   (b) $r_{YX_1|X_2} = 0.028$; $r_{YX_1|X_3} = 0.184$; $r_{YX_1|X_2,X_3} = -0.1265$. All these partial correlations are small, suggesting a spurious correlation of 0.35 between unemployment level and respiratory morbidity rate. The tests of hypotheses concerning the significance of these partial correlations are all nonsignificant:
       Testing H_0: $\rho_{XY_1|X_2} = 0$ yields $T = 0.131$ (22 df); $P > 0.80$.
       Testing H_0: $\rho_{YX_1|X_3} = 0$ yields $T = 0.878$ (22 df); $P > 0.20$.
       Testing H_0: $\rho_{YX_1|X_2,X_3} = 0$ yields $T = -0.584$ (21 df); $P > 0.40$.

8. (a) $r_{YX_2|X_1} = 0.764$, $r_{YX_3}|_{X_1} = -0.727$.

   (b) X_2 should be entered next, provided that it sjgnificantly adds to the prediction of Y.

   (c) Computed $T = 3.349$ (8 df); $0.01 < P < 0.02$ (two-sided).

   (d) $r^2_{Y(X_2,X_3)|X_1} = 0.7134$. Multiple–partial $F(X_2, X_3|X_1) = 8.713$ (2 and 7 df); $0.01 < P < 0.025$.

Chapter 12

1. (f) Multiple–partial $F(X_2, X_3|X_1) = 1.778$ (2 and 49 df); $0.10 < P < 0.25$.

   (g) For $\alpha = 0.05$, only $X_1 = \text{SBP1}$ should be used in the model.

   (h) Yes.

2. (a) Testing H_0: $\rho_{YX_1} = 0$ or H_0: $\beta_1 = 0$ in the model $Y = \beta_0 + \beta_1 X_1 + E$ yields $F(X_1) = 25.758$ (1 and 20 df); $P < 0.001$.
       Testing H_0: $\rho_{YX_2|X_1} = 0$ or H_0: $\beta_2 = 0$ in the model $Y = \beta_0 + \beta_1 X_1 + \beta_2 X_2 + E$ yields partial $F(X_2|X_1) = 6.322$ (1 and 19 df); $0.01 < P < 0.025$.
       Testing H_0: $\rho_{YX_3|X_1,X_2} = 0$ or H_0: $\beta_3 = 0$ in the model $Y = \beta_0 + \beta_1 X_1 + \beta_2 X_2 + \beta_3 X_3 + E$ yields partial $F(X_3|X_1, X_2) = 5.267$ (1 and 18 df); $0.025 < P < 0.05$.

   (b) $R^2_{Y|X_1} = 0.563$, $R^2_{Y|X_1,X_2} = 0.672$, $R^2_{Y|X_1,X_2,X_3} = 0.746$. All these R^2 values are high, and there are reasonably large increases in R^2 corresponding to the additions of X_2 and X_3 to the model.

(c) H_0: $\rho_{Y(X_2,X_3)|X_1} = 0$ or H_0: $\beta_2 = \beta_3 = 0$ in the model $Y = \beta_0 + \beta_1 X_1 + \beta_2 X_2 + \beta_3 X_3 + E$. Multiple–partial $F(X_2, X_3|X_1) = 6.505$ (2 and 18 df); $0.005 < P < 0.01$.

(d) Variables in the given ANOVA table are not entered in the necessary order. We require that X_3 enter first, and then that X_1 and X_2 enter in any order. Use the formula

multiple–partial $F(X_1, X_2|X_3)$

$$= \frac{[\text{SS regression } (X_1, X_2, X_3) - \text{SS regression } (X_1)]/2}{\text{MS residual } (X_1, X_2, X_3)}$$

which is distributed as an F with 2 and 18 df under H_0.

4. (c) H_0: $\rho_{Y(X_1X_3,X_2X_3)|X_1,X_2,X_3} = 0$.
Multiple–partial $F(X_1X_3, X_2X_3|X_1, X_2, X_3) = 0.362$ (2 and 36 df), $P \gg 0.25$.
Conclusion: The regression equation of Y on X_1 and X_2 when SPB is absent is "parallel" to the equation of Y on X_1 and X_2 when SPB is present (i.e., corresponding regression coefficients are the same).

(d) H_0: $\rho_{YX_3|X_1,X_2} = 0$. Partial $F(X_3|X_1, X_2) = 6.855$ (1 and 38 df), $0.01 < P < 0.025$.
Conclusion: The regression equation when SPB is absent is not coincident with the equation when SPB is present ($\alpha = 0.05$).

(e) SPB is associated with HDL measurement, but the other two independent variables are not and so should not be included in the model.

Chapter 13

1. (a) $\text{SBP} = \beta_0 + \beta_1(\text{QUET}) + \beta_2(\text{SMK}) + \beta_3(\text{QUET·SMK}) + E$, where

$$\text{SMK} = \begin{cases} 1 & \text{if smoker} \\ 0 & \text{if nonsmoker} \end{cases}$$

Smokers: $\text{SBP} = (\beta_0 + \beta_2) + (\beta_1 + \beta_3)(\text{QUET}) + E$.
Nonsmokers: $\text{SBP} = \beta_0 + \beta_1(\text{QUET}) + E$.

(b) Smokers: $\hat{\text{SBP}} = 79.253 + 20.118(\text{QUET})$.
Nonsmokers: $\hat{\text{SBP}} = 49.312 + 26.303(\text{QUET})$.

(c) H_0: $\beta_3 = 0$ (parallelism): partial $F(\text{QUET·SMK}|\text{QUET, SMK}) = 0.796$ (1 and 28 df); $P > 0.25$.
H_0: $\beta_2 = \beta_3 = 0$ (coincidence): multiple–partial $F(\text{SMK, QUET·SMK}|\text{QUET}) = 4.03$ (2 and 28 df); $0.025 < P < 0.05$.

(d) Test for parallelism gives the same result as previously obtained. Test for coincidence gives a different result for $\alpha = 0.05$.

3. (a) Males: $\hat{\text{SBPSL}} = (\hat{\beta}_0 + \hat{\beta}_2) + (\hat{\beta}_1 + \hat{\beta}_4)(\text{SBP1}) + (\hat{\beta}_3 + \hat{\beta}_6)(\text{RW})$
$+ (\hat{\beta}_5 + \hat{\beta}_7)(\text{SBP1·RW})$.
Females: $\hat{\text{SBPSL}} = (\hat{\beta}_0 - \hat{\beta}_2) + (\hat{\beta}_1 - \hat{\beta}_4)(\text{SBP1}) + (\hat{\beta}_3 - \hat{\beta}_6)(\text{RW})$
$+ (\hat{\beta}_5 - \hat{\beta}_7)(\text{SBP1·RW})$.

(b) Variables are entered in the given ANOVA table in an inappropriate order for making the required tests. To carry out either the test for parallelism or the test for coincidence, the correct order of entry should be X_1, X_3, and X_5 followed by X_2, X_4, X_6, and X_7.

In testing for parallelism (H_0: $\beta_4 = \beta_6 = \beta_7 = 0$), the test statistic to be used in the multiple–partial $F(X_4, X_6, X_7|X_1, X_2, X_3, X_5)$, which has the F distribution with 3 and 96 df under H_0.

In testing for coincidence (H_0: $\beta_2 = \beta_4 = \beta_6 = \beta_7 = 0$), the test statistic to be used is the multiple–partial $F(X_2, X_4, X_6, X_7 | X_1, X_3, X_5)$, which has the F distribution with 4 and 96 df under H_0.

6. (a) $R = 1$ and TD $= 1$: SB$\hat{P}$SL $= (\hat{\beta}_0 + \hat{\beta}_2 + \hat{\beta}_4 + \hat{\beta}_9) + \hat{\beta}_1$SBP1 $+ (\hat{\beta}_6 + \hat{\beta}_{11})$RW
$R = 0$ and TD $= 1$: SB$\hat{P}$SL $= (\hat{\beta}_0 + \hat{\beta}_3 + \hat{\beta}_4 + \hat{\beta}_7) + \hat{\beta}_1$SBP1 $+ (\hat{\beta}_6 + \hat{\beta}_{13})$RW
$R = -1$ and TD $= 1$: SB$\hat{P}$SL $= (\hat{\beta}_0 - \hat{\beta}_2 - \hat{\beta}_3 + \hat{\beta}_4 - \hat{\beta}_7 - \hat{\beta}_9) + \hat{\beta}_1$SBP1
$+ (\hat{\beta}_6 - \hat{\beta}_{11} - \hat{\beta}_{13})$RW
$R = 1$ and TD $= 0$: SB$\hat{P}$SL $= (\hat{\beta}_0 + \hat{\beta}_2 + \hat{\beta}_5 + \hat{\beta}_{10}) + \hat{\beta}_1$SBP1 $+ (\hat{\beta}_6 + \hat{\beta}_{12})$RW
$R = 0$ and TD $= 0$: SB$\hat{P}$SL $= (\hat{\beta}_0 + \hat{\beta}_3 + \hat{\beta}_5 + \hat{\beta}_8) + \hat{\beta}_1$SBP1 $+ (\hat{\beta}_6 + \hat{\beta}_{14})$RW
$R = -1$ and TD $= 0$: SB$\hat{P}$SL $= (\hat{\beta}_0 - \hat{\beta}_2 - \hat{\beta}_3 + \hat{\beta}_5 - \hat{\beta}_8 - \hat{\beta}_{10}) + \hat{\beta}_1$SBP1
$+ (\hat{\beta}_6 - \hat{\beta}_{12} - \hat{\beta}_{14})$RW
$R = 1$ and TD $= -1$: SB$\hat{P}$SL $= (\hat{\beta}_0 + \hat{\beta}_2 - \hat{\beta}_4 - \hat{\beta}_5 - \hat{\beta}_9 - \hat{\beta}_{10}) + \hat{\beta}_1$SBP1
$+ (\hat{\beta}_6 - \hat{\beta}_{11} - \hat{\beta}_{12})$RW
$R = 0$ and TD $= -1$: SB$\hat{P}$SL $= (\hat{\beta}_0 + \hat{\beta}_3 - \hat{\beta}_4 - \hat{\beta}_5 - \hat{\beta}_7 - \hat{\beta}_8) + \hat{\beta}_1$SBP1
$+ (\hat{\beta}_6 - \hat{\beta}_{13} - \hat{\beta}_{14})$RW
$R = -1$ and TD $= -1$: SB$\hat{P}$SL $= (\hat{\beta}_0 - \hat{\beta}_2 - \hat{\beta}_3 - \hat{\beta}_4 - \hat{\beta}_5 + \hat{\beta}_7 + \hat{\beta}_8 + \hat{\beta}_9 + \hat{\beta}_{10})$
$+ \hat{\beta}_1$SBP1 $+ (\hat{\beta}_6 + \hat{\beta}_{11} + \hat{\beta}_{12} + \hat{\beta}_{13} + \hat{\beta}_{14})$RW

(b) H_0: $\beta_{11} = \beta_{12} = \beta_{13} = \beta_{14} = 0$; multiple–partial $F = 1.409$ (4 and 89 df); $0.10 < P < 0.25$.

(c) Multiple–partial $F = 1.102$ (8 and 89 df); $P > 0.25$.

(d) H_0: $\beta_2 = \beta_3 = \beta_4 = \beta_5 = \beta_7 = \beta_8 = \beta_9 = \beta_{10} = \beta_{11} = \beta_{12} = \beta_{13} = \beta_{14} = 0$. Use

multiple–partial F

$$= \frac{[\text{SS regression (full model)} - \text{SS regression } (X_1, X_6)]/12}{\text{MS residual (full model)}}$$

which has 12 and 89 df under H_0.

7. (a) Location 1: $|\widehat{d_{ij}}| = -0.42 + 0.16(X_{ij1} + X_{ij2})$.
Location 2: $|\widehat{d_{ij}}| = 2.01 - 0.04(X_{ij1} + X_{ij2})$.

(b) Partial $F = 41.64$ (1 and 16 df); $P \ll 0.001$.

(c) Multiple–partial $F = 25.21$ (2 and 16 df); $P < 0.001$.

8. (a) $Y = \beta_0 + \beta_1 X + \beta_2 X^2 + \beta_3 Z + \beta_4 ZX + \beta_5 ZX^2 + E$, where

$$Z = \begin{cases} 0 & \text{if medium A} \\ 1 & \text{if medium B} \end{cases}$$

(b) Medium A: $\hat{Y}_A = 0.4946 + 0.00537X + 0.00022X^2$
Medium B: $\hat{Y}_B = 0.3509 + 0.00660X + 0.00021X^2$

(c) Parallelism: multiple–partial $F(ZX, ZX^2 | X, X^2, Z) = 0.006$ (2 and 34 df); $P \gg 0.25$.

(d) Coincidence: multiple–partial $F(Z, ZX, ZX^2 | X, X^2) = 4.08$ (3 and 34 df); $0.01 < P < 0.025$.

(e) No. A test to determine whether a quadratic term is needed in the model for each medium could be made by assessing whether the addition of X^2 and ZX^2 to a model containing X, Z, and ZX is significant. This would require results for a regression run based on a reduced model containing only X, Z, and ZX, but the results for such a run have not been presented.

Chapter 14

1. (a) $SBP = \beta_0 + \beta_1(QUET) + \beta_2(SMK) + E.$
   (b) Smokers: $\overline{SBP}$(adj.) = 148.547; $\overline{SBP}$(unadj.) = 147.823.
       Nonsmokers: $\overline{SBP}$(adj.) = 139.977; $\overline{SBP}$(unadj.) = 140.800.
   (c) $H_0: \beta_2 = 0$; partial $F(SMK|QUET) = 7.326$ (1 and 29 df); $0.01 < P < 0.025.$
   (d) $2.094 < \beta_2 < 15.048.$
   (e)

QUET CATEGORY NO. (i)	Z SCORES		$\bar{Y}_i$	S_i
	SMOKERS	NONSMOKERS		
1	0.024, 0.994, −0.218, −0.945, 1.479, −0.582, 1.236	0.145, −0.461, −1.673	133.80	8.25
2	0.841, 1.286	−1.714, −0.270, 0.508, −0.603, −0.048	137.43	9.00
3	1.413, 1.198, −0.953, 0.123	−0.092, −1.168, −0.523	148.86	9.30
4	1.522, 0.436, −1.736, 0.746	−0.650, 0.048, −0.650, 0.281	160.38	12.89
$\bar{Z}$:	0.4038	−0.4580		
$\overline{SBP}$ (with reference to cat. 3):	152.62	144.60		

Test of significance concerning the difference in mean (adjusted) scores:

$$Z = \frac{0.4038 - (-0.4580)}{\sqrt{0.05884 + 0.02901}} = 2.91, \qquad P = 0.0037$$

4. (a) $Y = \beta_0 + \beta_1 X + \beta_2 Z_1 + \beta_3 Z_2 + E$, where $\begin{cases} Z_1 = 1 & \text{if group 1 and 0 otherwise} \\ Z_2 = 1 & \text{if group 2 and 0 otherwise} \end{cases}$
   (b) $Y = \beta_0 + \beta_1 X + \beta_2 Z_1 + \beta_3 Z_2 + \beta_4(XZ_1) + \beta_5(XZ_2) + E.$
       $H_0: \beta_4 = \beta_5 = 0$; multiple–partial $F(XZ_1, XZ_2|X, Z_1, Z_2) = 4.1312$ (2 and 24 df); $0.025 < P < 0.05$; do not reject H_0 at $\alpha = 0.01.$
   (c) Group 1: $\bar{Y}$(adj.) = 4.40; $\bar{Y}$(unadj.) = 4.10.
       Group 2: $\bar{Y}$(adj.) = 4.80; $\bar{Y}$(unadj.) = 4.86.
       Group 3: $\bar{Y}$(adj.) = 4.74; $\bar{Y}$(unadj.) = 5.00.
       Testing for differences among the adjusted mean scores: $H_0: \beta_2 = \beta_3 = 0$; multiple–partial $F(Z_1, Z_2|X) = 0.637$ (2 and 26 df); $P > 0.25.$

5. (a) $VIAD = \beta_0 + \beta_1(IQM) + \beta_2(IQF) + \beta_3 Z + E.$
   (b) Male experimenter: $\overline{VIAD}$(adj.) = −3.295; $\overline{VIAD}$(unadj.) = −3.00.
       Female experimenter: $\overline{VIAD}$(adj.) = 1.901; $\overline{VIAD}$(unadj.) = 1.60.
   (c) $H_0: \beta_3 = 0$; partial $F(Z|IQM, IQF) = 2.751$ (1 and 16 df); $0.10 < P < 0.25.$
   (d) $-1.446 < \beta_3 < 11.838.$
   (e) (1), (2)

IQ CATEGORY NO (i)	Z SCORES		$\bar{Y}_i$	S_i
	GROUP 1	GROUP 2		
1	−0.876	−0.138, −0.415, 1.429	−5.50	10.85
2	−1.583, −0.287, 0.361	−0.503, 1.009, 1.009	−2.67	4.63
3	0.586, 0.152, −1.260	1.238, −0.717	3.60	9.21
4	−1.474, −0.473, 1.074	0.619, 0.255	1.20	10.99
$\bar{Z}$:	−0.3780	+0.3786		
$\overline{VIAD}$ (with reference to cat. 2):	−4.42	+0.917		

(e) (3) Test of significance concerning the difference in mean (adjusted) scores:

$$Z = \frac{-0.3780 - 0.3786}{\sqrt{0.08401 + 0.06204}} = -1.98, \qquad P = 0.0477.$$

7. (a) $\bar{Y}_{EXP}(adj.) = 13.75$, $\bar{Y}_{CON}(adj.) = 12.68$.
   (b) Partial $F(Z|X) = 0.3356$ (1 and 34 df); $P \gg 0.25$.
   (c) Alternative: two-sample t test comparing mean difference scores for each group. Computed $T = 1.324$ (35 df); $0.20 < P < 0.40$.
   (d) The two tests are *not* equivalent: regression model for covariance analysis is $Y = \beta_0 + \beta_1 X + \beta_2 Z + E$, whereas the regression model for the t test is $Y - X = \beta_0^* + \beta_1^* Z + E$.
   (e) Lack-of-fit $F = 6.357$ (30 and 4 df); $0.025 < P < 0.05$. Suggests (at $\alpha = 0.05$) that the covariance model is not completely appropriate.

Chapter 15

1. (a) Forward solution: $W\hat{G}T = 37.600 + 0.053(AGE \cdot HGT)$; $R^2 = 0.754$.
   (b) Backward solution: same answer as in part (a).
   (c) $W\hat{G}T = 6.553 + 0.722(HGT) + 2.05(AGE)$; $R^2 = 0.780$.
2. (a) Forward solution: $S\hat{B}P = 48.050 + 1.709(AGE) + 10.294(SMK)$.
   (b) Backward solution: same answer as in part (a).
   (c) All possible regressions solution: same answer as in part (a).

VARIABLES IN MODEL	QUET	AGE	SMK	QUET, AGE	QUET, SMK	AGE, SMK	QUET, AGE, SMK
R^2	0.551	0.601	0.061	0.641	0.641	0.730	0.761

(d) For $\alpha = 0.05$, add AQ only to get

$$S\hat{B}P = 79.096 + 0.482(AGE) + 9.819(SMK) + 0.186(AQ)$$

for which $R^2 = 0.773$.

4. (a) Stepwise solution: $\hat{Y} = -34.073 + 1.224X_2 + 4.399X_3$; $R^2 = 0.802$.
   (b) Backward solution: same answer as in part (a).
   (c) All-possible-regressions solution:
       (1) Best one-variable solution involves X_3, with $R^2 = 0.748$.

(2) Best two-variable solution involves X_2, X_3, with $R^2 = 0.802$.
(3) The three-variable solution yields $R^2 = 0.818$.

Chapter 17

1. (a) $\bar{Y}_1 = 7.5$, $\bar{Y}_2 = 5.0$, $\bar{Y}_3 = 4.333$, $\bar{Y}_4 = 5.167$, $\bar{Y}_5 = 6.167$
$S_1 = 1.6432$, $S_2 = 1.2649$, $S_3 = 1.0328$, $S_4 = 1.4720$, $S_5 = 2.0412$

(b)

SOURCE	df	SS	MS	F
Treatments	4	36.467	9.1167	3.90
Error	25	58.500	2.3400	
Total	29	94.967		

(c) $F = 9.1167/2.3400 = 3.90$ (4 and 25 df), $0.01 < P < 0.025$. Conclude that the treatments have significantly different effects at $\alpha = 0.025$ but not at $\alpha = 0.01$.

(d) Estimates of effects ($\bar{Y}_i - \bar{Y}$): 1.867 for treatment 1, -0.633 for 2, -1.300 for 3, -0.467 for 4, 0.533 for 5. And $\sum_{i=1}^{5}(\bar{Y}_i - \bar{Y}) = 0$.

(e) $Y = \beta_0 + \alpha_1 X_1 + \alpha_2 X_2 + \alpha_3 X_3 + \alpha_4 X_4 + E$. For

$$X_i = \begin{cases} 1 & \text{for treatment } i \ (i = 1, 2, 3, 4) \\ 0 & \text{otherwise} \end{cases}$$

the regression coefficients may be expressed as follows: $\beta_0 = \mu_5$, $\alpha_1 = \mu_1 - \mu_5$, $\alpha_2 = \mu_2 - \mu_5$, $\alpha_3 = \mu_3 - \mu_5$, $\alpha_4 = \mu_4 - \mu_5$. For

$$X_i = \begin{cases} -1 & \text{for treatment } 5 \\ 1 & \text{for treatment } i \ (i = 1, 2, 3, 4) \\ 0 & \text{otherwise} \end{cases}$$

the regression coefficients may be expressed as follows: $\beta_0 = \mu$, $\alpha_1 = \mu_1 - \mu$, $\alpha_2 = \mu_2 - \mu$, $\alpha_3 = \mu_3 - \mu$, $\alpha_4 = \mu_4 - \mu$, and $-(\alpha_1 + \alpha_2 + \alpha_3 + \alpha_4) = \mu_5 - \mu$.

(f) Ranking the sample means yields $\bar{Y}_1 > \bar{Y}_5 > \bar{Y}_4 > \bar{Y}_2 > \bar{Y}_3$, so the order of comparisons is 1 versus 3 (significant), 1 versus 2 (nonsignificant), and so on, with the remaining comparisons nonsignificant. The confidence intervals for the first two of these comparisons using an overall confidence level of 95% are:

Scheffé: $0.24 < (\mu_1 - \mu_3) < 6.10$; $-0.43 < (\mu_1 - \mu_2) < 5.43$
Tukey: $0.57 < (\mu_1 - \mu_3) < 5.76$; $-0.10 < (\mu_1 - \mu_2) < 5.10$
LSD: $0.34 < (\mu_1 - \mu_3) < 6.00$; $-0.33 < (\mu_1 - \mu_2) < 5.33$

Conclude with an overall $\alpha = 0.05$ that treatments 1 and 3 are significantly different, but that all other pairwise comparisons are nonsignificant. Schematically:

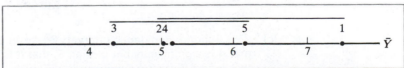

Of the three methods, Scheffé's method gives the widest interval, the LSD method the next-to-widest, and Tukey's method the narrowest interval.

2. (a)

SOURCE	df	SS	MS	F
Between cities	2	26	13.0	4.89
Within cities	6	16	2.67	
Total	8	42		

(f) $-6.59 < (\mu_I - \mu_{II}) < -1.41$.

3. (a)

SOURCE	df	SS	MS	F
Labs	2	18.7	9.35	1.75
Error	9	48.0	5.33	
Total	11	66.7		

(c) $Y_{ij} = \mu + T_i + E_{ij}$; $i = $ I, II, III; $j = 1, 2, 3, 4$: T_i and E_{ij} are independent random variables satisfying $T_i \frown N(0, \sigma_T^2)$ and $E_{ij} \frown N(0, \sigma^2)$.

(d) A measure of repeatability is σ^2, which is estimated by MSE $= 5.33$; a measure of reproducibility is $(\sigma^2 + 4\sigma_T^2)$, which is estimated by MS(labs) $= 9.35$.

SOURCE	df	SS	MS	F
Between groups	2	6,796.87	3,398.435	32.94
Within groups	197	20,323.00	103.162	
Total	199	27,119.87		

(c) Use a t statistic of the form

$$T = \frac{\bar{Y}_{combined} - \bar{Y}_{no\text{-}diff}}{\sqrt{MSE(\frac{1}{125} + \frac{1}{75})}} \frown t_{197} \text{ under } H_0$$

7. (b) The validity of the independence assumption might be questioned, since the same high schools were considered at different points in time (although different students were involved at each time point).

(c)

SOURCE	df	SS	MS	F
Years	2	551.33	275.67	2.28
Error	12	1,453.60	121.13	
Total	14	2,004.93		

8. (c) 95% Scheffé confidence interval: $(\bar{Y}_i - \bar{Y}_j) \pm 3.38$ for $i, j = A, B, C$. Conclude that A is significantly better than C, that B is significantly better than C, but that A and B are not significantly different.

(d) The tests performed in (b) and (c) address only the issue of whether or not the persons differ *from each other* in ESP ability, not whether one or more actually has significant ESP ability. Consideration of the latter issue would require a test of $H_0: \mu = 13$ versus $H_A: \mu > 13$ for each person.

10. (a)

SOURCE	df	SS	MS	F
Locations	3	30.00	10.00	13.33 (0.001 $< P <$ 0.005)
Error	8	6.00	0.75	
Total	11	36.00		

Conclusion: Significant differences (at $\alpha = 0.05$) among the four locations.

(b) $F \doteq 39.2$ (1 and 8 df); $P < 0.001$.

(c) $F = 44.55$ (1 and 10 df); $P < 0.001$.

(d) $-3\mu_1 - \mu_2 + \mu_3 + 3\mu_4 = -3[\beta_0 + \beta_1(0)] - [\beta_0 + \beta_1(10)] + [\beta_0 + \beta_1(20)]$
   $\qquad + 3[\beta_0 + \beta_1(30)] = 100\beta_1.$

(e) Different denominator mean squares were used; if there is no lack of fit, both are estimates of σ^2. Modification would involve using the mean square for pure error, which is equivalent to the ANOVA error mean square, in the denominator.

(f) $F = 0.40$ (2 and 8 df); $P > 0.25$.

(g) Same value as obtained in part (f).

Chapter 18

1. (a) Rats are blocks, chemicals are treatments.

(b)

SOURCE	df	SS	MS	F
Chemicals	2	25.0	12.500	7.00
Rats	7	18.5	2.643	1.48
Error	14	25.0	1.786	
Total	23	68.5		

(c) Test for difference in chemical effects: $F = 12.5/1.786 = 7.00$ (2 and 14 df), $0.005 < P < 0.01$; reject H_0.

(d) 98% CI for $(\mu_I - \mu_{II})$: $-3.0 < (\mu_I - \mu_{II}) < 0.5$.

(e) $R^2 = 0.635$.

(f) Fixed-effects ANOVA model: $Y_{ij} = \mu + \tau_i + \beta_j + E_{ij}$, $i = I$, II, III, $j = 1, \ldots, 8$; ($\tau_i = i$th chemical effect, $\beta_j = j$th rat effect).
   Regression model: $Y = \mu + \alpha_1 X_1 + \alpha_2 X_2 + \sum_{j=1}^{7} \beta_j Z_j + E$, where

$$X_1 = \begin{cases} 1 & \text{if chemical I} \\ 0 & \text{if chemical II} \\ -1 & \text{if chemical III} \end{cases} \qquad X_2 = \begin{cases} 0 & \text{if chemical I} \\ 1 & \text{if chemical II} \\ -1 & \text{if chemical III} \end{cases}$$

$$Z_i = \begin{cases} -1 & \text{if rat 8} \\ 1 & \text{if rat } j \ (j=1,2,\ldots,7) \\ 0 & \text{otherwise} \end{cases}$$

2. (a)   Let n denote the number of individuals in each group, let

$$G = \begin{cases} 1 & \text{if group 1} \\ -1 & \text{if group 2} \end{cases}$$

and let

$$B_i = \begin{cases} -1 & \text{if block } n \\ 1 & \text{if block } j \qquad (j=1,2,\ldots,n-1) \\ 0 & \text{otherwise} \end{cases}$$

Then the following six regression models correspond to approaches (1) through (6):

(1) $SP2 - SP1 = \beta_0 + \sum\limits_{j=1}^{n-1} \beta_j B_j + \tau G + E$

(2) $SP2 - SP1 = \beta_0 + \sum\limits_{j=1}^{n-1} \beta_j B_j + \gamma(SP1) + \tau G + E$

(3) $SP2 = \beta_0 + \sum\limits_{j=1}^{n-1} \beta_j B_j + \gamma(SP1) + \tau G + E$

(4) $SP2 - SP1 = \beta_0 + \alpha_1(AGE) + \alpha_2(SEX) + \tau G + E$, where

$$SEX = \begin{cases} 1 & \text{if male} \\ -1 & \text{if female} \end{cases}$$

(5) $SP2 - SP1 = \beta_0 + \alpha_1(AGE) + \alpha_2(SEX) + \gamma(SP1) + \tau G + E$

(6) $SP2 = \beta_0 + \alpha_1(AGE) + \alpha_2(SEX) + \gamma(SP1) + \tau G + E$

(b)   For each of the models above, the null hypothesis of interest is $H_0: \tau = 0$.

(1)

SOURCE	df	MS
Blocks	$n-1$	MSB
Group|blocks	1	MSG
Error	$n-1$	MSE
Total	$2n-1$	

Critical region: $F = MSG/MSE \geq F_{1,n-1,1-\alpha}$.

(2)

SOURCE	df	MS
Blocks	$n-1$	MSB
SP1|blocks	1	MS1
Group|blocks, SP1	1	MSG
Error	$n-2$	MSE
Total	$2n-1$	

Critical region: $F = \text{MSG/MSE} \geq F_{1,n-2,1-\alpha}$.

(3)

SOURCE	df	MS
Blocks	$n-1$	MSB
SP1|blocks	1	MS1
Group|blocks, SP1	1	MSG
Error	$n-2$	MSE
Total	$2n-1$	

Critical region: $F = \text{MSG/MSE} \geq F_{1,n-2,1-\alpha}$.

(4)

SOURCE	df	MS
AGE	1	MSA
SEX|AGE	1	MSS
Group|SEX, AGE	1	MSG
Error	$2n-4$	MSE
Total	$2n-1$	

Critical region: $F = \text{MSG/MSE} \geq F_{1,2n-4,1-\alpha}$.

(5)

SOURCE	df	MS
AGE	1	MSA
SEX|AGE	1	MSS
SP1|AGE, SEX	1	MS1
Group|SEX, AGE, SP1	1	MSG
Error	$2n-5$	MSE
Total	$2n-1$	

Critical region: $F = \text{MSG/MSE} \geq F_{1,2n-5,1-\alpha}$.

(6)

SOURCE	df	MS
AGE	1	MSA
SEX|AGE	1	MSS
SP1|AGE, SEX	1	MS1
Group|SEX, AGE, SP1	1	MSG
Error	$2n-5$	MSE

Critical region: $F = \text{MSG/MSE} \geq F_{1,2n-5,1-\alpha}$.

(c) Using either model (2) or model (3) would be preferable to using model (1), unless it can be assumed that (SP2 − SP1) is not significantly correlated with SP1. However, to be on the safe side, controlling for SP1 is suggested; in any case, if SP1 is not an important covariate, this will be detected by the stepwise regression runs for approaches (2) and (3).

3. (b) The choice of model depends primarily on how one prefers to treat the copper-ion-concentration variable. If treated nominally as in part (a), only differences among the five concentration levels treated as (qualitative) groups can be detected. If treated continuously as in part (b), a trend (i.e., a dose–response type of relationship) between copper-ion concentration and BOD level can possibly be quantified; in part (b), this dose–response relationship is modeled as a straight line for each sample. These remarks suggest that the model in part (b) is probably preferable, since copper-ion concentration is actually a continuous variable.

(c) There appears to be a decreasing trend in the means, suggesting that increasing the copper concentration magnifies the retardation effect on the bacterial action.

(d) Test for the effect of copper-ion concentration: $F = 2{,}299.17/114.22 = 20.13$ (4 and 8 df), $P < 0.001$; reject H_0.

(f) $T = 2.090$ (7 df), $0.05 < P < 0.10$ (two-sided test); do not reject H_0 at $\alpha = 0.05$.

(g) $F = 2{,}509.66/286.83 = 8.75$ (3 and 9 df), $0.001 < P < 0.005$; reject H_0.

6. (d) No, since any or all of the companies may be discriminatory even though they do not differ in the extent of that discrimination. One would need to test whether any given company has a rate discrepancy which is significantly different from some standard rate (which corresponds to no discrimination).

(e) Both the normality and variance homogeneity assumptions are in question. Alternative analysis procedures would entail using nonparametric ANOVA techniques [e.g., see Siegel (1956)] or making a suitable data transformation (see Chapter 16).

Chapter 19

1. (b) Both factors should be considered fixed.

(d)

	EPINEPHRINE		
LEVORPHANOL	−	+	MARGINALS
−	2.25	5.28	3.77
+	1.75	2.57	2.16
MARGINALS	2.00	3.93	2.96

The presence of levorphanol appears to reduce stress, whereas the presence of epinephrine appears to increase stress. In the presence of epinephrine, levorphanol reduces stress by an average of 2.71 units, whereas in the absence of epinephrine, levorphanol reduces stress by only 0.50 unit (suggesting the possibility of an interaction effect).

(e)

SOURCE	df	SS	MS	F(fixed effects)
Levorphanol	1	12.832	12.832	12.59
Epinephrine	1	18.586	18.586	18.24
Interaction	1	6.161	6.161	6.05
Error	16	16.298	1.019	

(f) Main effect of levorphanol: $F = 12.832/1.019 = 12.59$ (1 and 16 df), $0.001 < P < 0.005$; the presence of levorphanol significantly reduces stress.

Main effect of epinephrine: $F = 18.586/1.019 = 18.24$ (1 and 16 df), $P < 0.001$; the presence of epinephrine significantly increases stress.

Interaction: $F = 6.161/1.019 = 6.05$ (1 and 16 df), $0.025 < P < 0.05$; the interaction effect as described in part (d) is significant at $\alpha = 0.05$ but not at $\alpha = 0.025$.

2. (a) The classification depends on the purpose of the investigation, although the most reasonable classification is specialty-fixed and city-random.

(b)

	CITY 1	CITY 2	CITY 3	MARGINALS
Ped	83.58	83.98	83.83	83.79
Ob-Gyn	79.03	77.20	77.35	77.86
Db-Hyp	62.95	62.75	58.50	61.40
MARGINALS	75.18	74.64	73.22	74.35

There appears to be a highly significant main effect of specialty, with pediatric FNP's being most competent, Ob-Gyn FNP's next most competent, and Db-Hyp FNP's least competent. There is no main effect of city, and apparently there is no interaction effect.

(c)

SOURCE	F (both fixed)	F (both random)	F (SPEC fixed CITY random)	F (SPEC random CITY fixed)
Specialty	$15.02_{(2,27)}$	$187.04_{(2,4)}$	$187.04_{(2,4)}$	$15.02_{(2,27)}$
City	$0.11_{(2,27)}$	$1.42_{(2,4)}$	$0.11_{(2,27)}$	$1.42_{(2,4)}$
Interaction	$0.08_{(4,27)}$	$0.08_{(4,27)}$	$0.08_{(4,27)}$	$0.08_{(4,27)}$

(d) The tests for each factor classification scheme all give the same results, which support the interpretation given in part (b).

(e) Use $(\bar{Y}_{Ped} - \bar{Y}_{Ob\text{-}Gyn}) \pm \sqrt{(3-1)F_{2,27,0.95}MSE(1/12 + 1/12)}$, which gives $-.4554 < (\mu_{Ped} - \mu_{Ob\text{-}Gyn}) < 11.41$.

(f) $Y = \beta_0 + \beta_1 S_1 + \beta_2 S_2 + \beta_3 C_1 + \beta_4 C_2 + \beta_5 S_1 C_1 + \beta_6 S_1 C_2 + \beta_7 S_2 C_1 + \beta_8 S_2 C_2 + E,$

where

$$S_1 = \begin{cases} -1 & \text{if Db-Hyp Spec.} \\ 1 & \text{if Ped Spec.} \\ 0 & \text{otherwise} \end{cases} \qquad S_2 = \begin{cases} -1 & \text{if Db-Hyp Spec.} \\ 1 & \text{if Ob-Gyn Spec.} \\ 0 & \text{otherwise} \end{cases}$$

$$C_j = \begin{cases} -1 & \text{if city 3} \\ 1 & \text{if city } j \qquad (j = 1, 2) \\ 0 & \text{otherwise} \end{cases}$$

$\hat{\beta}_0 = 74.35,\ \hat{\beta}_1 = 9.44,\ \hat{\beta}_2 = 3.51,\ \hat{\beta}_3 = 0.833,\ \hat{\beta}_4 = 0.292.$

4. (a)

	MALE	FEMALE	MARGINALS
75PL male coed	39.3	29.7	34.49
LS75 male coed	26.7	26.9	26.80
Not coed	34.0	31.7	32.85
MARGINALS	33.33	29.43	31.38

The table of means suggests that males are generally more sexist than females and that coed colleges with a large proportion of females have less sexism than the other types of colleges considered. Furthermore, and perhaps most importantly, it appears that males are much more sexist than females in coed schools with relatively few females, that there is little difference in attitudes in coed schools with relatively many females, and that males are somewhat more sexist than females in colleges that are not coed (i.e., there is an interaction effect).

5. (a) Yes. There is an apparent "same-direction" interaction, which is reflected in the fact that the difference in mean waiting times between suburban and rural court locations is larger for state 1 than for state 2.

6. (c)

SOURCE	F (both fixed)	F (both random)	F (INST fixed METH random)	F (INST random METH fixed)
Instructor	$0.0004_{(1,96)}$	$0.0004_{(1,1)}$	$0.0004_{(1,1)}$	$0.0004_{(1,96)}$
Method	$5.996_{(1,96)}$	$5.742_{(1,1)}$	$5.996_{(1,96)}$	$5.742_{(1,1)}$
Interaction	$1.044_{(1,96)}$	$1.044_{(1,96)}$	$1.044_{(1,96)}$	$1.044_{(1,96)}$

(d) Student knowledge of subject matter at beginning of the experiment, IQ, mathematics aptitude, attitudes toward subject matter, and so on.

(e) Analysis-of-covariance model:

$$Y = \beta_0 + \beta_1 C + \beta_2 I + \beta_3 M + \beta_4 IM + E$$

where

$$I = \begin{cases} 1 & \text{if instructor A} \\ -1 & \text{if instructor B} \end{cases} \qquad M = \begin{cases} 1 & \text{if self-instruction method} \\ -1 & \text{if lecture method} \end{cases}$$

9. (a) Species fixed, locations random (unless, for locations, only the four chosen are of interest).

(b)

SOURCE	F(mixed)	F(both fixed)
Species	$9.140_{(2,6)}$	$7.972_{(2,48)}$
Locations	$0.944_{(3,48)}$	$0.944_{(3,48)}$
Interaction	$0.872_{(6,48)}$	$0.872_{(6,48)}$

Mixed: species $(0.01 < P < 0.025)$, locations $(P > 0.25)$, interaction $(P > 0.25)$.
Fixed: species $(0.001 < P < 0.005)$, locations $(P > 0.25)$, interaction $(P > 0.25)$.

Chapter 20

1. (b)

SOURCE	df	SS	MS
Modern rank	2	687.72	343.86
Traditional rank	2	33.123	16.561
Interaction	4	4,369.7	1,092.4
Error	21	1,592.5	75.833

(Note: Harmonic mean $= 3.1034$.)

F(modern) $= 343.86/75.833 = 4.53$ (2 and 21 df), $0.01 < P < 0.025$; significant at 0.025.
F(traditional) $= 16.561/75.833 = 0.218$ (2 and 21 df), $P > 0.25$; not significant.
F(interaction) $= 1,092.4/75.833 = 14.41$ (4 and 21 df), $P < 0.001$; highly significant.
Keep in mind that finding such a significant interaction effect somewhat negates the importance of any main-effect tests.

(c) $Y = \beta_0 + \alpha_1 X_1 + \alpha_2 X_2 + \beta_1 Z_1 + \beta_2 Z_2 + \gamma_{11} X_1 Z_1 + \gamma_{12} X_1 Z_2 + \gamma_{21} X_2 Z_1 + \gamma_{22} X_2 Z_2 + E,$

where

$$X_i = \begin{cases} -1 & \text{if LO modern rank} \\ 1 & \text{if modern rank } i \ (i = 1, 2), \\ 0 & \text{if otherwise} \end{cases}$$

$$Z_j = \begin{cases} -1 & \text{if LO trad. rank} \\ 1 & \text{if trad. rank } j \ (j = 1, 2) \\ 0 & \text{if otherwise} \end{cases}$$

($i = 1$ for HI modern rank and $i = 2$ for MED modern rank)
($j = 1$ for HI trad. rank and $j = 2$ for MED trad. rank)

(d) *"Modern" main effect*:

(1) $\quad F(X_1, X_2) = \dfrac{977.70/2}{235.5172} = 2.076$ (2 and 27 df), $0.10 < P < 0.25$;

not significant.

(2) $\quad F(X_1, X_2 | Z_1, Z_2) = \dfrac{871.92/2}{246.5377} = 1.768$ (2 and 25 df), $0.10 < P < 0.25$;

not significant.

"Traditional" main effect:

(1) $\quad F(Z_1, Z_2) = \dfrac{301.30/2}{260.5690} = 0.578$ (2 and 27 df), $P > 0.25$;

not significant.

(2) $\quad F(Z_1, Z_2 | X_1, X_2) = \dfrac{195.52/2}{246.5377} = 0.397$ (2 and 25 df), $P > 0.25$;

not significant.

Interaction:

$$F(X_1 Z_1, X_1 Z_2, X_2 Z_1, X_2 Z_2 | X_1, X_2, Z_1, Z_2) = \frac{4{,}570.85/4}{75.8333} = 15.069, \ P < 0.001;$$

highly significant.

Regression analysis agrees well with unweighted means analysis.

(e) $\quad Y = \beta_0 + \beta_1 X_1 + \beta_2 X_2 + \beta_3 X_1 X_2 + E$, where, for example,

$$X_1 = \begin{cases} 0 & \text{if LO modern rank} \\ 1 & \text{if MED modern rank} \\ 2 & \text{if HI modern rank} \end{cases} \qquad X_2 = \begin{cases} 0 & \text{if LO trad. rank} \\ 1 & \text{if MED trad. rank} \\ 2 & \text{if HI trad. rank} \end{cases}$$

The difficulty arises with regard to assigning numerical values to the categories of each factor. The coding scheme for X_1 and X_2 given here assumes that the categories are "equally spaced," which may not really be the case.

2. (a) $Y = \beta_0 + \beta_1 X_1 + \beta_2 X_2 + \beta_3 X_1 X_2 + E$.

   (b) $Y = \beta_0 + \alpha_1 A + \beta_1 C + \gamma_{11} AC + E$, where

$$A = \begin{cases} -1 & \text{if high attitude} \\ 1 & \text{if low attitude} \end{cases} \qquad C = \begin{cases} -1 & \text{if high communication} \\ 1 & \text{if low communication} \end{cases}$$

This corresponds to a two-way ANOVA model with fixed effects.

   (c) A choice between the models given in parts (a) and (b) should depend on the confidence one has in the sensitivity of the measurements. If the measurement methods are such that both the communication and attitude scores range from 0 to 100, for example, with meaningful interpretations for scores differing by 2 or 3 points, the model in part (a) would probably be best. On the other hand, if the measurement methods can only be expected to discriminate between "high" and "low" scores in a meaningful way for the attitude and communication factors, the model in part (b) is to be preferred.

3. (a)

DEPARTMENT	AGE GROUP			
	30–39	40–49	50–59	MARGINALS
A	1.600	1.500	1.643	1.586
B	0.625	1.167	0.750	0.917
MARGINALS	1.000	1.318	1.444	1.283

This table of means suggests that DEPT A has more illness (on the average) than DEPT B and that the mean number of illness absences increases with age. There also appears to be somewhat of an interaction effect, since the mean number of illness absences for DEPT A does not seem to depend much on the age group being considered, whereas for DEPT B the mean in age group 40–49 is almost twice as large as the means in the other two age groups.

   (b) Harmonic mean $= 7.231$, $\hat{\sigma}_{av}^2 = 0.14478$.

   (c) $F(\text{DEPT}) = 5.8393/1.0469 = 5.58$ (1 and 47 df), $0.01 < P < 0.025$; significant at $\alpha = 0.025$.

   $F(\text{AGE}) = 0.1797/1.0469 = 0.172$ (2 and 47 df), $P > 0.25$; not significant.

   $F(\text{Interaction}) = 0.44082/1.0469 = 0.421$ (2 and 47 df), $P > 0.25$; not significant.

   The conclusion here is that DEPT A has a significantly higher mean number of illness absences than DEPT B, and that there is no age effect and no interaction effect.

   (e) The assumption of normality.

7. (b) Two-way tables of sample means:

	SUMMER	WINTER	
Rural	2.72	4.81	3.76
Urban	3.39	4.07	3.73
	3.06	4.44	

	EAST	CENTRAL	WEST	
Rural	3.78	3.80	3.71	3.76
Urban	3.52	3.83	3.85	3.73
	3.65	3.82	3.78	

	EAST	CENTRAL	WEST	
Summer	3.07	2.95	3.15	3.06
Winter	4.23	4.68	4.42	4.44
	3.65	3.82	3.78	

These tables indicate that there is a seasonal main effect (i.e., more TV viewing in the winter), but that there are no main effects of the other two factors. There is also an apparent interaction between "season" and "residence" such that seasonal differences are much greater for rural residents than for urban residents. Furthermore, there appears to be no "residence"-by-"region" interaction (since appropriate mean differences are quite small). There is a slight hint of a "season"-by-"region" interaction, since the summer–winter difference is somewhat larger in the central part of the state than in the other two regions. Finally, there appears to be some evidence of a three-way interaction, since there is no apparent two-way interaction between "region" and "season" for rural residents but there is somewhat of a "region"-by-"season" interaction for urban residents.

(c)

SOURCE	df (num. den.)	F	P VALUE
RESID	(1, 468)	0.13	$P > 0.25$
REGION	(2, 468)	1.25	$P > 0.25$
SEASON	(1, 468)	224.73	$P < 0.001$
RESID × REGION	(2, 468)	1.63	$0.10 < P < 0.25$
RESID × SEASON	(1, 468)	59.20	$P < 0.001$
REGION × SEASON	(2, 468)	3.54	$0.025 < P < 0.05$
RESID × REGION × SEASON	(2, 468)	3.30	$0.025 < P < 0.05$

The main effect of SEASON and the RESID × SEASON interaction effect are both very highly significant. The REGION × SEASON interaction and the triple interaction are significant at the 0.05 level.

(d) $Y = \beta_0 + \beta_1 R + \beta_2 S + \beta_3 L_1 + \beta_4 L_2 + \beta_5 RS + \beta_6 RL_1 + \beta_7 RL_2 + \beta_8 SL_1 + \beta_9 SL_2$
$+ \beta_{10} RSL_1 + \beta_{11} RSL_2 + E$, where

$$R = \begin{cases} 1 & \text{if rural} \\ -1 & \text{if urban} \end{cases} \qquad S = \begin{cases} 1 & \text{if winter} \\ -1 & \text{if summer} \end{cases}$$

$$L_1 = \begin{cases} 1 & \text{if east} \\ 0 & \text{if central} \\ -1 & \text{if west} \end{cases} \quad \text{and} \quad L_2 = \begin{cases} 0 & \text{if east} \\ 1 & \text{if central} \\ -1 & \text{if west} \end{cases}$$

8. (a)   Factors are SEX (fixed) and TREATMENT (fixed).

(b)

	CONTROL	ESTROGEN	TOTALS
Male	$n_{11} = 12$	$n_{12} = 12$	$n_{1.} = 24$
Female	$n_{21} = 8$	$n_{22} = 8$	$n_{2.} = 16$
	$n_{.1} = 20$	$n_{.2} = 20$	$n_{..} = 40$

$n_{1.} n_{.1} / n_{..} = 12 = n_{11}$

$n_{1.} n_{.2} / n_{..} = 12 = n_{12}$

$n_{2.} n_{.1} / n_{..} = 8 = n_{21}$

$n_{2.} n_{.2} / n_{..} = 8 = n_{22}$

(c)

SOURCE	df	SS	MS	F
Treatment	1	536.5562	536.5562	36.429 $(P < 0.001)$
Sex	1	19.7227	19.7227	1.339 $(P > 0.25)$
Interaction	1	1.3500	1.3500	0.092 $(P \gg 0.25)$
Error	36	530.2412	14.7289	

(d)

SOURCE	df	SS	MS	F
Treatment	1	504.60	504.60	34.259 $(P < 0.001)$
Sex	1	19.723	19.723	1.339 $(P > 0.25)$
Interaction	1	1.350	1.350	0.092 $(P \gg 0.25)$
Error	36	530.2412	14.7289	

[*Note:* SS terms have each been multiplied by the harmonic mean of the n_{ij}'s (HM $= 9.60$).]

(e) $Y = \beta_0 + \beta_1 S + \beta_2 T + \beta_3 ST + E$, where

$$S = \begin{cases} -1 & \text{if male} \\ 1 & \text{if female} \end{cases} \qquad T = \begin{cases} -1 & \text{if control} \\ 1 & \text{if estrogen} \end{cases}$$

(f)   The regression results demonstrate that in the proportional-cell-frequency case, it makes no difference to the analysis which main-effect variable is entered first, so the methodology for equal-cell-number two-way ANOVA is appropriate.

Chapter 21

1.   (a) T   (b) T   (c) F   (d) F   (e) F   (f) F   (g) F   (h) F
      (i) F   (j) T   (k) T   (l) F   (m) F   (n) T   (o) F   (p) T
      (q) T   (r) F   (s) F   (t) T   (u) T   (v) T   (w) F   (x) T
      (y) F

2.   (a)   Small sample size.
      (b)   Rotated factors have much more of a simple structure than do the principal components.
      (c)   Possibly items 12 and 13.
      (d)   Factor 1: 1, 2, 3, 4, 6, 10, 11
             Factor 2: 1, 5, 7, 8, 9

3.   (a)   The three dimensions were theoretically conceived to be "overlapping" to some extent.
      (b)   Factor 1 = "value," factor 2 = "confidence," factor 3 = "interest." Suggested clusters:

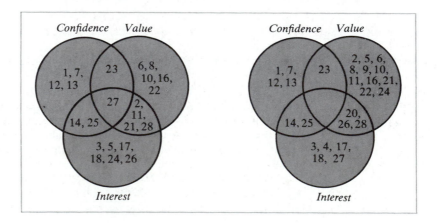

4.   (c)

FACTOR PML		FACTOR IML	
ITEM NO.	FACTOR WEIGHT	ITEM NO.	FACTOR WEIGHT
1	0.271	6	0.405
2	0.519	7	0.229
3	0.447	8	0.423
4	0.409	9	0.580
5	0.540	10	0.517

  (d) Person No. 1: PIML score $= (6.311) - (0.797) = 5.514$.
    Person No. 2: PIML score $= (1.841) - (1.957) = -0.116$.
    Person No. 3: PIML score $= (-1.034) - (6.046) = -7.080$.

5. *Factor 1:* hostility, suspiciousness, uncooperativeness.
  Factor 2: anxiety, guilt feelings, depressive mood.
  Factor 3: conceptual disorganization, hallucinatory behavior, unusual thought content.

Chapter 22

1. (a) Use the two-sample t test:

$$T = \frac{\bar{X}_1 - \bar{X}_2}{S_p \sqrt{1/n_1 + 1/n_2}} \frown t_{n_1 + n_2 - 2} \text{ (under } H_0\text{)}$$

   SBP: $T = 2.201$ with 23 df, $0.02 < P < 0.05$.
   CHOL: $T = 1.322$ with 23 df, $0.10 < P < 0.20$
   AGE: $T = 1.617$ with 23 df, $0.10 < P < 0.20$.
  (b) $Y = 0.4$ for noncases and $Y = -0.6$ for cases.
  (c) $c^* = 0.205849$.
  (d) $\ell = -0.0683\text{SBP} + 0.0118\text{CHOL} - 0.0677\text{AGE}$
   $b_0 = -\frac{1}{2}(\bar{\ell}_1 + \bar{\ell}_2) = 10.9545$
   $\ell^* = 10.9545 - 0.0683\text{SBP} + 0.0118\text{CHOL} - 0.0677\text{AGE}$
   (1) If $\ell^* > -0.4055$, assign to NCHD group; if $\ell^* < -0.4055$, assign to CHD group.
   (2) If $\ell^* > 0$, assign to NCHD group; if $\ell^* < 0$, assign to CHD group.
  (e) Noncase 1: $\ell^* = 1.3661$; assign to NCHD group.
   Case 10: $\ell^* = -1.5792$; assign to CHD group.
  (f) There is not a clear-cut choice; the first table has more incorrect classifications of actual CHD cases than the second table ($\frac{2}{15} = 0.1333$, $\frac{5}{10} = 0.5000$), whereas the second table has more incorrect classifications of noncases than the first table ($\frac{4}{15} = 0.2667$, $\frac{3}{10} = 0.3000$). The investigator would have to decide which type of error is most serious.
  (g) Apparent error rates: $\frac{4}{15} = 0.2667$, $\frac{3}{10} = 0.3000$. $\hat{P}_1 = \phi(-0.5061) = \hat{P}_2 = 0.3050$.
  (h) $F = 1.871$ with 3 and 21 df, $0.10 < P < 0.20$; conclude that there are no significant differences between the two groups based on a model involving SBP, CHOL, and AGE.
  (i) Only SBP makes a *significant* contribution to discrimination ($P < 0.05$); therefore, recommend a model involving only this variable.
  (j) $c^* = 0.21547$

   $b_1 = -0.06804$

   $b_0 = -\frac{1}{2}(\bar{\ell}_1 + \bar{\ell}_2) = 10.1288$

   Rule: If $\ell^* = (10.1288 - 0.06804\text{SBP})$ is greater than 0, assign to NCHD group; if $\ell^* < 0$, assign to CHD group. To compare this model with the one containing all three variables, examine the misclassification rates.

  (k) $P(\ell^*) = \dfrac{1}{1 + e^{-10.1288 + 0.06804\text{SBP}}}$

  where $1 - P(\ell^*)$ denotes the probability (or risk) of becoming a CHD case.

Risk for noncase 1 = 0.2802.
Risk for case 10 = 0.7369.

2. (a) $\ell = -1.7881$ (max. breadth) $+ 6.7966$ (basi-alv. length)
$+ 0.6289$ (nasal height) $+ 4.6846$ (basi-br. height)

(b) $b_0 = -1,074,5625$
Rule: Assign to period I if $\ell^* = -1,074.5625 + \ell > 0$; assign to period II otherwise.

(c) $F = 11.730$ with 4 and 5 df, $0.005 < P < 0.01$.

(d) Period I, No. 1: $\ell^* = 10.522$; assign to period I.
Period II, No. 1: $\ell^* = -18.659$; assign to period II.

(e) Perfect discrimination is achieved.

4.
	Both variables	MW only
	$c^* = 0.1273$	$c^* = 0.1294$
	$\ell = -0.0035\text{MW}$	$\ell = -0.0223\text{MW}$
	$+ 0.2442\text{MAGE}$	

5. $\ell^* = 4.1468 + 0.72106\text{DEF} - 1.70867\text{ENV} - 0.66671\text{RACE} - 0.1806\text{ECON}$

6. $\ell^* = 6.17532 - 3.22708\text{TD} + 1.60927\text{WR} + 1.16119\text{HS} - 0.51035\text{AD}$

8. (b) Order: X_1 (significant), X_3 (significant), X_2 (nonsignificant).

(c) Cutoff point: 1.0735.

(d) For $\ell = 1.25614X_1 + 0.79186X_2 + 2.63219X_3$, assign to species 1 if $\ell > 1.0735$ and assign to species 2 if $\ell < 1.0735$. It is not necessary to compute c^*, unless the a priori probabilities and/or costs of misclassification are such that $\log_e [(p_2 c_{21})/(p_1 c_{12})]$ is nonzero.

(e) Cutoff point: 0.6678. Using $\ell = 1.25118X_1 + 3.36758X_3$, assign to species 1 if $\ell > 0.6678$ and assign to species 2 if $\ell < 0.6678$.

(f) $F = 50.82$ (2 and 21 df), $P < 0.001$.

(g) The rule involving X_1 and X_3 only is more appropriate, since it is simpler and discriminates as well as the rule involving all three variables.

Chapter 23

1. (b) Two-response, no factor → independence

One-response, one-factor → homogeneity

	RESPONSE CATEGORIES ($r = 6$)						
	1	2	3	4	5	6	
	EAST AND APPROVE	EAST AND DISAPPROVE	EAST AND NO OPINION	SOUTH AND APPROVE	SOUTH AND DISAPPROVE	SOUTH AND NO OPINION	TOTALS
One subpopulation ($s = 1$)	$n_{11} = 500$	$n_{12} = 350$	$n_{13} = 150$	$n_{14} = 400$	$n_{15} = 300$	$n_{16} = 300$	$n_{1.} = 2000$
	Π_{11}	Π_{12}	Π_{13}	Π_{14}	Π_{15}	Π_{16}	

(e) H_0: $f_1(\Pi) = \log_e \Pi_{11} + \log_e \Pi_{16} - \log_e \Pi_{14} - \log_e \Pi_{13} = 0$

$f_2(\Pi) = \log_e \Pi_{12} + \log_e \Pi_{16} - \log_e \Pi_{15} - \log_e \Pi_{13} = 0$

so that $f_1(\Pi) = \sum_{\gamma=1}^{6} k_{1\gamma} \log_e a_\gamma(\Pi)$, where $a_\gamma(\Pi) = \Pi_{1\gamma}$, $\gamma = 1, 2, \ldots, 6$, $k_{11} = 1$, $k_{12} = 0$, $k_{13} = -1$, $k_{14} = -1$, $k_{15} = 0$, and $k_{16} = 1$; and $f_2(\Pi) = \sum_{\gamma=1}^{6} k_{2\gamma} \log_e a_\gamma(\Pi)$, where $a_\gamma(\Pi) = \Pi_{1\gamma}$, $\gamma = 1, 2, \ldots, 6$, $k_{21} = 0$, $k_{22} = 1$, $k_{23} = -1$, $k_{24} = 0$, $k_{25} = -1$, and $k_{26} = 1$.

(h) H_0: $a_1(\Pi) = (\Pi_{11} - \Pi_{21}) = 0$, $a_2(\Pi) = (\Pi_{12} - \Pi_{22}) = 0$, so that

$$a_1(\Pi) = \sum_{i=1}^{2} \sum_{j=1}^{3} a_{1,ij} \Pi_{ij}, \qquad \text{where } a_{1,11} = 1, a_{1,21} = -1, \text{ and all other } a_{1,ij} = 0,$$

$$a_2(\Pi) = \sum_{i=1}^{2} \sum_{j=1}^{3} a_{2,ij} \Pi_{ij}, \qquad \text{where } a_{2,12} = 1, a_{2,22} = -1, \text{ and all other } a_{2,ij} = 0.$$

(i) $\chi^2 = 64.96$ with 2 df, $P < 0.0005$; therefore, reject H_0 of "no association" (i.e., there is strong evidence of either dependence or nonhomogeneity, depending on the sampling scheme used).

(j) Usual χ^2 of 64.96 is close to GSK χ^2 values of 62.82 and 67.14.

4. (a)

SUBPOPULATIONS ($s = 12$)			CATEGORIES OF RESPONSE ($r = 2$)	
SUB-POPULATION NO.	YEAR OF GRADUATION	MD TYPE	RESPONDENT	NON-RESPONDENT
1	Before 1940	Infect. dis. ped.	Π_{11} ($n_{11} = 5$)	Π_{12} ($n_{12} = 5$)
2	Before 1940	General pract.	Π_{21} ($n_{21} = 38$)	Π_{22} ($n_{22} = 33$)
3	Before 1940	Family phys.	Π_{31} ($n_{31} = 56$)	Π_{32} ($n_{32} = 86$)
4	1940–1950	Infect. dis. ped.	Π_{41} ($n_{41} = 9$)	Π_{42} ($n_{42} = 10$)
5	1940–1950	General pract.	Π_{51} ($n_{51} = 85$)	Π_{52} ($n_{52} = 58$)
6	1940–1950	Family phys.	Π_{61} ($n_{61} = 42$)	Π_{62} ($n_{62} = 51$)
7	1951–1960	Infect. dis. ped.	Π_{71} ($n_{71} = 23$)	Π_{72} ($n_{72} = 14$)
8	1951–1960	General pract.	Π_{81} ($n_{81} = 102$)	Π_{82} ($n_{82} = 60$)
9	1951–1960	Family phys.	Π_{91} ($n_{91} = 81$)	Π_{92} ($n_{92} = 61$)
10	After 1960	Infect. dis. ped.	$\Pi_{10,1}$ ($n_{10,1} = 8$)	$\Pi_{10,2}$ ($n_{10,2} = 2$)
11	After 1960	General pract.	$\Pi_{11,1}$ ($n_{11,1} = 49$)	$\Pi_{11,2}$ ($n_{11,2} = 21$)
12	After 1960	Family phys.	$\Pi_{12,1}$ ($n_{12,1} = 27$)	$\Pi_{12,2}$ ($n_{12,2} = 35$)

(b) $a_\gamma(\Pi) = \Pi_{\gamma 1}$, $\gamma = 1, 2, \ldots, 12$.

(c) $a_1(p) = \frac{5}{10} = 0.5000$, $a_2(p) = \frac{38}{71} = 0.5352$, $a_3(p) = \frac{56}{142} = 0.3944$, $a_4(p) = \frac{9}{19} = 0.4652$, $a_5(p) = \frac{85}{143} = 0.5944$, $a_6(p) = \frac{42}{93} = 0.4516$, $a_7(p) = \frac{23}{37} = 0.6241$, $a_8(p) = \frac{102}{162} = 0.6296$, $a_9(p) = \frac{81}{142} = 0.5704$, $a_{10}(p) = \frac{8}{10} = 0.8000$, $a_{11}(p) = \frac{49}{70} = 0.7000$, $a_{12}(p) = \frac{27}{62} = 0.4355$.

(d) $\Pi_{\xi 1} = \mu + x_1 \alpha_1 + x_2 \alpha_2 + x_3 \beta_1 + x_4 \beta_2 + x_5 \beta_3 + x_1 x_3 \gamma_{13} + x_1 x_4 \gamma_{14} + x_1 x_5 \gamma_{15} + x_2 x_3 \gamma_{23}$
$+ x_2 x_4 \gamma_{24} + x_2 x_5 \gamma_{25}$, where

$$x_1 = \begin{cases} 1 & \text{if infect. dis. ped.} \\ 0 & \text{otherwise} \end{cases} \qquad x_2 = \begin{cases} 1 & \text{if gen. ped.} \\ 0 & \text{otherwise} \end{cases}$$

$$x_3 = \begin{cases} 1 & \text{if grad. before 1940} \\ 0 & \text{otherwise} \end{cases} \qquad x_4 = \begin{cases} 1 & \text{if grad. 1940–1950} \\ 0 & \text{otherwise} \end{cases}$$

$$x_5 = \begin{cases} 1 & \text{if grad. 1951–1960} \\ 0 & \text{otherwise} \end{cases}$$

(e)

SOURCE	df	χ^2	P VALUE
MD type	2	12.630	$0.001 < P < 0.005$
Year of graduation	3	9.669	$0.01 < P < 0.05$
Interaction	6	6.732	$0.30 < P < 0.40$

Conclude that "MD type" is strongly significant, that "year of graduation" is mildly significant, and that there is no interaction effect.

(f) MD type: $d = 2$, H_0: $\alpha_1 = 0$ and $\alpha_2 = 0$
Year of graduation: $d = 3$, H_0: $\beta_1 = 0$, $\beta_2 = 0$, $\beta_3 = 0$
Interaction: $d = 6$, H_0: $\gamma_{13} = 0$, $\gamma_{14} = 0$, $\gamma_{15} = 0$, $\gamma_{23} = 0$, $\gamma_{24} = 0$, $\gamma_{25} = 0$

(g) Modified model: $\Pi_{\xi 1} = \mu + x_1\alpha_1 + x_2\alpha_2 + x_3\beta_1 + x_4\beta_2 + x_5\beta_3$

5. (b) $a_\gamma(\Pi) = \Pi_{\gamma 1}$, $\gamma = 1, \ldots, 12$.

(e)

SOURCE	df	χ^2	P VALUE
W	1	7.356	$0.005 < P < 0.01$
E	2	0.156	$P > 0.90$
S	1	2.705	$P \doteq 0.10$
$W \times E$	2	2.832	$P \doteq 0.25$
$W \times S$	1	8.402	$0.001 < P < 0.005$
$E \times S$	2	2.285	$0.30 < P < 0.40$
$W \times E \times S$	2	1.611	$0.40 < P < 0.50$

Only the main effect of W and the $W \times S$ interaction effect are strongly significant; also, the main effect of S is significant at the $\alpha = 0.10$ level. Probably the most important finding is that the difference in the likelihood of getting byssinosis between dusty and not dusty work places is far greater for smokers than for nonsmokers, which is reflected in the highly significant $W \times S$ interaction effect.

(f) $\Pi_{\xi 1} = \mu + x_1\alpha + x_2\beta_1 + x_3\beta_2 + x_4\gamma + x_1x_2\delta_{12} + x_1x_3\delta_{13} + x_1x_4\delta_{14}$

(g)

SOURCE	df	χ^2	P VALUE
W	1	20.619	$P < 0.0005$
E	2	1.017	$P \doteq 0.60$
S	1	0.625	$0.40 < P < 0.50$
$W \times E$	2	12.758	$0.001 < P < 0.005$
$W \times S$	1	13.324	$P < 0.0005$
Goodness of fit	4	4.414	$0.30 < P < 0.40$

W and $W \times S$ are strongly significant, as before; but for this model $W \times E$ is also significant. Also, the fit of the model is adequate since the goodness-of-fit test is not significant.

(h)

	1	2	3	4	5	6
p_{i1}	0.0556	0.1288	0.1500	0.2388	0.1111	0.2715
$\hat{p}_{i1}$	0.0456	0.1402	0.1454	0.2400	0.1447	0.2393

	7	8	9	10	11	12
p_{i1}	0.0118	0.0103	0.0095	0.0121	0.0102	0.0196
$\hat{p}_{i1}$	0.0093	0.0118	0.0090	0.0115	0.0128	0.0153

(i) Based on the results in part (g), one might try the model

$$\Pi_{\xi 1} = \mu + x_1\alpha + x_1 x_2 \delta_{12} + x_1 x_3 \delta_{13} + x_1 x_4 \delta_{14}$$

Nevertheless, based on the results in part (h), the model defined in part (f) provides an adequate fit.

INDEX